Digital Communication Systems Design

Digital Communication Systems Design

MARTIN S. RODEN
California State University, Los Angeles

PRENTICE HALL, Englewood Cliffs, New Jersey 07632

Library of Congress Cataloging-in-Publication Data

Roden, Martin S.
Digital communication systems design/Martin S. Roden.
p. cm.
Bibliography: p.
Includes index.
ISBN 0-13-211574-3
1. Digital communications. I. Title.
TK5103.7.R633 1988
621.38'0413—dc19

Editorial-production supervision: *Raeia Maes*
Cover design: *20/20 Services, Inc.*
Manufacturing buyer: *S. Gordon Osbourne*

Printed in the United States of America

10 9 8 7 6 5 4 3 2 1

ISBN 0-13-211574-3 025

PRENTICE-HALL INTERNATIONAL (UK) LIMITED, *London*
PRENTICE-HALL OF AUSTRALIA PTY. LIMITED, *Sydney*
PRENTICE-HALL CANADA INC., *Toronto*
PRENTICE-HALL HISPANOAMERICANA, S.A., *Mexico*
PRENTICE-HALL OF INDIA PRIVATE LIMITED, *New Delhi*
PRENTICE-HALL OF JAPAN, INC., *Tokyo*
SIMON & SCHUSTER ASIA PTE. LTD., *Singapore*
EDITORA PRENTICE-HALL DO BRASIL, LTDA., *Rio de Janeiro*

Contents

Preface

This text is intended for senior- or graduate-level use in courses dealing with digital communications. It assumes a knowledge of basic system analysis and probability. It does not assume that the student has already had a course in analog communications.

While stressing the theory and the derivation of results, the text has a strong design emphasis. Students are taken through the derivations and though processes involved in system design. The performance of these systems is then evaluated, and practical implementations are discussed.

While a design emphasis is important so that the student sees the techniques of appplying theory to the real world, one must be cautioned not to stress *training* at the expense of *education.* Thus, the actual performance curves are far less important than the steps involved in developing these curves. As more and more systems evolve, the chances increase of applying a result to a situation for which it is not intended. An example of the danger of a heavily applied approach can be found in the integrated circuits discussed in the text. Section 3.2.5 presents practical PCM encoders. Actual IC chip numbers are included. There is some likelihood that these chips will be discontinued or changed before this text even reaches print. The justification for inclusion of this material is to give the student examples of application of theory—not to serve as a design handbook.

The text is divided into five broad categories. The first consists of one chapter covering background material, including review of probability, spectral analysis, and numbering systems. The block diagram of a digital communication system is presented in this chapter. Then Chapters 2 through 6 begin analyzing the various blocks in the system. Chapter 2 explores the characteristics of the channel. Distortion and practical conditioning systems for telephone channels are considered. A thorough discussion of random noise processes also appears in this chapter. Chapter 3 treats source encoding and includes an extensive discussion of PCM and delta modulation. Chapter 4 presents channel encoding, focusing upon entropy codes, block codes for forward error correction, and convolutional and cyclic codes. One section of this chapter discusses criteria for selection of the best code. Chapter 5 explores various topics common to all transmission and reception systems. It includes matched filtering, decision theory, and multiplexing. In contrast to analog systems, digital communication systems require attention to several levels of timing. This is the theme of Chapter 6, which deals with symbol and frame synchronization. One section of this chapter explores codes with desirable synchronization properties.

The third major section of the text, Chapters 7 through 10, deals with techniques for transmitting digital signals. Chapter 7 presents baseband transmission techniques, including a discussion of frequency spectra, intersymbol interference, equalization, and performance. The binary matched filter is presented and analyzed. We derive the important threshold effect, which relates quantization noise to transmission bit errors. Chapters 8, 9, and 10 present parallel treatments of the three major classes of carrier modulation—ASK, FSK, and PSK. Within each chapter we discuss frequency spectrum, modulators and demodulators, performance, and M-ary transmission.

The fourth section of the book, consisting of Chapter 11, pulls together the results of the previous chapters in order to derive criteria for system design. The various systems are compared with respect to performance, bandwidth, and transmission rate as a function of bandwidth.

The final section of the text, Chapter 12 and 13, deals with two contemporary topics, secure communications and computer communications. As more and more communication of data becomes part of our everyday life, security becomes an increasingly important consideration. Chapter 12 presents spread spectrum and cryptography, the two major techniques for providing communication security. Chapter 13 explores computer communications, an application of the theory presented throughout the text. The chapter focuses upon system architecture, protocols, and local area networks. ISDN is presented as a contemporary application of digital communication principles.

Problems are presented at the end of each chapter, with extensive solved examples being included within the chapters. Appendices include answers to selected problems, a table of error functions, a table of Marcum-Q functions, references, and a glossary of terms.

I gratefully acknowledge my students at Cal State L.A. for helping to point out the best pedagogy for approaching each topic. I acknowledge Dennis J. E. Ross for assistance throughout this project. Finally, my thanks go to Integrated Computer Systems for providing me a number of opportunities to teach their course, Digital Communication Systems Design. This gave me the opportunity to interact with many professionals and to become aware of the relevance of the theory to the real world.

Martin S. Roden

Introduction to the Student

I do not think anybody could have predicted the magnitude of the data revolution. The general public sees only the tip of the iceberg. Advances in integrated electronics are driving the price of digital systems down to a level where they represent the cheapest approach to many situations.

Vast numbers of people enjoy the convenience of automated-teller bank machines. Countless others give a credit card to a merchant, who then zips the card through a magnetic reader to get immediate confirmation of validity—the machine even types the credit slip. Telephones can be used to transfer funds, television quality (picture and sound, if not programming) improves with a move toward digital TV, compact discs are replacing analog recorded discs, office computing is becoming distributed, and computer communication is available to almost everybody.

Where universities traditionally taught *analog* communications with some PCM and digital communication thrown in at the end, the trend is now to teach *digital* communications first. This is happening not only because of the universality of applications, but also because, in many ways, digital communication is easier to analyze than is the analog counterpart. Except in the analysis of the noise performance of receivers and in A/D conversion, digital communication deals with discrete rather than continuous operations. Basic probability and digital analysis are all that is needed to understand these systems.

As you study the material in this text, I caution you to place greater emphasis upon understanding the derivations than upon applying the results. Performance curves are presented in several places, but if you do not understand the various assumptions made in deriving these curves, you are in danger of using them in situations for which they were never intended. While the excitement of arriving at meaningful numerical and practical results tends to peak your interest and motivate you toward the subject, it is critical that you stress *education* rather than *training*. The applications will change rapidly as the technology develops, but the underlying theory never changes.

If you have any comments about this book, whether positive or negative, I would be most appreciative if you would communicate them to me at California State University, Los Angeles, CA 90032. Thanks, and enjoy the subject!

Digital Communication Systems Design

1

Introduction and Background Material

1.1 INTRODUCTION AND HISTORY

Among the earliest forms of communication were vocal-cord sounds generated by animals and human beings, with reception via the ear. When greater distances were required, the sense of sight was used to augment that of sound. For example, in the second century B.C., Greek telegraphers used torch signals to communicate. Different combinations and positions of torches were used to represent the letters of the Greek alphabet. These early torch signals represented the first example of data communications! Later, drum sounds were used to communicate over greater distances, again calling upon the sense of sound. Increased distances were possible, since the drum sounds were more easily distinguished from the background than were human vocal-cord sounds.

In the eighteenth century, communication of letters was performed using semaphore flags. These semaphore flags, like the torches of ancient Greece, relied upon the human eye to receive the signal. This reliance, of course, severely limited the transmission distances.

In 1753, Charles Morrison, a Scottish surgeon, suggested an electrical transmission system using one wire (plus ground) for each letter of the alphabet. A system of pithballs and paper with letters printed on it was used at the receiver.

In 1835, Samuel Morse began experimenting with telegraph as we know it today. Two years later, in 1837, telegraph was invented by Morse in the United States and by Sir Charles Wheatstone in Great Britain. The first public telegram was sent in 1844, and electrical communication was established as a major component of life.

The communication schemes described above are essentially *digital* in nature. This is true since only a limited number of messages are used. It was not until Alexander Graham Bell invented the telephone in 1876 that *analog* electrical communication came into play. Following this invention, it appeared that analog communication would totally subsume digital communication. Indeed, Western Union Telegraph Company tried, in vain, to get into the telephone business.

It took another century to go full circle. By 1976, digital communication was replacing analog communication in many areas traditionally dominated by the analog format. This explosion of interest in digital communication was made possible by revolutionary advances in computers and solid state electronics.

Commercial applications of digital communications started as early as 1962. The *T1 transmission system,* introduced in that year by the *Bell System,* marked the beginning of a commercial digital revolution. By the end of that year about 250 digital communication circuits had been installed. Sometime in mid-1976 the number exceeded 3 million! And the surface had only been scratched.

By the mid 1980s, when computers were just celebrating their 40th birthday, and solid state electronics was younger still, computer-controlled digital networks were already commercially available. The information society had reached a new level of maturity which was to have a profound influence upon many phases of human life. The *cashless society* was just around the corner. Instant communication access, whether from an automobile, a commercial airplane, or the middle of an athletic field, became a reality.

Humanity seems to have an insatiable appetite for data. Home computers are becoming universal. The rate at which history is being written is increasing at an unfathomable speed.

It took 20 *centuries* to go from torch signals communicating data to electrical signals transmitting the same data. It took only 20 *years* to go from primitive electrical data transmission to highly advanced high-speed communications and processing. The end is not yet in sight.

1.2 ELEMENTS OF A DIGITAL SYSTEM

We begin our study of digital communications with an overview of an entire communication system. Figure 1.1 is a block diagram of a typical digital transmission and reception system. This is meant to be a comprehensive diagram, and not all of the illustrated blocks will be present in every practical system. We shall briefly describe the function of each block in this section. These functions will be expanded upon greatly in later chapters of the text.

The *source encoder* operates upon one or more analog signals to produce a periodic train of *symbols*. These symbols may be binary (1's and 0's) or may be members of a set with more than two elements. With channels which are used to communicate from more than one source at the same time, the source encoder also may contain a *multiplexer*. The source encoding operation is the subject of Chapter 3 of

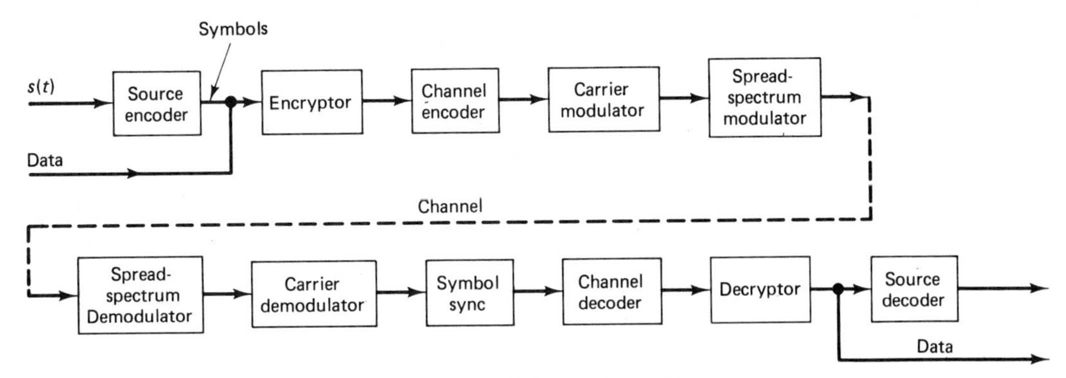

Figure 1.1 Digital transmission and reception system

this text, and multiplexing is covered in Chapter 5. Although we have indicated that the input to the source encoder is a time signal, $s(t)$, a *data* communication system may in fact start with a digital signal (e.g., the output obtained by pressing keys on a keyboard).

As electrical communication replaces written communication, security becomes increasingly important. We must assure that only the intended receiver can understand the message, and that only the authorized sender can transmit it. *Encryption* provides such security. As unauthorized receivers and transmitters become more sophisticated and computers become larger and faster, the challenges of secure communication become greater. Encryption is discussed in Chapter 12 of this text.

The *channel encoder* provides a different type of communication security than that provided by the encryptor. It increases efficiency and/or decreases the effects of transmission errors. Whenever noise is introduced into a communication channel, it is possible for one transmitted symbol to be interpreted as a different symbol at the receiver. We can decrease the effects of these errors by providing structure to the messages in the form of redundancy. In its simplest form, this would require repeating of the messages. The entire process is known as *forward error correction,* since proper encoding permits error correction without the necessity of the receiver asking the transmitter for additional information. The channel encoding process is discussed in detail in Chapter 4 of this text.

The output of the channel encoder is a digital signal composed of symbols. For example, in a binary system the output would be a train of 1's and 0's. An electrical channel can transmit signals only in the form of an electrical waveform. This is an important point. Do not be misled into thinking a digital signal can be transmitted in unmodified form. For example, if you use an audio channel to transmit "10101," you will speak five words thereby creating an audio pressure wave. When you speak the first word, "one," you are transmitting an analog waveform corresponding to the particular sounds in that word. You are therefore transmitting a digital signal using analog waveforms. This may appear to be a roundabout process, and indeed, it is. In order to send an analog signal, we will be changing it to a digital signal, sending that digital signal using analog waveforms, converting the analog waveforms at the receiver to the digital signal, and then changing the digital signal back to analog! This process will be shown to possess advantages over the strictly analog system under the conditions of certain types of distortion or in a noisy environment.

Returning now to the *carrier modulator,* its purpose is to produce analog waveforms corresponding to the discrete symbols at its input. This important process is the topic of the middle portion of this text, Chapters 7, 8, 9, and 10.

We saw earlier that the encryptor is present to make the transmission secure from unauthorized receivers or transmitters. In that sense, the encryptor produces a symbol sequence that is highly distinguishable only by the intended receiver. Additional security is often provided by use of *spread spectrum* techniques. One purpose of spread spectrum is to prevent an unauthorized listener from even distinguishing between the various symbols. Such a listener will mistake the signal for wideband noise. Spread spectrum is discussed in Chapter 12.

Viewing the second part of Fig. 1.1, we see that the receiver is simply a mirror image of the transmitter. It is necessary to "undo" each operation that was performed at the transmitter. The only variation from this one-to-one correspondence is that the *carrier modulator* of the transmitter has been replaced by two blocks in the receiver: the *carrier demodulator* and the *symbol synchronizer.* Once the analog waveforms are reproduced at the receiver, it is critical that the overall signal be properly partitioned into segments corresponding to each symbol and to each message. This partitioning is the function of the *synchronizer.* We spend an entire chapter, Chapter 6, on these and related timing considerations.

1.3 COMPARISON OF DIGITAL AND ANALOG COMMUNICATIONS

Before exploring the advantages of one communication system over the other, let us clarify some important definitions. An *analog communication system* is one which transmits analog signals—those time signals which can take on a continuum of values over a defined interval of time. If the analog time signals are sampled, one can think of the time samples as a list of numbers which must be transmitted. This list is still comprised of analog numbers—those which can take on an infinity of values within certain defined limits. The system is not yet digital. We refer to it as a *discrete time* or *sampled* system. If the sample values are now constrained to belong to a discrete set (e.g., the integers), the system becomes digital. The key concept is that the sample values can no longer take on every value within a continuous range.

Many systems are hybrid combinations of both digital and analog. As your eyes scan this page, your physiological system is operating as an analog receiver, since you are looking for gradations of images anywhere on this page. But the basic form of communication is digital, since you have programmed your mind to look for a limited number of signals—the alphanumerics plus a limited number of Greek and mathematical symbols. On a higher level, you are looking for words from our communication dictionary—a set of perhaps 30,000 possibilities. Getting way ahead of the game, if you saw the above phrase as "a set of perhaps 30,000 possiblities," you probably would correctly receive the message even though the last word in the string is not a member of your dictionary (note the typographical error, unless the computer typesetting equipment corrected it). You are able to correctly receive the message since we are communicating digitally—not every possible message is being used. In fact, if "possiblities" were part of our dictionary, you might have made a reception error in receiving it instead as "possibilities."

Now that we have agreed upon the definitions, we shall examine some of the advantages and disadvantages of digital communication when compared to analog communication. These topics are explored in greater detail later in this text. They are presented here to motivate the study of digital systems.

We begin by listing the major advantages and disadvantages.

Advantages of Digital Communications:

1. Errors often can be corrected.
2. Signal manipulation (e.g., encryption) is simple to perform.
3. Greater dynamic range (difference between largest and smallest values) is often possible.

Disadvantages of Digital Communications:

1. Generally requires more bandwidth than analog.
2. Synchronization is required.

We now expand upon these advantages and disadvantages. The major advantage of digital communication lies in its error-correcting capability. (For example, "possiblities" can be recognized as a typographical error.) This concept is critical! As long as all possible signal waveforms are not being transmitted, the receiver can often recognize when an error has been made. Sometimes this recognition is enough, but often the error is also corrected automatically at the receiver.

Figure 1.2 contrasts an analog communication system to a digital system. Note that in the analog system, amplifiers appear along the transmission path. Each amplifier introduces gain, but it amplifies both the signal and any additive noise that exists along the transmission path.

The lower part of Fig. 1.2 shows the digital communication system. Note that the amplifiers of the analog system have been replaced by *regenerative repeaters.* The repeaters not only perform the function of amplification but also "clean up" the signal.

We have illustrated the two systems using what we will later call "unipolar baseband" signals. The input signal can take on only one of two possible values—0

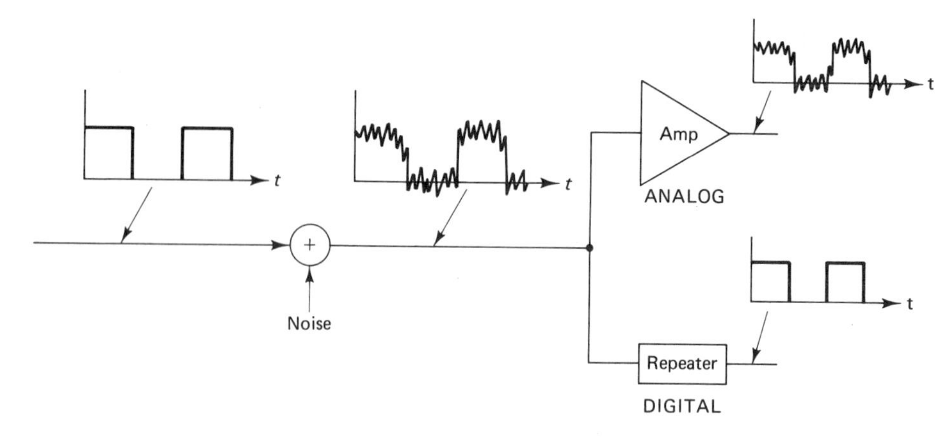

Figure 1.2 Comparison of analog to digital communication

or 1. Thus, the repeater must only decide which of these two values is present in any particular interval, then reproduce *that exact value* for transmission over the next leg of the channel.

The second advantage of digital communication relates to the fact that we are dealing with numbers rather than waveforms. These numbers can be manipulated by simple logic circuits or, if necessary, by microprocessors. Complex operations can easily be performed in order to accomplish signal-processing functions or security in transmission. Comparable analog operations would require complex hardware.

The third advantage relates to dynamic range. We illustrate this with an example. Traditional analog phonograph disc recording suffers from limited dynamic range. Very loud sounds require extreme variations in the shape of the groove on a record, and it is difficult for the needle to follow these variations. Digital recordings do not suffer from this problem, since all amplitude values, whether very small or very large, are transmitted using the same limited set of signals.

All is not ideal. There are disadvantages to digital communication when compared to analog communication. Digital systems generally require more bandwidth than do analog systems. For example, a single voice channel can be transmitted using *single-sideband AM* with a bandwidth of less than 5 kHz. Transmitting the same signal using digital techniques can require at least four times this bandwidth.

An additional disadvantage is the need to provide synchronization throughout the digital communication system. It is important for the system to know when each symbol is starting and stopping, and to properly associate each symbol with the correct transmission. This last point will become clear when we discuss multiplexing. Some would argue that the analog system, in its simplest form, is "more forgiving." Indeed, a single AM voice channel is extremely robust. The carrier frequency or bandwidth can be off and still the message can be understood.

1.4 NUMBERING SYSTEMS

The decimal system, although well suited to counting on one's fingers, is less than desirable for digital communications. The reasons should become clear as you proceed through this text.

The *base* of the decimal system is 10. Any decimal number, N, can be expressed in terms of $\log_{10}(N)$ (rounded up) digits. Thus, for example, the log of 20 is 1.3, which indicates that two digits are required to express this number. We express any number in decimal by assigning weighting factors to each digit based upon its position in the number. The rightmost or *least significant digit* has a weight of 10^0 or 1, the next digit has a weight of 10^1, and so on. Each digit is weighted by one higher power of 10 as we move from right to left, finally arriving at the *most significant digit*. As an example, the number $(1234)_{10}$ represents

$$(1 \times 10^3) + (2 \times 10^2) + (3 \times 10^1) + (4 \times 10^0)$$

Note that we have introduced the notation of parentheses around a number, with a subscript indicating the base of the numbering system.

A numbering system can have any nonzero base, although base 1 is not very exciting. The simplest choice is base 2, since each digit can then take on only one of two values. If base 2 is used, the numbering system is known as *binary*, and the digits are called binary digits, or *bits*. If we call the two values each digit can assume "0" and "1," then any number is written as a series of 0's and 1's. Each bit again has a weight associated with it, and the weights are powers of 2. For example, the binary number $(101101)_2$ is interpreted as

$$(1 \times 2^5) + (0 \times 2^4) + (1 \times 2^3) + (1 \times 2^2) + (0 \times 2^1) + (1 \times 2^0)$$

or, in decimal numbers, 45.

Binary arithmetic is performed in the same manner as decimal arithmetic. We simply need to keep careful track of carries and borrows, and to realize that no digit can exceed 1.

Example 1-1

Perform the following binary arithmetic operations.

(a) $1101011 + 1110$ (b) $1101011 - 1110$

(c) 1101011×1110 (d) $1101011 \div 1110$

Solution

(a)
$$\begin{array}{r} 1101011 \\ +\quad 1110 \\ \hline 1111001 \end{array}$$

(b)
$$\begin{array}{r} 1101011 \\ -\quad 1110 \\ \hline 1011101 \end{array}$$

```
(c)        1101011
      ×       1110
      ------------
          11010110
         1101011
        1101011
      ------------
       10111011010
```

```
                  111
(d) 1110)1101011
         1110
         -----
         11001
          1110
         -----
          10111
           1110
          -----
           1001  remainder
```

1.4.1 Modulo Arithmetic

A modulo counting system can be understood by viewing a periodic counter which recycles when it reaches the modulus of the system. Thus, for example, a modulo 6 counter generates the sequence

$$012345012345012345\ldots$$

The decimal number 7, in a modulo 6 system, becomes 1. To derive this, we subtract the largest multiple of 6 from the number and observe the remainder.

This concept leads to two distinct ways of viewing modulo arithmetic. To find the value of a number, N, in modulo m, we can:

(a) Divide N by m and observe the remainder, or
(b) Express N in a counting system using m as a base, and observe the least significant digit.

With the exception of the discussion of public key systems in cryptography, our use of modulo arithmetic will be confined to base 2. Modulo-2 addition is also known as the *exclusive-or* operation and is defined by the following equalities:

$$0 + 0 = 1 + 1 = 0$$

$$0 + 1 = 1 + 0 = 1$$

1.4.2 Other Numbering Systems

While binary systems were the earliest and most commonly used form of digital system (after decimal), many applications require that other numbering systems be used. In fact, communication systems have used base 4, 8, and 16, and computer systems have used 4, 8, 16, and 32.

In this section we concentrate upon the octal and hexadecimal systems. The *octal* system has base 8; the *hexidecimal* has base 16. Since 8 and 16 are powers of 2, numbers in these base systems can be related back to binary. We can go from binary to octal by combining bits in groups of 3. Transformation from octal to binary is performed by expanding each octal digit into 3 bits.

Example 1-2

(a) Convert $(1101101110)_2$ to octal.
(b) Convert $(17325)_8$ to binary.

Solution

(a) Grouping the bits by 3's, we have

$$(001)(101)(101)(110)$$

and converting each 3-bit group to octal, we have 1556, so

$$(1101101110)_2 = (1556)_8$$

(b) We convert each octal digit to 3 bits, so

$$(17325)_8 = (001)(111)(011)(010)(101)$$

or

$$(1111011010101)_2$$

The hexadecimal system can be arrived at by grouping bits in groups of four. Since each digit can now take on any one of 16 values, and we have common symbols for only 10 of these values (the digits 0 to 9), we must invent six new symbols. The first six letters of the alphabet, A through F, are usually used to represent 10 through 15, respectively. Thus, the hexadecimal number $(17EA)_{16}$ is equivalent to

$$(1 \times 16^3) + (7 \times 16^2) + (14 \times 16^1) + (10 \times 16^0) = (6122)_{10}$$

Converting this number to binary, we have

$$(0001)(0111)(1110)(1010)$$

or

$$(1011111101010)_2$$

Example 1-3

(a) Convert the hexadecimal number B3C10 to binary.
(b) Convert the binary number 101101110001011 to octal and to hexadecimal.

Solution

(a) To convert from hexadecimal to binary, we expand each hexadecimal symbol into 4 bits. Thus,

$$(B3C10)_{16} = (10110011110000010000)_2$$

(b) To convert from binary to octal, combine bits in groups of 3 to yield

$$(101)(101)(110)(001)(011) = (55613)_8$$

For hexadecimal, we combine in groups of 4 bits to yield

$$(101)(1011)(1000)(1011) = (5B8B)_{16}$$

Example 1-4

Perform the following arithmetic operations.
(a) In base 8: (13024 + 35171); (35171 × 4013); (35171 − 13024); and (35171 ÷ 571).
(b) In base 16: (A35C + 12D3); (A35C − 12D3); (A35C × 12D3); and (A35C ÷ 12D3).

Solution. All operations are similar to decimal, with caution exercised on carries and borrows.
(a) In base 8,

```
  35171      35171        35171           47
+ 13024    - 13024      ×  4013     571)35171
  -----      -----      -------         2744
  50215      22145       127553         ----
                         35171          5531
                       1647440          5117
                       ---------        ----
                       165445463         412 remainder
```

Note that the subtraction could have been performed using the $N-1$'s complement of 13024, which is 64753, and adding to get 35171 + 64753 = 122144. Then dropping the leading 1 and adding 1 yields 22145, the correct answer.
(b)

```
  A35C       A35C        A35C              8
+ 12D3     - 12D3      × 12D3     12D3)A35C
  ----       ----       -----          9698
  B62F       9089       1EA14          ----
                       84BAC            CC4 remainder
                      146B8
                      A35C
                      -------
                      C031CD4
```

Again, the subtraction could have been performed using complements as follows:

A35C + ED2C = 19088

Canceling the leading 1 and adding 1 yields 9089, as above.

1.4.3 Bits and Bauds

Throughout this text we will be talking about transmission of digital information, and the speed of that transmission is most important. Speed is measured in units of *bits per second,* sometimes written as bps.

Example 1-5

A digital thermometer samples temperature every second and displays an integer between 0 and 100 to represent the Celsius temperature. It is desired to transmit this as binary numbers. What is the speed of transmission?

Solution. A 7-bit binary number can be used to represent any decimal integer between 0 and 127. Thus, each sample temperature could be converted into a 7-bit binary number, and we would be sending at a rate of 7 bits per second. Note at this time that there are other ways to send this sampled temperature information using binary numbers. The technique outlined above is neither the fastest nor the most efficient way to send this information. As a hint of what is to come, we observe at this time that 27 of the possible bit combinations (the numbers from 101 to 127) are never used, since the problem postulated temperature readings between 0 and 100. This observation is a key to understanding why this is not the best transmission technique. More is said about this in Chapter 4.

When we start talking about the channel, another speed term is sometimes used, the *baud*. The baud is a unit of signaling speed and can be very loosely thought of as the number of pulses per second in the channel. It is actually the minimum time interval that must elapse between successive symbols. For example, if each sinusoidal burst gives 2 bits of information, the baud rate is one-half of the bit rate. The baud rate is sometimes (but rarely) the same as the bit rate, and there is some confusion in this regard. The two rates will be equal only if timing is uniform throughout and all pulses are used to send information (i.e., no extra pulses are used for other purposes such as forward error correction).

1.5 SOME SIMPLE CODES

Before getting to the theory and the details of application, let us pause to present some commonly used, and historically significant, codes for digitally transmitting text.

A *code* is a set of rules that assigns a *code word* to every message drawn from a *dictionary* of acceptable messages. The code words must consist of *symbols* from an acceptable *alphabet*. At present, we confine our attention to code words that consist of binary numbers, and the dictionary of messages will be the letters of the English alphabet together with some control functions. We do this because, historically, digital communication of messages was done letter by letter.

The *Baudot code* is one of the earliest, and now essentially obsolete, paper-tape codes used in Teletype machines. This code still finds limited application with radio amateurs and some communication systems for the deaf, as well as in printing telegraphy. The Baudot code assigns a 5-bit binary number to each letter of the alphabet. Since there are only 32 distinct 5-bit binary numbers (2^5), this code does not provide much flexibility. In fact, the 26 capital letters plus the space, line feed, and carriage return leave only three code words for all the other symbols, including the digits! This primitive code circumvents this shortcoming by providing a *shift* instruction, just as the shift lock on a typewriter introduces a whole new set of characters. One code word places all following words into the "shift" or "figures" mode, while another code word returns the system to the "letters" mode. Subtracting these two shift-instruction code words leaves 30 possible words in each mode, or a total of 60 words in the dictionary. This is enough for the uppercase alphabet, 10 digits, common symbols (e.g., $, #, and %), the line feed, the carriage return, and the friendly traditional Teletype BELL. Note that if a shift is required, it becomes necessary to send two code symbols in order to transmit only one information symbol. The Baudot code is illustrated in Table 1-1.

The *American Standard Code for Information Interchange,* or ASCII code, has become the standard for digital communication of individual alphabet symbols. This code is also used for very short range communications, such as from the keyboard to the processor of a computer. The basic code consists of code words of 7-bit length, thus providing 128 dictionary words (2^7). This is sufficient for the entire upper- and lowercase alphabet, plus a number of common digital control

TABLE 1-1 BAUDOT CODE

Letters Mode	Figures Mode	Code Word
Blank	Blank	00000
E	3	00001
Line feed	Line feed	00010
A	—	00011
Space	Space	00100
S	Bell	00101
I	8	00110
U	7	00111
Carriage return	Carriage return	01000
D	$	01001
R	4	01010
J	'	01011
N	,	01100
F	!	01101
C	:	01110
K	(	01111
T	5	10000
Z	"	10001
L	)	10010
W	2	10011
H	#	10100
Y	6	10101
P	∅	10110
Q	1	10111
O	9	11000
B	?	11001
G	&	11010
Figures	Figures	11011
M	.	11100
X	/	11101
V	;	11110
Letters	Letters	11111

messages. An eighth bit is often added as a *parity-check* bit for error detection. We discuss this process in Chapter 4.

The ASCII code is reproduced in Table 1-2. Examination of the table indicates that the code is arranged in a systematic way. For example, suppose that some application did not require any of the control symbols or any of the lowercase letters. We could then confine our attention to the middle four columns in the table. Examination should convince you that we could then drop the next-to-most significant bit, b_6, from each code word for each of the 64 possible messages.

TABLE 1-2 THE ASCII CODE

			b_7	0	0	0	0	1	1	1	1
			b_6	0	0	1	1	0	0	1	1
			b_5	0	1	0	1	0	1	0	1
b_1	b_2	b_3	b_4								
0	0	0	0	NUL	DLE	SP	0	@	P	′	p
0	0	0	1	BS	CAN	(	8	H	X	h	x
0	0	1	0	EOT	DC4	$	4	D	T	d	t
0	0	1	1	FF	FS	,	<	L	/	l	∫
0	1	0	0	STX	DC2	"	2	B	R	b	r
0	1	0	1	LF	SUB	*	:	J	Z	j	z
0	1	1	0	ACK	SYN	&	6	F	V	f	v
0	1	1	1	SO	RS	.	>	N	^	n	~
1	0	0	0	SOH	DC1	!	1	A	Q	a	q
1	0	0	1	HT	EM	)	9	I	Y	i	y
1	0	1	0	ENQ	NAK	%	5	E	U	e	u
1	0	1	1	CR	GS	−	=	M	[	m	}
1	1	0	0	ETX	DC3	#	3	C	S	c	s
1	1	0	1	VT	ESC	+	;	K	]	k	{
1	1	1	0	BEL	ETB	'	7	G	W	g	w
1	1	1	1	SI	US	/	?	O	_	o	DEL

Alternatively, if neither control symbols nor uppercase letters were required, a different bit could be dropped. Examination of the table should indicate which bit it is.

We define the more self-explanatory control symbols below. Some of these are discussed more fully in Section 5.8, where network protocols are introduced.

SOH start of heading

STX start of text

ETC end of text

EOT end of transmission

BS	backspace
HT	horizontal tab
LF	line feed
VT	vertical tab
FF	form feed
CR	carriage return
CAN	cancel
SUB	substitute

The final traditional code we present is the *Selectric code.* This is one of many specialized codes that have been widely used in the past. The Selectric typewriter was the standard of the industry before the days of electronic typewriters. The code is historically interesting, since it was specifically configured for the mechanical operation of that form of typewriter. The Selectric typewriter, and similar computer printers, uses a 7-bit code to control the position of the typing ball. Although this permits 128 distinct code symbols, only 88 of these are used. This would appear to be inefficient, and in fact the code is geared toward the specific typing-ball arrangement. One of the bits in the code controls the shift operation (this rotates the ball by 180 degrees). The other 6 bits control four modes of ball rotation and two modes of ball tilt. The Selectric typewriter can be instructed to type any letter or symbol by pushing one or more of the seven levers under the machine. The various controls, such as space and line feed, are handled completely separately from the symbol codes. The Selectric code is shown in Table 1-3. The seven code bits are presented as four controlling the 16 rows of the table and three controlling the 8 columns.

Conversion from one code to another (e.g., ASCII to Selectric) is trivial using integrated electronic hardware. It simply requires a table-lookup operation, which can be programmed into a read-only memory (ROM).

We close this section by again emphasizing that in the context of communications, the historical codes presented herein are not very efficient. As a hint of what we shall rigorously derive later, we observe that in the ASCII code, the symbol "%" uses a 7-bit code, while the letter "E" also uses a 7-bit code. Since the letter E occurs far more often in the English language than does the symbol %, it would be more desirable to use a shorter code word, even if this meant that the symbol % would need a code word longer than 7 bits. Practical applications of this concept are not hard to find. Intercontinental telephone dialing is one example, where a direct call from the United States to London requires 6 digits plus the local number, while a call to Galway, Ireland, requires 8 digits plus the local number. Indeed, we could have gone back over 100 years for an example. Samuel Morse realized this efficiency consideration when he coded the letter E into a single DIT, whereas the letter Q is coded into the much longer DAH-DAH-DI-DAH.

TABLE 1-3 SELECTRIC CODE

			T5	0	0	0	0	1	1	1	1
			T1	0	0	1	1	0	0	1	1
			T2	0	1	0	1	0	1	0	1
S	R2A	R2	R1								
0	0	0	0	–	b	w	9				
0	0	0	1	y	h	s	∅	/	l	o	4
0	0	1	0								
0	0	1	1								
0	1	0	0	q	k	i	6	,	c	a	8
0	1	0	1	p	e	′	5	;	d	r	7
0	1	1	0	=	n	.	2	f	u	v	3
0	1	1	1	j	t	$\frac{1}{2}$	z	g	x	m	1
1	0	0	0	—	B	W	(				
1	0	0	1	Y	H	S	)	?	L	O	$
1	0	1	0								
1	0	1	1								
1	1	0	0	Q	K	I	¢	,	C	A	*
1	1	0	1	P	E	″	%	:	D	R	&
1	1	1	0	+	N	.	@	F	U	V	#
1	1	1	1	J	T	$\frac{1}{4}$	Z	G	X	M	±

But we are getting way ahead of the game. The codes we have just presented are of great historical significance, and in the case of the ASCII and Selectric, they are of practical utility as well.

1.6 PRINCIPLES OF PROBABILITY

The reliability of various digital communication systems is of great importance to us throughout this text. Reliability considerations require that the effects of many types of external disturbances be considered. Since we wish to evaluate the performance of communication systems under a wide variety of scenarios, we will not restrict ourselves to specific signals. Nor can we completely characterize the exact form of the disturbances. It will thus become necessary to deal with *averages,* and this will enter us into the realm of probability.

Probability theory can be approached either on a strictly mathematical level or in an empirical setting. The mathematical approach embeds probability theory into the realm of abstract set theory and is itself somewhat abstract. The empirical approach satisfies one's intuition and is sufficient for our projected applications.

We use the "relative-frequency" definition of probability. Before we examine this definition, let us look at some related definitions.

An *experiment* is a set of rules governing an operation that is performed.

An *outcome* is the result realized after performing the experiment one time.

An *event* is a combination of outcomes.

For example, consider the experiment defined by flipping a single die. There are six possible outcomes, each being one of the six surfaces of the die facing upward after the performance of the experiment. There are many possible events (64 to be precise). For example, one event would be that of "an even number of dots showing." This event is a combination of the three outcomes: two dots, four dots, and six dots. Another possible event would be "one dot showing." This is called an *elementary event,* since it is actually equal to one of the outcomes.

We can now define what is meant by the probability of an event. Suppose that the experiment is performed N times, where N is very large. Suppose also that in n of these N experiments, the outcome belongs to a given event. If N is large enough, the probability of this event is given by the ratio, n/N. That is, it is the fraction of times that the event occurred. Formally, we define the probability of an event, A, as

$$\Pr\{A\} = \lim_{N\to\infty} \frac{n_A}{N}$$

where n_A is the number of times that the event A occurs in N performances of the experiment. This definition is intuitively satisfying. For example, if a coin were flipped many times, the ratio of the number of heads to the total number of flips would approach 1/2. We would therefore define the probability of a head to be $\frac{1}{2}$.

Suppose that we now consider two different events, A and B, with probabilities

$$\Pr\{A\} = \lim_{N\to\infty} \frac{n_A}{N} \quad \text{and} \quad \Pr\{B\} = \lim_{N\to\infty} \frac{n_B}{N}$$

If A and B cannot possibly occur at the same time, we call these two events *disjoint.* The events "an even number of dots" and "two dots" are not disjoint in the die-throwing example, while the events "an even number of dots" and "an odd number of dots" are disjoint.

The probability of event A *or* event B would be the ratio of the number of times A or B occurs divided by N. If A and B are disjoint, this is seen to be

$$\Pr\{A \text{ or } B\} = \lim_{N\to\infty} \frac{n_A + n_B}{N} = \Pr\{A\} + \Pr\{B\} \tag{1-1}$$

Equation (1-1) expresses the *additivity* concept. That is, if two events are disjoint, the probability of their "sum" is the sum of the probabilities.

Since each of the outcomes (elementary events) is disjoint from every other outcome, and each event is a sum of outcomes, we see that it would be sufficient to assign probabilities to only the outcomes. We could derive the probability of any event from these. For example, in the die-flipping experiment, the probability of an even outcome is the sum of the probabilities of a "two dots," "four dots," and "six dots" outcome.

Example 1-6

Consider the experiment of flipping a coin twice. List the outcomes, events, and their respective probabilities.

Solution. The outcomes of this experiment are (letting H denote heads and T tails)

HH, HT, TH, and TT

We shall assume that somebody has used some intuitive reasoning or has performed this experiment enough times to establish that the probability of each of the four outcomes is $\frac{1}{4}$. There are 16 events—i.e., 16 combinations of these outcomes. These are given by

{HH}, {HT}, {TH}, {TT}

{HH, HT}, {HH, TH}, {HH, TT}, {HT, TH}, {HT, TT}, {TH, TT}

{HH, HT, TH}, {HH, HT, TT}, {HH, TH, TT}, {HT, TH, TT}

{HH, HT, TH, TT}, and {0}

Note that the comma within the curly brackets is read as "or". Thus, the events {HH, HT} and {HT, HH} are identical. For completeness we have included the zero event, denoted {0}. This is the event made up of none of the outcomes and is called the *null* event.

Using the additivity rule, the probability of each of these events would be the sum of the probabilities of the outcomes comprising each event. Therefore,

$$\Pr\{HH\} = \Pr\{HT\} = \Pr\{TH\} = \Pr\{TT\} = \tfrac{1}{4}$$

$$\Pr\{HH, HT\} = \Pr\{HH, TH\} = \Pr\{HH, TT\} = \Pr\{HT, TH\} = \Pr\{HT, TT\}$$
$$= \Pr\{TH, TT\} = \tfrac{1}{2}$$

$$\Pr\{HH, HT, TH\} = \Pr\{HH, HT, TT\} = \Pr\{HH, TH, TT\} = \Pr\{HT, TH, TT\} = \tfrac{3}{4}$$

$$\Pr\{HH, HT, TH, TT\} = 1$$

$$\Pr\{0\} = 0$$

The last two probabilities indicate that the event made up of all four outcomes is the "certain" event. It has probability "1" of occurring, since each time the experiment is performed the outcome must belong to this event. Similarly, the zero event has probability zero of occurring, since each time the experiment is performed the outcome does not belong to the zero event.

1.6.1 Random Variables

We would like to perform several forms of analysis upon the probabilities. With this in mind, it is not too satisfying to have symbols such as "heads," "tails," and "two dots" floating around. We would much prefer to work with numbers. We therefore associate a real number with each possible outcome of an experiment. Thus, in the "single flip of the coin" experiment, we could associate the number "0" with "tails" and "1" with "heads." Similarly, we could just as well associate "3.1416" with heads and "2" with "tails."

The mapping (function) that assigns a number to each outcome is called a *random variable.* With a random variable so defined, many things can now be done that could not have been done before. For example, we can plot the various outcome probabilities as a function of the random variable. To make such a meaningful plot, we must first define something called the *distribution function,* $F(x)$. If the random variable is denoted by* X, then the distribution function, $F(x)$, is defined by

$$F(x_0) = \Pr\{X \leqslant x_0\}$$

We note that the set, $\{X \leqslant x_0\}$ defines an event, since it represents a combination of outcomes.

Example 1-7

Assign two different random variables to the "one flip of the die" experiment, and plot the two resulting distribution functions.

Solution. The first assignment we will choose is the one that is naturally suggested by this particular experiment. That is, we assign the number "1" to the outcome described by the face with one dot ending up in the top position. We assign "2" to "two dots," "3" to "three dots," and so on. We therefore see that the event $\{X \leqslant x_0\}$ includes the one-dot outcome if x_0 is between 1 and 2. If x_0 is between 2 and 3, the event includes the one-dot and two-dot outcomes. Thus, the distribution function is easily found; it is shown in Fig. 1.3(a).

Let us now choose another assignment of the random variable, this one representing a less natural choice.

Outcome	Random Variable
One dot	1
Two dots	π
Three dots	2
Four dots	$\sqrt{2}$
Five dots	11
Six dots	5

We have chosen strange numbers to indicate that the mapping is arbitrary. The resulting distribution function is plotted in Fig. 1.3(b). As an example, let us verify one point on the distribution function, the point for $x = 3$. The event $\{X \leqslant 3\}$ is the event made up of the three outcomes: one dot, three dots, and four dots. This is so because the value of the random variable assigned to each of these three outcomes is less than 3.

*We shall use capital letters for random variables, and lowercase letters for the values they take on. Thus, $X = x_0$ means that the random variable, X, is equal to the number, x_0.

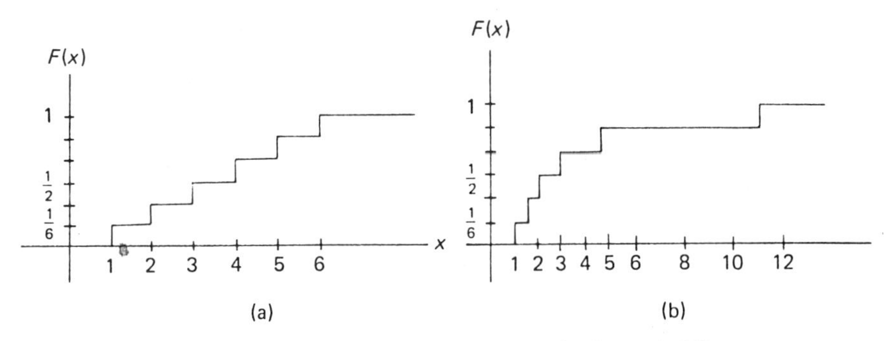

Figure 1.3 Distribution functions for Example 1-7

Note that a distribution function can never decrease with increasing argument—it is a *monotonic nondecreasing function.* This is so because an increase in argument can only add outcomes to the event. We also easily verify that

$$F(-\infty) = 0 \quad \text{and} \quad F(+\infty) = 1$$

We now define the derivative of the distribution as

$$p_x(x) = \frac{dF(x)}{dx}$$

This derivative, $p_x(x)$, is known as the *density function* of the random variable, X. If we integrate both sides of this expression and recognize that $F(-\infty) = 0$, we have

$$F(x_0) = \int_{-\infty}^{x_0} p_x(x)\, dx \tag{1-2}$$

The random variable can be used to define any event. For example, $\{x_1 < X \leqslant x_2\}$ defines an event. Since the events $\{X \leqslant x_1\}$ and $\{x_1 < X \leqslant x_2\}$ are disjoint, the additivity principle can be used to prove that

$$\Pr\{x_1 < X \leqslant x_2\} = \Pr\{X \leqslant x_2\} - \Pr\{X \leqslant x_1\}$$

This expression is now combined with Eq. (1-2) to yield

$$\begin{aligned}\Pr\{x_1 < X \leqslant x_2\} &= \int_{-\infty}^{x_2} p_x(x)\, dx - \int_{-\infty}^{x_1} p_x(x)\, dx \\ &= \int_{x_1}^{x_2} p_x(x)\, dx \qquad (1\text{-}3)\end{aligned}$$

The probability that X is between any two limits is found by integrating the density function between these two limits. This explains the reason for the terminology, "density function."

Since the distribution function can never have a negative slope, the density function can never go negative. Since $F(+\infty)$ is 1, the total area under the density function must be unity.

The examples given previously in this section (die and coin) would result in densities that contain impulse functions. This is so because the distribution contains discontinuities. A more common class of experiments gives rise to random variables whose density functions are continuous. This class is logically called the class of *continuous random variables.*

We shall now interrupt the development of probability theory to present four common examples of random variables, one discrete and three continuous. The discrete example is the *binomial distribution,* and the continuous examples include the *uniform, Gaussian,* and *Rayleigh* densities.

1.6.2 Binomial Distribution

The binomial distribution occurs in many cases of interest to us. We will use it to find the probability of a specified number of bit errors in a digital transmission.

Let us illustrate this distribution via the experiment of tossing a coin n times. Let the probability of heads be p and of tails be q. Of course,

$$q = 1 - p$$

Now suppose we wish to find the probability of k heads out of N tosses. If we take one possible sequence—say the first k tosses are heads and the remaining $N - k$ tosses are tails—the probability of that particular sequence is

$$p^k q^{N-k}$$

But there are other ways we could get k heads out of N tosses. For example, we could start with $N - k$ tails followed by k heads. The probability of this second sequence (and of any other ordering of k heads and $N - k$ tails) is still

$$p^k q^{N-k}$$

The overall probability of k heads (in any position) is found by adding the individual probabilities together, since the outcomes are disjoint. Since the probability of each outcome is the same, we need only multiply this probability by the number of ways we can distribute k heads among N positions. This number is given by the ratio of factorials as follows:

$$\binom{N}{k} = \frac{N!}{k!(N-k)!}$$

$\binom{N}{k}$ is the *binomial coefficient*. Thus the probability of k heads is given by

$$P(k) = \binom{N}{k} p^k q^{N-k} = \binom{N}{k} p^k (1-p)^{N-k} \tag{1-4}$$

Equation (1-4) is the binomial distribution function. This will apply to any binary situation where the individual trials are independent of each other. For example, if the flipping of the coin is replaced by the transmission of a binary 0 or 1, and p is the probability of bit error, Eq. (1-4) becomes the probability of k bit errors within a transmission of N bits.

1.6.3 Uniform Density

The simplest continuous density function is the *uniform density*. A typical version of such a density is shown in Fig. 1.4, where a and b are given parameters. Note that the value of the density function is a constant over a range of the x axis. That constant must be chosen to make the total area under the density function equal to unity.

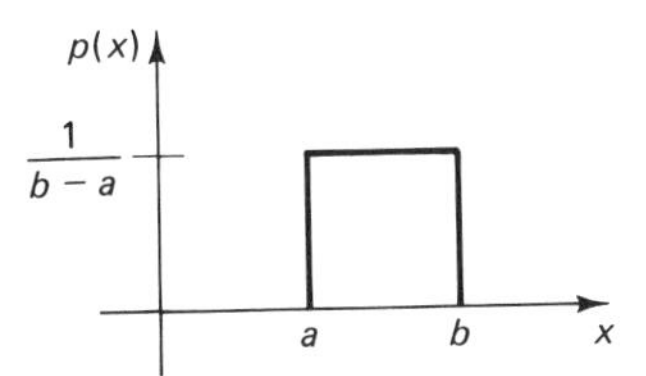

Figure 1.4 Uniform probability density function

As an example of a situation in which the uniform density arises, suppose that you were asked to turn on a sinusoidal generator. The output of the generator would be of the form

$$v(t) = A \cos(2\pi f_0 t + \phi)$$

Since the absolute time at which you turn on the generator can be considered to be random (actually, it depends upon your reaction time—if your reaction time is long compared to the period of the sinusoid, the angle ϕ can be considered to be random), it is reasonable to expect that ϕ is uniformly distributed between 0 and 2π. The density will therefore be as shown in Fig. 1.4 with $a = 0$ and $b = 2\pi$.

The uniform density is the continuous analogy of an experiment where each outcome is equally likely. Thus, for example, the single flip of the die with 6 outcomes, each with probability 1/6, represents a discrete version of a uniform density.

1.6.4 Gaussian Density

The most common density encountered in the real world is called the *Gaussian density function.* Its commonness is explained by the *central limit theorem,* a theorem we shall discuss in a few moments. The density is defined by the equation

$$p_X(x) = \frac{1}{\sqrt{2\pi}\sigma} \exp\left(\frac{-(x-m)^2}{2\sigma^2}\right) \tag{1-5}$$

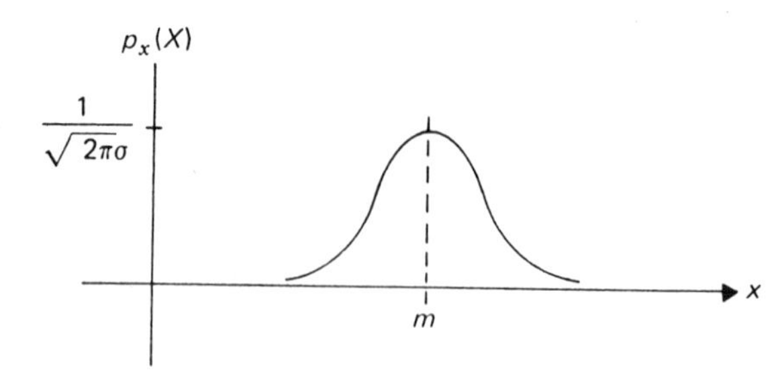

Figure 1.5 Gaussian probability density function

where m and σ are given constants. The density function is sketched in Fig. 1.5. Note that the parameter, m, dictates the center position or symmetry point of the density. The other parameter, σ, indicates the spread of the density. As σ increases, the "bell-shaped" curve gets wider and the peak decreases. Alternatively, as σ decreases, the density sharpens into a narrow pulse with a higher peak.

In order to evaluate probabilities that Gaussian variables are within certain ranges, we find it necessary to integrate this density. Unfortunately, the function of Eq. (1-5) cannot be integrated in closed form. The Gaussian density is sufficiently important that this integral has been computed and tabulated under the name *error function* (see Appendix II). It is defined by the following integral:

$$\text{erf}\,(x) = \frac{2}{\sqrt{\pi}} \int_0^x e^{-u^2}\,du \tag{1-6}$$

It can be shown that erf $(\infty) = 1$. Therefore,

$$\int_x^\infty e^{-u^2}\,du = \int_0^\infty e^{-u^2}\,du - \int_0^x e^{-u^2}\,du$$
$$= 1 - \text{erf}\,(x)$$

For convenience, this last expression is tabulated under the name *complementary error function.* Thus,

$$\text{erfc}\,(x) = 1 - \text{erf}\,(x)$$

The area under a Gaussian density with any values of m and σ can be expressed in terms of error functions. For example, if x_1 and x_2 are both larger than m, we have

$$\Pr\{x_1 < X \leqslant x_2\} = \frac{1}{\sqrt{2\pi}\sigma} \int_{x_1}^{x_2} \exp\left[\frac{-(x-m)^2}{2\sigma^2}\right] dx$$

We now make the following change of variables:

$$u = \frac{x-m}{\sqrt{2}\sigma}$$

to get

$$\Pr\{x_1 < x \leqslant x_2\} = \frac{1}{\sqrt{\pi}} \int_{(x_1-m)/\sqrt{2}\sigma}^{(x_2-m)/\sqrt{2}\sigma} e^{-u^2}\,du$$

$$= \frac{1}{2}\operatorname{erf}\left(\frac{x_2 - m}{\sqrt{2}\sigma}\right) - \frac{1}{2}\operatorname{erf}\left(\frac{x_1 - m}{\sqrt{2}\sigma}\right)$$

Now that we are familiar with the Gaussian density, let's consider why it occurs so frequently in the real world. It results whenever a large number of factors contribute to an end result, as in the case of radio static. Two conditions must be satisfied before the sum of many random variables starts to appear Gaussian. The first relates to the individual variances (spreads) and to their infinite sum. The sum must approach infinity as n, the number of variables added together, approaches infinity. The second condition is satisfied if the component densities go to zero outside some range (this is a sufficient but not a necessary condition). Since all quantities we deal with in the real world have bounded ranges, they satisfy this condition.

Although we do not prove the central limit theorem here, a simple example is often given to indicate its reasonableness. Suppose that we add together uniform random variables, where each is distributed between -1 and $+1$ as shown in Fig. 1.6(a). After the first two variables are added, the resulting density will be the convolution of the two original densities (this result is proven later in this section), as shown in Fig. 1.6(b). This doesn't yet look very much like the Gaussian curve, but we have added only two variables.

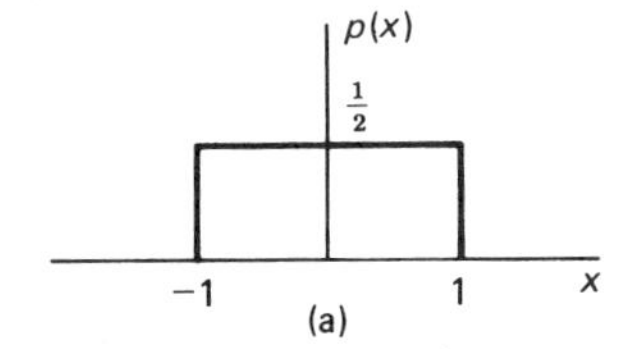

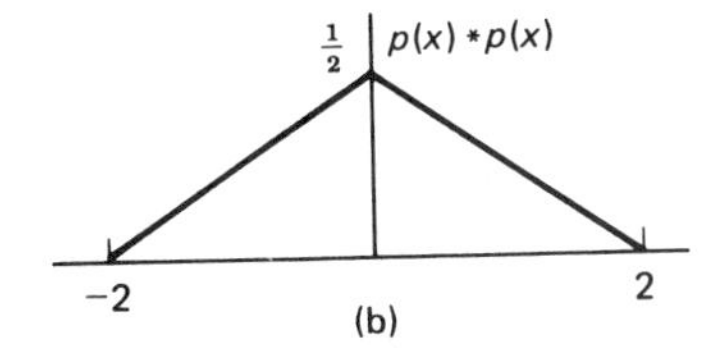

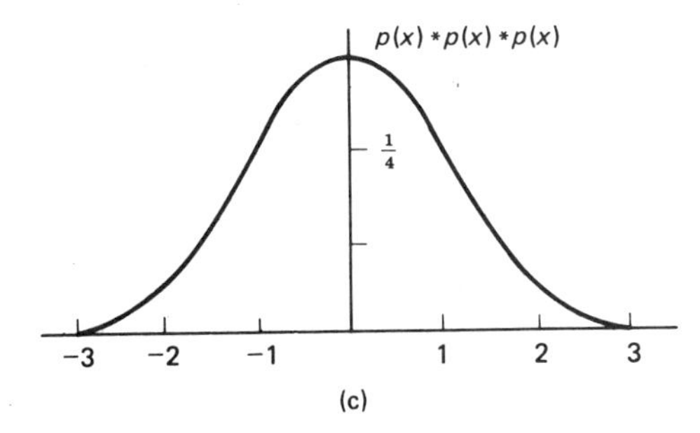

Figure 1.6 Illustration of central limit theorem

If a third variable is added, the ramp of Fig. 1.6(b) is convolved once more with the uniform density to get the parabolic curve of Fig. 1.6(c). At this point, the curve bears a faint resemblance to the bell-shaped Gaussian density. As more and more variables are added, the agreement becomes closer and closer. In fact, the series converges quite rapidly to a Gaussian density, even if the densities we start with are not uniform.

Example 1-8

A binary communication system is one that sends only two possible messages. The simplest form of binary system is one in which either *zero* or *one* volt is sent. Consider such a system in which the transmitted voltage is corrupted by additive atmospheric noise. If the receiver receives anything above $\frac{1}{2}$ volt, it assumes that a *one* was sent. If it receives anything below $\frac{1}{2}$ volt, it assumes that a *zero* was sent. Measurements have shown that if one volt is transmitted, the received signal level is random and has a Gaussian density with $m = 1$ and $\sigma = \frac{1}{2}$. Find the probability that a transmitted *one* will be interpreted as a *zero* at the receiver (i.e., a *bit error*).

Solution. The received signal level has a Gaussian density with $m = 1$ and a $\sigma^2 = (\frac{1}{2})^2$. Thus, if we designate the random variable as V, we have

$$p_V(v) = \frac{\sqrt{2}}{\sqrt{\pi}} \exp\left[\frac{-(v-1)^2}{2(0.5)^2}\right]$$

Assuming that anything received above a level of 0.5 is called "1," the probability that a transmitted "1" will be interpreted as a "0" at the receiver is simply the probability that the random variable V is less than 0.5. This is given by the integral

$$\int_{-\infty}^{0.5} p_V(v)\, dv = \frac{\sqrt{2}}{\sqrt{\pi}} \int_{-\infty}^{0.5} \exp\left[-2(v-1)^2\right] dv$$

In order to reduce this to a form that can be found in a table of error functions, we make the change of variable

$$u = \sqrt{2}(v-1)$$

to get

$$\Pr(\text{error}) = \frac{1}{\sqrt{\pi}} \int_{-\infty}^{-\sqrt{2}/2} \exp(-u^2)\, du$$

$$= \frac{1}{\sqrt{\pi}} \int_{\sqrt{2}/2}^{\infty} \exp(-u^2)\, du$$

The second equality is true since the integrand is even. This is now seen to be related to the complementary error function

$$\Pr(\text{error}) = 0.5 \operatorname{erfc}(\sqrt{2}/2) = 0.16$$

Thus, on the average, one would expect 16 out of every 100 transmitted 1's to be misinterpreted as 0's at the receiver. This is an extremely poor level of performance.

1.6.5 Rayleigh Density Function

A third commonly encountered density function is the *Rayleigh density function.* It is given by the formula

$$p_X(x) = \begin{cases} \dfrac{x}{K^2} \exp\left(\dfrac{-x^2}{2K^2}\right), & x > 0 \\ 0, & x < 0 \end{cases}$$

where K is a given constant. Figure 1.7 shows two representative sketches of the density function for different values of K.

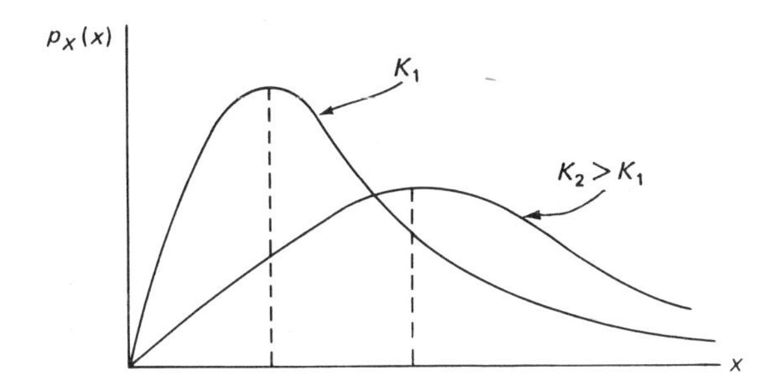

Figure 1.7 Rayleigh probability density function

The Rayleigh density function is related to the Gaussian density function. In fact, the square root of the sum of the squares of two Gaussian distributed random variables is, itself, Rayleigh. Thus, if we transform from rectangular to polar coordinates, the radius is given by

$$r = \sqrt{x^2 + y^2}$$

Therefore, if x and y are both Gaussian, in most cases (the restriction is one of *independence,* a term we have not yet defined) r will be Rayleigh. As an ideal example, suppose that you were throwing darts at a target on a dartboard and that the horizontal and vertical components of your error were Gaussian distributed with zero mean. The *distance* from the center of the target would then be Rayleigh distributed.

The *chi-square* distribution is closely related to the Rayleigh. If we were to consider r^2 instead of r, we would find that variable to be chi-square distributed. That is, a chi-square distribution results from summing the squares of Gaussian variables. If we sum two such variables, the result is chi-square with 2 *degrees of freedom.* In general, if

$$z = x_1^2 + x_2^2 + x_3^2 + \cdots + x_n^2$$

and x_i are Gaussian with $m = 0$, z will be chi-square with n degrees of freedom. If the Gaussian variables all have $\sigma = 1$, the actual form of the density is given by

$$p(z) = \begin{cases} \dfrac{(z)^{(n/2)-1}}{2^{n/2}\,(\frac{n}{2}-1)!} e^{-z/2}, & z > 0 \\ 0, & z < 0 \end{cases} \tag{1-7}$$

The "!" in Eq. (1-7) indicates the *factorial* operation.

1.6.6 Conditional Probabilities

We ask now whether we can tell if one random variable has any effect upon another. In the die experiment, if we knew the time at which the die hit the floor, would it tell us anything about which face was showing? This question leads naturally into a discussion of *conditional probabilities.*

Let us abandon the density functions for a moment and speak of two events, A and B. The probability of event A *given that event B has occurred* is defined by

$$\Pr\{A \mid B\} = \frac{\Pr\{A \text{ and } B\}}{\Pr\{B\}} \tag{1-8}$$

For example, if A represented two dots appearing in the die experiment and B represented an even number of dots, the probability of A given B would be the probability of two dots assuming that we know the outcome is either two, four, or six dots. Thus, the conditioning statement has reduced the scope of possible outcomes from six to three. We would intuitively expect the answer to be $\frac{1}{3}$. Now, using Eq. (1-8), the probability of A and B is the probability of getting two *and* an even number of dots simultaneously (in set theory, this is known as the *intersection*). This is simply the probability of two dots, or $\frac{1}{6}$. The probability of B is the probability of two, four, or six dots, which is $\frac{1}{2}$. The ratio is $\frac{1}{3}$ as expected.

Similarly, we could have defined event A as "an even number of dots" and event B as "an odd number of dots." The event "A AND B" would therefore be the zero event, and $\Pr\{A \mid B\}$ would be zero. This is reasonable, since the probability of an even outcome given that an odd outcome occurred is clearly zero.

We are now ready to define independence. Two events, A and B, are said to be *independent* if

$$\Pr\{A \mid B\} = \Pr\{A\}$$

The probability of A given that B occurred is simply the probability of A. Knowing that B has occurred tells us nothing about A. Substituting this into Eq. (1-8) shows that independence implies

$$\Pr\{A \text{ and } B\} = \Pr\{A\}\Pr\{B\}$$

You have used this fact before in simple experiments. For example, we assumed that the probability of flipping a coin and having it land with heads facing up was $\frac{1}{2}$. The probability of flipping the coin twice and getting two heads is $\frac{1}{2} \times \frac{1}{2} = \frac{1}{4}$. This is so because the events are independent of each other.

Example 1-9

A coin is flipped twice. Four different events are defined.

A is the event of getting a head on the first flip.
B is the event of getting a tail on the second flip.
C is the event of a match between the two flips.
D is the elementary event of a head on both flips.

Find $\Pr\{A\}$, $\Pr\{B\}$, $\Pr\{C\}$, $\Pr\{D\}$, $\Pr\{A|B\}$, and $\Pr\{C|D\}$. Are A and B independent? Are C and D independent?

Solution. The events are defined by the following combination of outcomes.

$$A = \text{HH, HT}$$

$$B = \text{HT, TT}$$

$$C = \text{HH, TT}$$

$$D = \text{HH}$$

Therefore, as in Example 1-6 we have

$$\Pr\{A\} = \Pr\{B\} = \Pr\{C\} = \tfrac{1}{2}$$

$$\Pr\{D\} = \tfrac{1}{4}$$

In order to find $\Pr\{A|B\}$ and $\Pr\{C|D\}$ we use Eq. (1-8):

$$\Pr\{A|B\} = \frac{\Pr\{A \text{ and } B\}}{\Pr\{B\}}$$

and

$$\Pr\{C|D\} = \frac{\Pr\{C \text{ and } D\}}{\Pr\{D\}}$$

The event $\{A \text{ and } B\}$ is $\{\text{HT}\}$. The event $\{C \text{ and } D\}$ is $\{\text{HH}\}$. Therefore,

$$\Pr\{A|B\} = \frac{\frac{1}{4}}{\frac{1}{2}} = 0.5$$

$$\Pr\{C|D\} = \frac{\frac{1}{4}}{\frac{1}{4}} = 1$$

Since $\Pr\{A|B\} = \Pr\{A\}$, the event of a head on the first flip is independent of that of a tail on the second flip. Since $\Pr\{C|D\} \neq \Pr\{C\}$, the event of a match and that of two heads are not independent.

1.6.7 Functions of a Random Variable

"Everybody talks about the weather, but nobody does anything." We, as communication engineers, would be open to the same type of accusation if all we ever did was make statements such as "there is a 42% probability that the noise will cause a bit error." A significant part of communication engineering involves itself with changing noise from one form to another in the hope that the new form will be less annoying than the old. We must therefore study the effects of processing upon random phenomena.

Consider a function of a random variable, $y = g(x)$, where X is a random variable with known density function. Since X is random, Y is also random. We therefore ask what the density function of Y will be. The event $\{x_1 < X \leqslant x_2\}$ corresponds to the event $\{y_1 < Y \leqslant y_2\}$,* where

$$y_1 = g(x_1) \quad \text{and} \quad y_2 = g(x_2)$$

That is, the two events are identical, since they include the same outcomes. We are assuming for the moment that $g(x)$ is a single-valued function. Since the events are identical, their probabilities must also be equal:

$$\Pr\{x_1 < X \leqslant x_2\} = \Pr\{y_1 < Y \leqslant y_2\}$$

and in terms of the densities,

$$\int_{x_1}^{x_2} p_X(x)\,dx = \int_{y_1}^{y_2} p_Y(y)\,dy$$

If we now let x_2 get very close to x_1, this equation becomes

$$p_X(x_1)\,dx = p_Y(y_1)\,dy$$

and finally

$$p_Y(y_1) = \frac{p_X(x_1)}{dy/dx}$$

If, on the other hand, $y_1 > y_2$, we find (you should prove this result) that

$$p_Y(y_1) = -\frac{p_X(x_1)}{dy/dx}$$

We can account for both of these cases by writing

$$p_Y(y_1) = \frac{p_X(x_1)}{|dy/dx|}$$

Finally, writing $x_1 = g^{-1}(y_1)$, and realizing that y_1 can be set equal to any value, we have

$$p_Y(y) = \frac{p_X(g^{-1}(y))}{|dy/dx|} \tag{1-9}$$

*We are assuming that $y_1 < y_2$ if $x_1 < x_2$. That is, $g(x)$ is monotonic increasing and has a positive derivative. If this is not the case, the inequalities will have to be reversed.

If the function $g(x)$ is not monotone, the event $\{y_1 < Y \leqslant y_2\}$ corresponds to several intervals of the variable X. For example, if $g(x) = x^2$, then the event $\{1 < Y \leqslant 4\}$ is the same as the event $\{1 < X \leqslant 2\}$ or $\{-2 < X \leqslant -1\}$. Therefore,

$$\int_1^2 p_X(x)\, dx + \int_{-2}^{-1} p_X(x)\, dx = \int_1^4 p_Y(y)\, dy$$

In terms of the density functions, this would mean that $g^{-1}(y)$ has several values. Denoting these values as x_a and x_b, then

$$p_Y(y) = \left.\frac{p_X(x)}{|dy/dx|}\right|_{x=x_a} + \left.\frac{p_X(x)}{|dy/dx|}\right|_{x=x_b} \tag{1-10}$$

Example 1-10

A random voltage, v, is put through a full-wave rectifier. The voltage is uniformly distributed between -2 volts and $+2$ volts. Find the density of the output of the full-wave rectifier.

Solution. Calling the output y, we have $y = g(v)$, where $g(v)$ and the density of V are sketched in Fig. 1.8. Note that we have made the "natural" assignment of the random variable by letting it be equal to the value of voltage. At every value of V, $|dg/dv| = 1$. For $y > 0$, $g^{-1}(y) = \pm y$. For $y < 0$, $g^{-1}(y)$ is undefined. That is, there are no values of v for which $g(v)$ is negative. Equation (1-10) is then used to find

$$p_Y(y) = p_X(y) + p_X(-y), \qquad y > 0$$

$$p_Y(y) = 0, \qquad y < 0$$

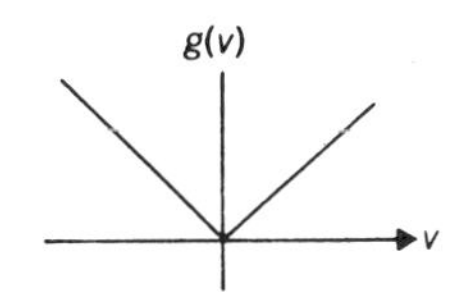

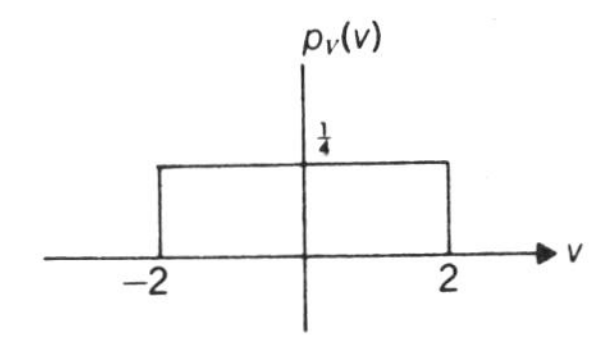

Figure 1.8 Function and probability density for Example 1-10

This result is shown in Fig. 1.9.

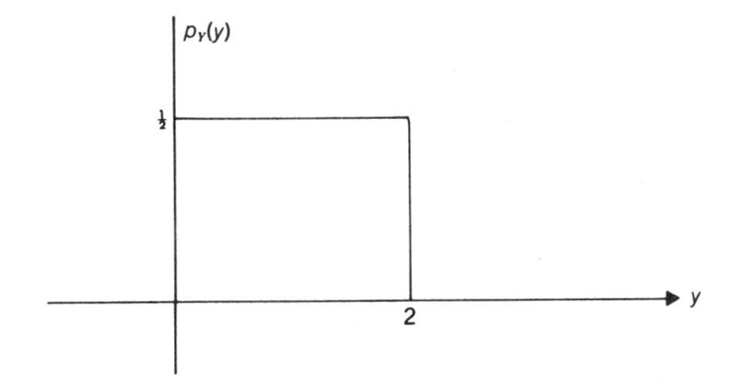

Figure 1.9 Output probability density function for Example 1-10

1.6.8 Expected Values

Expected values, or averages, are extremely important in communications. As one example, the average of the square of a voltage is closely related to the power generated by that voltage. The power of the noise voltage is an important measure of the disturbance caused by that voltage.

Several classes of expected values come up often enough to be given names. These are the mean, variance, and moments of a random variable. We shall define these terms in the following paragraphs.

Picture yourself as a teacher who has just given an examination. How would you average the resulting grades? You would probably add them all together and divide by the number of grades. Thus, if an experiment is performed many times, the average of the random variable resulting is found in the same way.

Let x_i, $i = 1, 2, \ldots, M$, represent the possible values of the random variable, and let n_i represent the number of times x_i occurs as the random variable assigned to the outcome. The average of the random variable after N performances of the experiment would therefore be

$$X_{\text{avg}} = \frac{1}{N} \sum_{i=1}^{M} n_i x_i$$
$$= \sum_{i=1}^{M} \frac{n_i}{N} x_i$$

Since x_i has ranged over all possible values of the random variable, we observe that

$$\sum_{i=1}^{M} n_i = N$$

As N approaches infinity, n_i/N becomes the probability, $\Pr\{x_i\}$. Therefore,

$$X_{\text{avg}} = \sum_{i=1}^{M} x_i \Pr\{x_i\} \tag{1-11}$$

This average value is sometimes called the *mean* or *expected value* of X and is often given the symbol m_x.

Now suppose that we wish to find the average value of a continuous random variable. We can use the previous result of Eq. (1-11) if we first round off the continuous variable to the nearest multiple of Δx. Thus, if X is between $k\,\Delta x - \frac{1}{2}\Delta x$ and $k\,\Delta x + \frac{1}{2}\Delta x$, we shall round it off to $k\,\Delta x$. The probability of X being in this range is given by

$$\int_{k\Delta x-(1/2)\Delta x}^{k\Delta x+(1/2)\Delta x} p_X(x)\,dx$$

which, if Δx is small, approximately equals $p_X(k\,\Delta x)\,\Delta x$. Therefore, Eq. (1-11) can be rewritten as

$$\sum_{k=-\infty}^{\infty} k\,\Delta x\, p_X(k\,\Delta x)\,\Delta x \approx X_{\text{avg}}$$

As Δx approaches zero, this becomes

$$X_{\text{avg}} = m_x = \int_{-\infty}^{\infty} x p_X(x)\,dx \tag{1-12}$$

The same reasoning can be used to find the average of any function of the random variable. Suppose that y were defined as $y = g(x)$. The expected value of Y is given by Eq. (1-12), where we simply substitute y for x:

$$Y_{\text{avg}} = \int_{-\infty}^{\infty} y p_Y(y)\,dy$$

We now express $p_Y(y)$ in terms of $p_X(x)$ in the form given by Eq. (1-9):

$$Y_{\text{avg}} = \int_{-\infty}^{\infty} y\, p_X(g^{-1}(y))\, \frac{dy}{|dy/dx|}$$

which, after substituting for y, becomes

$$[g(x)]_{\text{avg}} = \int_{-\infty}^{\infty} g(x)\, p_X(x)\,dx \tag{1-13}$$

This is an extremely useful equation. It tells us that in order to find the expected value of a function of X, we simply integrate that function weighted by the *density of X*. It is not necessary to first find the density of the new random variable.

We shall switch between several different standard notational forms for the average value operator. These are

$$[g(x)]_{\text{avg}} = \overline{g(x)} = E\{g(x)\}$$

Often we find the expected value of a power of the random variable. This is called the *moment*. Thus, the expected value of x^n is known as the nth *moment* of the random variable X.

If we first shift the random variable by its mean and then take a moment of the resulting variable, we obtain the *central moment*. Thus, the nth central moment is given by the expected value of $(x - m_x)^n$. The *second central moment* is extremely important as it relates to power. It is called the *variance* and represented by the symbol σ^2. Thus, the variance is given by

$$\sigma^2 = E\{(x - m_x)^2\}$$

$$= \int_{-\infty}^{\infty} (x - m_x)^2 p_X(x)\,dx$$

The variance gives a measure of how far we can expect the variable to deviate from its mean value. As the variance gets larger, the density function tends to "spread out."

Example 1-11

Find the average value of a binomial distributed variable.

Solution. The distribution function is given by

$$P(k) = \binom{N}{k} p^k(1-p)^{N-k}$$

The mean value is found by adding together the various values of k weighted by their probabilities. Thus,

$$\begin{aligned} E(K) &= \sum_{k=1}^{N} k \binom{N}{k} p^k(1-p)^{N-k} \\ &= \sum_{k=1}^{N} k \frac{N!}{k!(N-k)!} p^k(1-p)^{N-k} \\ &= \sum_{k=1}^{N} \frac{N!}{(k-1)!(N-k)!} p^k(1-p)^{N-k} \\ &= \sum_{k=1}^{N} \frac{(N-1)!}{(k-1)!(N-k)!} Np\, p^{k-1}(1-p)^{N-k} \end{aligned}$$

The binomial theorem shows that

$$(x+y)^r = \sum_{N=0}^{r} \binom{r}{N} x^N y^{r-N}$$

so the mean value becomes

$$E(K) = Np[p + (1-p)]^{N-1} = Np$$

This makes intuitive sense. For example, suppose the experiment is one of tossing a coin, and p is the probability of heads, or $\frac{1}{2}$. This states that if we toss the coin 100 times, the average number of heads is 0.5×100, or 50 heads.

Example 1-12

X is uniformly distributed as shown in Fig. 1.10. Find $E\{x\}$, $E\{x^2\}$, $E\{\cos x\}$ and $E\{(x-m_x)^2\}$.

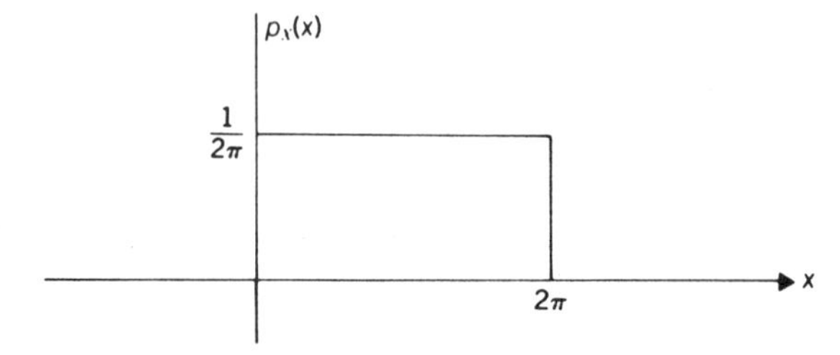

Figure 1.10 Probability density function of X for Example 1-12

Solution.

$$E\{x\} = \int_{-\infty}^{\infty} x p_X(x)\, dx = \frac{1}{2\pi}\int_0^{2\pi} x\, dx = \pi$$

$$E\{x^2\} = \int_{-\infty}^{\infty} x^2 p_X(x)\, dx = \frac{1}{2\pi}\int_0^{2\pi} x^2\, dx = \frac{4}{3}\pi^2$$

$$E\{\cos x\} = \int_{-\infty}^{\infty} \cos x p_X(x)\, dx = \frac{1}{2\pi}\int_0^{2\pi} \cos x\, dx = 0$$

$$E\{(x-\pi)^2\} = \int_{-\infty}^{\infty} (x-\pi)^2 p_X(x)\, dx = \frac{1}{2\pi}\int_0^{2\pi} (x-\pi)^2\, dx = \frac{\pi^2}{3}$$

Example 1-13

X is a Gaussian-distributed random variable with density function

$$p_x(x) = \frac{1}{\sqrt{2\pi}\sigma} e^{-(x-m)^2/2\sigma^2}$$

(a) Show that $p_X(x)$ integrates to unity.
(b) Find $E\{x\}$.
(c) Find $E\{(x - m_x)^2\}$.

Solution

(a) We first write the integral directly:

$$\frac{1}{\sqrt{2\pi}\sigma}\int_{-\infty}^{+\infty} e^{-(x-m)^2/2\sigma^2}\, dx$$

Making a change of variables, $y = (x - m)/\sigma$, we have

$$\frac{1}{\sqrt{2\pi}\sigma}\int_{-\infty}^{+\infty} e^{-(x-m)^2/2\sigma^2}\, dx = \frac{1}{\sqrt{2\pi}}\int_{-\infty}^{+\infty} e^{-y^2/2}\, dy$$

This integral cannot be evaluated in closed form for general limits of integration. In the case of infinite limits, the integral can be evaluated. Although the infinite limit integral can be found in most any table of integrals, it is instructive to show a "trick" for evaluating it. Define

$$I = \int_{-\infty}^{+\infty} e^{-y^2/2}\, dy$$

Then

$$I^2 = \int_{-\infty}^{+\infty}\int_{-\infty}^{+\infty} e^{-x^2/2} e^{-y^2/2}\, dx\, dy$$

$$= \int_{-\infty}^{+\infty}\int_{-\infty}^{+\infty} e^{-(x^2+y^2)/2}\, dx\, dy$$

Transforming to polar coordinates, we have

$$x^2 + y^2 = r^2$$

and

$$dx\,dy = r\,dr\,d\theta$$

$$I^2 = \int_0^{2\pi}\int_0^{\infty} e^{-r^2/2}\, r\,dr\,d\theta$$

Since the integrand is not a function of θ, this becomes

$$I^2 = 2\pi \int_0^{\infty} re^{-r^2/2}\,dr$$

The integrand in this equation is the derivative of $-e^{-r^2/2}$. Therefore,

$$I^2 = 2\pi\left(-e^{-r^2/2}\Big|_0^{\infty}\right) = 2\pi$$

and

$$I = \sqrt{2\pi}$$

Finally,

$$\frac{1}{\sqrt{2\pi}} I = 1$$

which is the desired result.

(b)

$$\begin{aligned}
E\{x\} &= \frac{1}{\sqrt{2\pi}\sigma}\int_{-\infty}^{+\infty} xe^{-(x-m)^2/2\sigma^2}\,dx \\
&= \frac{1}{\sqrt{2\pi}\sigma}\int_{-\infty}^{+\infty} (x-m)e^{-(x-m)^2/2\sigma^2}\,dx \\
&\quad + \frac{1}{\sqrt{2\pi}\sigma}\int_{-\infty}^{+\infty} e^{-(x-m)^2/2\sigma^2}\,dx \\
&= \frac{\sigma}{\sqrt{2\pi}} e^{-(x-m)^2/2\sigma^2}\Big|_{-\infty}^{+\infty} + m \\
&= 0 + m = m
\end{aligned}$$

Therefore, the m in the formula for the Gaussian density is actually the mean value of the random variable. This should have been obvious from the symmetry displayed in Fig. 1.5.

(c)

$$E\{(x-m)^2\} = \frac{1}{\sqrt{2\pi}\sigma}\int_{-\infty}^{+\infty} (x-m)^2 e^{-(x-m)^2/2\sigma^2}\,dx$$

This can be integrated by parts. Let

$$u = x - m$$

$$dv = (x - m)e^{-(x-m)^2/2\sigma^2}\,dx$$

Then

$$du = dx$$

$$v = -\sigma^2 e^{-(x-m)^2/2\sigma^2}$$

$$E\{(x - m)^2\} = \frac{\sigma^2}{\sqrt{2\pi}\sigma}\int_{-\infty}^{+\infty} e^{-(x-m)^2/2\sigma^2}\,dx$$

$$- (x - m)\sigma^2 e^{-(x-m)^2/2\sigma^2}\Big|_{-\infty}^{+\infty}$$

$$= \sigma^2$$

Therefore, σ^2 in the formula for the Gaussian density is the variance of the random variable.

Example 1-14

X is a Rayleigh distributed random variable with density function

$$p_X(x) = \frac{x}{4}\exp\left(-\frac{x^2}{8}\right)U(x)$$

Find the mean and variance of the random variable.

Solution. The mean value is given by

$$\int_{-\infty}^{\infty} x\,p_X(x)\,dx = \int_0^{\infty}\frac{x^2}{4}\exp\left(-\frac{x^2}{8}\right)dx = \sqrt{2\pi}$$

The variance is given by

$$\int_{-\infty}^{\infty}(x - \sqrt{2\pi})^2 p_X(x)\,dx = 1.7$$

We have consulted tables of integrals to evaluate the mean and variance. Alternatively, integration by parts could be used.

1.6.9 Multidimensional Random Variables

What if more than one parameter were required to describe the outcome of an experiment? For example, in the die-tossing experiment, consider the outcome as the face that is showing *and* the time at which the die hits the table. We would therefore have to define two random variables. The density function would become two-dimensional and would be defined by

$$\Pr\{x_1 < X \leqslant x_2 \text{ and } y_1 < Y \leqslant y_2\} = \int_{y_1}^{y_2}\int_{x_1}^{x_2} p(x, y)\,dx\,dy$$

Example 1-15

X and Y are each Gaussian random variables and they are independent of each other. What is their joint density?

Solution

$$P_X(x) = \frac{1}{\sqrt{2\pi}\sigma_1} \exp\left[-\frac{(x-m_1)^2}{2\sigma_1^2}\right]$$

$$p_Y(y) = \frac{1}{\sqrt{2\pi}\sigma_2} \exp\left[-\frac{(y-m_2)^2}{2\sigma_2^2}\right]$$

$$p(x, y) = \frac{1}{2\pi\sigma_1\sigma_2} \exp\left[-\frac{(x-m_1)^2}{2\sigma_1^2}\right] \exp\left[-\frac{(y-m_2)^2}{2\sigma_2^2}\right]$$

The previous example found the joint density of *independent* Gaussian variables by simply multiplying the individual densities. We will need an expression for multidimensional Gaussian random variables in order to evaluate the performance of various digital receivers. The variables will not always be independent.

If two zero-mean random variables are jointly Gaussian, the joint density can be expressed as

$$p(x, y) = \frac{1}{2\pi\sqrt{\sigma_x^2\sigma_y^2\sigma_{xy}^2}} \exp\left(\frac{-x^2\sigma_y^2 + 2\sigma_{xy}xy - y^2\sigma_x^2}{2(\sigma_x^2\sigma_y^2 - \sigma_{xy}^2)}\right) \tag{1-14}$$

σ_x^2 and σ_y^2 are the variances of x and y, respectively, and σ_{xy} is the *covariance.* It is equal to the expected value of the product, xy. Note that if x and y are independent, the expected value of the product is the product of the expected values. Since the variables are zero mean, this product would be zero. You should verify that, with $\sigma_{xy} = 0$, Eq. (1-14) reduces to the product of two Gaussian densities.

Suppose we now relax the zero-mean restriction and increase the number of variables to N. We can then express the multidimensional Gaussian density in matrix form as

$$p(\bar{X}) = \frac{1}{(2\pi)^{N/2}\sqrt{|R|}} \exp\left[\frac{-(\bar{X} - \bar{M})^T[R]^{-1}(\bar{X} - \bar{M})}{2}\right]$$

where

$$\bar{X} = \begin{bmatrix} x_1 \\ x_2 \\ \cdot \\ \cdot \\ \cdot \\ x_N \end{bmatrix}, \quad \bar{M} = \begin{bmatrix} m_1 \\ m_2 \\ \cdot \\ \cdot \\ \cdot \\ m_N \end{bmatrix}$$

and

$$[R] =$$

$$\begin{bmatrix} E\{x_1 - m_1)^2\} & E\{(x_1 - m_1)(x_2 - m_2)\} & \cdots & E\{(x_1 - m_1)(x_N - m_N)\} \\ E\{(x_2 - m_2)(x_1 - m_1)\} & E\{(x_2 - m_2)^2\} & & E\{(x_2 - m_2)(x_N - m_N)\} \\ \cdot & \cdot & & \cdot \\ \cdot & \cdot & \cdots & \cdot \\ \cdot & \cdot & \cdots & \cdot \\ E\{(x_N - m_N)(x_1 - m_1)\} & E\{(x_N - m_N)(x_2 - m_2)\} & \cdots & E\{(x_N - m_N)^2\} \end{bmatrix}$$

We now examine functions of multidimensional random variables. Suppose that a new set of random variables, y_1 through y_N, is defined according to the functions,

$$\begin{aligned} x_1 &= f_1(y_1, y_2, \ldots, y_N) \\ x_2 &= f_2(y_1, y_2, \ldots, y_N) \\ &\cdot \\ &\cdot \\ &\cdot \\ x_N &= f_N(y_1, y_2, \ldots, y_N) \end{aligned}$$

The probability that x is within some N-dimensional volume, A, is given by

$$\int \underset{A}{\cdots} \int p_X(x_1, x_2, \ldots, x_N)\, dx_1\, dx_2 \cdots dx_N$$

If we now make a formal change of variables, this integral becomes

$$\int \underset{B}{\cdots} \int p_X[f_1(y_1 \cdots y_N), f_2(y_1 \cdots y_N), \ldots, f_N(y_1 \cdots y_N)]\, |J|\, dy_1 \cdots dy_N \tag{1-15}$$

where J is the Jacobian of the transformation, given by

$$J = \begin{bmatrix} \dfrac{\partial f_1}{\partial y_1} & \cdots & \dfrac{\partial f_N}{\partial y_1} \\ \cdot & & \cdot \\ \cdot & & \cdot \\ \cdot & & \cdot \\ \dfrac{\partial f_1}{\partial y_N} & \cdots & \dfrac{\partial f_N}{\partial y_N} \end{bmatrix}$$

and B is the volume in y-space corresponding to A in x-space. The integrand in Eq. (1-15) is the joint density of $(y_1, y_2, \ldots, y_N)$.

Example 1-14

x_1 and x_2 are joint Gaussian, zero mean, with $\sigma_1^2 = \sigma_2^2 = 1$. $\sigma_{12} = \frac{1}{2}$. Two new variables are defined as

$$y_1 = x_1 + x_2$$

$$y_2 = 2x_1 + x_2$$

Find the joint density, $p_Y(y_1, y_2)$.

Solution

$$p_X(x_1, x_2) = \frac{1}{\pi} \exp\left(\frac{-x_1^2 + x_1 x_2 - x_2^2}{3/2}\right)$$

Solving for x_1 and x_2, we have

$$x_1 = -y_1 + y_2$$

$$x_2 = 2y_1 - y_2$$

and the Jacobian is given by

$$J = \begin{vmatrix} -1 & 2 \\ 1 & -1 \end{vmatrix} = -1$$

and

$$|J| = +1$$

The new density is proportional to

$$p_y(y_1, y_2) = \frac{1}{\pi} \exp\left\{\frac{2}{3}[-(-y_1 + y_2)^2 + (-y_1 + y_2)(2y_1 - y_2) - (2y_1 - y_2)^2]\right\}$$

$$= \frac{1}{\pi} \exp\left(\frac{-7y_1^2 + 9y_1 y_2 - 3y_2^2}{3/2}\right)$$

Thus, y_1 and y_2 are joint Gaussian with

$$\sigma_{y_2}^2 = 7, \qquad \sigma_{y_1}^2 = 3, \qquad \sigma_{y_1 y_2} = \tfrac{9}{2}$$

In Example 1-14 the new variables turned out to be joint Gaussian. This result is no accident. We now sketch the proof that any linear function of joint Gaussian variables results in joint Gaussian variables. This result will prove most useful to us later.

Let

$$\bar{X} = \begin{bmatrix} x_1 \\ \cdot \\ \cdot \\ \cdot \\ x_N \end{bmatrix}$$

be joint Gaussian.

Define a new set of variables, $\bar{Y}$, by

$$\bar{Y} = [L]\bar{X}$$

where $[L]$ is an $N \times N$ matrix. Then

$$\bar{X} = [L]^{-1}\bar{Y}$$

The Jacobian of this transformation can be shown (using the rules for matrix inversion) to be $1/|L|$. The density of $\bar{Y}$ becomes

$$p(\bar{Y}) = \frac{1}{(2\pi)^{N/2}(|L|^2|R|)^{1/2}} \exp\left[\frac{-(L^{-1}\bar{Y} - \overline{M})^T R^{-1}(L^{-1}\bar{Y} - \overline{M})}{2}\right]$$

We can rewrite this as

$$p(\bar{Y}) = \frac{1}{(2\pi)^{N/2}|F|^{1/2}} \exp\left[\frac{-(\bar{Y} - L\overline{M})^T F^{-1}(\bar{Y} - L\overline{M})}{2}\right] \tag{1-16}$$

where

$$[F] = [L][R][L]^T$$

Equation (1-16) describes a multidimensional Gaussian density. Therefore, *any linear combination of Guassian variables is itself Gaussian.*

1.6.10 Density Function of the Sum of Two Random Variables

Suppose that $z = x + y$, where x and y are random variables which are statistically independent of each other. We wish to find the density function of z in terms of the densities of x and y. The result will be useful in later work.

We start by finding the distribution function of the sum variable, z. The probability that z is less than some quantity, z_0, is the probability that the sum, $x + y$, is less than z_0. This is the probability that x and y fall within the shaded region of Fig. 1.11. Since we have assumed independence, the probability of falling in this shaded region is given by,

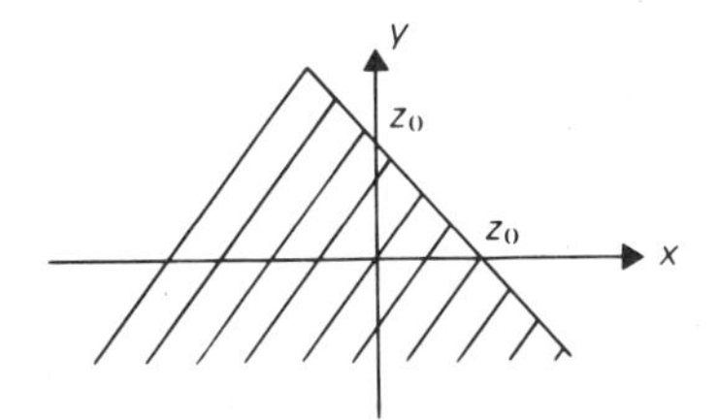

Figure 1.11 Region of integration for the sum of two random variables

$$F_Z(z_0) = \int_{-\infty}^{\infty}\int_{-\infty}^{z_0-y} p(x, y)\, dx\, dy$$

$$= \int_{-\infty}^{\infty}\int_{-\infty}^{z_0-y} p_X(x)p_Y(y)\, dx\, dy$$

$$= \int_{-\infty}^{\infty} p_Y(y) \int_{-\infty}^{z_0-y} p_X(x)\, dx\, dy$$

We now differentiate to find the density function.

$$p_Z(z_0) = \frac{dF_Z(z_0)}{dz_0}$$

$$= \int_{-\infty}^{\infty} p_Y(y)p_X(z_0 - y)\, dy \tag{1-17}$$

Equation (1-17) has a very simple interpretation. That is, the density of the sum of two independent random variables is equal to the convolution of the two individual densities. We used this result earlier to examine the central limit theorem.

1.6.11 Characteristic Functions

The convolution of functions is equivalent to multiplication of their Fourier transforms. We define

$$C_x(f) = E\{e^{j2\pi fx}\} = \int_{-\infty}^{\infty} p(x)e^{j2\pi fx}\, dx \tag{1-18}$$

$C(f)$ is simply the Fourier transform of $p(x)$. If we sum two random variables to form $z = x + y$, we have, from Eq. (1-18) and the convolution theorem of Fourier transforms,

$$p_Z(z) = \int_{-\infty}^{\infty} p_X(z - x)p_Y(x)\, dx$$

$$C_z(f) = C_x(f)C_y(f)$$

The function defined in Eq. (1-18) is known as the *characteristic function* of the random variable. Besides being useful for finding the distribution of a sum of random variables, the characteristic function can also be used for finding the expected value of powers of the random variable, the moments.

Differentiating both sides of Eq. (1-18) with respect to f yields

$$\frac{dC_x(f)}{df} = \int_{-\infty}^{\infty} j\,2\pi x p(x)e^{j2\pi fx}\, dx$$

and evaluating this at $f = 0$ we have

$$\left. \frac{dC_x(f)}{df} \right|_{f=0} = j\,2\pi E(x)$$

Thus, the mean of the random variable is found by evaluating the derivative of its characteristic function at $f = 0$ and dividing by $2\pi j$. Taking additional derivatives, we find that

$$E\{x^n\} = (j\,2\pi)^{-n} \left. \frac{d^n C_x(f)}{df^n} \right|_{f=0} \tag{1-19}$$

Example 1-15

Find the characteristic function of a Gaussian random variable with zero mean and unit variance. Use this to verify the values of the mean and variance.

Solution. The density and characteristic functions are given by

$$p(x) = \frac{1}{\sqrt{2\pi}} \exp\left(-\frac{x^2}{2}\right)$$

$$C_x(f) = \frac{1}{\sqrt{2\pi}} \int_{-\infty}^{\infty} \exp\left(-\frac{x^2}{2}\right) \exp\,(j\,2\pi f x)\,dx$$

We complete the square in the exponent to find

$$C_x(f) = \frac{1}{\sqrt{2\pi}} \int_{-\infty}^{\infty} \exp\left[-\frac{(x - j\,2\pi f)^2}{2}\right] \exp\left[\frac{(j\,2\pi f)^2}{2}\right] dx$$

$$= \exp\,(-2\pi^2 f^2) \frac{1}{\sqrt{2\pi}} \int_{-\infty}^{\infty} \exp\left[-\frac{(x - j\,2\pi f)^2}{2}\right] dx$$

But the integrand is a unit-variance, nonzero-mean Gaussian variable, so the value of the integral over the infinite limits is unity. Therefore,

$$C_x(f) = \exp\,(-2\pi^2 f^2)$$

We now find the first two moments.

$$E(x) = \frac{1}{j\,2\pi} \left. \frac{dC_x(f)}{df} \right|_{f=0}$$

$$= \frac{1}{j\,2\pi} \left[-4\pi^2 f \exp\,(-2\pi^2 f^2)\,|_{f=0}\right] = 0$$

$$E(x^2) = \frac{1}{(j\,2\pi)^2} \left. \frac{d^2 C_x(f)}{df^2} \right|_{f=0}$$

$$= 1$$

1.7 SPECTRAL ANALYSIS

We encounter two types of time signals in digital communication—continuous and discrete. These signals can be analyzed in either the time or frequency domain. The Fourier transform is used as a tool in frequency analysis.

1.7.1 The Discrete Fourier Transform

A continuous band-limited time signal can be represented by time samples. Practical system limitations often make it desirable to work with the sample values rather than with continuous functions of time. This is particularly true in data acquisition systems where the parameters being measured do not change rapidly.

We shall analyze a time record consisting of N samples. If the sampling period is T seconds, the total time record will be NT seconds long. We shall calculate the Fourier transform of this time-limited function. The continuous version of the Fourier transform is defined by

$$S(f) = \int_0^{NT} s(t)e^{-j2\pi ft}\,dt \tag{1-20}$$

If the time function is now discrete instead of continuous, we replace the integral of Eq. (1-20) with a summation.

$$S(f) = \frac{1}{N}\sum_{n=0}^{N-1} s(nT)e^{-j2\pi fnT}$$

Suppose we now assume that the limited time record of $s(t)$ represents one period of a periodic function. If the period is NT seconds, the fundamental frequency is the reciprocal of this, $f_0 = 1/NT$. The Fourier transform will have nonzero values only for this fundamental and its harmonics. We therefore sample f at multiples of the fundamental to get

$$S(mf_0) = \frac{1}{N}\sum_{n=0}^{N-1} s(nT)e^{-j2\pi nm/N} \tag{1-21}$$

We now define W as the Nth complex root of unity. That is,

$$W = \exp(-j2\pi/N)$$

With this definition, Eq. (1-21) becomes

$$S(mf_0) = \frac{1}{N}\sum_{n=0}^{N-1} s(nT)W^{mn} \tag{1-22}$$

This is the usual form of the *discrete Fourier transform* (DFT). It should be a simple matter to verify that this transform is periodic in m with period N. That is,

$$S[(N+m)f_0] = S(mf_0)$$

For this reason, it is necessary to calculate the transform for only N frequency points.

Example 1-16

Find the DFT of the following sampled time function:

t	0	T	$2T$	$3T$	$4T$	$5T$	$6T$	$7T$
$s(t)$	1	1	1	1	1	1	0	0

This is a sampled step as shown in Fig. 1.12.

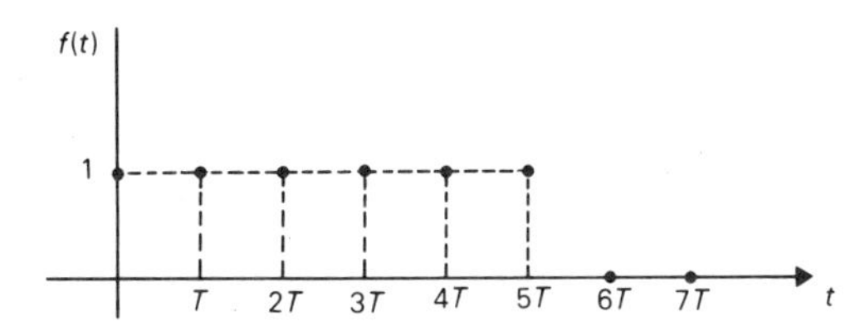

Figure 1.12 Sampled step function for Example 1-16

Solution. The number of sample points $N = 8$. Therefore,

$$S(m\Omega) = \frac{1}{8}\sum_{0}^{7} s(nT)W^{+mn}$$

$$= \frac{1}{8}(1 + W^{+m} + W^{+2m} + W^{+3m} + W^{+4m} + W^{+5m})$$

where

$$W = e^{-j2\pi/8} = \frac{1-j}{\sqrt{2}}$$

and

$$\Omega = \frac{1}{8T}$$

Therefore,

$$S(0) = \frac{6}{8}$$

$$S(\Omega) = \frac{1}{8}(1 + e^{-j\pi/4} + e^{-2j\pi/4} + e^{-3j\pi/4} + e^{-4j\pi/4} + e^{-5j\pi/4})$$

$$= \frac{1}{8}\left(-j - \frac{j}{\sqrt{2}} - \frac{1}{\sqrt{2}}\right)$$

$$S(2\Omega) = \frac{1}{8}(1 - j)$$

$$S(3\Omega) = \frac{1}{8}\left(j + \frac{1}{\sqrt{2}} - \frac{j}{\sqrt{2}}\right)$$

$$S(4\Omega) = 0$$

$$S(5\Omega) = \frac{1}{8}\left(-j + \frac{1}{\sqrt{2}} + \frac{j}{\sqrt{2}}\right)$$

$$S(6\Omega) = \frac{1}{8}(1 + j)$$

$$S(7\Omega) = \frac{1}{8}\left(j + \frac{j}{\sqrt{2}} - \frac{1}{\sqrt{2}}\right)$$

Because of the periodicity of the DFT, there is no need to calculate any additional samples of the transform. We note from the solution above that the real part of the DFT is symmetric about the midpoint and that the imaginary part of the DFT is antisymmetric about the midpoint. This is true in general and is derived from the property of continuous Fourier transforms; that is, $S(-f) = S^*(f)$. One can easily see from the DFT defining equation [Eq. (1-21)] that

$$S(-m\Omega) = S^*(m\Omega)$$

Since the DFT is periodic,

$$S[(m + N)\Omega] = S(m\Omega)$$

we have

$$S[(N - m)\Omega] = S^*(m\Omega)$$

which is a mathematical statement of the observation above. The DFT can be shown to have other properties similar to those of the continuous Fourier transform. Some of these properties are explored in the problems at the end of the chapter.

1.7.2 The Fast Fourier Transform

The discrete Fourier transform is a convenient way to approximate the Fourier transform when sampled waves are involved, and it lends itself readily to computation on a digital computer. Computation of the DFT as in Eq. (1-21) would involve summing N products (complex) for each of N frequency sample points. If N is large, this can become quite tedious.

The fast Fourier transform (FFT) is an algorithm that greatly simplifies and shortens the calculations that must be done to find the DFT. We begin by rewriting Eq. (1-22) in matrix form:

$$S(m\Omega) = [W]s(nT) \tag{1-23}$$

We have multiplied the DFT by N. Any constant multiplication can be accounted for by including the reciprocal multiple when taking the inverse transform. $\overline{S(m\Omega)}$ is a column vector,

$$\overline{S(m\Omega)} = \begin{bmatrix} S(0) \\ S(1\Omega) \\ S(2\Omega) \\ S(3\Omega) \\ \cdot \\ \cdot \\ \cdot \\ S\{(N-1)\Omega\} \end{bmatrix}$$

and $\overline{s(nT)}$ is a column vector,

$$\overline{s(nT)} = \begin{bmatrix} s(0) \\ s(1T) \\ s(2T) \\ s(3T) \\ \cdot \\ \cdot \\ \cdot \\ s\{(N-1)T\} \end{bmatrix}$$

$[W]$ is a matrix of the form

$$[W] = \begin{bmatrix} W^0 & W^0 & W^0 & W^0 & \cdots & W^0 \\ W^0 & W^1 & W^2 & W^3 & \cdots & W^N \\ W^0 & W^2 & W^4 & W^6 & \cdots & W^{2N} \\ W^0 & W^3 & W^6 & W^9 & \cdots & W^{3N} \\ \cdot & \cdot & \cdot & \cdot & \cdots & \cdot \\ W^0 & W^{1N} & W^{2N} & W^{3N} & \cdots & W^{N^2} \end{bmatrix}$$

The FFT algorithm is derived by factoring the matrix $[W]$ into a product of matrices. This can be done in such a way that each of the new matrices contains many zeros, thereby simplifying the calculations.

Let us first take a more intuitive approach. We wish to calculate

$$S(m) = \sum_{n=0}^{N-1} s(n)W^{mn}$$

Assume that N is even. We can then separate the sequence $s(n)$ into two sequences, each having $N/2$ points. Call these two sequences $s_e(q)$ and $s_o(r)$, where s_e is composed of the even points (every other point) of s, and s_o is composed of the odd points. This splitting is illustrated for a representative sequence in Fig. 1.13. Each of the two new sequences has a discrete Fourier transform formed by the addition of $N/2$ terms.

$$S_e(m) = \sum_{q=0}^{N/2-1} s_e(q) W^{2mq}$$

$$0 \leqslant m < N/2 - 1$$

$$s_o(m) = \sum_{r=0}^{N/2-1} s_o(r) W^{2mr}$$

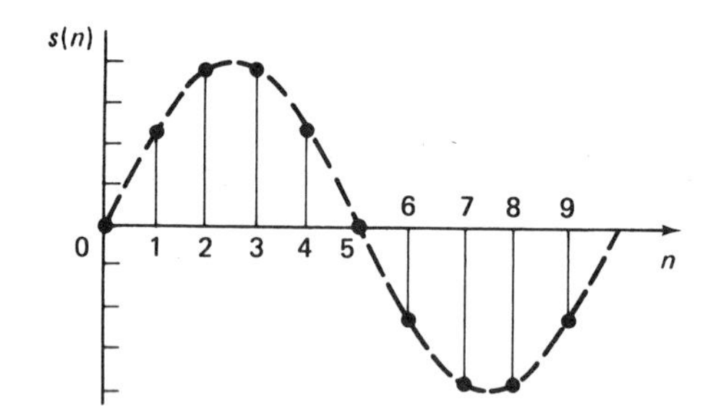

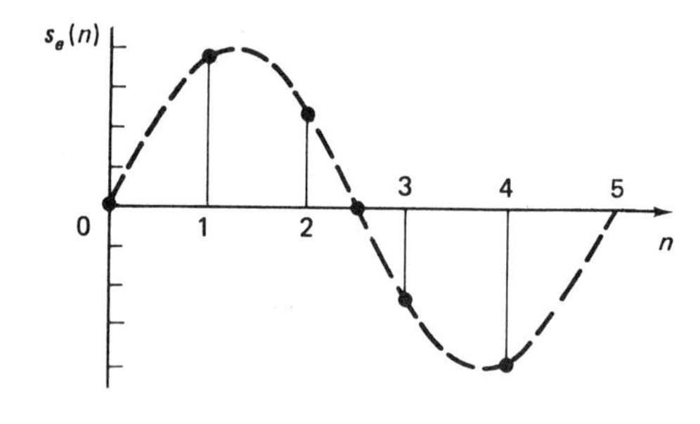

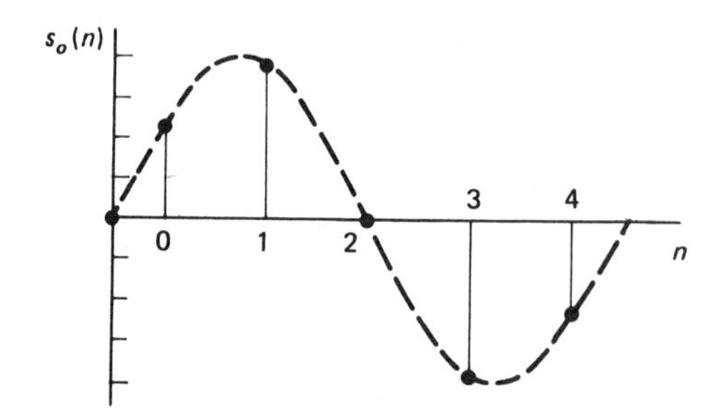

Figure 1.13 Sample splitting for fast Fourier transform

The original discrete Fourier transform can be related to these separate DFTs by noting that $s_e(q) = s(2q)$ and $s_o(r) = s(2r + 1)$. Therefore, the original DFT is given by

$$S(m) = \sum_{k=0}^{N/2-1} [s_e(k)M^{2mk} + s_o(k)W^{m(2k+1)}] \qquad 0 \leqslant m < N/2 - 1$$

$$= S_e(m) + W^m S_o(m) \tag{1-24}$$

This gives $S(m)$ for the first half of frequency points. To complete the picture, we note that $S_e(m)$ and $S_o(m)$ are periodic with period $N/2$. Therefore, to find $S(m)$ for $N/2 \leqslant m < N$, we plug in $m = m + N/2$ and

$$S(m + N/2) = S_e(m + N/2) + W^{m+N/2}S_o(m + N/2)$$

$$S(m + N/2) = S_e(m) + W^m W^{N/2} S_o(m)$$

but $W^{N/2} = -1$, so

$$S(m + N/2) = S_e(m) - W^m S_o(m) \tag{1-25}$$

Finally putting this together, we have

$$\begin{aligned} S(m) &= S_e(m) + W^m S_o(m) \\ S(m + N/2) &= S_e(m) - W^m S_o(m) \end{aligned} \qquad 0 \leqslant m \leqslant N/2 - 1$$

Calculation of $S_e(m)$ requires summation of $N/2$ complex products for each of the $N/2$ frequency points. We thus do $N^2/4$ complex arithmetic operations to find $S_e(m)$. The same is true of $S_o(m)$. Therefore, finding $S(m)$ requires $N + (N^2/2)$ operations. This compares to N^2 operations if we had simply proceeded to find $S(m)$ using Eq. (1-21).

But why stop there? Let us take s_e and s_o and split each of them in the same manner. We can continue this process as long as the number of samples at each step is still divisible by 2.

If the original number of points in the discrete function (samples of time function) is a power of 2, we can continue splitting in half until the problem has been reduced down to sequences of one sample each. We then note that the discrete Fourier transform of a single sample is the sample value itself. That is,

$$S(0) = \sum_{n=0}^{0} s(n)W^n = s(0)$$

Example 1-17

Find the discrete Fourier transform of the function shown in Fig. 1.14.

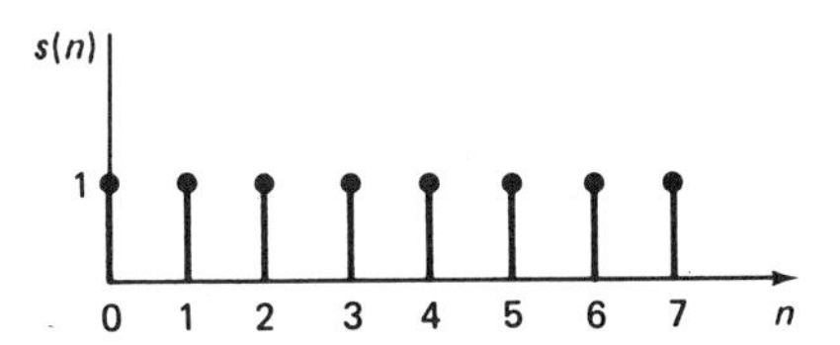

Figure 1.14 Function for Example 1-17

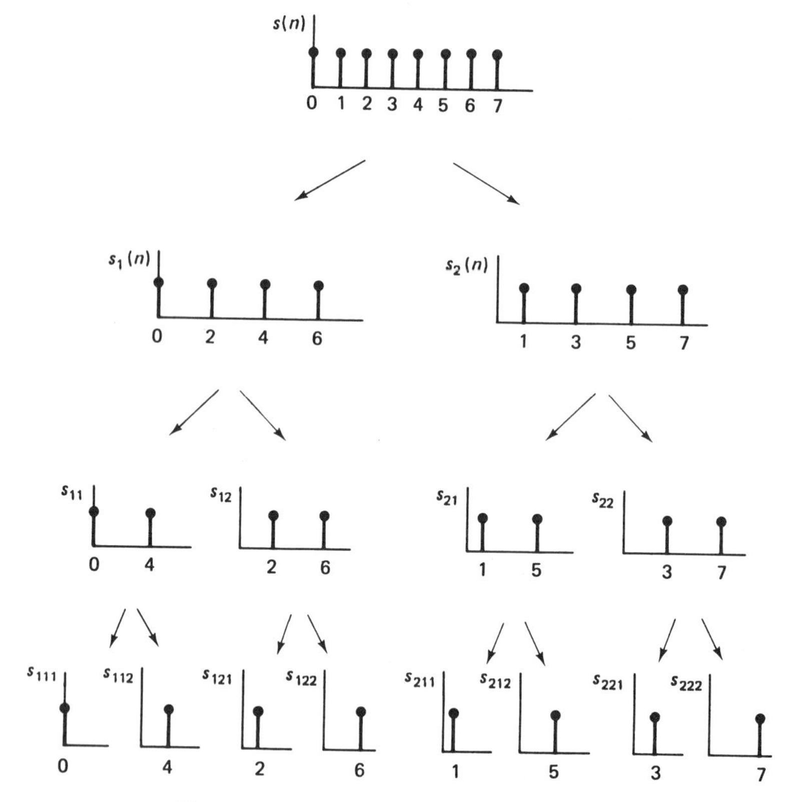

Figure 1.15 Sample splitting for Example 1-17

Solution. We start by successively splitting the samples. $s(n)$ is divided to form $s_1(n)$ and $s_2(n)$. Each of these is then divided two more times. The discrete functions are shown in Fig. 1.15. The DFT of $s(n)$ is then the sum of the DFT of s_1 with W^m times the DFT of s_2. But the DFT of s_1 can be expressed in terms of the DFTs of s_{11} and s_{12}, and so on. The required multiplications are most conveniently illustrated by the flowchart shown in Fig. 1.16. We start at the left of the chart with the eight reordered sample values. We then form eight intermediate variables, denoted A_0 through A_7. Each of these is the sum of two preceding variables, one of which is weighted by a power of W. The weighting factor is shown adjacent to the arrow on the chart. If no number appears next to an arrow, multiplication by unity is assumed. Note that W^0 is included to make the result more general, even though this could be replaced by 1.

After calculating the eight values of the A variable, we proceed to calculate B_0 through B_7 in a similar manner, and finally $S(0)$ through $S(7)$ in the same way. Once a new set of eight values is calculated, the preceding set is no longer used. Therefore, in addition to providing tremendous computational savings, the FFT can also significantly cut down on memory requirements when a computer is being used to calculate DFTs.

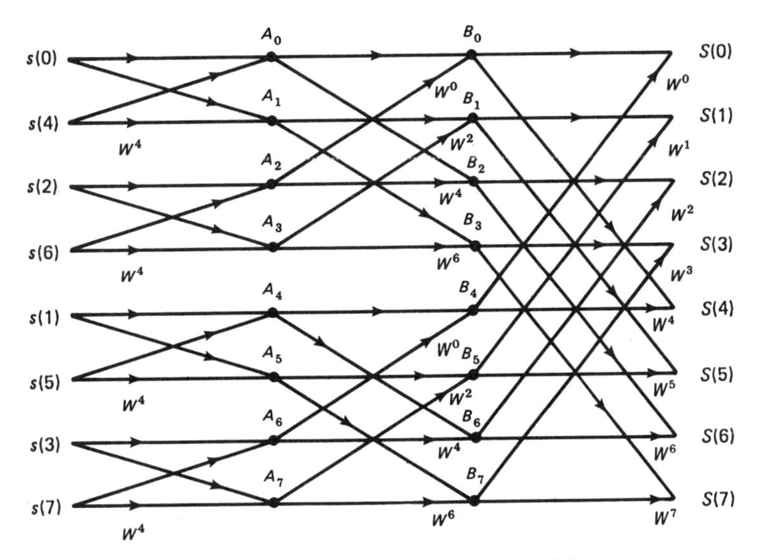

Figure 1.16 FFT flowchart for Example 1-17

Applying this flowchart to the example, we find that

$$A_0 = 2$$

$$A_1 = 1 + W^4$$

$$A_2 = 2$$

$$A_3 = 1 + W^4$$

$$A_4 = 2$$

$$A_5 = 1 + W^4$$

$$A_6 = 2$$

$$A_7 = 1 + W^4$$

$$B_0 = 4$$

$$B_1 = 1 + W^2 + W^4 + W^6$$

$$B_2 = 2 + 2W^4$$

$$B_3 = 1 + W^4 + W^6 + W^{10} = 1 + W^2 + W^4 + W^6$$

$$B_4 = 4$$

$$B_5 = 1 + W^2 + W^4 + W^6$$

$$B_6 = 2 + 2W^4$$

$$B_7 = 1 + W^4 + W^6 + W^{10} = 1 + W^2 + W^4 + W^6$$

and

$$S(0) = 8$$
$$S(1) = 1 + W^1 + W^2 + W^3 + W^4 + W^5 + W^6 + W^7$$
$$S(2) = 2(1 + W^2 + W^4 + W^6)$$
$$S(3) = 1 + W^1 + W^2 + W^3 + W^4 + W^5 + W^6 + W^7$$
$$S(4) = 4 + 4W^4$$
$$S(5) = 1 + W^1 + W^2 + W^3 + W^4 + W^5 + W^6 + W^7$$
$$S(6) = 2(1 + W^2 + W^4 + W^6)$$
$$S(7) = 1 + W^1 + W^2 + W^3 + W^4 + W^5 + W^6 + W^7$$

The student should substitute the values of the various powers of W to find the Fourier transform values. After this laborious task, you will realize that the computation is not really necessary. If the various vectors are sketched, it should become obvious that $S(0) = 8$ and all the other transform values are zero. Not terribly surprising when you realize we are taking the DFT of a sampled constant.

Examination of Example 1-17 would show that the number of arithmetic operations required at each level was N, where N is the number of samples of the original function. The number of levels is $\log_2 N$. Therefore, the total number of arithmetic operations required to form the FFT is $N \log_2 N$. Figure 1.17 shows this function together with N^2, which was the number of operations required to find the DFT. The figure is included to illustrate the dramatic computational savings offered by the FFT.

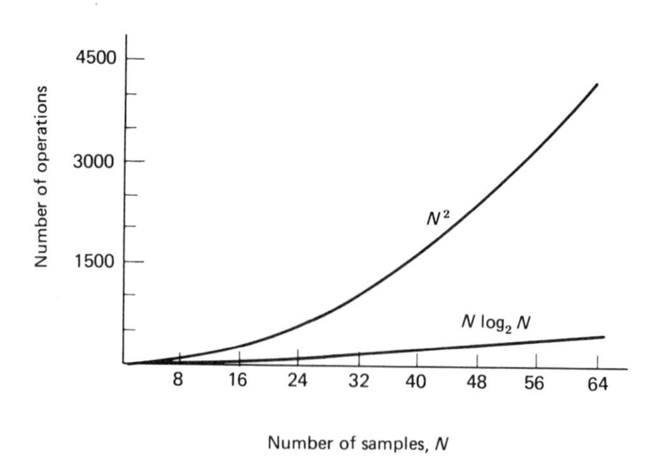

Figure 1.17 Comparison of number of operations to perform DFT vs. FFT

Example 1-18

Redo the discrete Fourier transform of Example 1-17, but this time use the fast Fourier transform. The function was given by

$$s(n) = \begin{cases} 1 & 0 \leqslant n \leqslant 5 \\ 0 & 6 \leqslant n \leqslant 8 \end{cases}$$

Solution. We solve for the FFT using the flowchart of Fig. 1.16. Figure 1.18 is a repeat of this chart, where the specific values of the variable at each node have been indicated. The Fourier transform is indicated by the eight values along the right side of the diagram. These are identical to those found earlier using the DFT.

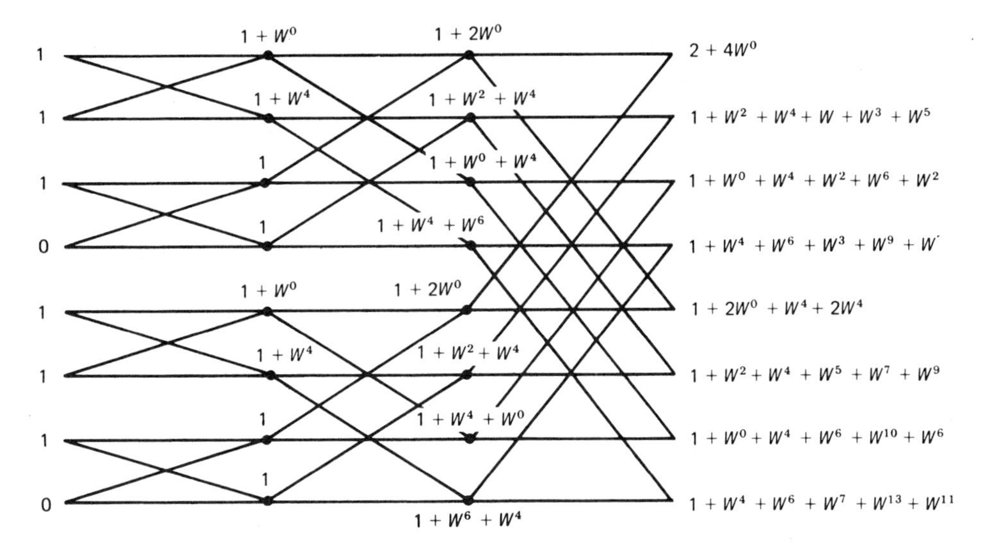

Figure 1.18 FFT flowchart for Example 1-18

We now briefly turn our attention to the matrix factorization formulation of the FFT. The flowchart of Fig. 1.16 is sufficient for finding Fourier transforms. We examine the matrix factorization only to gain more insight into the workings of the algorithm. Equation (1-23) is repeated below, where for simplicity we have assumed that $N = 4$.

$$\begin{bmatrix} S(0) \\ S(1) \\ S(2) \\ S(3) \end{bmatrix} = \begin{bmatrix} W^0 & W^0 & W^0 & W^0 \\ W^0 & W^1 & W^2 & W^3 \\ W^0 & W^2 & W^4 & W^6 \\ W^0 & W^3 & W^6 & W^9 \end{bmatrix} \begin{bmatrix} s(0) \\ s(1) \\ s(2) \\ s(3) \end{bmatrix}$$

The matrix can be factored into the product of two matrices, each of which has two zeros in each row.

$$\begin{bmatrix} W^0 & W^0 & W^0 & W^0 \\ W^0 & W^1 & W^2 & W^3 \\ W^0 & W^2 & W^4 & W^6 \\ W^0 & W^3 & W^6 & W^9 \end{bmatrix} = \begin{bmatrix} W^0 & W^0 & W^0 & W^0 \\ W^0 & W^1 & W^2 & W^3 \\ W^0 & W^2 & W^0 & W^2 \\ W^0 & W^3 & W^2 & W^1 \end{bmatrix}$$

$$= \begin{bmatrix} 1 & W^0 & 0 & 0 \\ 0 & 0 & 1 & W^1 \\ 1 & W^2 & 0 & 0 \\ 0 & 0 & 1 & W^3 \end{bmatrix} \begin{bmatrix} 1 & 0 & W^0 & 0 \\ 0 & 1 & 0 & W^0 \\ 1 & 0 & W^2 & 0 \\ 0 & 1 & 0 & W^2 \end{bmatrix}$$

This factorization leads to a two-step computation as illustrated in the flowchart of Fig. 1.19. This is the same flowchart that we would have derived using the previous splitting procedure. Note that if we started with an 8 × 8 matrix for $[W]$, we would have factored this into the product of three matrices, each of which had only two nonzero values in each row.

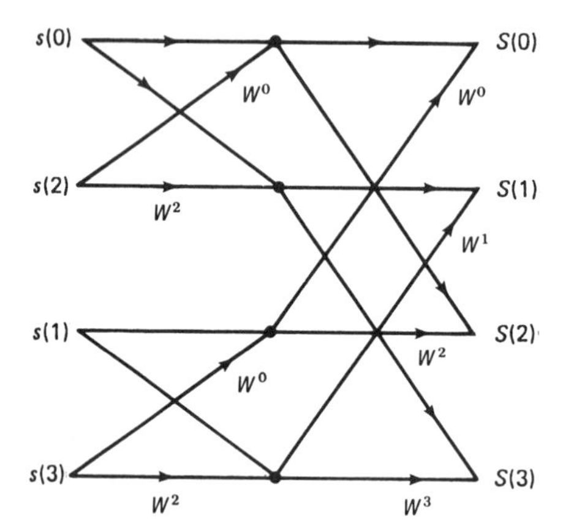

Figure 1.19 FFT flowchart for four samples

1.8 SUMMARY

This chapter lays the groundwork for our study of digital communication systems. Following a brief history presented in the first section, Section 1.2 develops the block diagram of a digital transmission and reception system. This block diagram will be referred to many times throughout your studies, since the remainder of this text expands upon the various blocks.

Section 1.3 presents the principal advantages and disadvantages of digital communications as compared to analog. The major advantages discussed are error-correction capability, ease of performing signal processing and encryption, and increased dynamic range. The disadvantages are increased bandwidth and the need for providing various levels of synchronization.

Section 1.4 contains a very basic review of numbering systems. This is included to make the reader comfortable in working with binary operations. Section 1.5 presents simple codes, many of which are important for primarily historical reasons. Some of the codes presented are still being used in contemporary systems.

Section 1.6 contains a thorough treatment of discrete and continuous probability theory. This material will prove critically important in analyzing the digital systems presented throughout this text. In particular, calculations of bit error rate require use of many results derived in this section.

We close the chapter with a discussion of frequency analysis using the discrete Fourier transform (DFT) and the fast Fourier transform (FFT). While this material is not used extensively in the analysis within the text, it is of great significance in digital signal processing.

PROBLEMS

1.1. Perform the following binary arithmetic operations.
- **(a)** 10110 × 11101
- **(b)** 110011 ÷ 11
- **(c)** 1011101 + 10001111
- **(d)** 100110110 − 1111

1.2. Perform the following binary subtractions.
- **(a)** 11001 − 1101
- **(b)** 11111 − 11011
- **(c)** 111 − 101010
- **(d)** 101011 − 1111

1.3. Perform the following octal arithmetic operations.
- **(a)** 5231 − 434
- **(b)** 7112 + 357
- **(c)** 17135 ÷ 325
- **(d)** 315 × 604

1.4. Perform the following hexadecimal arithmetic operations.
- **(a)** FA3B4 + 138C
- **(b)** FA3B4 − 59D7
- **(c)** 1F3B9 ÷ 2C18
- **(d)** 9ABC × 17F

1.5. Perform the following octal subractions.
- **(a)** 7777 − 2222
- **(b)** 13714 − 7325
- **(c)** 3172 − 5211
- **(d)** 7311 − 6422

1.6. Perform the following hexadecimal subtractions.
- **(a)** 178AB − FACE

(b) 905C − A328
(c) 731111 − AF82
(d) 99999 − 77777

1.7. Suppose that in the ASCII coding problem, uppercase letters are not required. Which bit could be dropped? Dropping this bit, give the code for the word "hello."

1.8. Four coins are tossed at the same time. List all possible outcomes of this experiment. List five representative events. Find the probabilities of the following events:
(a) All tails
(b) One head only
(c) Three matched

1.9. If a perfectly balanced die is rolled, find the probability that the number of spots facing upward is greater than or equal to 2.

1.10. An urn contains three white balls and seven black balls. An experiment is performed in which three balls are drawn out in succession without replacing any. List all possible outcomes and assign probabilities to each.

1.11. In Problem 1.10 you are told that the first two draws are white balls. Find the probability that the third is also a white ball.

1.12. An urn contains four red balls, seven green balls, and five white balls. Another urn contains five red balls, nine green balls, and two white balls. One ball is drawn from each urn. What is the probability that both balls will be of the same color?

1.13. Three people, A, B, and C, live in the same neighborhood and use the same busline to go to work. Each of the three has a probability of $\frac{1}{4}$ of making the 6:10 bus, a probability of $\frac{1}{2}$ of making the 6:15 bus, and a probability of $\frac{1}{4}$ of making the 6:20 bus. Assuming independence, what is the probability that they all take the same bus?

1.14. The probability density function of a certain voltage is given by

$$p_V(v) = ve^{-v}U(v)$$

where $U(v)$ is the unit step function.
(a) Sketch this probability density function.
(b) Sketch the distribution function of v.
(c) What is the probability that v is between 1 and 2 V?

1.15. Find the density of $y = |x|$ given that $p(x)$ is as shown below. Also find the mean and variance of both y and x.

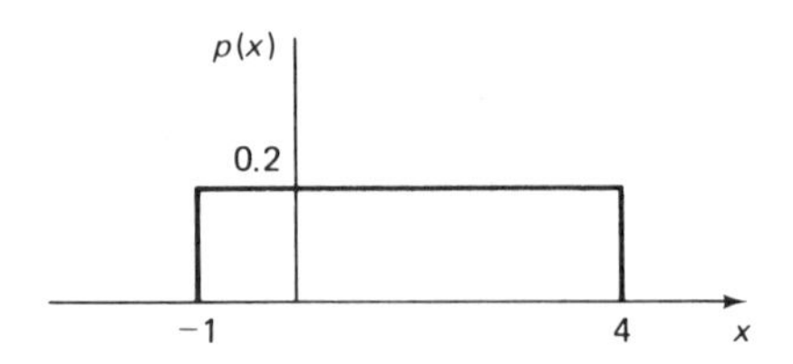

1.16. Find the mean and variance of x where the density of x is as shown below.

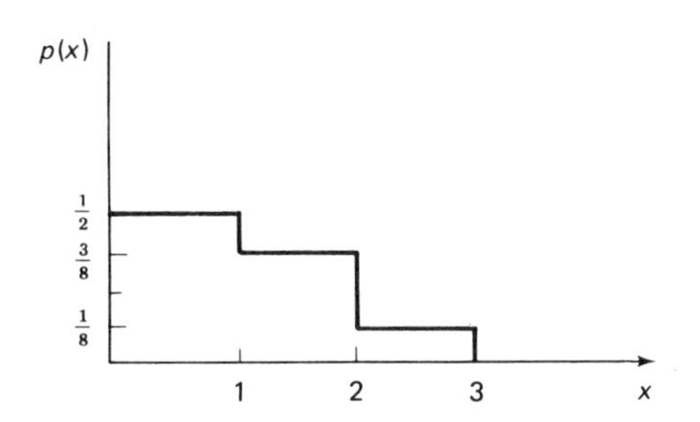

1.17. The density of x is shown below. A random variable, y, is related to x as shown in the figure. Determine the density of y.

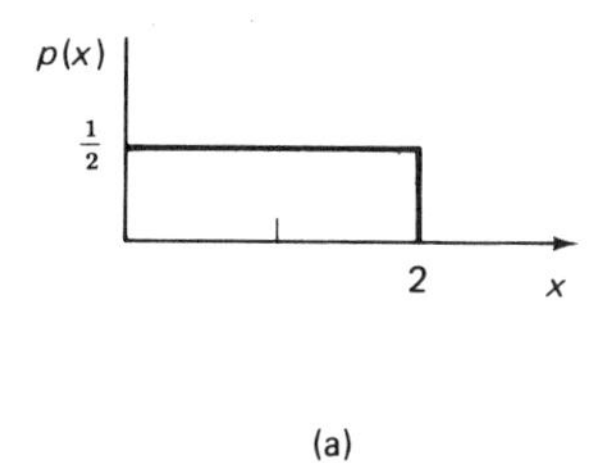

(a)

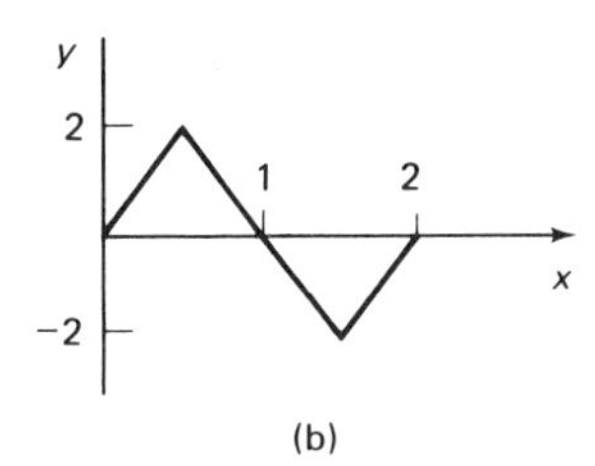

(b)

1.18. Find the expected value of $y = \sin x$ if x is uniformly distributed between 0 and 2π.

1.19. The random variable x is uniformly distributed between $x = -1$ and $x = +1$. The random variable y is uniformly distributed between $y = 0$ and $y = +5$. x and y are independent of each other. A new variable, z, is formed as the sum of x and y.

(a) Find the probability density function of z.

(b) Find the mean and variance of x, y, and z.

(c) Find a general relationship among the means of x, y, and z.

(d) Find a general relationship among the variances of x, y, and z.

1.20. A Gaussian density has mean 3 and variance 16.

(a) Write the equation for this density.

(b) Find the probability that the random variable is between 5 and 10.

1.21. You are told that the integral of a zero mean Gaussian density from $x = 5$ to ∞ is 0.01. Find the variance of the random variable.

1.22. x_1 and x_2 are joint Gaussian zero-mean variables.

$$\sigma_1^2 = 1, \qquad \sigma_2^2 = 2, \qquad \sigma_{12} = 1$$

Define

$$y_1 = x_1 + 2x_2$$

$$y_2 = 2x_1$$

Find the joint density, $p(y_1, y_2)$.

1.23. Using characteristic functions, find the first three moments of

$$p(x) = 2e^{-2x}, \qquad x > 0$$

1.24. A Gaussian random variable has a variance of 9. The probability that the variable is greater than 5 is 0.1. Find the mean of the random variable.

1.25. A zero-mean Gaussian random variable has a variance of 4. Find x_0 such that

$$\Pr(|x| > x_0) < 0.001$$

1.26. A random variable is Rayleigh distributed as follows.

$$p(x) = \frac{x}{4} \exp\left(\frac{-x^2}{8}\right) U(x)$$

Find the probability that the random variable is between 1 and 3.

1.27. Find the density of $y = |x|$ given that $p(x)$ is as shown below. Also find the mean and variance of both y and x.

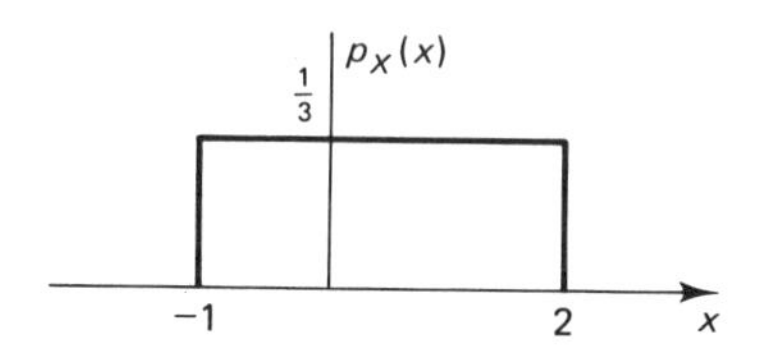

1.28. x is a Gaussian random variable with zero mean and unit variance. A new variable is defined by

$$y = |x|$$

(a) Find the density of y.
(b) Find the mean value of y.
(c) Find the variance of y.

1.29. A function is defined by

$$y = e^{-2x}$$

Find the density of y, $p(y)$, if
(a) x is uniformly distributed between 0 and 3.
(b) $p(x) = e^{-x}U(x)$.

1.30. Find the mean and variance of x where the density of x is shown below.

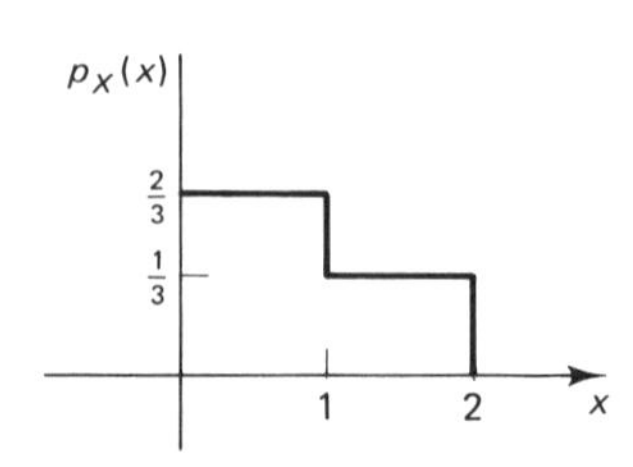

1.31. The density function of x is shown below. A random variable, y, is related to x as shown. Determine $p_Y(y)$.

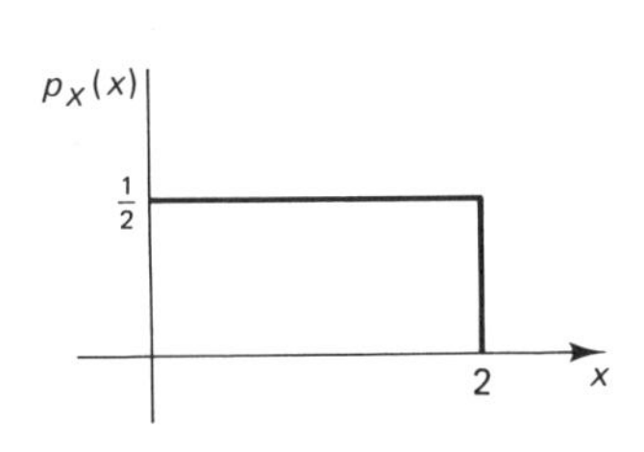

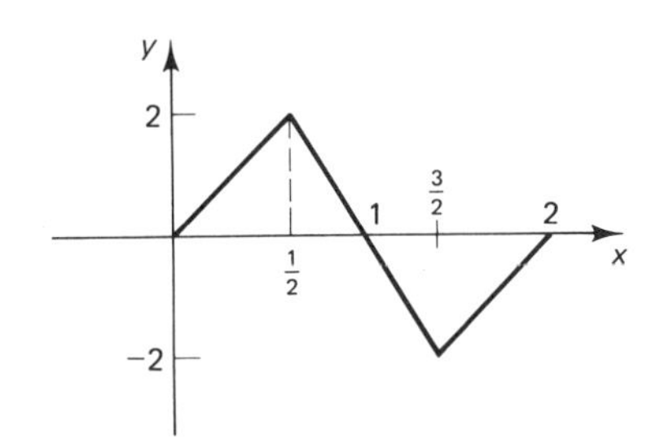

1.32. Find the expected value of $y = \sin x$ if x is uniformly distributed between 0 and 2π.

1.33. You are told that the mean rainfall per year is 10″ in California and that the variance in the amount of rain per year is 1. Can you tell, from this information, what the density of rainfall is? If so, roughly sketch this density function.

1.34. The random variable x is Rayleigh distributed. Find the mean, second moment, and variance.

1.35. **(a)** Find and sketch the density of the sum of two independent random variables as follows. One of the variables is uniformly distributed between −1 and +1. The second variable is triangularly distributed between −2 and +2.

(b) Compare the result to a Gaussian density function, where the variance should be chosen to make the Gaussian variable as close as possible to your answer to part (a). (You will have to start by defining "closeness.")

1.36. A random variable x has the following density function:

$$p_x(x) = Ae^{-2|x|}$$

(a) What is the value of A?

(b) Find the probability that x is between −3 and +3.

(c) Find the variance of x.

1.37. A voltage is known to be Gaussian distributed with mean value of 4. When this voltage is impressed across a 4-ohm resistor, the average power dissipated is 8 watts. Find the probability that the voltage exceeds 2 volts at any instant of time.

1.38. Find the DFT of the sample function shown below.

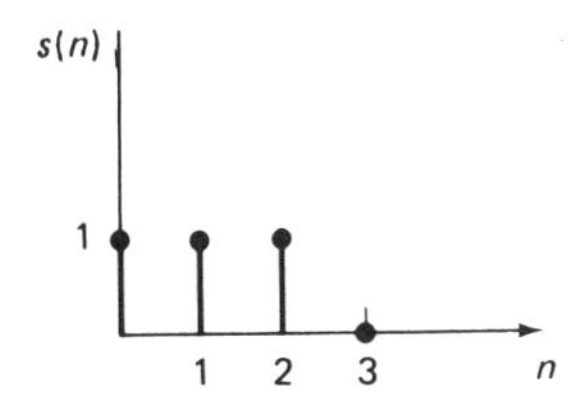

1.39. Show that

(a) the DFT of an even function is real.

(b) the DFT of an odd function is imaginary.

1.40. What is the effect upon the DFT of a function, $s(t)$, if the function is delayed by n sampling periods? That is, find the DFT of $s(t - nT_s)$ in terms of the DFT of $s(t)$.

1.41. Find the DFT of the convolution $f(t) * g(t)$ in terms of the DFT of $f(t)$ and the DFT of $g(t)$.

1.42. Find the flowchart for a 16-sample FFT.

1.43. Find the FFT of the sampled signal from Problem 1.38.

1.44. Find the FFT of the sampled ramp shown below.

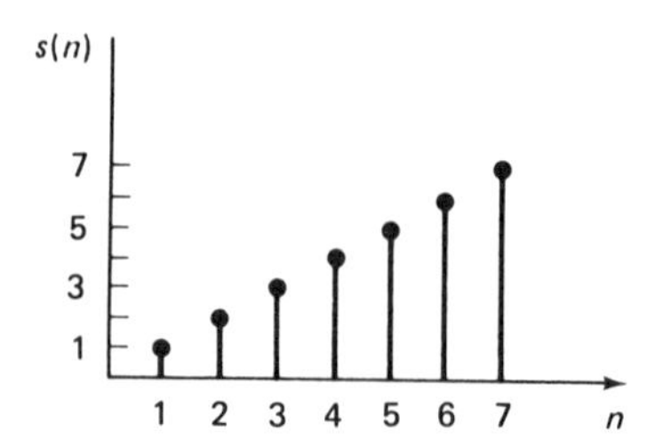

1.45. Find the DFT of the following function.

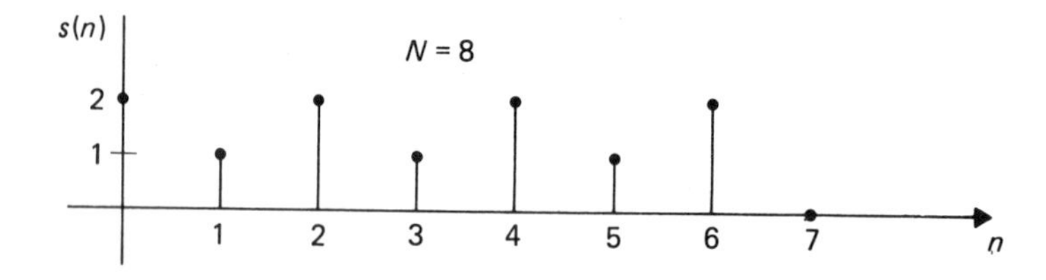

1.46. Find the DFT of the following function.

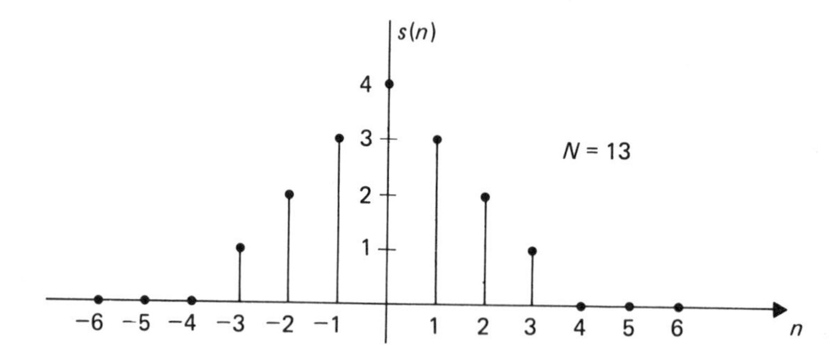

1.47. Find the FFT of the signals in Problem 1.45 and compare your answer with those found using the DFT.

2

The Channel

The *channel* is what separates the transmitter from the receiver in a communication system. Channels affect communication in two ways. First, a channel can alter the form of a signal during its movement from transmitter to receiver. If this alteration includes anything other than multiplication by a constant and/or shifting of the signal in time, we say that *distortion* occurs. Second, the channel can add *noise* waveforms to the original transmitted signal.

Data communication often takes place on telephone channels. Many of these channels are carryovers from a time when they were designed primarily for analog voice transmission. It is important for us, therefore, to examine the characteristics of the telephone channel.

2.1 THE MEMORYLESS CHANNEL

In order to characterize the channel's behavior for the digital configuration, we must introduce the concept of *memory*. A channel is *memoryless* if each element of the output sequence depends only upon the corresponding input sequence element and upon the channel characteristics. Figure 2.1 shows a block diagram of a channel with input sequence $s_{in}(n)$ and output sequence $s_{out}(n)$. If this channel is memoryless, $s_{out}(n)$ depends only upon $s_{in}(n)$, and not upon $s_{in}(n-1)$ or any other input sample values. This may seem a bit distressing, since in continuous system analysis we find that almost any interesting system has memory. In a linear time-invariant system, the output is expressed as a convolution of the input with the im-

Figure 2.1 Communication channel

pulse response. Unless the impulse response is itself an impulse, the output depends upon past values of the input, and the system has memory.

In analog electric circuits, the capacitors and inductors are the memory devices. In discrete systems, delay blocks constitute memory.

The memoryless channel can be characterized by a *transition matrix* composed of conditional probabilities. As an example consider the binary channel, where s_{in} can take on either of two values, 0 or 1. For a particular input bit, s_{out} can equal either 0 or 1. We can thus completely characterize the effects of the channel with four conditional probabilities defined as follows:

$$P\{s_{out} = 0 \mid s_{in} = 0\} = P_{00}$$

$$P\{s_{out} = 1 \mid s_{in} = 0\} = P_{10}$$

$$P\{s_{out} = 0 \mid s_{in} = 1\} = P_{01}$$

$$P\{s_{out} = 1 \mid s_{in} = 1\} = P_{11}$$

The transition probability matrix is then

$$[T] = \begin{bmatrix} P_{00} & P_{01} \\ P_{10} & P_{11} \end{bmatrix}$$

Note that in the absence of noise and distortion, one would expect $[T]$ to be the identity matrix. That is, $P_{ij} = 1$ if $i = j$ and 0 if $i \neq j$. Further note that the sum of entries in any column of the transition matrix must be unity, since, given the value of the input, the output must be one of the two possibilities.

If the input alphabet contains more than two symbols, the order of the transition matrix will increase accordingly.

Example 2-1

A digital communication system has a symbol alphabet composed of four entries, and a transition matrix given by the following:

$$[T] = \begin{bmatrix} \frac{1}{4} & \frac{1}{2} & \frac{1}{6} & \frac{1}{6} \\ \frac{1}{4} & \frac{1}{6} & \frac{1}{2} & \frac{1}{6} \\ \frac{1}{4} & \frac{1}{6} & \frac{1}{6} & \frac{1}{3} \\ \frac{1}{4} & \frac{1}{6} & \frac{1}{6} & \frac{1}{3} \end{bmatrix}$$

(a) Find the probability of a single transmitted symbol being in error assuming that all four input symbols are equally probable at any time.

(b) Find the probability of a correct symbol transmission.

(c) If the symbols are denoted A, B, C, and D, find the probability that the transmitted sequence BADCAB will be received as DADDAB.

Solution

(a)

$$P_e \,|\, 0 \text{ sent} = P_{10} + P_{20} + P_{30} = \tfrac{1}{4} + \tfrac{1}{4} + \tfrac{1}{4} = \tfrac{3}{4}$$

$$P_e \,|\, 1 \text{ sent} = P_{01} + P_{21} + P_{31} = \tfrac{1}{2} + \tfrac{1}{6} + \tfrac{1}{6} = \tfrac{5}{6}$$

$$P_e \,|\, 2 \text{ sent} = P_{02} + P_{12} + P_{32} = \tfrac{1}{6} + \tfrac{1}{2} + \tfrac{1}{6} = \tfrac{5}{6}$$

$$P_e \,|\, 3 \text{ sent} = P_{03} + P_{13} + P_{23} = \tfrac{1}{6} + \tfrac{1}{6} + \tfrac{1}{3} = \tfrac{2}{3}$$

Total probability of error is given by the average of the four quantities.

$$P_e = \sum_{i=0}^{3} (P_e \,|\, i \text{ sent})P(i)$$

$$= \frac{1}{4}\left(\frac{3}{4} + \frac{5}{6} + \frac{5}{6} + \frac{2}{3}\right) = \frac{37}{48}$$

(b)

$$P_c = P_{00}P(0) + P_{11}P(1) + P_{22}P(2) + P_{33}P(3)$$

$$= \frac{1}{4}\left(\frac{1}{4} + \frac{1}{6} + \frac{1}{6} + \frac{1}{3}\right) = \frac{11}{48}$$

Note that the sum,

$$P_e + P_c = 1$$

so we could have found the probability of correct transmission without solving the formula given above.

(c) Since each symbol is independent of the other symbols, the probability is given by the product

$$P(\text{DADDAB}) = P_{31}P_{00}P_{33}P_{32}P_{00}P_{11}$$

$$= \frac{1}{6} \times \frac{1}{4} \times \frac{1}{3} \times \frac{1}{6} \times \frac{1}{4} \times \frac{1}{6} = \frac{1}{10{,}368}$$

An alternative way of displaying transition probabilities is by use of the transition diagram. Figure 2.2(a) illustrates this diagram for the binary channel. The transition probabilities are indicated next to the arrow on the appropriate branch from input to output. The summation of probabilities leaving any node must be unity.

Figure 2.2(b) illustrates a special case of the binary memoryless channel, one in which the two conditional error probabilities are equal. The channel is completely characterized by a single parameter, p. Such a channel is known as the *binary symmetric channel* (*BSC*) and is a reasonably good model for many practical situations.

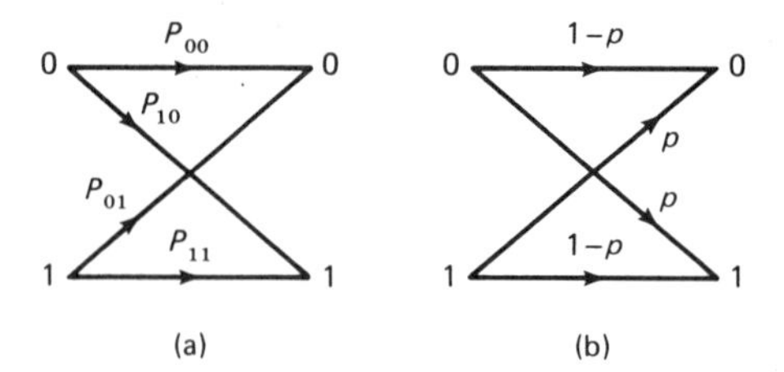

Figure 2.2 Transition diagram to model channel

We can use the BSC in order to derive some important transmission-error probabilities. This analysis will reinforce the comments made in Chapter 1 regarding the advantages of digital transmission. A transmission channel can be modeled as a number of tandem binary symmetric channels. As an example, suppose in transmitting a digital signal over a long distance the signal path includes a number of repeaters. Further suppose that the path between each repeater and the following repeater can be modeled as a BSC. The overall channel can then be viewed as a tandem connection of BSCs. Figure 2.3 illustrates two such interconnections.

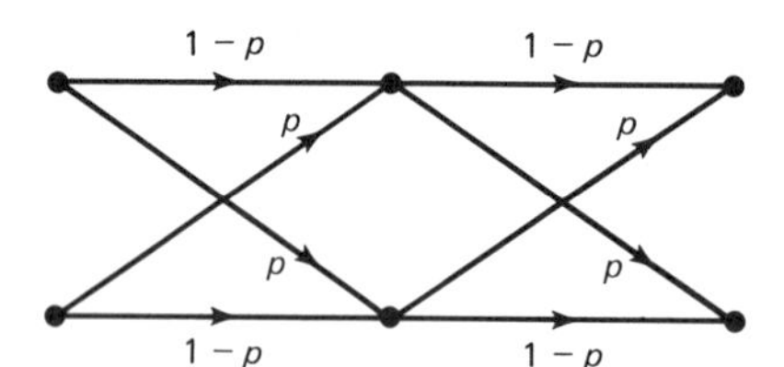

Figure 2.3 Tandem connections of two BSCs

The probability of error, or *bit error rate,* following one hop is given by

$$\begin{aligned} P_e &= P(0 \mid 1)P(1) + P(1 \mid 0)P(0) \\ &= (p)P(1) + (p)P(0) \\ &= (p)[P(1) + P(0)] = p \end{aligned}$$

We have used the shorthand notation

$$P(i \mid j) = P(s_{\text{out}} = i \mid s_{\text{in}} = j)$$

We now examine the system which includes two hops. The overall probability of correct transmission is

$$P_c = P(1 \mid 1)P(1) + P(0 \mid 0)P(0)$$

A transmitted 1 will be received as a 1 provided that no errors occur in either hop. Alternatively, if an error occurs in each of the two hops, the 1 will be correctly received. The probability of no errors is $(1 - p)^2$, while the probability of two errors is $(p)^2$. Therefore, the probability of correct transmission is given by

$$P_c = [p^2 + (1 - p)^2][P(1) + P(0)]$$
$$= p^2 + (1 - p)^2$$

The probability of error is given by

$$P_e = 1 - P_c = 1 - p^2 - (1 - p)^2$$
$$= 1 - p^2 - 1 + 2p - p^2$$
$$= 2p - 2p^2$$
$$= 2p(1 - p)$$

If we assume that $1 - p$ is close to 1, as one would hope in a binary channel, this can be approximated by

$$P_e = 2p$$

Comparing this with the result for a single hop, we see that the probability of bit error for two hops is approximately twice that of a single hop. In fact, in general (see Problem 2.4), the probability of error goes up linearly with the number of hops. Thus, for n binary symmetric channels in tandem, the overall probability of bit error is n times the bit error rate for a single BSC. This contrasts sharply with the noise behavior of an analog system as the transmission distance increases.

As a hint of where we are heading, let's examine the result a little further. Suppose you were required to design a transmission system to cover a distance of 500 kilometers. You decide to install a repeater station every 10 kilometers, so you require 50 such segments in your overall transmission path. You find that the bit error rate for each segment is $p = 10^{-6}$. Therefore, the overall bit error rate is given by 50×10^{-6} or 5×10^{-5}. You can therefore expect an average of 1 bit error for every 20,000 transmitted bits. Suppose that is not good enough. There are many techniques at your disposal, and in fact, the remainder of this text concentrates upon this problem. But let us say that you intuitively reason that reducing the distance between repeaters will decrease the error rate. You decide to double the number of repeaters. That is, you decrease the spacing from 10 kilometers to 5 kilometers. You then have 100 tandem hops instead of only 50, so the overall error is 100 times the probability of error for a single hop. Provided that the single-hop error probability is less than half of the previous 10^{-6}, you are ahead of the game. In fact, we will see that the single-hop error rate can be expected to decrease much faster than at a rate linearly related to the spacing. Thus, we have arrived at the essence of the value of digital communication and the concept of regeneration of signals. We hasten to point out that decreasing repeater spacing is not without cost. You must build additional repeater stations, so the true design problem contains a number of tradeoff decisions. We will say far more about this later.

2.2 BANDWIDTH CONSIDERATIONS

Whenever a waveform is sent through a channel, the bandwidth of the channel proves important in determining potential errors and maximum transmission rates.

System *bandwidth* relates to the system function—that is, the Fourier transform of the impulse response of the channel. All channels have a maximum frequency beyond which input components are almost entirely attenuated. This is due to distributed capacitance and inductance. As frequencies increase, the parallel capacitance tends to "short out" the signal and the series inductance "open-circuits."

Some channels also exhibit a low-frequency cutoff due to the dual of the effects mentioned above. If there is a low-frequency cutoff, the channel can be modeled as a band-pass filter. If there is no lower-frequency cutoff, the channel model becomes a low-pass filter.

Data communication channels are categorized according to bandwidth. There are three generally used grades of channel: the narrowband, voiceband, and wideband.

Bandwidths up to 300 hertz (Hz) are in the *narrowband* or *telegraph grade*. These can be used for slow data transmission, of the order of 600 bits per second (bps). They cannot reliably be used for voice transmissions.

Voiceband channels have bandwidths between 300 Hz and 4 kHz and can transmit data at rates of the order of 10 kilobits per second (kbps). The public telephone (subscriber loop) circuits are voiceband, and we shall have more to say about these in Section 2.4.

Wideband channels have bandwidths greater than 4 kHz. These can be leased from a carrier (e.g., telephone company).

We begin our analysis by considering the response of the channel to a pulse, since this is a form of digital communication we will be discussing in detail. Let us assume that the channel resembles an ideal low-pass filter with cutoff frequency f_m. The impulse response of this low-pass filter is then of the form

$$A\frac{\sin 2\pi f_m t}{\pi t} \tag{2-1}$$

We have assumed that there is no delay in the channel. That is, the phase characteristic of the channel system function is assumed to be zero. If there were a delay, the impulse response would be shifted in time by the appropriate amount.

The step response, $a(t)$, is the time integral of the impulse response. The step and impulse responses are sketched in Fig. 2.4.

The response of the system to a square pulse can be found from the sum of two step responses, and it is sketched in Fig. 2.5. In drawing this pulse response, we have assumed that T is sufficiently large that the oscillations of $a(t)$ have died out before the negative step occurs. This requires that

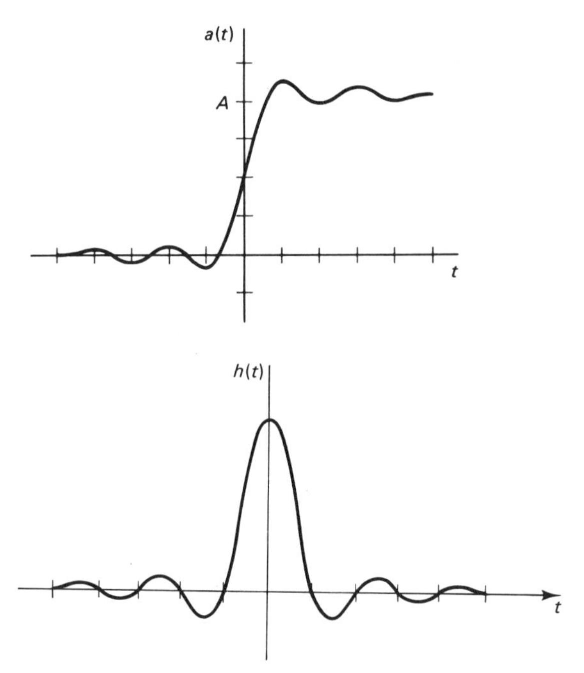

Figure 2.4 Step and impulse response of ideal low-pass filter

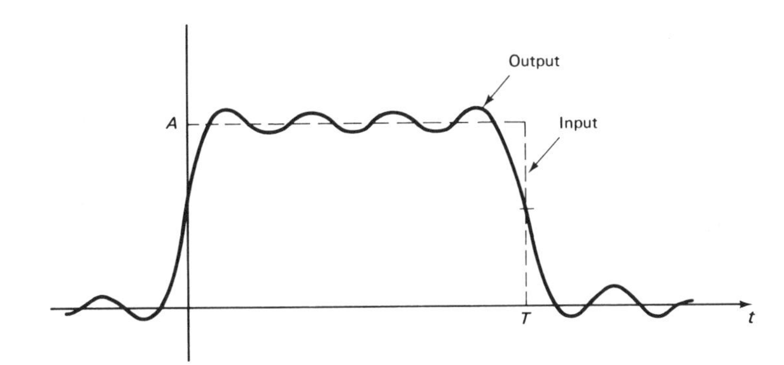

Figure 2.5 Response of low-pass filter to square pulse

$$T \gg 1/f_m$$

It should be noted that the ideal low-pass filter is noncausal (it predicts) and therefore cannot exist in the real world. Specific channel models will be presented as needed later in this text. For now, we briefly consider the case of the simple first-order low-pass filter, one composed of a single resistor and energy storage element. The pulse response of the *RC* circuit is shown in Fig. 2.6.

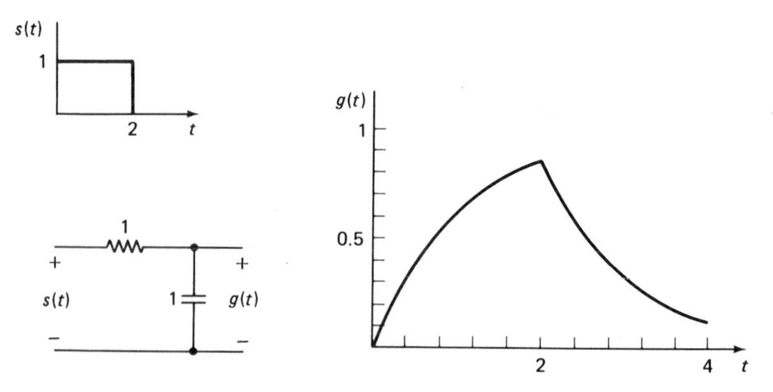

Figure 2.6 Pulse response of *RC* circuits

The bandwidth of this first-order circuit can be found by first finding the system function, $H(f)$:

$$H(f) = \frac{1}{1 + j2\pi fRC}$$

A commonly used definition of bandwidth uses the half-power point as reference. This is the frequency at which the amplitude drops to 0.707 of the maximum. Using this definition for cutoff frequency yields

$$f_m = \frac{1}{2\pi RC}$$

for the first-order low-pass filter.

In both the ideal low-pass filter and the first-order approximation, we see that the bandwidth is inversely related to the amount of pulse spreading. When we use pulses to send digital information, this spreading affects how closely together the pulses can be placed. This, in turn, specifies the maximum rate of communication.

If the square pulse now modulates a sinusoidal carrier and the low-pass filter is replaced by a band-pass filter with center frequency equal to the carrier frequency, the modulated pulse response is a modulated version of the signal shown in Fig. 2.5. These relationships are sketched in Fig. 2.7.

If the carrier frequency is not the same as the center frequency of the channel passband, the result must be modified. We deal with this case in Chapter 7.

2.3 DISTORTION

Anything that the channel does to a signal other than pure delay and constant multiplication is considered to be *distortion*. Let us assume that the channels we will encounter are linear and therefore cannot change the frequencies of the input. Some

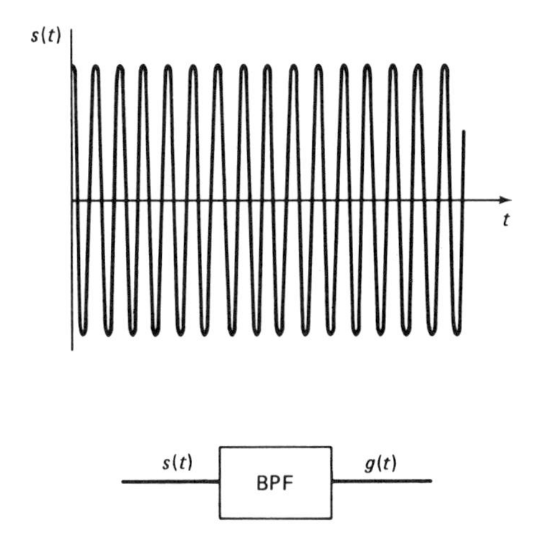

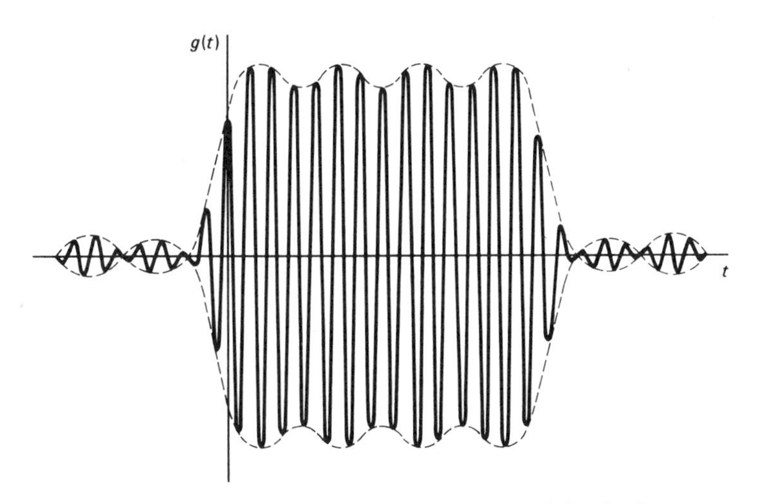

Figure 2.7 Response of band-pass filter to modulated pulse

nonlinear forms of distortion are significant at higher transmission frequencies. Indeed, the higher frequencies are affected by air turbulence, which causes a frequency modulation. This phenomenon is capitalized upon in *Doppler radar systems* used for weather monitoring.

Linear distortion is characterized by *time dispersion* (spreading) due either to multipath effects or to the characteristics of the channel. We deal with multipath in Chapter 5. For now, we look at the effects that can be readily characterized by the system function of the channel. The channel can be characterized by a system transfer function of the form

$$H(f) = A(f)e^{-j\theta(f)} \tag{2-2}$$

The *amplitude factor* is $A(f)$ and the *phase factor* is $\theta(f)$.

Distortion arises from the two frequency-dependent terms in Eq. (2-2). If $A(f)$ is not a constant, we have what is known as *amplitude distortion.* If $\theta(f)$ is not linear in f, we have *phase distortion.*

2.3.1 Amplitude Distortion

Let us first assume that $\theta(f)$ is linear with frequency. The transfer function is therefore of the form

$$H(f) = A(f)e^{-j2\pi f t_0}$$

where the proportionality constant has been denoted as t_0, since it represents the channel delay.

One general way to analyze this for a variety of amplitude variations is to expand $A(f)$ into a series. As an example, it may be possible to expand $A(f)$ in a Fourier series. This expansion will be true if $A(f)$ is band-limited to a certain range of frequencies. In such cases, we can write

$$H(f) = \sum_n H_n(f)$$

where the terms in the summation are of the form

$$H_n(f) = a_n \cos \frac{n\pi f}{f_m} e^{-j2\pi f t_0}$$

These terms can be related to the cosine filter whose amplitude characteristic follows a cosine wave in the passband. This is shown in Fig. 2.8. The impulse response for this filter is

$$\begin{aligned} H(f) &= \left(A + a \cos \frac{N\pi}{f_m} f\right) e^{-j2\pi f t_0} \\ &= Ae^{-j2\pi f t_0} + \frac{a}{2}\left\{\exp\left[j2\pi f\left(\frac{N}{2f_m} - t_0\right)\right] + \exp\left[j2\pi f\left(-\frac{N}{2f_m} - t_0\right)\right]\right\} \end{aligned}$$

If the input, $s(t)$, were bandlimited, the output of the cosine filter would be

$$g(t) = As(t - t_0) + \frac{a}{2} s\left(t + \frac{N}{2f_m} - t_0\right) + \frac{a}{2} s\left(t - \frac{N}{2f_m} - t_0\right) \tag{2-3}$$

Returning to the general filter case, we see that the output of a system with amplitude distortion is a sum of shifted inputs. Thus with

$$H(f) = \sum_n a_n \cos\left(\frac{n\pi f}{f_m}\right) e^{-j2\pi f t_0}$$

the output due to an input, $s(t)$, is given by

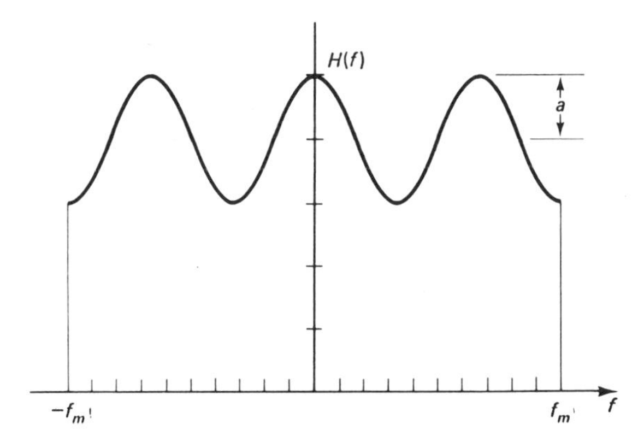

Figure 2.8 Cosine filter characteristic

$$g(t) = \sum_n \frac{a_n}{2}\left[s\left(t + \frac{n}{2f_m} - t_0\right) + s\left(t - \frac{n}{2f_m} - t_0\right)\right] \tag{2-4}$$

Equation (2-4) can be computationally difficult to evaluate. This approach is therefore usually used only when the Fourier series contains relatively few significant terms.

Example 2-2

Consider the triangular filter characteristic shown in Fig. 2.9. Assume that the phase characteristic is linear, with slope $-2\pi t_0$. Find the output of this filter when the input signal is

$$s(t) = \frac{\sin 400\pi t}{\pi t}$$

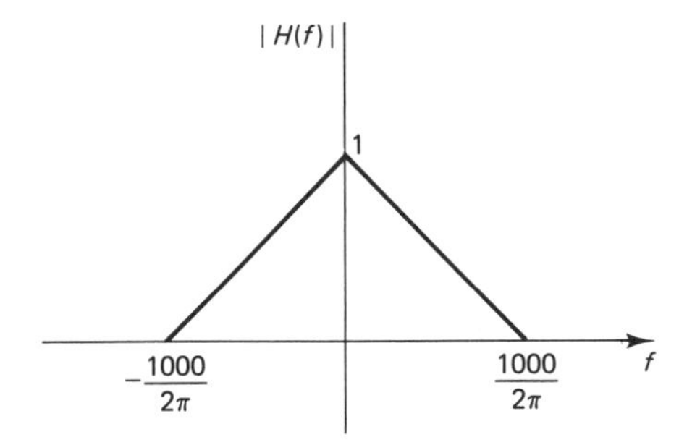

Figure 2.9 Filter characteristic for Example 2-2

Solution. It is first necessary to expand $H(f)$ in a Fourier series to get

$$H(f) = \frac{1}{2} + \frac{4}{\pi^2}\cos\frac{\pi f}{1000} + \frac{4}{9\pi^2}\cos\frac{3\pi f}{1000} + \frac{4}{25\pi^2}\cos\frac{5\pi f}{1000} + \cdots$$

$s(t)$ is bandlimited so that all frequencies are passed by the filter. This is so because

$S(f)$ is zero at frequencies above $200/2\pi$ and the filter cuts off at $f = 1000/2\pi$. If we retain the first three nonzero terms in the series, the output becomes

$$g(t) = \frac{1}{2}s(t - t_0) + \frac{2}{\pi^2}\left[s\left(t - \frac{\pi}{1000} - t_0\right) + s\left(t + \frac{\pi}{1000} - t_0\right)\right]$$

$$+ \frac{2}{9\pi^2}\left[s\left(t - \frac{3\pi}{1000} - t_0\right) + s\left(t + \frac{3\pi}{1000} - t_0\right)\right]$$

This result is sketched as Fig. 2.10 for $t_0 = 0.05$ s.

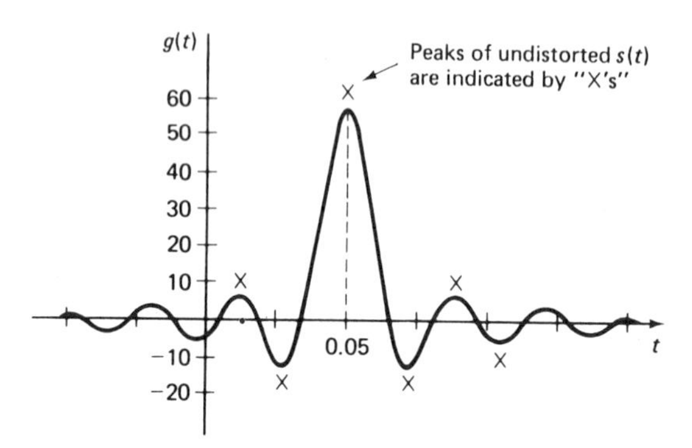

Figure 2.10 Output function for Example 2-2

Note that the output does not appear significantly different from the undistorted function. The figure shows "X's" at the peaks of the undistorted function. Examining the differences, we see that the distortion is manifest as an attenuation and a slight time delay. That is, the peaks of the output are smaller and slightly to the right of the corresponding peaks of the undistorted waveform.

2.3.2 Phase Distortion

Phase variation away from the distortionless (linear-phase) case can be characterized by variations in the slope of the phase characteristic and in the slope of a line from the origin to a point on the curve. We define *group delay* (also known as *envelope delay*) and *phase delay* as follows:

$$t_{ph}(f) = \frac{\theta(f)}{2\pi f}, \qquad t_{gr}(f) = \frac{d\theta(f)}{df} \tag{2-5}$$

Figure 2.11 illustrates these definitions for a representative phase characteristic. For an ideal distortionless channel, the phase characteristic is linear and the group and phase delays are both constant for all f. In fact, both of these delays would be equal to the time delay of the input signal, t_0.

In many cases, the phase characteristic can be approximated as a piecewise linear curve. As an example, examine the phase characteristic of Fig. 2.11. If we

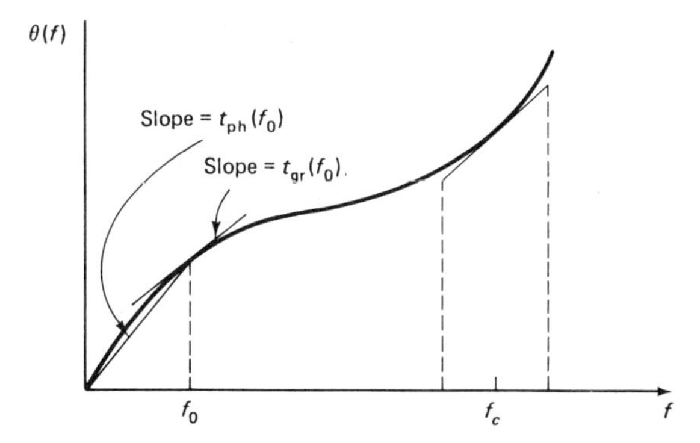

Figure 2.11 Definition of group delay and phase delay

were to operate in a relatively narrow band around f_c, the phase could be approximated as the first two terms in a Taylor-series expansion:

$$\theta(f) = \theta(f_c) + \frac{d\theta(f_c)}{df}(f - f_c)$$

$$= t_{\text{ph}}(f_c)f_c + (f - f_c)t_{\text{gr}}(f_c)$$

This equation applies for positive frequency, and the negative of this equation applies for negative frequency. This is so because the phase function for a real system must be an odd function of frequency.

Suppose that the amplitude factor is constant, $A(f) = A$, and a wave of the form $s(t) \cos 2\pi f_c t$ is the input to the system. This could represent a sinusoidal burst used to send a binary digit. The Fourier transform of the input is

$$\tfrac{1}{2}[S(f - f_c) + S(f + f_c)]$$

The Fourier transform of the output, $G(f)$, is given by the product of the input transform with the system function. Thus,

$$G(f) = \tfrac{1}{2}[S(f - f_c)]A \exp[j t_{\text{ph}}(f_c)2\pi f_c] \exp[j 2\pi(f - f_c)t_{\text{gr}}(f_c)] \tag{2-6}$$

Equation (2-6) has been written for positive f. For negative f, the correct expression is the complex conjugate of Eq. (2-6). We must now find the time function corresponding to this Fourier transform. The transform can be simplified by taking note of the following three Fourier-transform relationships:

$$s(t - t_0) \cos 2\pi f_c t \leftrightarrow \tfrac{1}{2}[S(f - f_c)e^{-j2\pi(f-f_c)t_0} + S(f + f_c)e^{-j2\pi(f+f_c)t_0}]$$

$$s(t - t_1) \cos 2\pi f_c(t - t_1) \leftrightarrow \tfrac{1}{2}[S(f - f_c) + S(f + f_c)]e^{-j2\pi t_1}$$

$$s(t - t_0) \cos 2\pi f_c(t - t_1) \leftrightarrow \tfrac{1}{2}[S(f - f_c)e^{-j2\pi(f-f_c)(t_0-t_1)} + S(f + f_c)e^{-j2\pi(f+f_c)(t_0-t_1)}]e^{-j2\pi t_1}$$

Using these relationships, we find from Eq. (2-6) that

$$g(t) = As[t - t_{\text{gr}}(f_c)] \cos 2\pi f_c[t - t_{\text{ph}}(f_c)]$$

This result indicates that the amplitude (envelope) of the burst is delayed by an amount equal to the group delay, and the sinusoidal portion is delayed by an amount equal to the phase delay. Both group and phase delay are evaluated at the frequency of the sinusoid. This result will prove significant later. Getting ahead of the game, we will work with two types of receiver: the coherent and incoherent. Incoherent receivers operate only upon the envelope of the received signal, so the envelope delay is critical to the receiver operation. On the other hand, coherent receivers use all the waveform information, so phase delay will also be important.

2.4 TELEPHONE CHANNELS

We mentioned earlier that the vast majority of present-day data communication is done on telephone lines. This channel is not ideally suited to efficient data communication. In the present section we examine some of the reasons for the unsuitability. As a hint toward the problems we will encounter, we note that the human ear is not very sensitive to phase variations, so many distorting effects that cause extreme hardship to data could be tolerated for voice transmissions.

There are two types of phone lines in use today. The *dial-up line* is routed through voice switching offices. The switching operations add impulse noise, which is heard as occasional clicks during a phone conversation, not terribly devastating to a conversation. But it should not be too surprising to learn that this wreaks havoc with data communication. The alternative is the *leased line,* which is a permanent circuit that is not subject to the public type of switching. Since the same line is used every time, some types of distortion can be predicted and compensated.

In the dial-up circuit, assorted problems arise due to procedures adopted for voice channels. For example, the system has many bridge taps. A *bridge tap* is a "jumper" which is installed when a phone is removed, or when extra phone jacks have been installed for possible later expansion. These are not serious causes of distortion for voice transmission, but their capacitance leads to delays that can destroy data.

The phone system is specifically tailored to audio signals with an upper frequency in the vicinity of 4 kHz. When these lines are used for data, this upper cutoff is often stretched to provide data rates above 10 kbps. Loading coils in the line improve performance in the voice band, but cause additional amplitude distortion above 4 kHz, thus making higher bit rates more difficult to achieve.

Long-distance phone channels contain *echo suppressors* which are voice activated. These prevent a speaker from receiving an echo due to reflections from transitions in the channel. The time delay activating these suppressors can make certain types of data operation impossible. Many telephone-line data sets contain a provision to disable to echo suppressors, using a tone of about 2000 Hz.

Phone lines have amplitude characteristics which are not constant with frequency, and they therefore contribute amplitude distortion. Figure 2.12 shows a

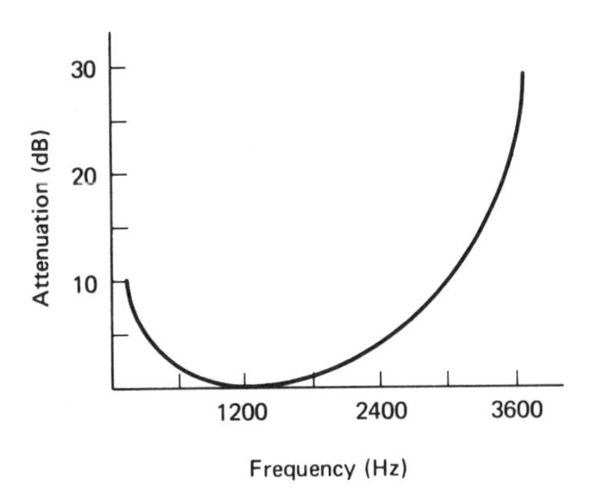

Figure 2.12 Attenuation characteristic for typical telephone channel

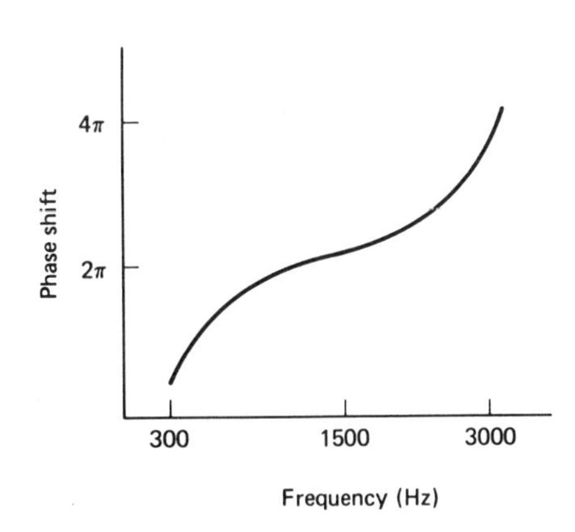

Figure 2.13 Phase characteristic for typical telephone channel

typical attenuation or loss curve. The loss is given in decibels (dB) and is relative to about 1000 Hz, where the minimum loss occurs.

Phase distortion also occurs in the phone line. A typical phase characteristic for about 7 kilometers of phone line is shown in Fig. 2.13.

Let us now assume that a data signal is sent through a telephone channel. We simplify the data signal by assuming it to be a 1-kHz square wave. The square wave has components at the fundamental frequency and at all odd harmonics of this frequency. Figure 2.14 shows the square wave, its Fourier-series approximation using only the 1-kHz and 3-kHz components, and the output of a phone line assuming the amplitude and phase variations of Figs. 2.12 and 2.13. Note the pulse spreading caused by the network. This spreading allows one pulse to interact with neighboring pulses, thereby leading to increased errors.

We have been concentrating upon envelope and phase distortion due to a nonlinear phase characteristic and a nonconstant amplitude characteristic. Ad-

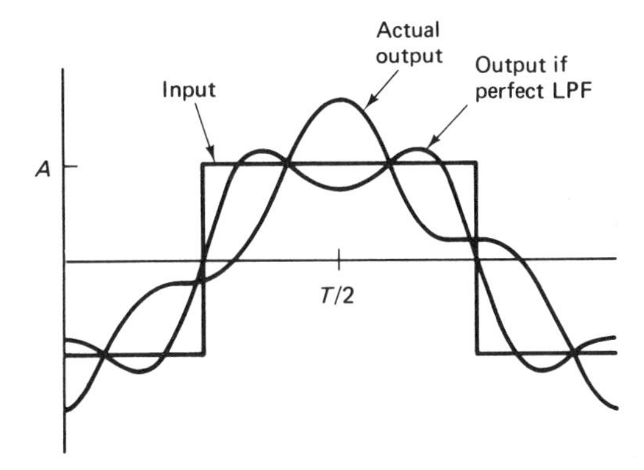

Figure 2.14 Response of typical telephone line to square wave

ditional problems exist when using phone lines for data transmission. The various amplifiers and electronic equipment introduce nonlinear distortion. Sometimes there are also frequency offsets introduced by the modulation equipment. These offsets are typically less than 5 Hz.

Voice-grade channels are classified according to the maximum amount of attenuation distortion and maximum envelope delay distortion within a particular frequency range. Telephone companies can provide conditioning to reduce the effects of particular types of distortion. In purchasing a telephone channel, a particular conditioning is specified and the company guarantees a certain performance level. Naturally, the better the channel, the higher the cost. Table 2-1 lists typical channel conditioning characteristics for representative types of conditioning.

TABLE 2-1 TYPICAL TELEPHONE CHANNEL PARAMETERS

Conditioning	Attenuation Frequency Range	Distortion Variation	Envelope Delay Frequency Range	Distortion Variation, μs
Basic	500–2500	−2 to +8	800–2600	1750
	300–3000	−3 to +12		
C1	1000–2400	−1 to +3	1000–2400	1000
	300–2700	−2 to +6	800–2600	1750
	300–3000	−3 to +12		
C2	500–2800	−1 to +3	1000–2600	500
	300–3000	−2 to +6	600–2600	1500
			500–2800	3000
C4	500–3000	−2 to +3	1000–2600	300
	300–3200	−2 to +6	800–2800	500
			600–3000	1500
			500–3000	3000

As an example, let us examine the C2 channel. If you purchase such a channel and use it within the band of frequencies between 500 and 2800 Hz, you are guaranteed that the attenuation will not vary beyond the range of −1 to +3 dB (relative to response at 1004 Hz). If you use it in the wider band between 300 Hz and 3000 Hz, the guaranteed attenuation range increases to −2 to +6 dB. Similarly, the envelope delay variation will be less than 3000 microseconds if you operate within the band between 500 Hz and 2800 Hz. In addition to the parameters given in Table 2-1, the various lines have specified losses, loss variations, maximum frequency error, and phase jitter. For example, the C1 channel specifies a loss of 16 dB ± 1 dB at 1000 Hz. The variation in loss is limited to 4 dB over long periods of time. The frequency error is limited to 5 Hz and the phase jitter to 10 degrees.

2.5 NOISE

Communication would be trivial if there were no distortion and no noise. Section 2.4 considered various ways in which the channel changes the shape of the signal between transmitter and receiver. We called this distortion. In addition to distorting the signal, the channel also adds an undesired function to the transmitted signal. We call this undesired term *noise.* This added term arises from many sources—among them, other transmitters, electrical storms, radiation from wires carrying high-frequency or pulsed signals, thermal motion of electrons, and reflections of signals from other sources. Some of these added signals, such as some types of jamming signals in a radar system, may be deterministic. Other types are composed of components from so many sources that we find it more convenient to analyze them as *random processes.*

We usually assume that additive noise is a random process that is *Gaussian.* This means that if the process is sampled at any group of points in time, the joint probability density of the samples is a joint Gaussian density. This assumption proves important, because if we start with a Gaussian random process and put it through any linear system, the result will be a Gaussian process. This fact will be used many times in examining the performance of various digital communication systems.

Besides knowing that the noise process is Gaussian, it will be necessary to know something about the correlations of various samples. We do this by examining the *autocorrelation function* and its Fourier transform, known as the *power spectral density.*

Until this point, we have considered primarily single random variables. All averages (and related parameters) were simply numbers. Now we shall, in effect, add another dimension to the study—the dimension of time. Instead of talking of numbers only, we will now be able to characterize random functions. The advantages of such a capability should be obvious. Our approach to random functions' analysis will begin with the consideration of discrete-time functions (i.e., sampled), since these will prove to be a simple extension of random variables.

Imagine a single die being flipped 1000 times. Let X_i be the random variable assigned to the outcome of the ith flip. Now list the 1000 values of the random variables

$$x_1, x_2, x_3, \ldots, x_{999}, x_{1000}$$

For example, if the random variable is assigned to be equal to the number of dots on the top face after flipping the die, a typical list might resemble that shown in Eq. (2-7),

$$4, 6, 3, 5, 1, 4, 2, 5, 3, 1, 4, 5, \ldots \tag{2-7}$$

Suppose that all possible sequences of random variables are now listed. We will have a collection of 6^{1000} entries, each one resembling that shown in Eq. (2-7).

This collection will be known as the *ensemble* of possible outcomes. This ensemble, together with associated statistical properties, forms a *stochastic process.* In this particular example, the process is discrete-valued and discrete-time.

If we were to view one digit, say the third entry, a random variable would result. In this example the random variable would represent that assigned to the outcome of the third flip of the die.

We can completely describe the above stochastic process by specifying the 1000-dimensional probability density function

$$p(x_1, x_2, x_3, \ldots, x_{999}, x_{1000})$$

This 1000-dimensional probability density function is used in the same way as a one-dimensional probability density function. Thus, the 1000-dimensional integral of the probability density function is unity.

$$\int_{-\infty}^{\infty}\int_{-\infty}^{\infty} \cdots \int_{-\infty}^{\infty} p(x_1, x_2, \ldots, x_{1000})\, dx_1\, dx_2 \ldots dx_{1000} = 1$$

The probability that the variables fall within any specified volume is the integral of the probability density function over that volume. The expected value of any single variable can be found by integrating the product of that variable with the probability density function. Thus, for example, the expected value of x_1 is

$$\int_{-\infty}^{\infty}\int_{-\infty}^{\infty} \cdots \int_{-\infty}^{\infty} x_1 p(x_1, x_2, \ldots, x_{1000})\, dx_1\, dx_2 \ldots dx_{1000}$$

We will call this a *first-order average.*

In a similar manner, the expected value of any multidimensional function of the variables is found by integrating the product of that function with the density function. Thus, for example, the expected value of the product x_1x_2 is given by

$$E\{x_1x_2\} = \int_{-\infty}^{\infty}\int_{-\infty}^{\infty} \cdots \int_{-\infty}^{\infty} x_1x_2p(x_1, x_2, \ldots, x_{1000})\, dx_1\, dx_2\, dx_{1000}$$

This would be known as a *second-order average.*

In many instances it will prove sufficient to specify only the first- and second-order averages (moments). That is, one would specify

$$E\{x_i\} \quad \text{and} \quad E\{x_ix_j\}$$

for all i and j.

We would now like to extend this example to the case of an infinite number of random variables in each list and an infinite number of lists in the ensemble. It may be helpful to refer back to the above simple example from time to time.

The most general form of the stochastic process would result if our simple experiment could yield an infinite range of values as the outcome, and if the sampling period approached zero (i.e., the die is flipped faster and faster).

Another way to arrive at a stochastic process is to perform a discrete experiment, but to each outcome assign a *time function* instead of a number. As a simple example, consider the experiment defined by picking a 6-volt DC generator from an infinite inventory in a warehouse. The generator voltage is then measured as a function of time on an oscilloscope. This waveform, $v(t)$, will be called a *sample function* of the process. The waveform will not be a perfect constant of 6 volts value but will wiggle around due to imperfections in the generator construction and due to radio pickup when the wires act as an antenna. There are an infinite number of possible sample functions of the process. This infinite number of samples forms the ensemble. Each time we choose a generator and measure its voltage, a sample function from this infinite ensemble results.

If we were to sample the voltage at a specific time, say $t = t_0$, the sample, $v(t_0)$, can be considered as a random variable. Since $v(t)$ is assumed to be continuous with time, there are an infinity of random variables associated with the process.

Summing this up and changing notation slightly, let $x(t)$ represent a stochastic process. Then $x(t)$ can be thought of as being an infinite ensemble of all possible sample functions. For every specific value of time, $t = t_0$, $x(t_0)$ is a random variable.

As in the simple example which opened this section, we can first hope to characterize the process by a joint density function of the random variables. Unfortunately, since there are an infinite number of random variables, this density function would be infinite-dimensional. We therefore resort to first- and second-moment characterization of the process.

The first moment is given by the mean value:

$$m(t) \triangleq E\{x(t)\} \tag{2-8}$$

The second moments are given by

$$R_{xx}(t_1, t_2) \triangleq E\{x(t_1)x(t_2)\} \tag{2-9}$$

for all t_1 and t_2.

At the moment we must think of these averages as being taken over the ensemble of possible time-function samples. That is, in order to find $m(t_0)$, we would average all members of the ensemble at time t_0. In practice, we could measure the voltage of a great number of the generators at time t_0 and average the resulting numbers. In the case of the DC generators, one would intuitively expect this average not to depend upon t_0. Indeed, most processes we will consider have mean values which are independent of time.

A process with overall statistics that are independent of time is called a *stationary* process. If only the mean and second moment are independent of time, the process is *wide-sense stationary*. Given a stationary process, $x(t)$, then the process described by $x(t - T)$ will have the same statistics independent of the value of T. Clearly, for a stationary process, $m(t_0)$ will not depend upon t_0.

If the process is stationary,

$$R_{xx}(t_1, t_2) = E\{x(t_1)x(t_2)\} = E\{x(t_1 - T)x(t_2 - T)\} \tag{2-10}$$

Since Eq. (2-10) applies for all values of T, let $T = t_1$. Then

$$R_{xx}(t_1, t_2) = E\{x(0)x(t_2 - t_1)\}$$

This indicates that the second moment, $R_{xx}(t_1, t_2)$ does not depend upon the actual values t_1 and t_2, but only upon the difference, $t_2 - t_1$. That is, the left-hand time point can be placed anywhere, and as long as the right-hand point is separated from this by $t_2 - t_1$, the second moment will be $R_{xx}(t_1, t_2)$. We state that mathematically as

$$R_{xx}(t_1, t_2) = R_{xx}(t_2 - t_1) \tag{2-11}$$

for a stationary process.

The function $R_{xx}(t_2 - t_1)$ is called the *autocorrelation* of the process, $x(t)$.

In the work to follow, the argument, $t_2 - t_1$ will be replaced by "τ".

$R_{xx}(t_2 - t_1)$ tells us most of what we need to know about the process. In particular, it gives us some idea as to how fast a particular sample function can change with time. If t_2 and t_1 are sufficiently far apart such that $x(t_1)$ and $x(t_2)$ are independent random variables, the autocorrelation reduces to

$$\begin{aligned} R_{xx}(t_2 - t_1) &= E\{x(t_1)x(t_2)\}, \\ &= E\{x(t_1)\}E\{x(t_2)\} = m^2 \end{aligned} \tag{2-12}$$

For most processes encountered in communications, the mean value is zero ($m = 0$). In this case, the value of "t" at which $R_{xx}(t)$ goes to zero represents the time over which the process is "correlated." If two samples are separated by more than this length of time, one sample has no effect upon the other.

Example 2-3

You are given a stochastic process, $x(t)$, with mean value m_x and autocorrelation, $R_{xx}(\tau)$. Obviously, the process is at least wide-sense stationary. If not, the mean and autocorrelation could not have been given in this form. Find the mean and autocorrelation of a process $y(t)$, where

$$y(t) = x(t) - x(t - T)$$

Solution. To solve problems of this type, we need only recall the definition of the mean and autocorrelation and the fact that taking expected values is a linear operation. Proceeding, we have

$$\begin{aligned} m_y &= E\{y(t)\} = E\{x(t) - x(t - T)\} \\ &= E\{x(t)\} - E\{x(t - T)\} \\ &= m_x - m_x = 0 \end{aligned}$$

$$\begin{aligned} R_{yy}(\tau) &= E\{y(t)y(t + \tau)\} \\ &= E\{[x(t) - x(t - T)][x(t + \tau) - x(t + \tau - T)]\} \\ &= E\{x(t)x(t + \tau)\} - E\{x(t)x(t + \tau - T)\} \\ &\quad - E\{x(t - T)x(t + \tau)\} + E\{x(t - T)x(t + \tau - T)\} \\ &= R_{xx}(\tau) - R_{xx}(\tau - T) - R_{xx}(\tau + T) + R_{xx}(\tau) \end{aligned}$$

$$= 2R_{xx}(\tau) - R_{xx}(\tau - T) - R_{xx}(\tau + T)$$

We note that the process $y(t)$ is wide-sense stationary. If this were not the case, both m_y and R_{yy} would be functions of t.

Example 2-4

Consider the experiment of starting a sinusoidal generator of deterministic frequency f_0 and amplitude A. The exact time of start is random. Thus,

$$x(t) = A \sin (2\pi f_0 t + \theta)$$

where the phase, θ, can be considered a random variable with a uniform density as shown in Fig. 2.15.

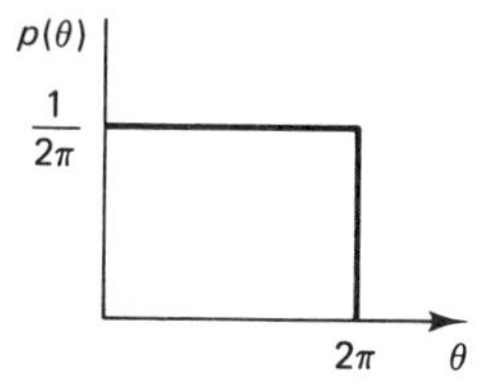

Figure 2.15 Density function of phase for Example 2-4

Find the autocorrelation of this random process.

Solution. The autocorrelation is found directly from the definition,

$$R(\tau) = E\{x(t)x(t + \tau)\}$$

$$= E\{A^2 \sin (2\pi f_0 t + \theta) \sin (2\pi f_0(t + \tau) + \theta)\}$$

By using trigonometric identities, this becomes

$$E\{\tfrac{1}{2} A^2[\cos 2\pi f_0\tau - \cos (2\pi f_0(2t + \tau) + \theta)]\}$$

The first term in this result is not random, so its expected value is the function itself. The expected value of the second term is found by integrating its product with the density of θ.

$$R(\tau) = \frac{1}{2}A^2 \cos 2\pi f_0\tau - \frac{1}{2\pi}\int_0^{2\pi} \cos [2\pi f_0(2t + \tau) + \theta]\, d\theta$$

The second term represents the integral of a perfect cosine function over two entire periods. This integral is equal to zero. Thus,

$$R(\tau) = \tfrac{1}{2} A^2 \cos 2\pi f_0\tau$$

Suppose you were asked to find the average value of the voltage in the DC generator example. You would have to measure the voltage (at any given time) of many generators and then compute the average. Once you were told that the process is stationary, you would probably be tempted to take one generator and average its voltage over a large time interval. You would expect to get 6 volts as the result of either technique. That is, you would reason that

$$m_v = E\{v(t)\} = \lim_{T\to\infty} \frac{1}{T} \int_{-T/2}^{T/2} v_i(t)\, dt$$

Here, $v_i(t)$ is one sample function of the ensemble. Actually this reasoning will not always be correct. Suppose, for example, that one of the generators was burned out. If you happened to choose this particular generator, you would find that $m_v = 0$, which is certainly not correct.

If any single sample of a stochastic process contains all of the information (statistics) about the process, we call the process *ergodic*. Most processes which we deal with are ergodic. The generator example is ergodic as long as none of the generators is exceptional (e.g., burned out).

Clearly a process which is ergodic must also be stationary. This is so because, once we allow that the averages can be found from one time sample, these averages can no longer be a function of the time at which they are computed. Conversely, a stationary process need not be ergodic (e.g., the burned-out generator example . . . convince yourself this process would still be stationary).

If a process is ergodic, the autocorrelation can be found from any time sample:

$$R_{xx}(\tau) = E\{x(t)x(t + \tau)\} = \lim_{T\to\infty} \frac{1}{T} \int_{-T/2}^{T/2} x(t)x(t + \tau)\, dt \tag{2-13}$$

The power spectral density of a stochastic process is given by

$$G(f) = \mathscr{F}[R(t)] = \int_{-\infty}^{\infty} R(t)e^{-j2\pi ft}\, dt \tag{2-14}$$

and the total average power by

$$P_{\text{avg}} = 2\int_{0}^{\infty} G(f)\, df \tag{2-15}$$

Here, the word "average" takes on more meaning than it did for deterministic signals. Note that since

$$R(t) = \int_{-\infty}^{\infty} G(f)e^{j2\pi ft}\, df$$

then

$$P_{\text{avg}} = \int_{-\infty}^{\infty} G(f)\, df = R(0) \tag{2-16}$$

The observation that $R(0)$ is the average power makes a great deal of sense, since

$$R(0) = E\{x^2(t)\}$$

If a stochastic process now forms the input to a linear system as shown in Fig. 2.16, the output will also be a stochastic process. That is, each sample function of

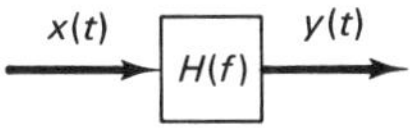

Figure 2.16 Stochastic process as input to a linear system

the input process yields a sample function of the output process. The autocorrelation of the output process is

$$G_{yy}(f) = G_{xx}(f)\,|H(f)|^2 \tag{2-17}$$

Taking inverse transforms,

$$R_{yy}(t) = R_{xx}(t) * h(t) * h(-t) \tag{2-18}$$

We therefore have a way of finding the average power of the output of a system when the input is a stochastic process. Since most noise waveforms can be thought of as being samples of a stochastic process, the above analysis is critical to any noise reduction scheme analysis.

Example 2-5

A received signal is made up of two components, signal and noise.

$$r(t) = s(t) + n(t)$$

The signal can be considered as a sample of a random process, since random amplitude fluctuations are introduced by turbulence in the air. You are told that the autocorrelation of the signal process is

$$R_s(\tau) = 2e^{-|\tau|}$$

The noise is a sample function of a random process with autocorrelation

$$R_n(\tau) = e^{-2|\tau|}$$

You are told that both processes have zero mean value, and that they are independent of each other.

Find the autocorrelation and total power of $r(t)$.

Solution. From the definition of autocorrelation,

$$\begin{aligned} R_r(\tau) &\triangleq E\{r(t)r(t+\tau)\} \\ &= E\{[s(t)+n(t)][s(t+\tau)+n(t+\tau)]\} \\ &= E\{s(t)s(t+\tau)\} + E\{s(t)n(t+\tau)\} \\ &\quad + E\{s(t+\tau)n(t)\} + E\{s(t+\tau)n(t+\tau)\} \end{aligned} \tag{2-19}$$

Since the signal and noise are independent,

$$E\{s(t+\tau)n(t)\} = E\{s(t+\tau)\}E\{n(t)\} = 0$$

and

$$E\{s(t)n(t+\tau)\} = E\{s(t)\}E\{n(t+\tau)\} = 0$$

Finally, Eq. (2-19) becomes

$$R_r(\tau) = R_s(\tau) + R_n(\tau)$$
$$= 2e^{-|\tau|} + e^{-2|\tau|}$$

The total power of $r(t)$ is simply $R_r(0) = 3$.

2.5.1 White Noise

Let $x(t)$ be a stochastic process with a constant power spectral density (*see* Fig. 2.17). This process contains all frequencies to an equal degree. Since white light is composed of all frequencies (colors), the process described above is known as *white noise.*

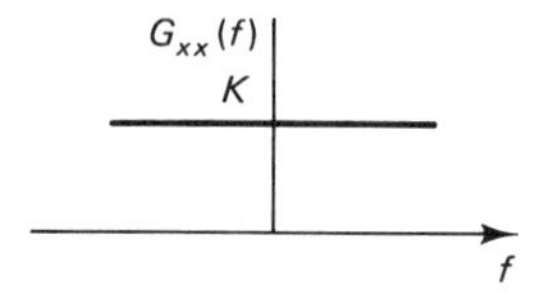

Figure 2.17 Power spectral density of white noise

The autocorrelation of white noise is the inverse Fourier transform of K, which is simply $R_{xx}(t) = K\,\delta(t)$.

The average power of white noise, $R(0)$, is infinity. It therefore cannot exist in real life. However, many types of noise encountered can be assumed to be approximately white.

Example 2-6

White noise forms the input to the RC circuit of Fig. 2.18 (approximation to a low-pass filter). Find the autocorrelation and power spectral density of the output of this filter.

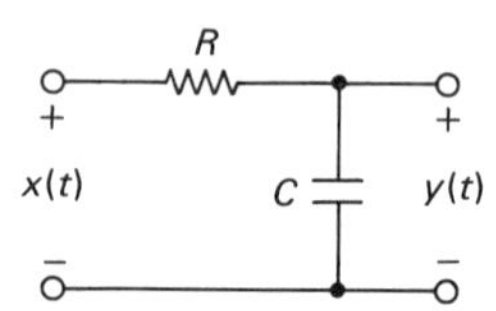

Figure 2.18 Circuit for Example 2-6

Solution

$$G_{yy}(f) = G_{xx}(f)\,|H(f)|^2 = \frac{K}{1 + (2\pi f)^2C^2R^2}$$

$$R_{yy}(\tau) = \frac{K}{2RC}\exp\left(\frac{-|\tau|}{RC}\right)$$

Example 2-7

Repeat Example 2-6 for an ideal low-pass filter with cutoff frequency f_m, as shown in Fig. 2.19.

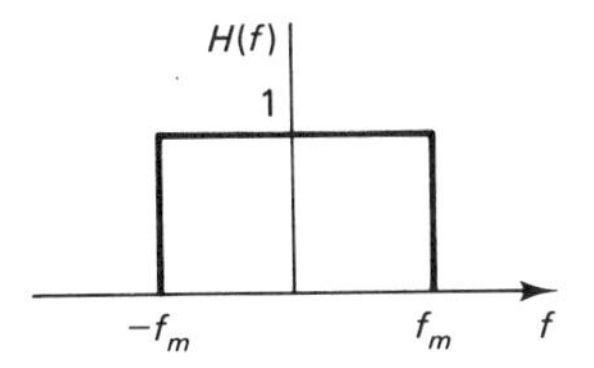

Figure 2.19 Low-pass filter characteristic for Example 2-7

Solution

$$G_{yy}(f) = G_{xx}(f)\,|H(f)|^2$$

$$= \begin{cases} K & |f| < f_m \\ 0 & \text{otherwise} \end{cases}$$

This is shown in Fig. 2.20. The autocorrelation is the inverse transform of the power spectral density and is given by

$$R_{yy}(\tau) = K\,\frac{\sin 2\pi f_m \tau}{\pi\tau}$$

A process with a power spectral density as shown in Fig. 2.20 is known as *bandlimited white noise.*

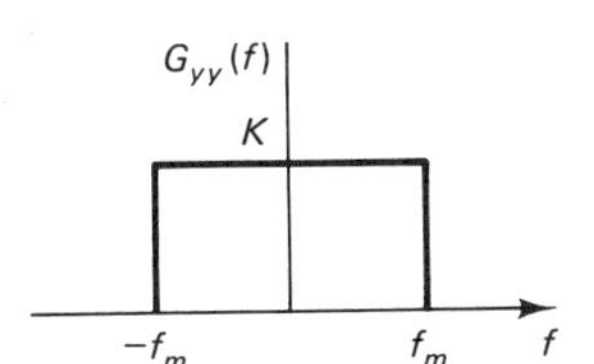

Figure 2.20 Output power spectral density for Example 2-7

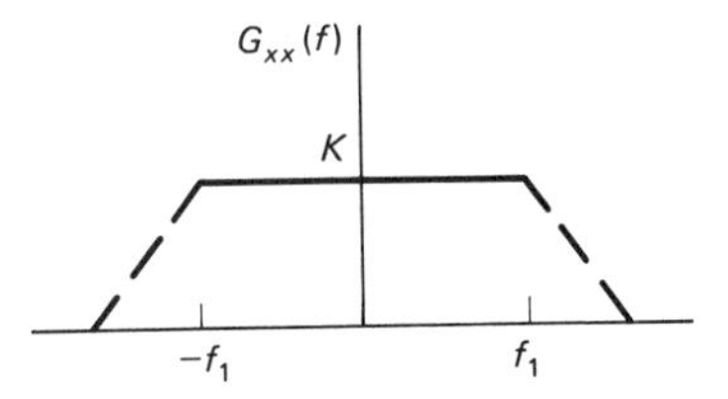

Figure 2.21 Power spectral density of input to filter

In the previous example, suppose that the input process had a power spectrum as shown in Fig. 2.21 with $f_1 \gg f_m$. The output process would be identical to that found in Example 2-7. Therefore, if a processing system exhibits an upper cutoff frequency (all real systems approximately do) and the input noise has a flat spectrum up to this cutoff frequency, no error will be made by considering the input noise to be white. This will simplify the required mathematics considerably.

Since $R(t)$ is zero for $t \neq 0$, two samples of white noise are independent of each other even if they are taken very closely together.* That is, knowing the sample value of white noise at one instant of time tells us absolutely nothing about its value

*Actually the zero correlation [$x(t)$ and $x(t + \tau)$ are "uncorrelated"] is a necessary, but not sufficient, condition for independence of the two random variables. The only complete definition of independence relates to conditional probabilities or, equivalently, to the joint density function. However, one can show that two random variables which are jointly Gaussian are independent as long as they are uncorrelated. Therefore as long as we deal with Gaussian processes, we can use the words "uncorrelated" and "independent" interchangeably.

an instant later. From a practical standpoint, this is a sad situation. It would seem to make the elimination of noise more difficult. Although this is true, the "independence of samples" assumption serves to greatly simplify the analysis of many complex processing systems.

Up to this point we have said nothing about the actual probability distributions of the process. We have talked only about the first and second moments. Each random variable, $x(t_0)$, has a certain probability distribution. For example, it can be either Gaussian or uniformly distributed. By considering only the means and second moments we are not telling the whole story. Two completely different random variables may possess the same mean and variance.

Perhaps this last idea will become clearer if we draw an analogy with simple dynamics. The equations for the first and second moment of a random variable are

$$E\{x\} = \int_{-\infty}^{\infty} xp(x)\,dx \tag{2-20}$$

$$E\{x^2\} = \int_{-\infty}^{\infty} x^2p(x)\,dx \tag{2-21}$$

These are of the same form as the equations used to find the center of gravity (centroid) and the moment of inertia of a body. Certainly two bodies can be completely different in shape yet have the same center of gravity and moment of inertia.

We are rescued from this dilemma by the observation that most processes encountered in real life are Gaussian. This means that all relevant densities (joint and single) are Gaussian. Once we know that a random variable has a Gaussian distribution, the density function is completely specified by giving its mean and second moment.

Thermal and Shot Noise. The assumption of *white noise* is not too far off base for several common forms of noise. *Thermal noise* and *shot noise* are two physically encountered types of noise.

Thermal noise is produced by the random motion of electrons in a medium. The intensity of this motion increases with increasing temperature and is zero only at a temperature of absolute zero. If a resistor is hooked directly to the input of a sensitive oscilloscope, a random pattern will be displayed on the screen. The power spectral density of this random process can be shown to be of the form

$$G(f) = \frac{A\,|f|}{e^{B|f|} - 1} \tag{2-22}$$

where A and B are constants that depend upon temperature and other physical constants. Figure 2.22 is a sketch of $G(f)$. For low frequencies, $G(f)$ is almost constant. We can therefore usually approximate thermal noise as white noise.

Figure 2.22 Power spectral density of thermal noise

Figure 2.23 Shot noise current waveform

Shot noise occurs since, although we think of current as being continuous, it is actually a discrete phenomenon. In fact, current occurs in discrete pulses each time an electron moves between two points. A plot of current as a function of time would show a waveform which wiggles around what we conventionally think of as the current plot. An example is shown in Fig. 2.23. This variation of current around the average value is known as shot noise. An analysis (see the references) would show that, for frequencies of interest to us, shot noise can usually be thought of as being white noise.

2.5.2 Narrowband Noise—Quadrature Components

In studies of noise in communication systems, we encounter noise that is band-limited to some range of frequencies. Such a noise waveform would result if any noise signal (white, for example), were passed through a band-pass filter. It will prove useful to us to have a trigonometric expansion for such so called narrowband noise signals. The actual form will be

$$n(t) = x(t) \cos 2\pi f_0 t - y(t) \sin 2\pi f_0 t \tag{2-23}$$

where $n(t)$ is the noise waveform and f_0 is a frequency (often the center) within the band occupied by the noise. Since sine and cosine vary by 90 degrees, $x(t)$ and $y(t)$ are known as the "quadrature components" of the noise waveform.

Because the formula of Eq. (2-23) will be used many times in later studies, we shall spend some time examining the properties of the various time functions. The result of Eq. (2-23) can be derived by starting with the exponential notation

$$n(t) = Re\{r(t)e^{j2\pi f_0 t}\}$$

where $r(t)$ is a complex function with a low-frequency bandlimited Fourier transform, Re stands for "real part of," and the exponential has the effect of shifting the frequencies of $r(t)$ by f_0. Expanding the exponential using Euler's identity and letting $x(t)$ be the real part of $r(t)$ and $y(t)$ be the imaginary part, we obtain

$$\begin{aligned} n(t) &= Re\{(x(t) + jy(t))(\cos 2\pi f_0 t + j \sin 2\pi f_0 t)\} \\ &= x(t) \cos 2\pi f_0 t - y(t) \sin 2\pi f_0 t \end{aligned}$$

as in Eq. (2-23).

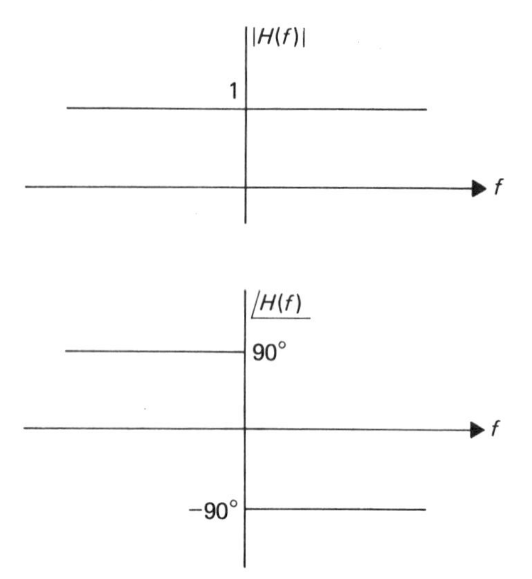

Figure 2.24 System function of Hilbert transformer

Explicit solution of Eq. (2-23) for $x(t)$ and $y(t)$ is not simple. One way to solve it is by using Hilbert transforms. Since this concept will be useful to us again in the discussion of envelopes and in single-sideband AM, we shall spend a few moments developing it.

Hilbert Transform. If all components of a time signal are shifted by −90 degrees in phase (and the amplitude is not changed), the result is a new time function known as the *Hilbert transform.* Since the impulse response of a real system, $h(t)$, is real, the real part of $H(f)$ must be even and the imaginary part odd. Alternatively, the amplitude of $H(f)$ must be even and the phase odd. Thus, the system function of Figure 2.24 results. $H(f)$ is then given by

$$H(f) = -j \operatorname{sgn}(f) \tag{2-24}$$

where sgn is the "sign" function. The impulse response of this system is found by evaluating the inverse Fourier transform of $H(f)$. A limiting process similar to that used in finding the Fourier transform of the unit step function yields

$$h(t) = \frac{1}{\pi t}$$

Thus the Hilbert transform of a signal, $s(t)$, is given by the convolution of the signal with $h(t)$:

$$\hat{s}(t) \triangleq \frac{1}{\pi} \int_{-\infty}^{\infty} \frac{s(\tau)}{t - \tau} d\tau$$

where the caret over the $s(t)$ is the notation used for the Hilbert transform.

If we take the Hilbert transform of a Hilbert transform, the effect in the frequency domain is to multiply the signal transform by $H^2(f)$. We see from Eq. (2-24) that $H^2(f) = -1$. Thus, the inverse Hilbert transform is given by

$$s(t) = -\frac{1}{\pi}\int_{-\infty}^{\infty}\frac{\hat{s}(\tau)}{t-\tau}\,d\tau$$

Example 2-8

Find the Hilbert transform of the following time signals.

(a) $s(t) = \cos(2\pi f_0 t + \theta)$

(b) $s_m(t) = \dfrac{\sin t}{t}\cos 2\pi x 100t$

(c) $s_m(t) = \dfrac{\sin t}{t}\sin 2\pi x 100t$

Solution. It is almost always easier to avoid time convolution by working with Fourier transforms.

(a) The Fourier transform of $s(t)$ is given by

$$S(f) = [\tfrac{1}{2}\,\delta(f - f_0) + \tfrac{1}{2}\,\delta(f + f_0)]e^{-j\theta f/f_0}$$

Note that the phase shift of θ radians is equivalent to a time shift of $\theta/2\pi f_0$ seconds. We now multiply this by $-j$ sgn (f) to get

$$\hat{S}(f) = [-\tfrac{1}{2}j\delta(f - f_0) + \tfrac{1}{2}j\delta(f + f_0)]e^{-j\theta f/f_0}$$

The quantity in square brackets is the Fourier transform of a sine wave, so the result is

$$\hat{s}(t) = \sin(2\pi f_0 t + \theta)$$

This is not surprising, since the Hilbert transform is a 90-degree phase-shifting operation.

(b) We shall denote

$$s(t) = \frac{\sin t}{t}$$

The Fourier transform of $s_m(t)$ is then given by

$$S_m(f) = \tfrac{1}{2}S(f - 100) + \tfrac{1}{2}S(f + 100)$$

Since $S(f)$ is bandlimited to $f = \pm 1$, the first term in $S_m(f)$ occupies frequencies between 99 and 101 Hz while the second term occupies frequencies between -101 and -99 Hz. When $S_m(f)$ is multiplied by $-j$ sgn (f) to get the Fourier transform of the Hilbert transform, we find

$$\hat{S}_m(f) = -\tfrac{1}{2}jS(f - 100) + \tfrac{1}{2}jS(f + 100)$$

The inverse transform of this yields

$$\hat{s}_m(t) = s(t) \sin 2\pi x 100t = \frac{\sin t}{t} \sin 2\pi x 100t$$

(c) We can use the fact that the Hilbert transform of a Hilbert transform is the negative of the original function. Thus by inspection, the result is

$$-s(t) \cos 2\pi x 100t = -\frac{\sin t}{t} \cos 2\pi x 100t$$

We are now ready to return to the solution of Eq. (2.23). If $x(t)$ and $y(t)$ are assumed to be bandlimited to frequencies below f_0, we can take the Hilbert transform of both sides of Eq. (2-23) to get

$$\hat{n}(t) = x(t) \sin 2\pi f_0 t + y(t) \cos 2\pi f_0 t \tag{2-25}$$

If Eq. (2-23) is multiplied by $\cos 2\pi f_0 t$ and Eq. (2-25) is multiplied by $\sin 2\pi f_0 t$, when the two expressions are added together, $y(t)$ is eliminated. This yields

$$n(t) \cos 2\pi f_0 t + \hat{n}(t) \sin 2\pi f_0 t = x(t)[\cos^2 2\pi f_0 t + \sin^2 2\pi f_0 t]$$

$$= x(t) \tag{2-26}$$

Similarly, if Eq. (2-23) is multiplied by $\sin 2\pi f_0 t$ and Eq. (2-25) by $\cos 2\pi f_0 t$, taking the difference eliminates $x(t)$. This yields

$$y(t) = \hat{n}(t) \cos 2\pi f_0 t - n(t) \sin 2\pi f_0 t \tag{2-27}$$

Example 2-9

Show that the system of Figure 2.25 yields the quadrature components at the outputs.

Solution. The inputs to the low-pass filters are given by

$$n_1(t) = 2x(t) \cos^2 2\pi f_0 t - 2y(t) \sin 2\pi f_0 t \cos 2\pi f_0 t$$

$$= x(t) + x(t) \cos 4\pi f_0 t - y(t) \sin 4\pi f_0 t$$

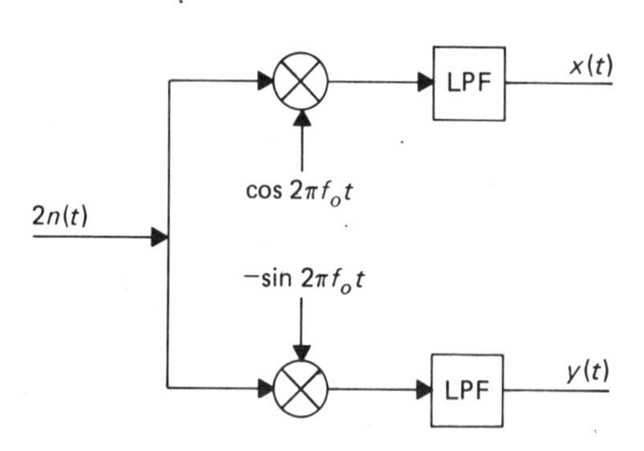

Figure 2.25 Generator for quadrature components

$$n_2(t) = -2x(t)\cos 2\pi f_0 t \sin 2\pi f_0 t + 2y(t)\sin^2 2\pi f_0 t$$
$$= -x(t)\sin 4\pi f_0 t + y(t) - y(t)\cos 4\pi f_0 t$$

Using the modulation theorem, we see that the Fourier transform of $x(t)\cos 4\pi f_0 t$ occupies a range around a frequency of $2f_0$, as is true of the other terms with $4\pi f_0$ in the above expression. The low-pass filter is designed to pass the frequencies of $x(t)$ and $y(t)$, so it will reject this high-frequency term. The outputs are therefore as shown on the diagram.

The autocorrelation of $x(t)$ and $y(t)$ can now be derived from Eqs. (2-25) and (2-27):

$$R_x(\tau) = R_y(\tau) = R_n(\tau)\cos 2\pi f_0\tau + \left[R_n(\tau) * \frac{1}{\pi t}\right]\sin 2\pi f_0\tau \qquad (2\text{-}28)$$

Finally, from Eq. (2-28) and using the modulation theorem, we get the power spectral density relationship:

$$G_x(f) = G_y(f) = G_n(f - f_0) + G_n(f + f_0)$$
$$f_0 - f_m < |f| < f_0 + f_m \qquad (2\text{-}29)$$

Equation (2-29) is the key result which will enable us to calculate the effects of noise upon AM and FM communication systems.

Example 2-10

Express the three narrowband noise processes of Fig. 2.26 in quadrature form using f_0 as the center frequency.

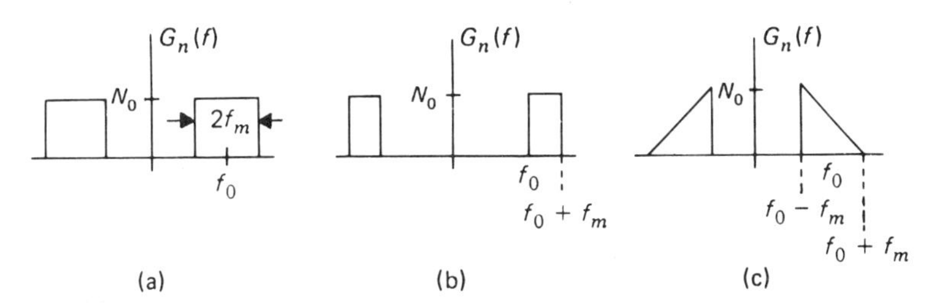

Figure 2.26 Noise spectral densities for Example 2-10

Solution. Using Eq. (2-29), we can immediately sketch the power spectral densities of $x(t)$ and $y(t)$, as in Fig. 2.27, where the noise is expressed as

$$n(t) = x(t)\cos 2\pi f_0 t - y(t)\sin 2\pi f_0 t$$

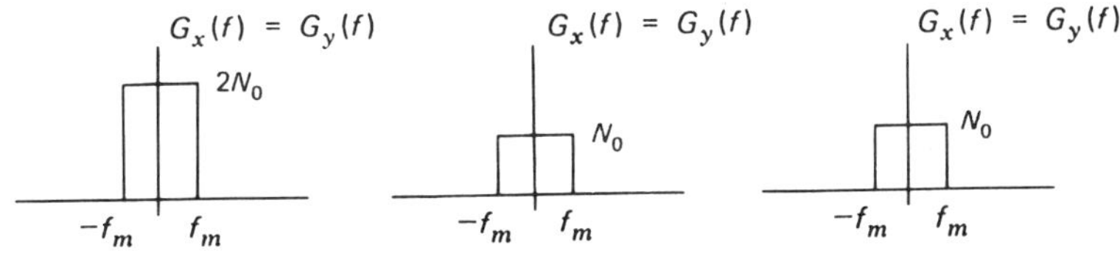

Figure 2.27 Power spectral densities of quadrature components for Example 2-10

2.6 SUMMARY

Before examining transmission and reception techniques, it is important to know the characteristics of the medium used to transport the signal from transmitter to receiver. This chapter explores the characteristics of the channel and introduces concepts and analysis techniques that will be used in later chapters.

We begin with a theoretical abstraction known as the memoryless channel. Although this type of channel does not exist in the real world, it forms a good starting point in order to establish the concepts and terminology and to introduce the important binary symmetric channel. The channel model provides for bit errors in each leg of the transmission path, and we are able to use the model for multisegment transmission analysis.

Section 2.2 explores bandwidth considerations, and we introduce the concepts of pulse spreading and intersymbol interference. This leads naturally to a discussion of distortion in Section 2.3. We develop analysis techniques for amplitude and phase distortion and examine the manifestations of these two factors.

Section 2.4 presents the practical aspects of telephone channels. This is important because telephone lines are the transmission channel for a significant percentage of digital communications. We examine the various grades of telephone line and the conditioning available to achieve a specified amplitude and phase characteristic.

Section 2.5 introduces the concepts of random noise and extends the probability analysis of Chapter 1 to the case of random processes. We present the important subjects of autocorrelation and power spectral density. These results will prove very important in finding bit error rates of various digital communication systems.

PROBLEMS

2.1. A binary channel is specified by the following two input/output relationships:

When the input is a binary 0, the received signal is a Gaussian random variable with mean zero and variance of unity.

When the input is a binary 1, the received signal is a Gaussian random variable with mean unity and variance of unity.

Assume that the channel output is a binary 1 if the received signal exceeds 0.5, and is 0 otherwise. Characterize this channel as a discrete memoryless channel and give the transition matrix.

2.2. A digital communication system has a symbol alphabet consisting of six entries with transition matrix as shown below.

$$\begin{bmatrix} \frac{1}{6} & 0.1 & 0.1 & 0.1 & 0.1 & 0.1 \\ \frac{1}{6} & 0.5 & 0.1 & 0.1 & 0.1 & 0.1 \\ \frac{1}{6} & 0.1 & 0.5 & 0.1 & 0.1 & 0.1 \\ \frac{1}{6} & 0.1 & 0.1 & 0.5 & 0.1 & 0.1 \\ \frac{1}{6} & 0.1 & 0.1 & 0.1 & 0.5 & 0.1 \\ \frac{1}{6} & 0.1 & 0.1 & 0.1 & 0.1 & 0.5 \end{bmatrix}$$

(a) Find the probability that exactly two symbols out of a ten-symbol message are in error.

(b) Find the probability of correct transmission of a five-symbol message.

2.3. You are given a binary symmetric channel with a bit error rate of 0.1.

(a) Find the probability that a transmitted message, 11111, will be received as 01101.

(b) Find the probability that a transmitted message, 11111, will be received as 10110.

2.4. A channel consists of n segments, each of which can be modeled as a BSC with crossover (error) probability p, where p is small. Show that the probability of error is approximately proportional to n. [*Hint:* You may wish to use an inductive proof. That is, start with the text result for two hops, and show that each additional hop increases the error probability by a fixed amount.]

2.5. You wish to design a communication system in order to link two stations separated by 5000 km. The system transmits 1 Mbps for 10 hours each day. Each bit error costs \$1 and each repeater station costs \$1 million (including maintenance over its life). A repeater station last 5 years. The bit error rate per hop is given by

$$\text{BER} = 10^{-6}e^{-10/X}$$

where X is the spacing in km. Develop formulas and discuss how you would find the optimum number of repeaters.

2.6. A given channel has a model as shown below. Find the bandwidth of this channel and also find the time response to a square pulse. Compare these to the bandwidth and pulse response of a single RC network.

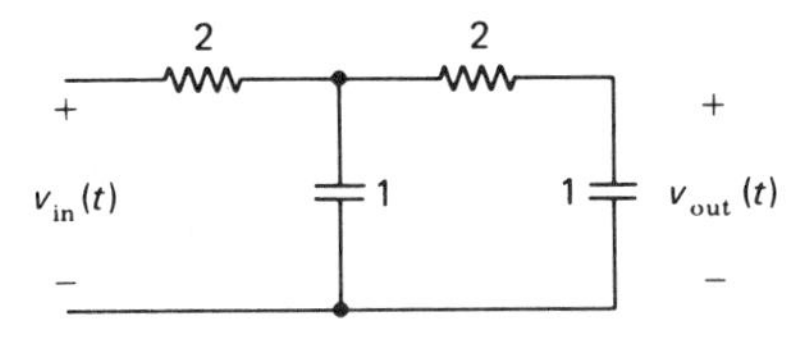

2.7. A filter has a sinusoidal amplitude response as shown below and a linear phase response with slope $-t_0$.

(a) Find the system response due to an input of $\cos t$.

(b) Find the system response due to an input of $(\sin t)/t$.

(c) Find the system response due to an input of $(\sin 10t)/t$.

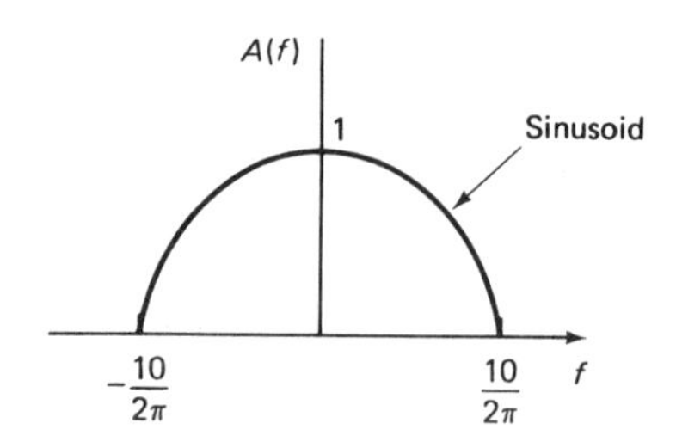

2.8. Repeat Problem 2.7 for the amplitude response shown below.

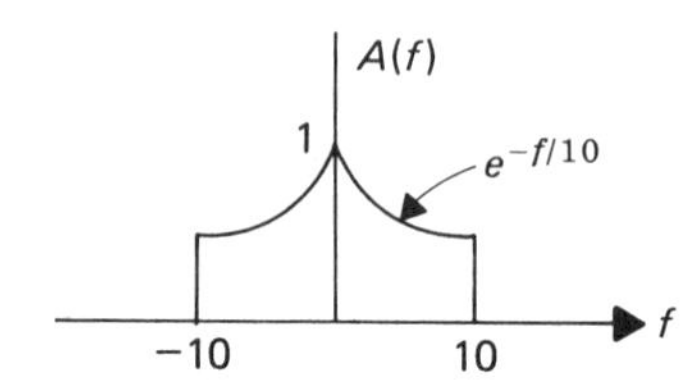

2.9. A system is as shown below.

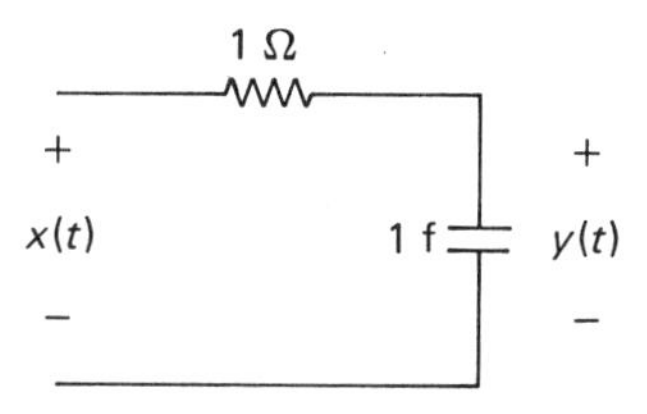

Assume that $\theta(f) = -2\pi f t_0$. Expand $A(f)$ in a series and use this to find the response due to a unit-amplitude unit-width square pulse.

2.10. You are given the system shown below.

The output of the system is $i(t)$. Find the phase distortion when the input is given by

$$v_{\text{in}}(t) = \frac{\sin t}{t} \cos 200t$$

2.11. Starting with Eq. (2.6), show that the time function corresponding to this Fourier transform is

$$s(t) = Ar(t - t_{gr}(f_0)) \cos 2\pi f_0(t - t_{ph}(f_0))$$

[*Hint:* You may find it useful to prove the following relationships first.]

$$r(t - t_0) \cos 2\pi f_0 t \leftrightarrow \tfrac{1}{2}[R(f - f_0)e^{-j2\pi(f-f_0)t_0} + R(f + f_0)e^{-j2\pi(f+f_0)t_0}]$$

$$r(t - t_1) \cos 2\pi f_0(t - t_1) \leftrightarrow \tfrac{1}{2}[R(f - f_0) + R(f + f_0)]e^{-j2\pi f t_1}$$

$$r(t - t_0) \cos 2\pi f_0(t - t_1) \leftrightarrow \tfrac{1}{2}[R(f - f_0)e^{-j2\pi(f-f_0)(t_0-t_1)} + R(f + f_0)e^{-j2\pi(f+f_0)(t_0-t_1)}]e^{-j2\pi f t_1}$$

2.12. Find the output of a typical telephone line when the input is
(a) $\cos 2\pi \times 500t + \cos 2\pi \times 1000t$.
(b) a periodic triangle of frequency 1 kHz.

2.13. White noise forms the input to a telephone line with amplitude and phase as shown in Figs. 2.12 and 2.13. The height of the two-sided noise power spectral density is K. Find the output power.

2.14. Find the cross correlation of narrowband noise with its Hilbert transform. That is, evaluate

$$E\{n(t)\hat{n}(t + \tau)\}$$

in terms of the center frequency, f_c, and the autocorrelation of $n(t)$.

2.15. Show that the cross correlation between the in-phase and quadrature terms in a narrowband noise expansion is given by

$$R_{xy}(\tau) = R_n(\tau) \sin 2\pi f_c\tau - \hat{R}_n(\tau) \cos 2\pi f_c\tau$$

2.16. Find the autocorrelation and power spectral density of the waveform

$$v(t) = A \cos (2\pi f_c t + \theta)$$

where A and f_c are not random, and θ is uniformly distributed between 0 and 2π.

2.17. Two narrowband noise processes have power spectral densities as shown below. Express each of these in quadrature form. For each process, choose two different center frequencies and sketch the power spectral density of $x(t)$ and $y(t)$.

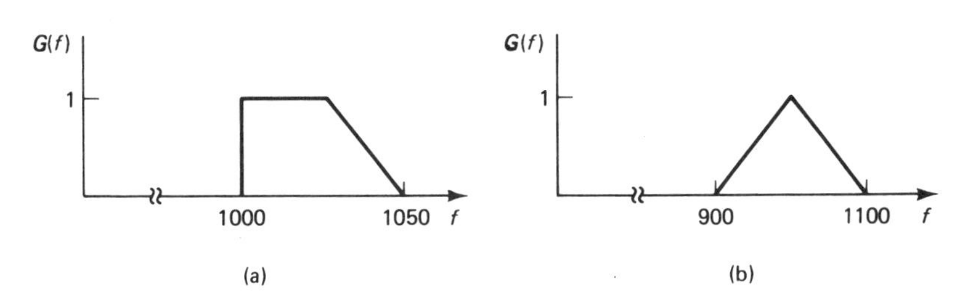

2.18. In a given communication system, the signal is

$$s(t) = 20 \cos t$$

Noise of power spectral density

$$G_n(f) = e^{-3|f|}$$

is added to the signal, and the resulting sum forms the input to a filter, $H(f)$.
(a) Find the signal-to-noise ratio (S/N) at the input to the filter.

(b) If the filter is an ideal low-pass filter with cutoff at $f = 2$, find the signal-to-noise ratio at the output of the filter.

2.19. $x(t)$ is white noise with autocorrelation $R_x(\tau) = \delta(\tau)$. It forms the input to an ideal low-pass filter with cutoff frequency f_m. Find the average power of the output, $E\{y^2(t)\}$.

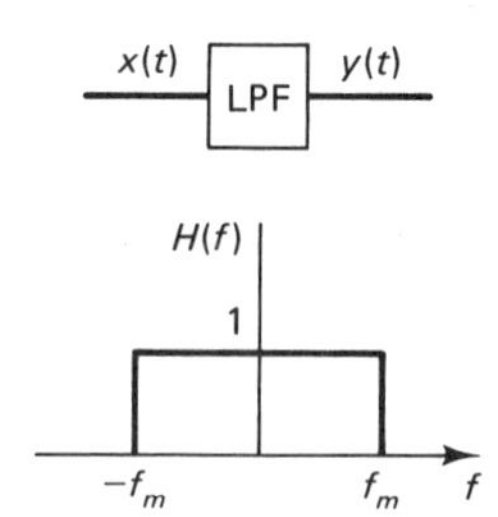

3

Source Encoding

The block diagram of a digital communication system presented in Fig. 1.1 contains a block entitled "Source Encoder." This block is not necessary if we begin with a data signal, such as might occur as the output of a digital computer. It is necessary when the original signal to be communicated is analog in format. That is, since our communication system can transmit only digital signals, it is necessary to first transform the analog waveform into a digital signal. This transformation, the job of the *source encoder,* is explored in the current chapter.

We start with a discussion of *sampling,* which is a technique of transforming from a continuous time axis to discrete points in time. Later sections then concentrate upon the amplitude of the signal and explore techniques for changing this continuous amplitude into a digital format.

3.1 SAMPLING

We need to find ways of converting an analog information waveform into a digital signal. To make this transformation requires two distinct operations. First the time axis must be discretized; then the resulting list of numbers must be reformatted so that each number is drawn from a discrete alphabet.

The sampling theorem. The changing of the continuous time axis into a discrete axis is accomplished by time sampling. The *sampling theorem,* sometimes known as Shannon's theorem or Kotelnikov's theorem, states that, if the Fourier transform of a time function is zero for $f > f_m$ and the values of the time function are known for $t = nT_s$ (for all integer values of n), then the time function is exactly known for *all* values of t provided that the samples are close enough together. The restriction is that $T_s < 1/2f_m$. Stated in an alternate way, $s(t)$ can be uniquely determined from its values at a sequence of equidistant points in time. The upper limit of T_s, $1/2f_m$, is known as the *Nyquist sampling rate.*

The upper limit on T_s can be expressed in a more meaningful way by taking the reciprocal of T_s to obtain the sampling frequency, denoted $f_s = 1/T_s$. The restriction then becomes

$$f_s > 2f_m$$

Thus, the sampling frequency must be at least twice the highest frequency of the signal being sampled. For example, if a voice signal has 3 kHz as a maximum frequency, it must be sampled at least 6000 times per second to comply with the conditions of the sampling theorem.

Before going further, we should observe that the spacing between the sample points is inversely related to the maximum frequency, f_m. This is intuitively satisfying, since the higher f_m is, the faster we would expect the function to vary. The faster the function varies, the closer together the sample points should be in order to permit reproduction of the function.

There are at least three common approaches toward proving the sampling theorem. We shall present only one proof. The proof we choose only requires a knowledge of basic amplitude-modulation theory. We choose it because the sampling system incorporated into this proof resembles a practical system we discuss later in this section.

Proof of the sampling theorem. Figure 3.1 shows a pulse train multiplying the original signal, $s(t)$. If the pulse train consists of narrow pulses, one would say that the output of the multiplier is a sampled version of the original waveform. In actuality, the output depends not only upon the sample values of the input, but upon a small range of values around each sample point. The theory does not require these extra values, which represent added information. However, practical systems usually sample over a small range of time surrounding the actual sample points. As we prove the theorem, it should become obvious that the multiplying function need not consist of perfect square pulses. In fact, the function can be any periodic signal.

Multiplying $s(t)$ by a $p(t)$ of the type shown in Fig. 3.1 is really a form of time gating. It can be viewed as the opening and closing of a gate, or switch.

Our goal is to show that the original signal can be recovered from the sampled waveform, $s_s(t)$. We shall do this by examining the Fourier transform of $s_s(t)$. The sampling theorem requires that we assume $s(t)$ has no energy above a frequency of f_m. The Fourier transform of $s(t)$, $S(f)$, therefore cuts off at f_m. Figure 3.2 shows a representative shape for this Fourier transform.

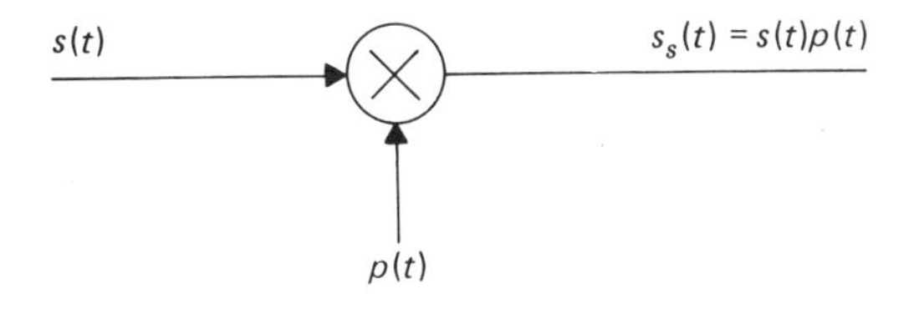

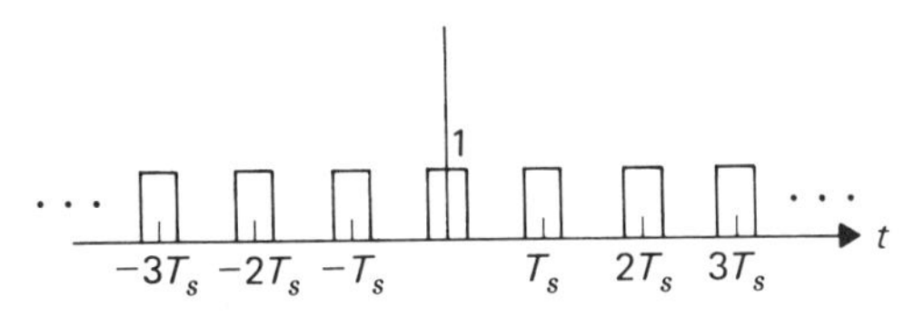

Figure 3.1 Sampling of $s(t)$ with a pulse train

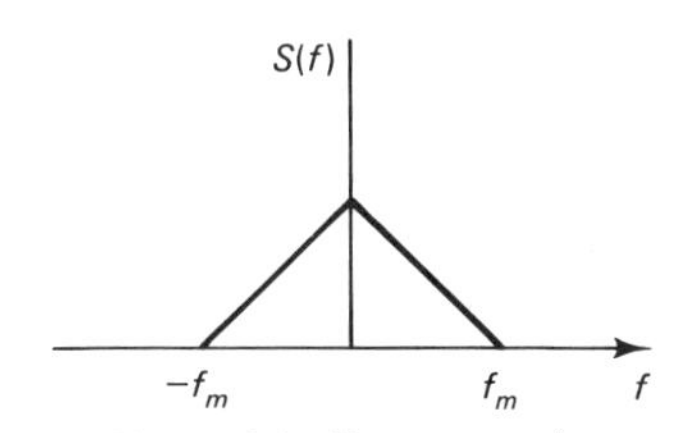

Figure 3.2 Representative frequency-limited Fourier transform

Since the multiplying pulse train is assumed to be periodic, it can be expanded in a Fourier series. The $p(t)$ shown is an even function, so we can use a trigonometric series containing only cosine terms. Thus,

$$\begin{aligned} s_s(t) &= s(t)p(t) \\ &= s(t)\left[a_0 + \sum_{n=1}^{\infty} a_n \cos 2\pi n f_s t\right] \\ &= a_0 s(t) + \sum_{n=1}^{\infty} a_n s(t) \cos 2\pi n f_s t \end{aligned} \tag{3-1}$$

where

$$f_s = \frac{1}{T_s}$$

The goal is to isolate the first term, which is proportional to the original $s(t)$. We can then undo the effects of any constant multiplier with an amplifier or attenuator.

Each term in the summation of Eq. (3-1) represents an AM waveform, where the information signal is $s(t)$ and the carrier frequency is nf_s. The frequency content is then centered around that carrier frequency. We can now find the Fourier transform of $s_s(t)$; it is sketched in Fig. 3.3. The shape centered at the origin is the transform of $a_0s(t)$, and the shifted versions represent the transforms of the various modulation terms. We see that the various terms will not overlap in frequency provided that $f_s > 2f_m$. But this is simply the condition given in the sampling theorem. Since the various terms occupy different bands of frequency, they can be separated from each other using linear filters. A low-pass filter with a cutoff frequency of f_m can be used to recover the $a_0s(t)$ term.

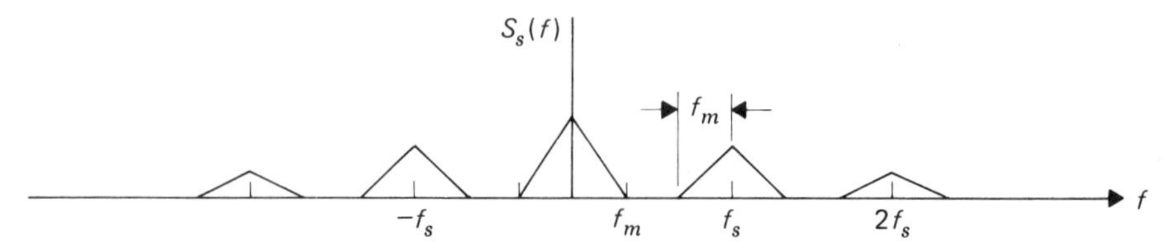

Figure 3.3 Fourier transform of sampled waveform

3.1.1 Errors in Sampling

The sampling theorem indicates that $s(t)$ can be perfectly recovered from its samples. If sampling is attempted in the real world, errors result from three major sources. Round-off errors occur when the various sample values are rounded off in the communication system. We will examine this type of error later in this chapter when we study the specific rules for rounding off (we will later call this error "quantization noise"). *Truncation errors* occur if the sampling is done over a finite time. That is, the sampling theorem requires that samples be taken for all time in the infinite interval, and every sample is used to reconstruct the value of the original function at any particular time. In a real system, the signal is observed for a limited time. We can define an error function as the difference between the reconstructed time function and the original function, and upper bounds can be placed upon the magnitude of this error function. Such bounds involve sums of the rejected time sample values, and some examples are included in the exercises at the end of this chapter.

A third error results if the sampling rate is not high enough. This situation can be intentional or accidental. For example, if the original time signal has a Fourier transform which asymptotically approaches zero with increasing frequency, a conscious decision is made to define a maximum frequency beyond which signal energy is negligible. In order to minimize this problem, the input signal is usually lowpass filtered prior to sampling. On the other hand, we may design a system with a high enough sampling rate, yet an unanticipated high-frequency signal appears at the input. In either case, the error caused by sampling too slowly is known as *aliasing,* a name derived from the fact that the higher frequencies disguise themselves in the form of lower frequencies. This is the same phenomenon that occurs if a rotating device is viewed as a sequence of individual frames, as in a television picture. As the device rotation speed increases, a point is reached where the perceived angular velocity starts decreasing. Eventually, a speed is reached (matched to the frame rate) where the device appears to be standing still. Further increases make the device appear to rotate in the reverse direction.

Analysis of aliasing is most easily performed in the frequency domain. Before doing that, we will illustrate the problem in the time domain. Figure 3.4 shows a sinusoid at a frequency of 3 Hz. Suppose we sample this sinusoid at 4 samples per

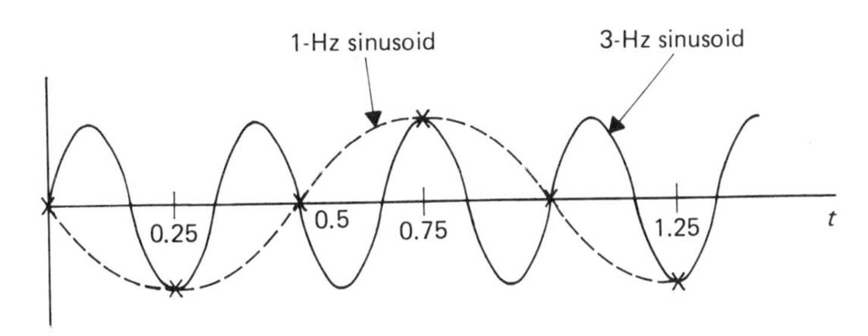

Figure 3.4 Illustration of aliasing error

second. The sampling theorem tells us that the minimum sampling rate for unique recovery is 6 samples per second, so 4 samples per second is not fast enough. The samples at the slower rate are indicated on the figure. But alas, these are the same samples that would result from a sinusoid at 1 Hz, as shown by the dashed curve. The 3-Hz signal is disguising itself (an alias) as a 1-Hz signal.

The Fourier transform of the sampled wave is found by periodically repeating the Fourier transform of the original signal. If the original signal has frequency components above one-half of the sampling rate, these components will fold back into the frequency band of interest. Thus, in Fig. 3.4, the 3-Hz signal folded back to fall at 1 Hz.

Figure 3.5 illustrates the case where a representative signal is being sampled by an ideal train of impulses (we use this as the ideal theoretical limit of narrow pulses). Note that the transform at the output of the low-pass filter is no longer the same as the transform of the original signal. If we denote the filter output as $s_0(t)$, the error would be defined as

$$e(t) = s_0(t) - s(t) \tag{3.2}$$

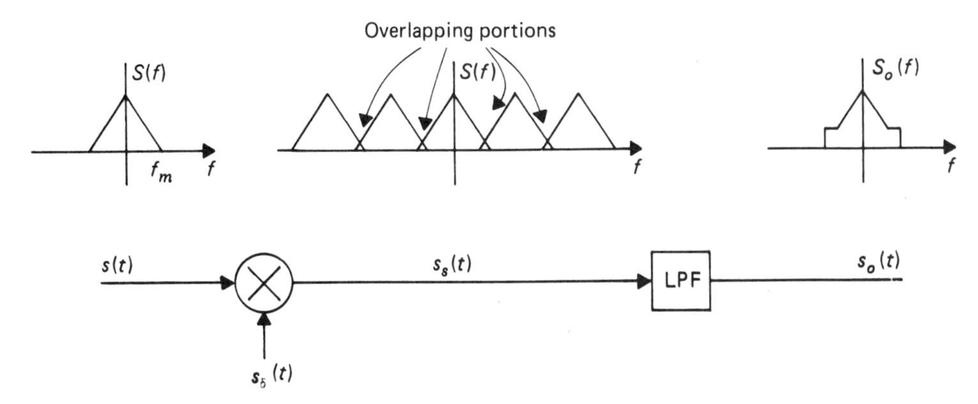

Figure 3.5 Aliasing error in sampling

If we take the Fourier transform of both sides, Eq. (3-2) yields

$$E(f) = S_0(f) - S(f)$$
$$= S(f - f_s) + S(f + f_s) \qquad \text{for } f < f_m$$

Note that if $S(f)$ were limited to frequencies below $f_s/2$, the error transform would be zero. Without assuming a specific form for $S(f)$, we cannot carry this example further. In general, various bounds can be placed upon the magnitude of the error function based upon properties of $S(f)$ for $f > f_s/2$.

Example 3-1

Assume that the bandlimited function,

$$s(t) = \frac{\sin 20\pi t}{\pi t}$$

is sampled at 19 samples/second. The impulse sampled function forms the input to a low-pass filter with cutoff frequency of 10 Hz, as illustrated in Fig. 3.6. Find the output of the low-pass filter and compare this with the original $s(t)$.

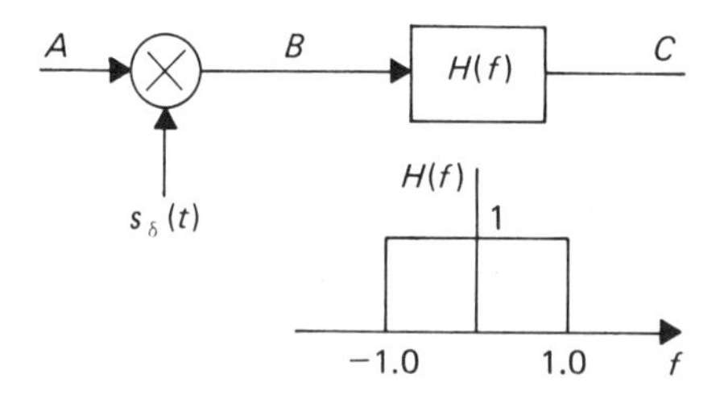

Figure 3.6 Low-pass filtering of sampled wave for Example 3-1

Solution. The Fourier transforms of the signals at points A, B, and C in Fig. 3.6 are shown in Fig. 3.7. The output time function is the inverse Fourier transform of $S_0(f)$ and is given by

$$s_0(t) = \frac{\sin 20\pi t}{\pi t} + 2\frac{\sin \pi t}{\pi t}\cos 19\pi t \tag{3-3}$$

The second term in Eq. (3-3) represents the aliasing error. It has a maximum amplitude of 2 at $t = 0$, at which point the signal has amplitude of 20. Do not be tempted to

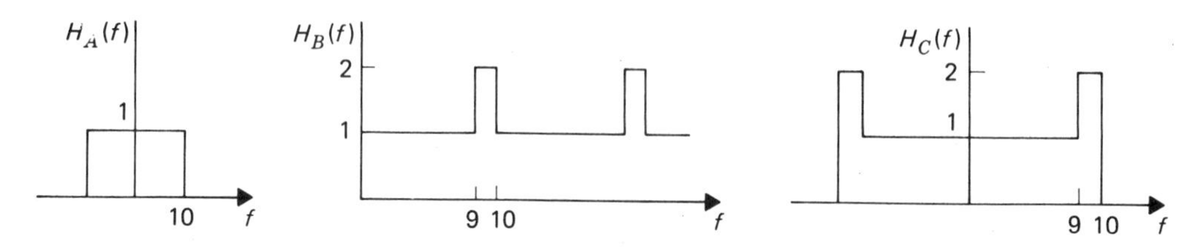

Figure 3.7 Fourier transform of signal at various points in system of Fig. 3.6

calculate a percentage error by taking the ratio of the error amplitude to the desired signal amplitude. Since the first term in Eq. (3-3) goes to zero at periodic points, and the second term is not necessarily zero at these same points, the percentage error can approach infinity. Errors are often analyzed by looking at the energy of the time function representing the error. Energy is the area under the square of the function. We could therefore find the energy of the second term in Eq. (3-3). Once this energy is found, it is divided by the total energy of the desired signal (the first term) to get a percentage of error.

We conclude this section with the idea that the restriction upon $S(f)$ imposed by the sampling theorem is not very severe in practice. All signals of interest in real life do possess Fourier transforms which are approximately zero above some frequency. No physical device can transmit infinitely high frequencies, since all channels contain series inductance and parallel (parasitic) capacitance. The inductance opens and the capacitance shorts as frequencies increase.

3.2 PULSE CODE MODULATION

Pulse code modulation is a technique for rounding off the amplitudes of samples of a waveform. This is the second of the two operations required to change an analog signal into the digital signal. The rounding-off operation is known as *quantization,* and the round-off error is known as *quantization noise.* We can view this in a simplistic manner as taking the original analog signal and approximating it as a staircase function, where the heights of the stairs are the permissible round-off values. Figure 3.8 shows an example of an analog signal and the staircase approximation. The more quantization round-off levels used, the closer the staircase function resembles the desired signal. The number of levels then determines the signal *resolution*—that is, how small a change in signal level can be detected by looking at the quantized version of the signal.

Once each sample value is rounded to the appropriate quantization level, we need transmit only enough information so that the receiver knows which level is being sent. PCM codes the various levels into binary numbers and sends the binary code corresponding to the particular round-off level. Thus, for example, if there

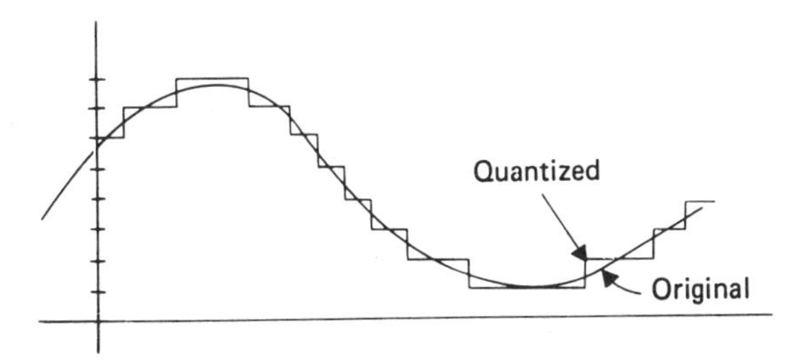

Figure 3.8 Staircase approximation to analog signal

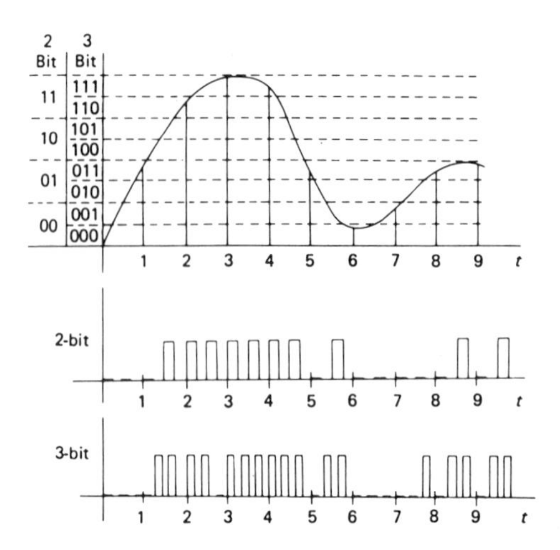

Figure 3.9 Analog-to-digital conversion

were eight quantization levels, the values could be coded into 3-bit binary numbers.

The resulting binary numbers can be transmitted using a variety of techniques. Such techniques form the major part of this text. For now, we shall illustrate the process using pulses—a positive pulse to represent a binary 1 and a zero pulse to represent a zero (we will later call this "baseband unipolar" transmission).

Figure 3.9 illustrates an analog signal which is being sampled at nine points in time. Also shown in the figure are the resulting binary pulse trains that would occur with four quantization levels and with eight levels. Note that two bits are needed for each sample in the case of four levels, and three bits for each sample in the case of eight levels. We refer to this as "2-bit PCM" and "3-bit PCM," respectively.

3.2.1 PCM Modulators

A *PCM modulator* is nothing more than an analog-to-digital converter. The converter first samples the waveform and then quantizes each sample value. We shall present several variations of the A/D converter. Before doing so, let us examine the quantization process in more detail.

Figure 3.10 illustrates 3-bit quantization. We are assuming that the sample values have been normalized to lie between 0 and 1. Figure 3.10(a) shows the range of functional values divided into eight regions. Each of these regions is assigned a 3-bit binary number. We have chosen eight regions because 8 is a power of 2. All 3-bit binary combinations are used, leading to greater efficiency. Figure 3.10(b) illustrates the quantization process as a functional relationship between input and output. We shall return to this figure when we analyze quantization noise (round-off error) later in this section.

Before leaving this figure, we shall use it to make a general observation about binary counting. As you examine the binary numbers along the ordinate, notice

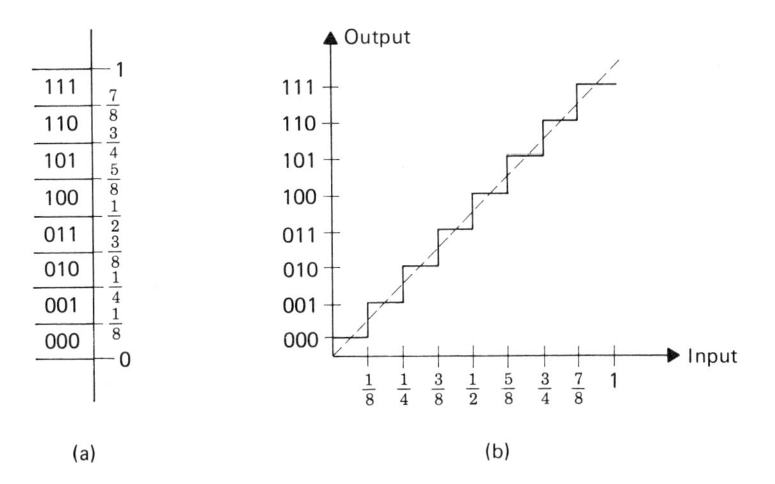

Figure 3.10 Three-bit PCM

that the first bit is equal to 1 for the top half of the range and to 0 for the bottom half. It oscillates with a period equal to the total range. The next bit alternates with a period equal to half the range and is equal to 1 for the top half *of each half* and to 0 for the bottom half *of each half*. This pattern continues with each successive bit subdividing the region by 2 and indicating which half of the new subregion the sample value is in.

There are three generic forms for the quantizer:

1. *Counting quantizers* serially count through each quantizing level.
2. *Serial quantizers* generate a code word, bit by bit. That is, they start with the most significant bit and work their way to the least significant bit.
3. *Parallel quantizers* generate all bits of a complete code word simultaneously.

3.2.2 Counting Quantizers

Figure 3.11 illustrates a *counting quantizer.* The ramp generator starts at each sampling point, and a binary counter is simultaneously started. The output of the sample-and-hold system is a staircase approximation to the original function, with stair steps that stay at the sample values throughout each sampling interval. A typical waveform is shown in the figure. The time duration of the ramp, and therefore the duration of the count, T_s, is proportional to the sample value. This is so because the ramp slope is kept constant. If the clock frequency is such that the counter has enough time to count to its highest count (all 1's) for a ramp duration corresponding to the maximum possible sample, then the ending counts on the counter will correspond to the quantizing levels.

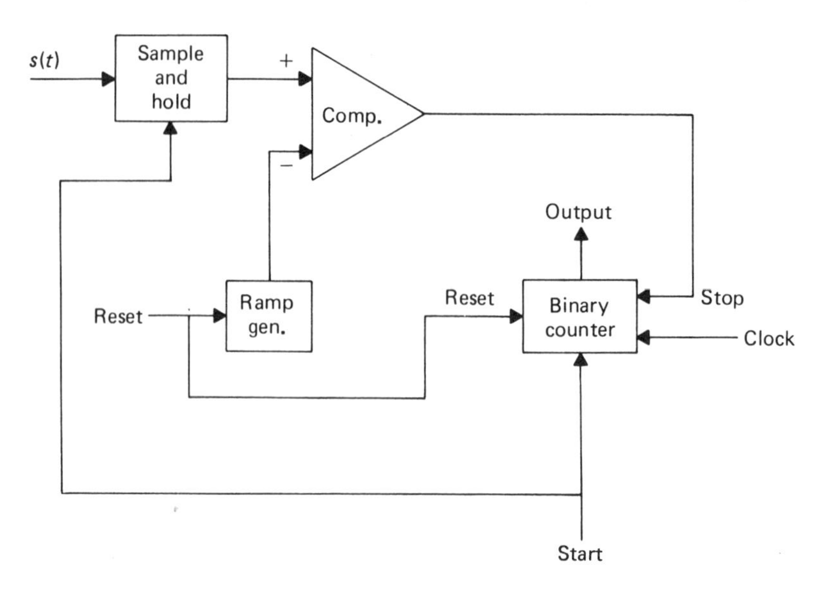

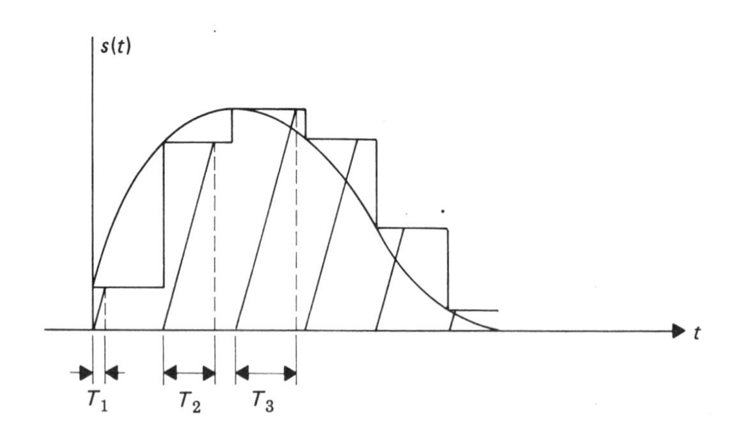

Figure 3.11 Counting quantizer

Example 3-2

You are designing a counting quantizer. Given that the ramp slope is 10^6 V/sec, signal amplitudes range from 0 to 10 V, and a 4-bit counter is used, what should the clock frequency be for a voice signal? Assume a maximum frequency of 3 kHz.

Solution. The only reason for worrying about the maximum frequency of the signal is to see if the ramp slope is sufficient to reach the maximum possible values within one sampling period. With a maximum signal frequency of 3 kHz, the minimum sampling rate for recovery is 6 kHz, so the maximum sampling period is $\frac{1}{6}$ msec. Since the ramp can reach the 10-V maximum in 0.01 msec, it is sufficiently fast to avoid overload problems. Therefore, the counter must be capable of counting from 0000 to 1111 in 0.01 msec. The clock frequency must be 1.6 MHz, since up to 16 counts are required in this sampling period.

Example 3-3

Design a counting ADC to convert $s(t) = \sin 2\pi t$ into a 4-bit digital signal. Choose appropriate parameters for the ADC.

Solution. To comply with the sampling theorem, the sampling rate must be greater than 2 samples/sec. Let us choose a rate 25% above this value, or 2.5 samples/sec. Many practical trade-off considerations go into this choice of sampling rate. If we use a rate very close to the minimum, very precise low-pass filtering is required at the receiver to reconstruct the original. On the other hand, if we use a much higher rate, the bandwidth of the transmitted waveform increases. We will also see that increasing the rate decreases the number of channels that can be simultaneously transmitted.

The individual sample values will range between -1 V and $+1$ V. The quantizer discussed in this section operates upon positive samples. We must therefore shift the signal by 1 V to assure that samples never go negative. The shifted samples range from 0 V to 2 V. The ramp must be capable of reaching the maximum sampling value within the sampling period, 0.4 second. The slope must therefore be at least $2/0.4 = 5$ volts/sec. In practice, we would choose a value larger than this to account for slight jitter in the timing of the system. We might choose a much larger slope value if we wanted to convert the sample in a small fraction of the period. This applies if a converter is being shared among a number of signals. At the minimum slope, it will take the ramp function 0.4 sec to reach the maximum sample value. The counter should therefore count from 0000 to 1111 in 0.4 sec. This requires a counting rate of 40 counts/sec.

3.2.3 Serial Quantizers

We now turn our attention to *serial quantizers.* The serial quantizer successively divides the ordinate into two regions. It first divides the axis in half and observes whether the sample is in the upper or lower half. The result of this observation generates the most significant bit in the code word.

The half-region in which the sample lies is then subdivided into two regions, and a comparison is again performed. This generates the next bit. The process continues a number of times equal to the number of bits of encoding.

Figure 3.12 shows a block diagram of this encoder for 3 bits of encoding and for inputs in the range of 0 to 1. The diamond-shaped boxes are comparators. They

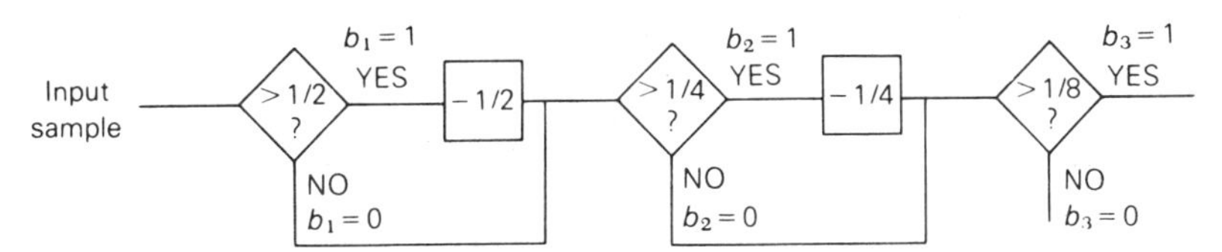

Figure 3.12 Serial quantizer

compare the input to some fixed value, giving one output if the input exceeds the fixed value and another output if the reverse is true. The block diagram indicates these two possibilities as two possible output paths labeled YES and NO. The figure is shown for 3-bit code words and for a range of input values between 0 and 1 V. If the input range of signal sample values were not 0 to 1, the signal could be normalized (shifted and then amplified or attenuated) to achieve values within this range. If more (or fewer) bits are required, the appropriate comparison blocks can be added (or removed).

Note that b_1 is the first bit of the coded sample value, known as the *most significant bit* or *msb*. b_3 is the third and final bit of the coded sample and is known as the *least significant bit* or *lsb*. The reason for this terminology is that the weight associated with b_1 is 2^2, or 4, while the weight associated with b_3 is 2^0, or 1.

Example 3-4

Illustrate the operation of the system of Fig. 3.12 for the following two input sample values: 0.2 V and 0.8 V.

Solution. For 0.2 V, the first comparison, with $\frac{1}{2}$, would yield a NO answer. Therefore, $b_1 = 0$. The second comparison, with $\frac{1}{4}$, would yield a NO answer, so $b_2 = 0$. The third comparison, with $\frac{1}{8}$, would yield a YES answer, so $b_3 = 1$. The binary code for 0.2 V is therefore 001.

For the 0.8-V input, the first comparison, with $\frac{1}{2}$, would yield a YES answer, so $b_1 = 1$. We then subtract $\frac{1}{2}$, leaving 0.3. The second comparison, with $\frac{1}{4}$, results in a YES answer, so $b_2 = 1$, and we subtract $\frac{1}{4}$, leaving 0.05. The third comparison, with $\frac{1}{8}$, yields NO, so $b_3 = 0$. The final code for 0.8 V is 110.

A simplified system can be realized if, at the output of the block marked "$-\frac{1}{2}$" in Fig. 3.12, a multiplication by 2 is performed and the result is fed back into the comparison with $\frac{1}{2}$. All blocks to the right can then be eliminated, as in Fig. 3.13. The signal sample can be cycled through as many times as desired to achieve any number of bits of code-word length.

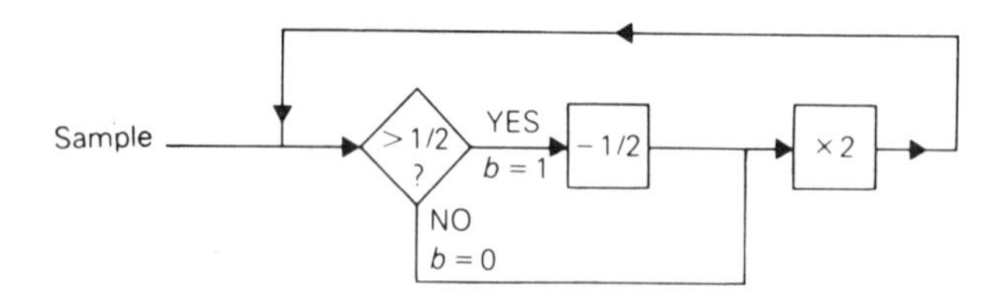

Figure 3.13 Simplified serial quantizer

3.2.4 Parallel Quantizers

The *parallel quantizer* (or *flash coder*) is the fastest in operation, since it develops all bits of the code word simultaneously. It is also the most complex, requiring a number of comparators that is only one less than the number of levels of quantization. We illustrate this with an example using 3 bits of encoding.

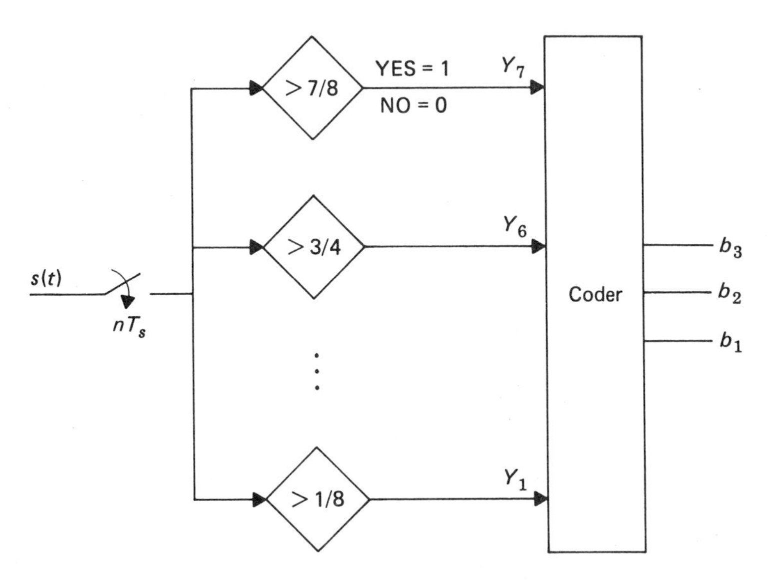

Figure 3.14 Parallel quantizer

Figure 3.14 shows a block diagram of the parallel 3-bit encoder. The block labeled "coder" observes the output of the seven comparators. It is a relatively simple combinational logic circuit. If all seven outputs are 1 (YES), the coder output is 111, since the sample value had to be greater than $\frac{7}{8}$. If comparator outputs 1 through 6 are 1 and output 7 is 0, the coder output is 110, since the sample had to be between $\frac{6}{8}$ and $\frac{7}{8}$. We continue through all levels and finally, if all comparator outputs are low, the sample had to be less than $\frac{1}{8}$, so the coder output is 000. Thus, only 8 of the 128 (2^8) possible coder inputs are legal, and the other 120 inputs represent illegal possibilities (e.g., there is no way a sample could be greater than $\frac{7}{8}$ and not greater than $\frac{5}{8}$). Thus, the combinational logic circuit contains a high percentage of *don't care* conditions, and the design is simplified accordingly.

While the serial quantizer takes advantage of the structure of binary numbers when counted in sequence, the parallel quantizer does not require such structure. In fact, the code for the quantization regions can be assigned in any useful manner. A problem with the sequential assignment is that transmission bit errors cause nonuniform reconstruction errors. A bit error in the msb causes a much greater error than one in the lsb.

In 1947, F. Gray, who was working with electronic coding devices, invented a "reflected binary code" in which adjacent numbers differ in only one bit position. We illustrate a 4-bit version at the top of page 108.

A change of one in the digit causes only one bit change. This code is easily implemented in the flash-coder logic. It can also be used in the other types of quantizers. In the counting quantizer, we simply vary the count sequence. In the serial quantizer, we follow the decision operations with a simple combinational logic circuit to convert the sequential code to the Gray code.

Digit	Binary	Gray
0	0000	0000
1	0001	0001
2	0010	0011
3	0011	0010
4	0100	0110
5	0101	0111
6	0110	0101
7	0111	0100
8	1000	1100
9	1001	1101
10	1010	1111
11	1011	1110
12	1100	1010
13	1101	1011
14	1110	1001
15	1111	1000

3.2.5 Practical PCM Encoders

Practical PCM encoders are constructed to follow the block diagrams presented in the previous three sections. Most forms are packaged as single integrated circuits. We shall briefly discuss one practical quantizer for each type of conversion. Since the technology is changing at a rapid pace, it should be emphasized that these practical implementations are presented for instructional purposes only. The design engineer should consult current specification sheets and handbooks to ensure that the best devices currently available are being incorporated into systems.

Counting quantizers are also known as *dual-slope* A/D converters. The input sample is first applied to an integrator for a fixed length of time, thus yielding an integrated output which is proportional to that sample value. The input to the integrator is then switched to a reference voltage (which is opposite in sign from the signal sample), the counter is started, and the output of the integrator is compared to zero. The counter is stopped when the integrator output ramp reaches zero.

The Intersil ICL7126 is a CMOS integrated circuit which simulates a counting quantizer. The IC package contains 40 pins and is illustrated in Fig. 3.15. Details of the operation of this chip can be found either in manufacturer's specification sheets or in the references covering analog integrated circuits. We shall give an overview of the chip inputs and outputs.

Pins 2 through 25 are used for the output display. The chip is configured to directly drive LCDs, as it includes seven-segment decoders and LCD drivers. The display is $3\frac{1}{2}$ digit, which means it can indicate numbers with magnitudes as high as 1999. The seven segment outputs for the *units* display are indicated as A1 to G1.

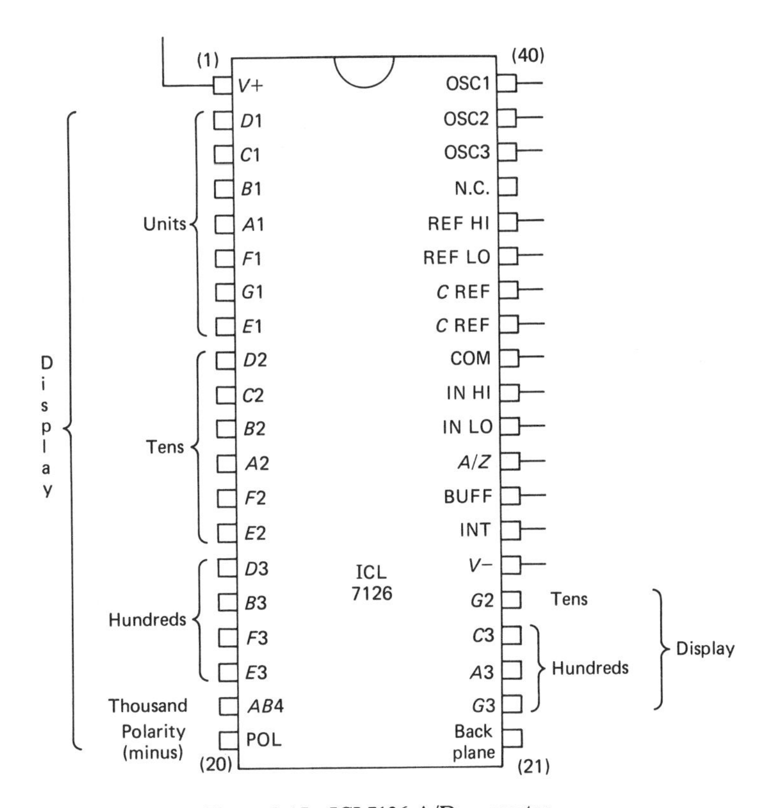

Figure 3.15 ICL7126 A/D converter

Those for the *tens* display use a suffix of 2, and the *hundreds* display uses 3. The *thousands* display is indicated as AB4, and only one lead is needed, since (for a $3\frac{1}{2}$ digit display) the thousands digit is either 0 or 1.

The analog input is applied to pins 30 and 31. The actual operation of the chip proceeds in three phases. The first is *auto-zero,* where the analog inputs are disconnected and internally shorted to a common, pin 32. The output of the comparator is shorted to the inverting input of the integrator.

The second phase occurs when the input signal value is integrated for a time corresponding to 1000 clock pulses. Finally, in the third phase, the reference voltage stored on a capacitor which is externally connected between pins 34 and 35 is used to cause the second ramp. The range of input values determines the required value of the reference, which is input to the REF HI pin (pin 36). If this input is 1 V, the chip is capable of converting voltages with magnitudes as high as 1.999. The clock can be derived from pins 38, 39 and 40. We can use either an external oscillator, a crystal connected between pins 39 and 40, or an *RC* circuit configured ac-

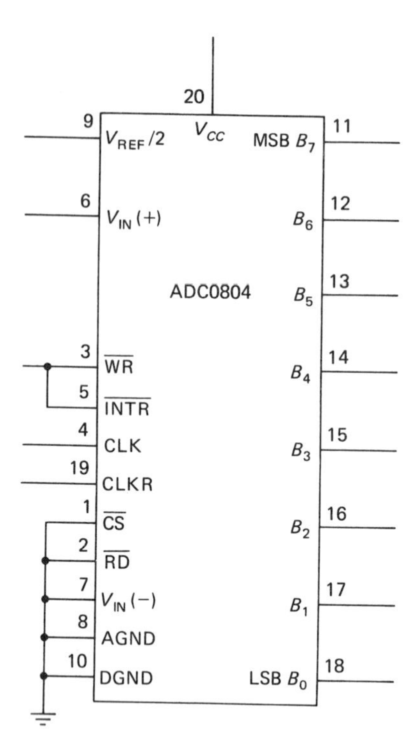

Figure 3.16 ADC0804 A/D converter

ross these pins. A complete A/D conversion of a single sample requires 4000 counts. The signal is integrated for one-fourth of this period, or 1000 counts. The second integration and the auto-zero require the remaining 3000 counts. The internal clock is developed by dividing the oscillator input by 4. Thus, for example, if we wish to perform 10 conversions per second, the oscillator input must be 160 kHz. This device is not capable of high-speed conversion; it should be used for slowly varying signals (low sampling rates) or DC inputs.

The National Semiconductor ADC 0804 is an example of an integrated circuit that performs serial A/D conversion, sometimes known as *successive-approximation* conversion. The chip is illustrated in Fig. 3.16. This is an 8-bit device. Its internal construction consists of a number of flip-flops, shift registers, a decoder, and a comparator. The full conversion takes eight internal clock pulses. The internal clock is provided by dividing the clock signal at pins 4 and 19 by 8. Thus, for example, with a 64-kHz signal on these pins, the chip can perform one conversion in 1 msec. The chip is capable of converting a sample in about 120 microseconds, so we still cannot use it for high-speed sampling.

The digital outputs, B_0 through B_7, appear on pins 11 through 18 of the chip. The chip is compatible with a microprocessor, which is the reason for much of the terminology labeling the pins on the chip. These lines are specified as follows:

Pin	Label	Function
1	CS (chip select)	Set LOW to initialize, HIGH to start conversion
2	RD (read)	Goes LOW to indicate microprocessor is ready to receive data
3	WR (write)	LOW to initialize, HIGH to start conversion
4	CLK	Input external oscillator or connect resistor between 4 and 19 to set oscillation frequency
5	INTR (interrupt)	Goes LOW to tell microprocessor that data are available
6	V_{IN}	Part of differential input (with pin 7 for negative input)
9	$V_{REF}/2$	Reference voltage (one-half of full-scale voltage)

The *parallel quantizer* is also known as a *flash A/D converter.* The RCA CA3308 is an example of an integrated circuit which accomplishes the conversion. The layout of this 24-pin chip is shown in Fig. 3.17.

The chip is capable of converting a sample in 66.7 nsec. The chip contains a bank of comparators. The analog input is on pins 16 and 21, and reference voltages are applied to pins 10, 15, 20, 22, and 23. The digital output is read from pins 1 to 8.

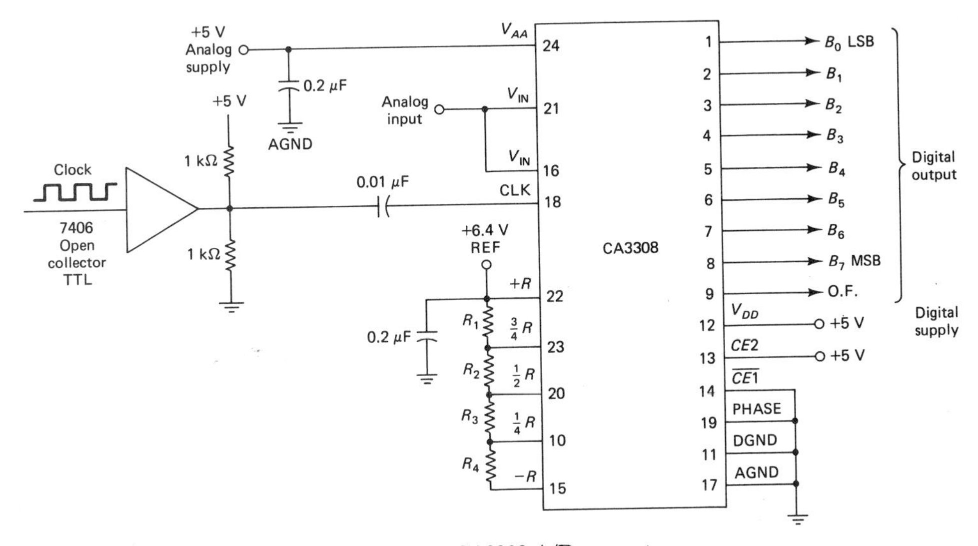

Figure 3.17 CA3308 A/D converter

3.2.6 PCM Decoders

We now shift our attention to the conversion of a digital signal to an analog signal. This is performed by a *digital-to-analog* converter (D/A converter, or DAC). To perform this conversion, we need simply associate a value with each binary code word. If the code word represents a region of sample values, the actual value chosen for the conversion is usually the center point of the region. If the A/D conversion is performed as previously described, the reverse operation is equivalent to assigning a weight to each bit position.

Let us illustrate the procedure for a 4-bit binary word. Let us further assume that the original analog sample fell in the range between 0 and 1 V and that sequential coding (as opposed to Gray coding) was used. We can see that conversion to the analog sample value is accomplished by converting the binary number to decimal, dividing by 16, and adding $\frac{1}{32}$. Thus, for example, the code 1101 represents the decimal number 13, so we convert this to $\frac{13}{16} + \frac{1}{32} = \frac{27}{32}$. The addition of the $\frac{1}{32}$ takes us from the bottom of the $\frac{1}{16}$-wide region to the middle.

Figure 3.18 illustrates a conceptual circuit to perform the conversion. If a 1 appears in the msb position, a $\frac{1}{2}$-V battery is switched into the circuit. The second bit controls a $\frac{1}{4}$-V battery, and so on.

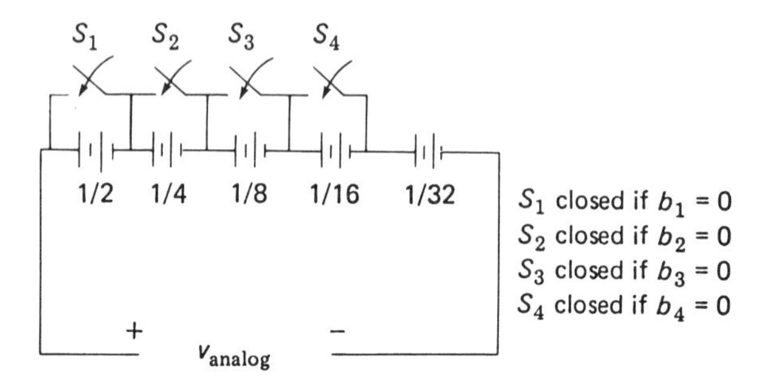

Figure 3.18 Conceptual diagram of D/A converter.

The ideal decoder of Fig. 3.18 is analogous to the serial quantizer, since each bit is being associated with a particular component of the sample value.

A more complex decoder results when an analogy to the counting operation is attempted. Figure 3.19 shows the counter decoder. A clock feeds a staircase generator and, simultaneously, a binary counter. The output of the binary counter is compared to the binary digitized input. When a match occurs, the staircase generator is stopped. The output of the generator is sampled and held until the next sample value is achieved. The final staircase approximation result is smoothed by a low-pass filter to recover the original signal.

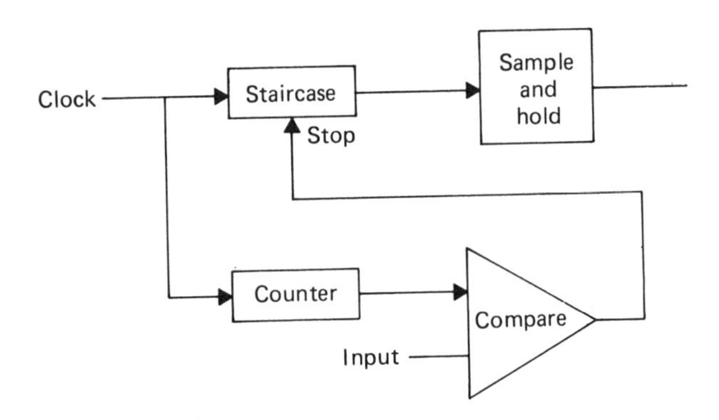

Figure 3.19 Block diagram of D/A converter

3.2.7 Practical PCM Decoders

Figure 3.20 shows a practical version of the ideal decoder of Fig. 3.19. The gain of the operational amplifier shown in the figure is $R/5R_{in}$, where R_{in} is the parallel combination of the resistors included in the input circuit (i.e., those with associated switches closed). In terms of input conductance, the gain is $RG_{in}/5$, where G_{in} is the sum of the conductances associated with closed switches. Therefore, closing S_1 contributes $\frac{8}{5}$ to the gain. Closing S_2 contributes $\frac{4}{5}$ to the gain; S_3 contributes $\frac{2}{5}$; S_4, $\frac{1}{5}$.

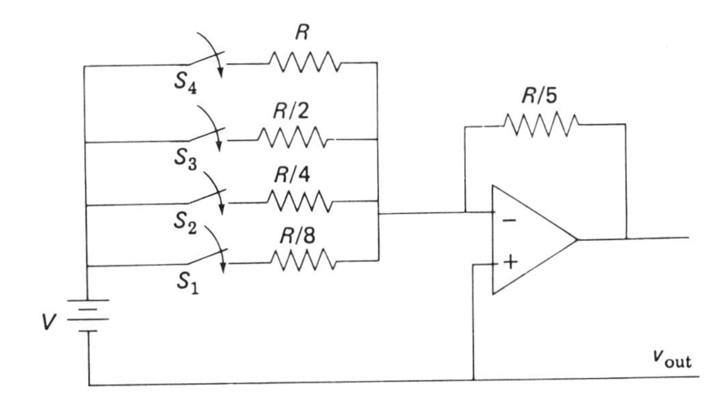

Figure 3.20 Practical D/A converter

In order to calculate the gain due to more than one switch being closed, we simply add the associated gains. Therefore, if the input voltage source, V, is made equal to 5 V, we see that S_1 contributes 8 V to the output; S_2 contributes 4 V, S_3 contributes 2 V, and S_4 contributes 1 V. The output voltage is an integer between 0 and 15 corresponding to the decimal equivalent of the binary number controlling the switches. To convert this system into a DAC analogous to that of Fig. 3.16, we simply divide the output by 16 and add $\frac{1}{32}$V. The division can be performed by scaling V or by scaling the resistor values. The addition of $\frac{1}{32}$V can be accomplished by adding a fifth resistor (unswitched) to the input. The value of this resistor would be 2R.

3.2.8 Nonuniform Quantization

Figure 3.21 illustrates the quantization process as an input-vs.-output function. In that figure, and in all of the earlier examples of this section, the range of sample values was divided into quantization regions, each of the same size as every other region. Thus, for example, with 3-bit quantization, we divided the entire range of sample values into eight equal regions. Because all of the regions are the same size, we refer to this as *uniform quantization.*

Under certain circumstances it will prove advantageous to use quantization intervals which are not all the same size. That is, we could replace the quantization function in Fig. 3.21 with that shown in Fig. 3.22. The function of Fig. 3.22 has the property that the spacing between quantization levels is no longer uniform, and the output levels are no longer in the center of each interval. Let us first see intuitively why nonuniform quantization may be preferable to uniform.

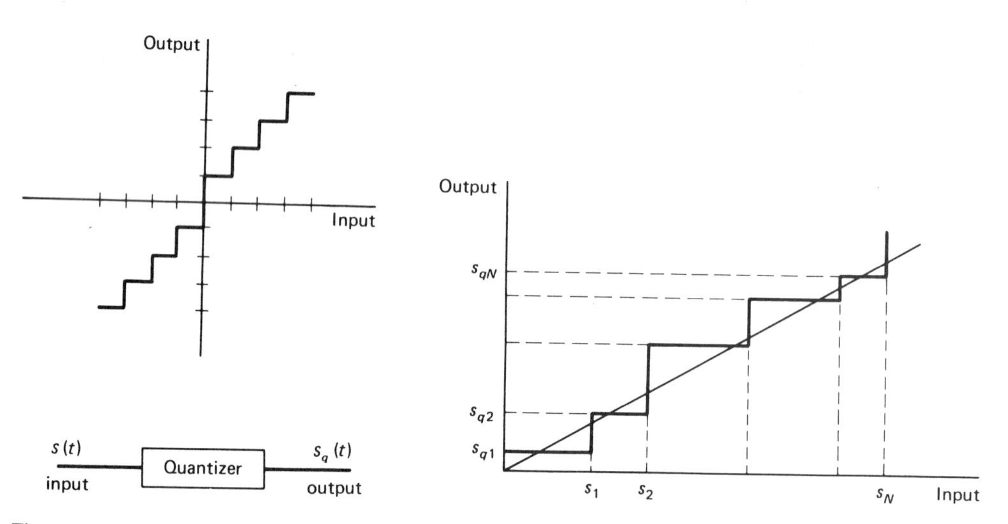

Figure 3.21 Input-output function for uniform quantizer

Figure 3.22 Input-output function for nonuniform quantizer

Consider a musical piece where the voltage waveform ranges from −2 to +2 volts. Suppose further that 3-bit uniform quantization is used. Therefore all voltages between 0 and $\frac{1}{2}$ volt will be coded into the same code word, 100, corresponding to the output reconstructed value of $\frac{1}{4}$ V. Likewise, all samples between 1.5 and 2 volts will be coded into a single code word, 111, corresponding to an output reconstructed value of $\frac{7}{4}$ V. During soft music passages, where for long periods the signal may not exceed $\frac{1}{2}$ volt, a great deal of music definition would be lost. The quantization provides the *same resolution* at high levels as at low, even though the human ear is less sensitive to changes at higher levels. The response of the human ear is nonlinear. It would appear to be desirable to use small quantization steps at the lower levels and larger steps at the higher levels.

As an alternative justification for examining nonuniform quantization, suppose the signal spent a far greater percentage of time at the low levels than at the higher levels. It would seem preferable to provide higher resolution at these lower levels at the expense of lower resolution at the high levels. Since the signal spends much less time at the higher levels, the average quantization error may well decrease using this approach.

The quantization error, or quantization noise, is a measure of the effectiveness of a quantization scheme. We shall examine that in detail in Section 3.2.11. For now, we will take a qualitative approach.

The average quantization error is a function of the quantization regions (s_i's in Fig. 3.22), the round-off values (s_{qi}'s in Fig. 3.22), and the probability density function of the sample values. We will show later that once quantization regions have been chosen (the s_i's in Fig. 3.22), the s_{qi}'s are chosen to be the center of gravity of the corresponding portion of the probability density. Figure 3.23 shows a representative example of a probability density function that resembles a Gaussian density. We have divided this into eight equal regions (this will prove to be a poor choice), indicated with boundaries of s_0 to s_8 on the figure. Given this choice of regions, the round-off levels would be approximately as shown in the figure. That is, we have shown the s_{qi} to be at the approximate center of gravity of each quantization region.

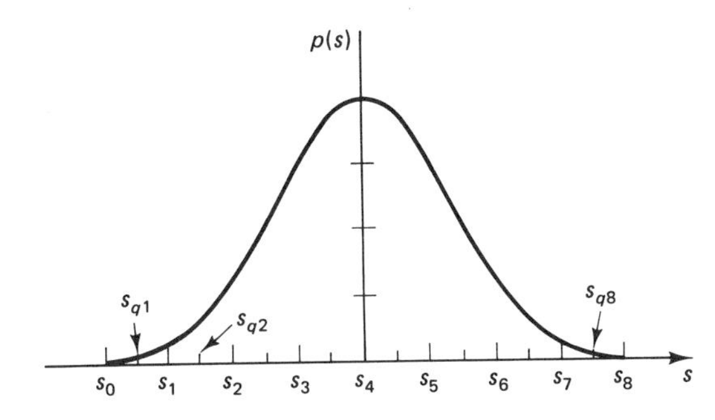

Figure 3.23 Probability density function of input signal

If, in addition to the specific quantization levels, we also make the actual intervals variable, a complicated optimization problem results and a complex system would be required to implement the result. Indeed, the coders (quantizers) presented in Section 3.2.5 would become extremely complicated, with each individual quantization level requiring separate comparators. For this reason, and also because systems should be applicable to a variety of input signals, suboptimum systems are almost always used. This leads us to a study of companding.

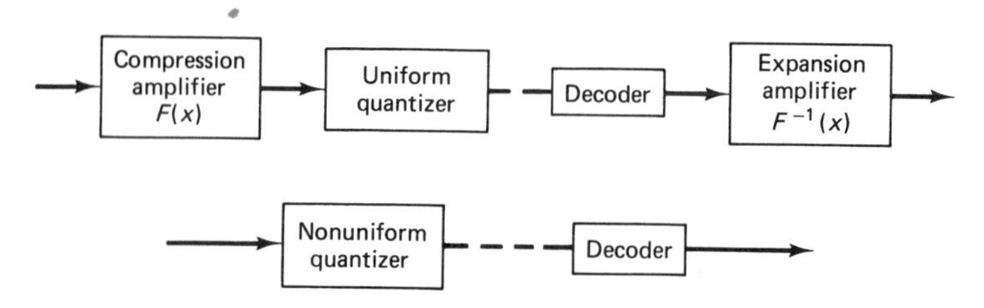

Figure 3.24 The concept of companding

3.2.9 Companding

The most common form of nonuniform quantization is known as *companding.* The name is derived from the words "compressing–expanding." The process is illustrated in Fig. 3.24. We see that the original signal is compressed using a memoryless nonlinear device. The compressed signal is then uniformly quantized. Following transmission, the decoded waveform must be expanded using a nonlinear function which is the inverse of that used in compression.

Prior to quantization, the signal is distorted by a function similar to that shown in Fig. 3.25. This operation compresses the extreme values of the waveform while enhancing the small values, much as the logarithm is used to permit viewing of very large and very small values on the same set of axes. If the analog signal forms the input to this compressor and the output is uniformly quantized, the result is equivalent to quantizing with steps that start out small and get larger for higher signal levels. This is shown in Fig. 3.25. We have divided the output of this compressor into eight equal quantization regions. The function is used to translate the boundaries of these regions to the abscissa, which represents the uncompressed input signal. Note that the regions in the s-axis start out small and get larger with increasing values of s.

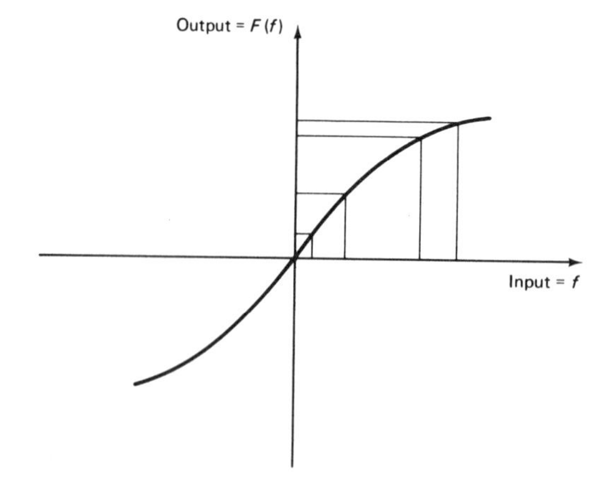

Figure 3.25 Typical compression curve

In order to realize hardware efficiencies, it is desirable to agree to some standards for companding. That is, if you have any hope of purchasing an off-the-shelf PCM encoder with companding, there is a need to agree to a limited number of standard compression formulae.

The most common application of companding is in voice transmission. North America and Japan have adopted a standard compression curve known as μ-law companding. Europe has adopted a different, but similar, standard known as A-law companding.

The μ-law compression formula is given by

$$F(s) = \text{sgn}(s) \frac{\ln(1 + \mu|s|)}{\ln(1 + \mu)}$$

This function is sketched for selected values of μ in Fig. 3.26. The parameter, μ, defines the degree of curvature of the function. A commonly used value is μ = 255.

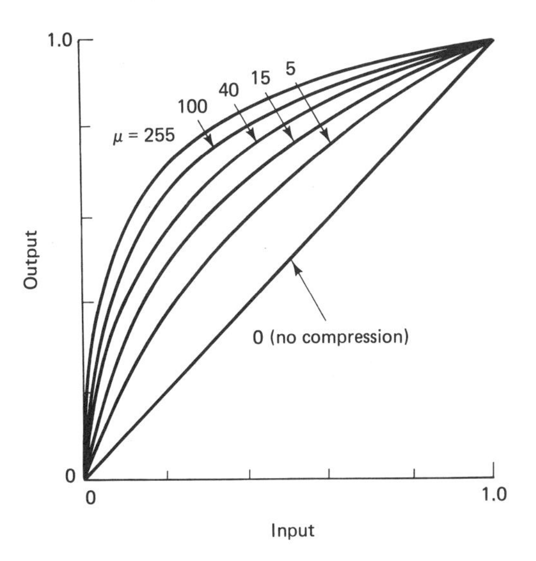

Figure 3.26 μ-law compression curves

We shall examine a possible implementation of a μ-255 encoder for 8-bit PCM encoding. One approach is to simulate a nonlinear system which follows the μ-255 input/output curve, then place the sample values through this system and uniformly quantize the output using an 8-bit A/D converter. An alternate approach is to approximate the μ-255 curve by a piecewise linear curve, as shown in Fig. 3.27. We are illustrating the positive input portion only, since the curve is an odd function. Note that we have approximated the positive portion of the curve by eight straight line segments. We divide the positive output region into eight equal segments, which effectively divides the input region into eight *unequal* segments.

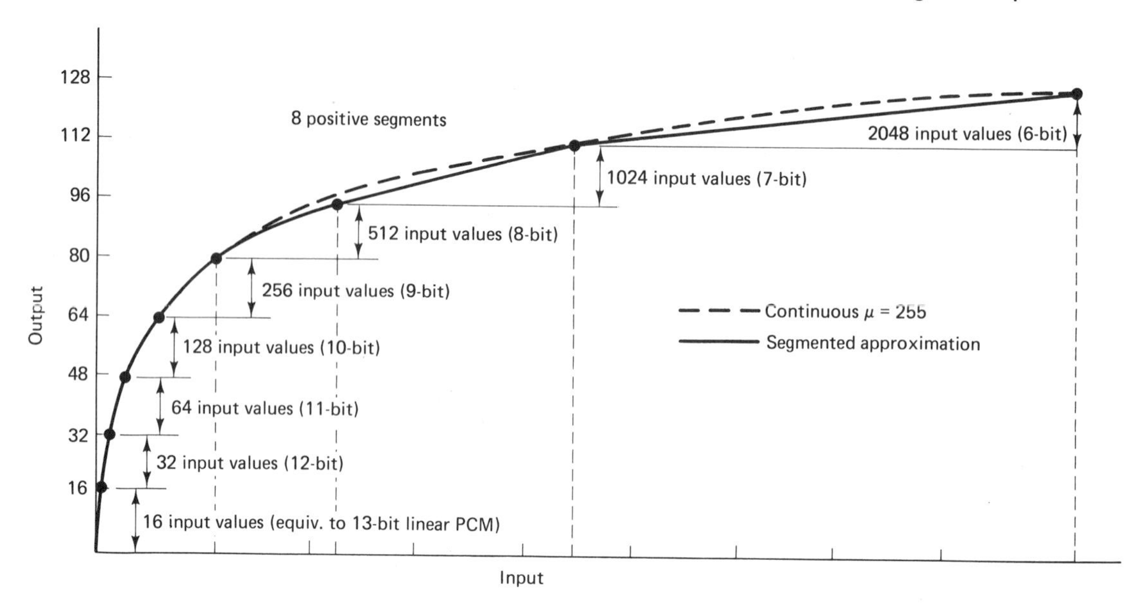

Figure 3.27 μ-255 compression curve

Within each of these segments, we uniformly quantize the value of samples using 4 bits of quantization. Thus, each of these eight regions is divided into 16 subregions, for a total of 128 regions on each side of the axis. We therefore have a total of 256, or 2^8, regions, which corresponds to 8 bits of quantization. The specific technique in sending a sample value is to send 8 bits coded as follows:

1 bit	gives the polarity of the sample: 1 for positive, 0 for negative.
3 bits	identify which piecewise segment the sample lies in.
4 bits	identify the quantization level within each sample region.

The logarithmic relationship of the μ-255 law leads to an interesting relationship among the eight segments. Each segment on the input axis is twice as wide as the segment to its left. The resolution of the first segment to the right of the origin is therefore twice that of the next segment, and so on. The sixth region to the right of the origin will cover an interval on the input axis which is one-sixteenth of the total swing. Thus, the resolution for samples in this particular interval is the same as that of uniform quantization using 8 bits of A/D. The resolution of the region just to the left of this will be the same as that of 9-bit uniform quantization. Similarly, as we move to the left, each region has the resolution of a uniform quantizer with one more bit of quantization than the previous one. These levels are marked on the figure.

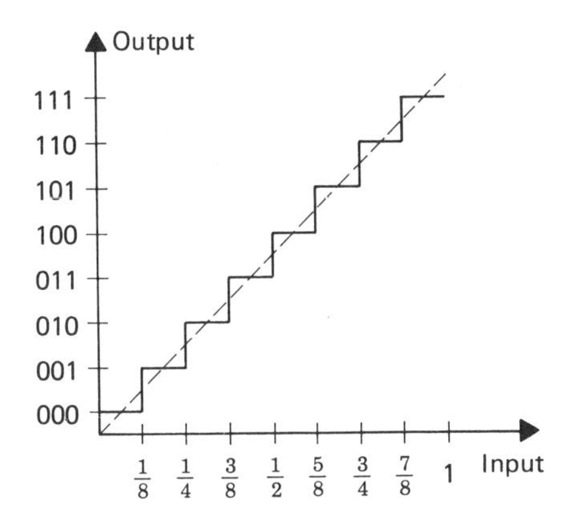

Figure 3.28 Input-output relationship of quantizer

3.2.10 Quantization Noise—Uniform Quantization

We begin our study of quantization noise in PCM by reexamining the quantization input-output relationship of Fig. 3.28. Quantization noise, or error, is defined as the time function which is the difference between $s_q(t)$, the quantized waveform, and $s(t)$. The error is given by

$$e(nT_s) = s(nT_s) - s_q(nT_s)$$

Figure 3.29(a) illustrates a representative time function, $s(t)$, and a quantized version of this time function, $s_q(t)$.

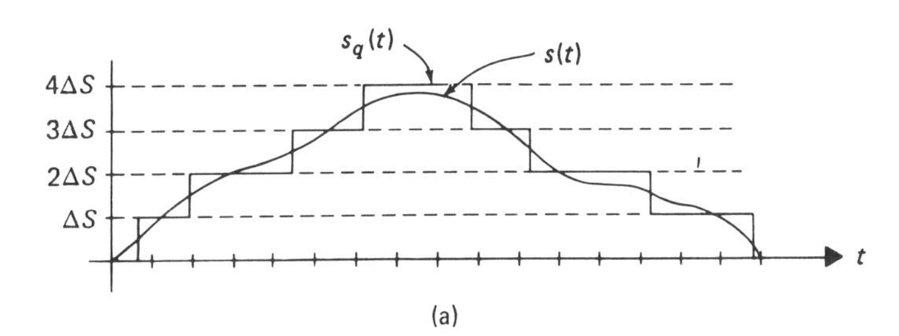

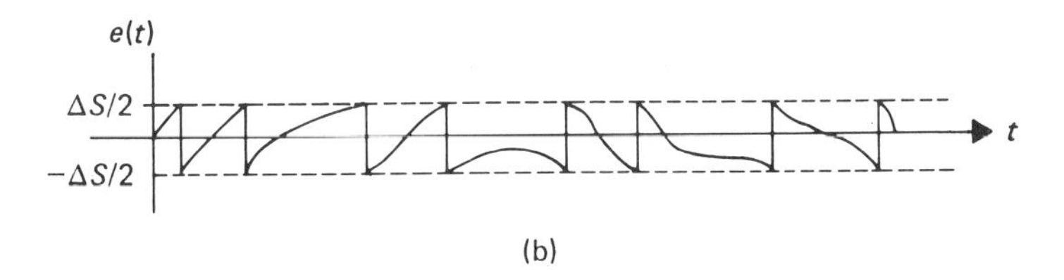

Figure 3.29 Typical time function and quantized version

While we are illustrating time functions, it is important to note that it is the *sample values* that are rounded off, not the analog time function. Therefore, the only significant values of $s_q(t)$ are the values at the time sample points, nT_s. Figure 3.29(b) shows the quantization error, $e(t)$, as the difference between $s(t)$ and $s_q(t)$. Again note that we are interested only in the values of this function at the sample points. The magnitude of the error term never exceeds one-half of the spacing between quantization levels.

We do not wish to restrict ourselves to a particular $s(t)$. It is therefore more important to find the average statistics of the error. In order to approach this problem, we need to find the probability density function of the error. The first step in the process is to plot the error as a function of the input sample value. This is shown in Fig. 3.30.

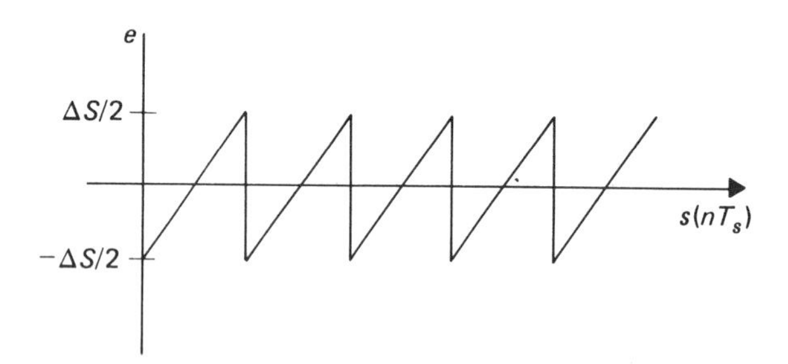

Figure 3.30 Quantization error as a function of input sample value

The error starts at $-\Delta S/2$ at the lower edge of each quantization interval and increases linearly to reach $+\Delta S/2$ at the top edge. If we now know the probability density function of the samples, $s(nT_s)$, it becomes a simple matter to find the probability density funciton of e. This is an application of "functions of a random variable." The result is shown in Eq. (3-4).

$$p(e) = \sum_i \frac{p(s_i)}{|de_i/df_i|} \tag{3-4}$$

The s_i are the various values of s corresponding to e. That is, if we set e to a certain value in Fig. 3.30, a number of values of s (equal to the number of quantization regions) yield this value of e. The magnitude of the slope of the function is always unity, so Eq. (3-4) can be simplified to

$$p(e) = \sum_i p(s_i) \tag{3-5}$$

The first value of s_i to the right of the origin is given by

$$s_i = e + \frac{\Delta S}{2}$$

All of the other s_i can be found by successively adding and subtracting ΔS from this value.

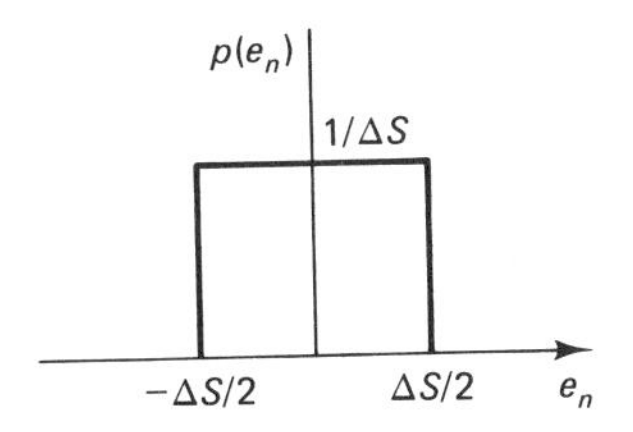

Figure 3.31 Probability density of quantization error

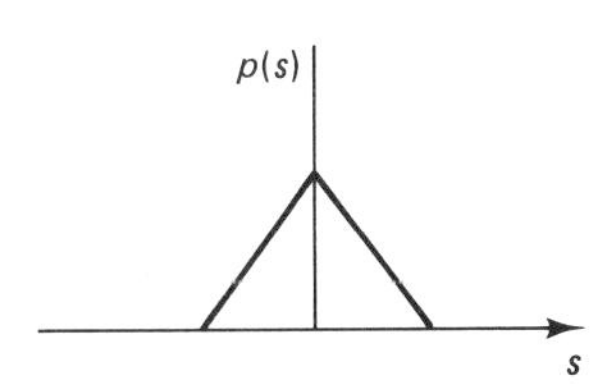

Figure 3.32 Triangular probability density function

If the original function is uniformly distributed over the range of values, each term in the summation of Eq. (3-5) will be a constant, the height of the original density function. The result will be a uniform density, as shown in Fig. 3.31. If the original density function is now triangular as shown in Fig. 3.32, the result will still be uniform. You can convince yourself of this by taking some representative values. In fact, as e increases, all of the s_i of Eq. (3-5) increase by the same amount. In a triangular density, for every value in the sum that is decreasing, another value is increasing by the same amount. A similar argument can be used for any density that is approximately piecewise linear over the range of a single quantization region. For this reason, the uniform density of Fig. 3.31 is assumed to be approximately correct over a wide range of input signals.

Now that we know the density of the error, we can find the average of its square, or the *mean square error:*

$$\begin{aligned} \text{mse} &= \int_{-\infty}^{\infty} p(e_n)e_n^2\, de_n \\ &= \frac{1}{\Delta S}\int_{-\Delta S/2}^{\Delta S/2} e_n^2\, de_n = \frac{\Delta S^2}{12} \end{aligned} \tag{3-6}$$

This result gives the average of the square of the error in a single sample of the time function. In order to assess the "annoyance" caused by this error, it is necessary to compare this value to the average of the square of the unperturbed time sample. This yields the very important signal-to-quantization-noise ratio. Thus,

$$\text{SNR} = \frac{\overline{s^2}}{\overline{e_n^2}} = \frac{12\overline{s^2}}{\Delta S^2} \tag{3-7}$$

Equation (3-7) is an extremely important result that will be used many times in the work to follow.

Example 3-5

Consider an audio signal comprised of the sinusoidal term

$$s(t) = 3 \cos 500t$$

(a) Find the signal-to-quantization-noise ratio if this is quantized using 10-bit PCM.

(b) How many bits of quantization are needed to achieve a signal-to-quantization-noise ratio of at least 40 dB?

Solution. (a) Equation (3-7) is used to find the signal-to-quantization-noise ratio. The only parameters which need to be evaluated are the signal power and the quantization region size. The total swing of the signal is 6 volts, so the size of each interval is $6/2^{10} = 5.86 \times 10^{-3}$. The signal power is $3^2/2 = 4.5$. The signal-to-quantization-noisc ratio is then given by

$$\text{SNR} = \frac{12 \times 4.5}{(5.86 \times 10^{-3})^2}$$

$$= 1.57 \times 10^6$$

If we wish to express this in decibels, we take its log and multiply by 10. Therefore,

$$\text{SNR} = 10 \times \log(1.57 \times 10^6) = 62 \text{ dB}$$

(b) The signal-to-noise ratio is specified as 40 dB. This corresponds to a ratio of 10^4. We use Eq. (3-7), where ΔS is unknown. Therefore,

$$\frac{12 \times 4.5}{\Delta S^2} = 10^4$$

and solving for ΔS yields

$$\Delta S = 7.35 \times 10^{-2}$$

We now note that

$$\Delta S = \frac{6}{2^N}$$

where N is the number of bits of quantization. We need to choose an N such that ΔS is no larger than 7.35×10^2. Thus,

$$\frac{6}{2^N} < 7.35 \times 10^{-2}$$

and

$$2^N > 81.6$$

We could use logarithms to solve this, but doing so really should not be necessary. If N is 6, the left side is 64. If N is 7, the left side is 128. Therefore, we require 7 bits of quantization to achieve a signal-to-noise ratio of at least 40 dB.

It should be clear that each additional bit of quantization reduces ΔS by a factor of 2. This will increase the signal-to-quantization-noise ratio by a factor of 4. A factor of 4 corresponds to 6 dB, since

$$10 \log 4 \approx 6$$

Hence, each additional bit of quantization increases the SNR by 4 dB.

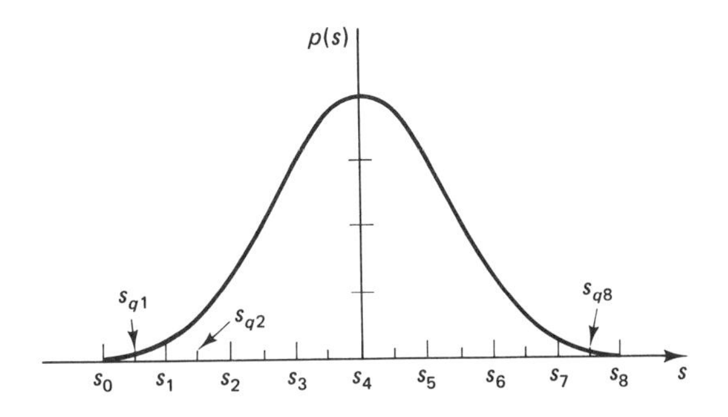

Figure 3.33 Probability density function of input signal

3.2.11 Quantization Noise—Nonuniform Quantization

Suppose that the input, $s(t)$, is distributed according to a probability density, $p(s)$, as shown in Fig. 3.33. This is approximately Gaussian. We have illustrated 3 bits of quantization, with the eight regions marked by the s_i and the round-off levels indicated by the s_{qi}. The mean-square quantization error is given by

$$\begin{aligned} \text{mse} &= E\{[s(nT_s) - s_q(nT_s)]^2\} \\ &= \int_s (s - s_q)^2 p(s)\, ds \\ &= \sum_i \int_{s_i}^{s_{i+1}} (s - s_{qi})^2 p(s)\, ds \end{aligned} \tag{3-8}$$

In Eq. (3-8), the s_{qi} are the various quantization round-off levels, and $p(s)$ is the probability density function of the signal. We shall be returning to this equation in a few moments when we examine companded systems. For now, we shall use the expression to prove a statement made earlier about the best location for the round-off levels. Let us assume that the regions have been specified (the s_i are given) and we wish to find the "optimum" location of the round-off values, s_{qi}. We interpret the word "optimum" to mean that we minimize the mean-square error. To do so, we differentiate Eq. (3-8) with respect to the s_{qi} and set the derivative to zero. This yields

$$\int_{s_i}^{s_{i+1}} (s - s_{qi}) p(s)\, ds = 0 \qquad \text{for all } i \tag{3-9}$$

Equation (3-9) indicates that once the quantization regions have been chosen, the round-off levels are selected to be the center of gravity of the corresponding portion of the probability density. Thus, the quantization levels, rather than being in the center of each interval, are skewed toward the higher-probability end of each interval. This is intuitively satisfying.

Example 3-6

Assume that the density function of $s(t)$ is a zero-mean Gaussian density with variance of $\frac{1}{9}$. Since the probability of a sample's exceeding a magnitude of 1 is less than 1%, assume that you quantize the region between -1 and $+1$ (i.e., values above a magnitude of 1 will saturate at either 000 or 111). Three-bit quantization is used.
(a) Find the mean-square quantization error, assuming uniform quantization is used.
(b) A scheme is proposed wherein the quantization regions are chosen to yield *equal area* under the probability density function over each region. That is, the probability of the function's being within any particular interval is the same as for any other interval. Choose the best location for the round-off values, and find the mean-square error.

Solution. (a) We shall use the approximate formula of Eq. (3-6) to find the mean-square error in the uniform case. The size of each interval is $\frac{2}{8} = \frac{1}{4}$. The error is therefore given by

$$\text{mse} = \frac{\Delta S^2}{12} = \frac{1}{192} = 5.2 \times 10^{-3}$$

(b) We must first find the quantization regions. We are dividing this into eight equal area segments, so the area under the density within each region must be $\frac{1}{8}$. Reference to a table of error functions will indicate that the s_i are given by

$$-1,\ -0.38,\ -0.22,\ -0.1,\ 0,\ 0.1,\ 0.22,\ 0.38,\ 1$$

Equation (3-9) is now used to find the round-off levels, s_{qi}. The equation reduces to

$$s_{qi} = 8 \int_{s_i}^{s_{i+1}} s\, p(s)\, ds \qquad \text{for all } i$$

This can be evaluated in closed form, or approximated. The resulting s_{qi} are given by

$$-0.54,\ -0.3,\ -0.16,\ -0.05,\ 0.05,\ 0.16,\ 0.3,\ 0.54$$

Finally, the mean-square error is found from Eq. (3-8) to be

$$\text{mse} = 5.3 \times 10^{-3}$$

It appears that the uniform quantizer is better than this particular nonuniform quantizer. However, with the Gaussian density and only 3 bits of quantization, Eq. (3-6) is not a very good approximation for the mean-square error. Recall that this equation required the density to be approximately piecewise linear over the various regions. The exact answer to part (a) could be found by applying Eq. (3-9). The result is approximately 6.2×10^{-3}, and therefore the nonuniform quantizer does provide an improvement in performance.

This example has suggested one possible algorithm for selection of quantization regions. In fact, this algorithm is not optimum and in some cases will even cause degradation of performance when compared to uniform quantization. The equation for mean-square error weights the probability by the square of the deflection from the quantized value before integration. In general, the problem is to minimize the error of Eq. (3-8) as a function of two variables, s_i and s_{qi}. The s_{qi} are constrained to satisfy Eq. (3-9). Unless the probability density can be expressed in closed form, this problem is computationally difficult.

The *mean-value theorem* can be used in Eq. (3-8) to get an approximation which improves as the number of bits of quantization increases. Using this approximation, the following rule for choosing the quantization regions results.

Choose the quantization regions to satisfy the uniform moment constraint

$$(s_{i+1} - s_i)^2\, p(\text{midpoint}) = \text{constant} \tag{3-10}$$

We will explore this more fully in the problems at the end of this chapter.

Companded systems. We shall derive an approximate expression for the improvement (or degradation) offered by a particular compression function as compared to uniform quantization. The result will be approximate, and the approximation improves as the number of bits of quantization increases—that is, as the quantization regions get smaller. In fact, we begin by assuming that the round-off values are in the middle of each interval. This will be correct only if the density can be assumed to be a constant over the width of the interval. Using this assumption, and further assuming that the density function can be approximated over the interval by its value at the round-off level, Eq. (3-8) reduces to

$$\begin{aligned} \text{mse} &= \sum_i \int_{s_i}^{s_{i+1}} (s - s_{qi})^2\, p(s)\, ds \\ &= \sum_i p(s_{qi}) \left. \frac{(s - s_{qi})^3}{3} \right|_{s_i}^{s_{i+1}} \end{aligned} \tag{3-11}$$

Now, using the assumption that s_{qi} is in the middle of the interval,

$$s_{qi} = \frac{s_i + s_{i+1}}{2}$$

and

$$\text{mse} = \sum_i p(s_{qi}) \frac{(s_{i+1} - s_i)^3}{12} \tag{3-12}$$

It is comforting to occasionally check our work against the known result for uniform quantization. In fact, if uniform step sizes of ΔS are substituted into Eq. (3-12), the result reduces to $\Delta S^2/12$, as we found earlier for uniform quantization. If this were not the case, we would have to recheck the derivation to find the error(s).

We can relate the interval size, $s_{i+1} - s_i$, to the slope of the compression curve. That is, if the compressed output is uniformly quantized with a step size of ΔS, the corresponding step sizes of the uncompressed waveform are approximately given by (see Fig. 3.23):

$$s_{i+1} - s_i \approx \frac{\Delta S}{F'(s_i)}$$

where $F'(s)$ is the derivative of the compression function.

We wish to take the limit of this summation as the intervals get smaller and smaller. To do so, we separate the square of the interval from the cube term in Eq. (3-12) and rewrite this squared term using the compression-function derivation. The remaining interval multiplier becomes the differential term. Thus,

$$\text{mse} = \frac{1}{12} \sum_i p(s_{qi})(s_{i+1} - s_i)^2(s_{i+1} - s_i) \tag{3-13}$$

$$= \frac{1}{12} \sum_i \frac{p(s_{qi})\,\Delta S^2}{[F'(s_i)]^2}(s_{i+1} - s_i)$$

which, in the limit, becomes

$$\text{mse} = \frac{\Delta S^2}{12} \int_s \frac{p(s)}{[F'(s)]^2}\,ds \tag{3-14}$$

The mean-square error for a uniform quantizer appears explicitly in Eq. (3-14). If the integral in Eq. (3-14) is less than unity, the compander has provided an improvement over the uniform quantizer.

The equations derived above can be used to evaluate a companding scheme, given any particular probability density of the signal samples. You will have ample opportunity to fine-tune your analysis skills in the problems at the back of the chapter. For now, we will present one set of results in a slightly different way.

Suppose we first look again at the uniform quantizer and attempt to evaluate the SNR as a function of the input signal power. If the input signal fills the entire region, the signal-to-noise ratio is

$$\text{SNR} = \frac{12\overline{s^2}}{\Delta S^2}$$

and if we assume that the signal is uniformly distributed between $-s_{\max}$ and $+s_{\max}$, and N bits of quantization, then

$$\overline{s^2} = \frac{s_{\max}^2}{3}$$

and

$$\Delta S = \frac{2s_{\max}}{2^N}$$

The signal-to-quantization-noise ratio is then

$$\text{SNR} = 2^{2N}$$

Thus, for example, for 8 bits of quantization, the signal-to-noise ratio is 2^{16} or 48 dB.

Suppose now that the signal power decreases, but the quantization levels are not changed. As long as the signal fills at least one quantization region ($s_{max}/128$), the quantization noise will vary over its entire range ($-\Delta S/2$ to $+\Delta S/2$), and the noise power can be expected to remain unchanged. Therefore, as the signal power decreases, the SNR would decrease in the same proportion. If we plot SNR as a function of input signal power, we get a linear curve as in Fig. 3.34. As soon as the signal increases beyond the range of the quantization levels, the noise increases rapidly. This is so because larger samples will saturate the system, and the noise is no longer limited in magnitude to $\Delta S/2$. For any specified SNR, the portion of this curve above that level represents the *dynamic range* of the quantizer. Thus, if we desire an SNR of at least 28 dB, the dynamic range is from -20 dB to about $+3$ dB relative to full-load.

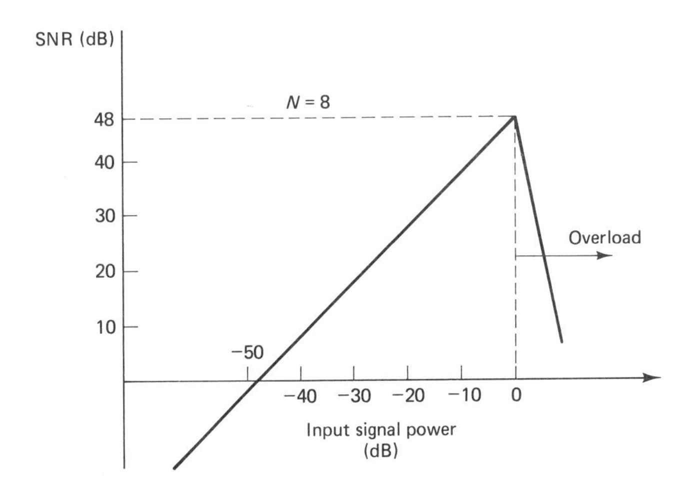

Figure 3.34 Signal-to-quantization-noise ratio as a function of input signal power

The companded system performs better than the uniform quantization system for small signals. This is so because the intervals get smaller as the sample size decreases.

The μ-255 system can be analyzed for various probability density functions of the signal. In particular, we can evaluate this for uniform, Laplace, or Gaussian densities. The Laplace density is an exponential density which closely approximates human speech. It is sketched in Fig. 3.35.

The performance of the companded system does not vary greatly over the various probability density functions of the input. Once we find the performance, we can compare it to that of the uniform quantizer. Figure 3.36 does this for a un-

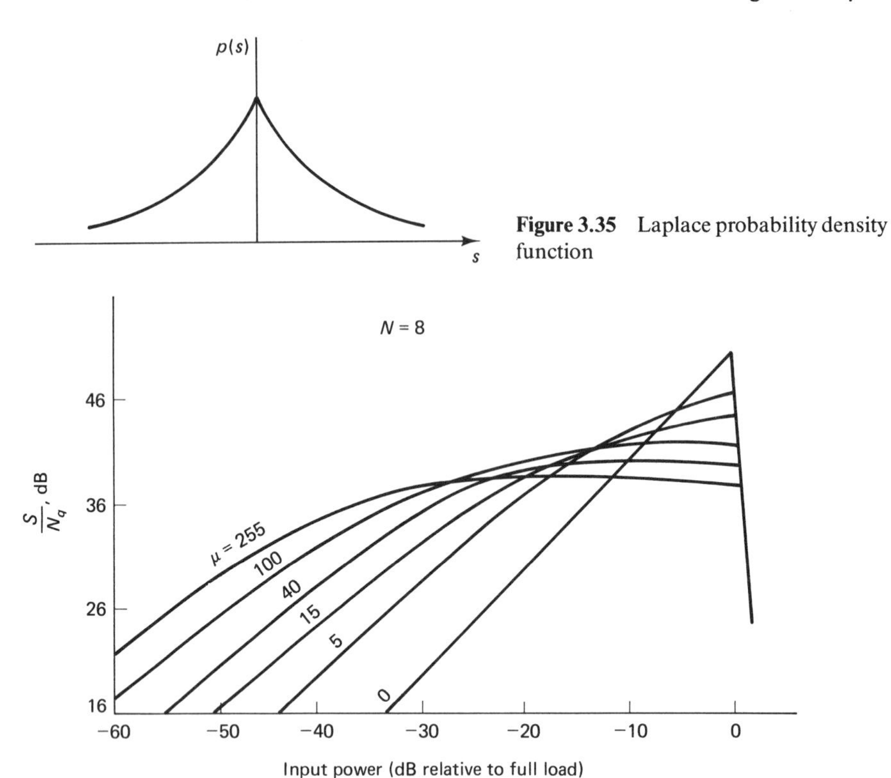

Figure 3.35 Laplace probability density function

Figure 3.36 Performance comparison of uniform quantizer to μ-law compander

iform density and μ-255 companding. The curve of Fig. 3.34 is repeated in this figure for comparison. Note that, as expected, the companded system outperforms the uniform quantizer for low signal levels. As the signal starts to fill the entire range of quantization intervals, the uniform quantizer performs better than the companded system. Thus, the companded system enhances performance in the presence of low-level signals, as expected.

3.2.12 Threshold Effect

The analyses of the previous two sections indicate that the signal-to-quantization-noise ratio can sometimes be increased by using nonuniform quantization. The ratio can *always* be improved by increasing the number of bits of quantization. That is, as the resolution improves, the SNR also improves. This might indicate that in designing a digital communication system, you should select the number of bits of quantization in order to achieve the desired SNR. However, this ignores a second important source of error. As we transmit the 1's and 0's resulting from the quantization process, some of these bits will be received in error. That is, a 1 will occasionally change to a 0, and a 0 to a 1. This is due to additive noise in the transmission path. It does not make much sense to significantly improve the resolution of the quantizer if, indeed, system performance is being limited by the additive noise.

We will see later (Section 7.8) how to tell when the performance is limited by quantization noise and when it is limited by additive noise. This will prove important in indicating where additional effort should be placed to improve performance. When the system is "quantization-noise limited," increasing the resolution (or going to nonuniform quantization) may prove cost-effective to improve performance. However, when the system is "additive-noise limited," effort should be expended to reduce the effects of additive noise. We could try shielding, using a different transmission path or different frequency range, or doing coding to reduce the effect of bit errors.

3.3 DELTA MODULATION

We studied pulse code modulation in the previous section. In that method of source encoding, each sample value is coded into a binary number. The resulting binary number must be capable of providing a measure of the sample value over the entire dynamic range. For example, if we start with a signal that ranges from -5 V to $+5$ V, the digital code must be capable of indicating sample values over a 10-V range. The resulting quantization noise is dependent upon this dynamic range.

If we could somehow reduce the dynamic range of the numbers we are trying to communicate, the noise performance would improve. Various alternate forms of source encoding operate on this principle.

Delta modulation is a simple technique for reducing the dynamic range of the numbers to be coded. Instead of sending each sample value, we send the *difference* between a sample and the previous sample. If sampling is being performed at the Nyquist rate, this difference has a dynamic range equal to that of the original samples. That is, at the Nyquist rate, each sample is independent of the previous sample. Two adjacent samples could be at the minimum and maximum signal amplitudes, respectively. If we sample at a rate higher than the Nyquist, the samples are dependent upon each other (Section 3.4 expands upon this concept). The dynamic range of the difference between two samples taken at a rate higher than the Nyquist rate can be less than that of the samples themselves. If the net result of sampling at a faster rate, but reducing the dynamic range, is that we can send the information using fewer binary digits (for the same quantization noise), we will have realized an improvement.

Delta modulation quantizes this difference using only one bit of quantization. Thus, for example, a "1" is sent if the difference is positive, and a "0" is sent if the difference is negative. Since there is only one bit of quantization, the differences are being coded into only two levels. We shall refer to these two possibilities as

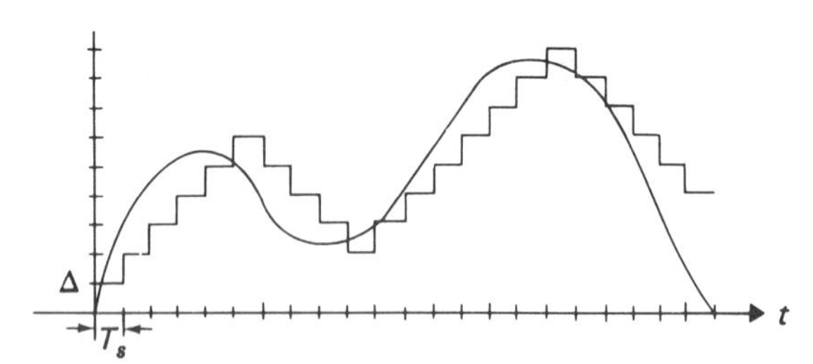

Figure 3.37 Analog waveform and staircase approximation

either $+\Delta$ or $-\Delta$. At every sample point, the quantized waveform can only either increase or decrease by Δ.

Figure 3.37 shows a typical analog waveform. Since the quantized waveform can only either increase or decrease by Δ at each sample point, we shall attempt to fit a staircase approximation to the analog waveform. This approximation is also indicated in the figure. We shall examine the choice of sampling rate and step size later in this section. In fitting the staircase to the function, we need only make a simple decision at each sample point. If the staircase is below the analog sample value, the decision is to increment positively (i.e., an UP step). If the staircase is above, we increment negatively (i.e., a DOWN step). The transmitted bit train for the example shown in Fig. 3.37 would be

1 1 1 1 1 0 0 0 0 1 1 1 1 1 1 1 0 0 0 0 0

The above description leads to a simple implementation of the quantizer, using a comparator and a staircase generator. The resulting A/D converter is shown in Fig. 3.38.

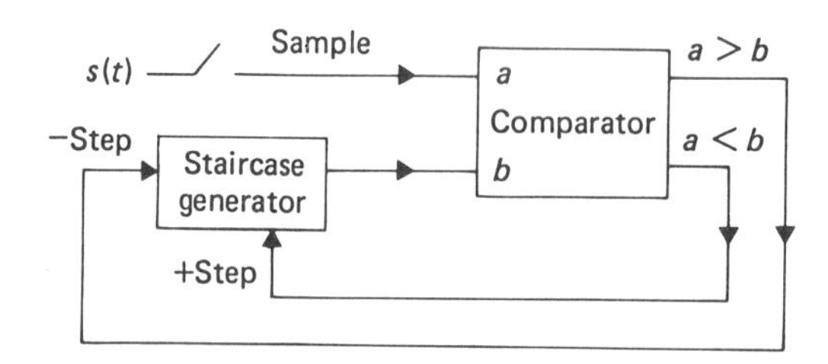

Figure 3.38 Delta A/D converter

The key to effective use of delta modulation is the intelligent choice of the two parameters, *step size* and *sampling rate*. These parameters must be chosen such that the staircase signal is a close approximation to the actual analog waveform. Since the signal has a definable upper frequency, we know the fastest rate at which it can change. However, to account for the fastest possible change in the signal, the sampling frequency and/or the step size must be increased. Increasing the sampling frequency means that the delta-modulated waveform requires a larger bandwidth. Increasing the step size increases the quantization error.

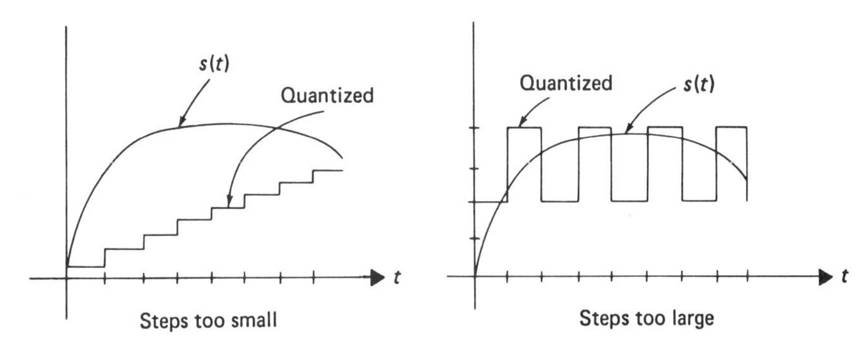

Figure 3.39 Consequences of poor choice of delta step size

Figure 3.39 shows the consequences of incorrect step size. If the steps are too small, we can experience a *slope overload* condition where the staircase cannot track rapid changes in the analog signal. If, on the other hand, the steps are too large, considerable overshoot will occur during periods when the signal is not changing rapidly. In that case, we have significant quantization noise, known as *granular noise.*

Delta modulation uses a source encoding process which has memory in order to reduce dynamic range. Under certain circumstances, it is possible to achieve the same level of performance as PCM with fewer bits transmitted each second. However, because the system has memory, bit transmission errors propagate. In PCM, a single bit error will cause an error in reconstructing the associated sample value. The error will affect *only* that single reconstructed sample. If a bit error occurs in delta modulation, the D/A converter in the receiver will go up instead of down (or vice versa), and all later values will contain an *offset error* of twice the step size. If a subsequent bit error occurs in the opposite direction, the offset error will be canceled. If the offset error poses a significant problem, the system can be periodically restarted from a reference level (usually 0).

3.3.1 Adaptive Delta Modulation

We have shown that the appropriate step size to use in delta modulation depends upon how rapidly the signal is changing from one sample to the next. When the signal is changing rapidly, a larger step size will help avoid overload. When the signal is changing slowly, a smaller step size will reduce the amount of overshoot, and therefore the quantization noise.

Adaptive delta modulation is a scheme which permits adjustment of the step size depending upon the characteristics of the analog signal. It is, of course, critical that the receiver be able to adapt step sizes in exactly the same manner as the transmitter. If this were not the case, the receiver could not uniquely recover the original quantized signal (the staircase function). Since all that is being transmitted is a series of binary digits, the step size must be derived from this bit train.

If a given-length string of bits contains an almost equal number of 1's and 0's, we can assume that the staircase is oscillating about a slowly varying analog signal. In such cases, we should reduce the step size. On the other hand, an excess of either 1's or 0's within a string of bits would indicate that the staircase is trying to catch up with the function. In such cases, we should increase the step size.

In one implementation, the step-size control is performed by a digital integrater. The integrator sums the bits over some fixed period. If the sum deviates from what it would be for an equal number of 1's and 0's, the step size is increased. In practice, the bit sum is translated into a voltage which is then fed into a variable-gain amplifier. The amplification is a minimum when the input voltage corresponds to an equal number of 1's and 0's in the period. The amplifier controls the step size.

Several adaptive delta-modulation algorithms are simpler to implement than that discussed above. Two such rules are the Song algorithm and the Space Shuttle algorithm.

The *Song algorithm* compares the transmitted bit with the previous bit. If the two are the same, the step size is increased by a fixed amount, Δ. If the two bits are different, the step size is reduced by the fixed amount, Δ. Thus, the step size is always changing, and it can get larger and larger, without limit, if necessary. We illustrate this for a step input function in Fig. 3.40. This is an extreme case and would not occur in the real world, since a step function possesses very high frequency components. Note that a damped oscillation occurs following the rapid change in the signal.

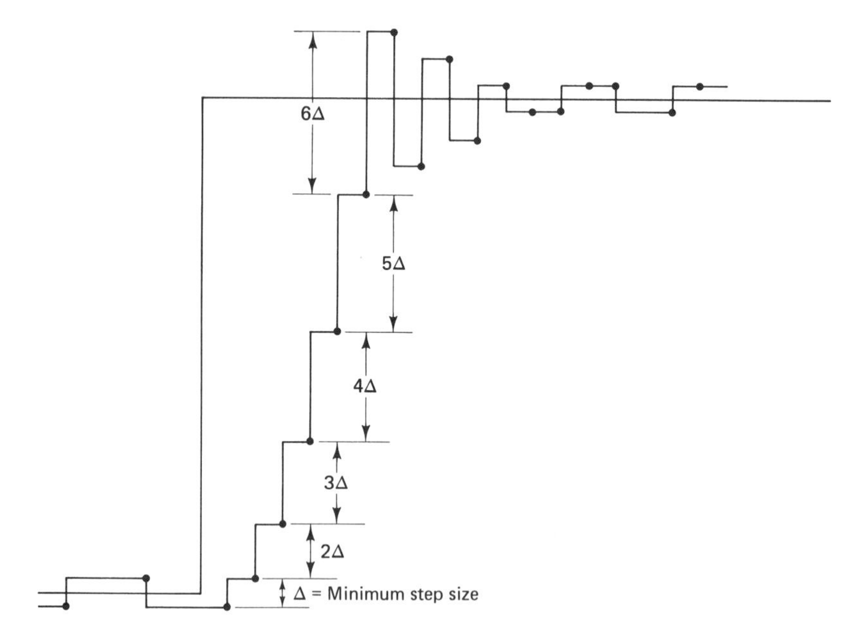

Figure 3.40 Song algorithm

If an analog signal is expected to have many abrupt transitions resembling the step function, the damped oscillations observed after the transition in the Song algorithm might be troublesome. Photographs of distinct and detailed objects could have many such transitions as they are scanned for transmission (i.e., a rapid change from white to black). The *Space Shuttle algorithm* is a modification of the Song algorithm and it eliminates the damped oscillations. When the present bit is the same as the previous bit, the step size increases by a fixed amount, Δ, just as with the Song algorithm. However, when the bits disagree, the step size reverts immediately to its minimum size, Δ. This is in contrast to the Song algorithm where the step size decreases toward zero at a rate of Δ every sampling period. The Space Shuttle algorithm is illustrated in Fig. 3.41 for the same step function input as illustrated in Fig. 3.40.

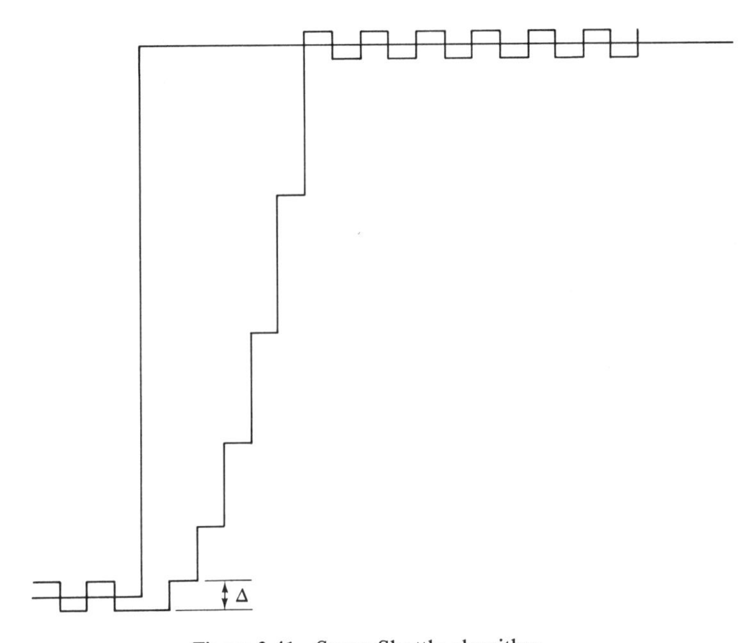

Figure 3.41 Space Shuttle algorithm

3.3.2 Quantization Noise in Delta Modulation

We once again define the quantization error as the difference between the original signal and the quantized (staircase) approximation. Thus,

$$e(t) = s(t) - s_q(t)$$

We will assume that the sampling rate and step size are chosen in order to avoid overload. Under this assumption, the magnitude of the quantization noise

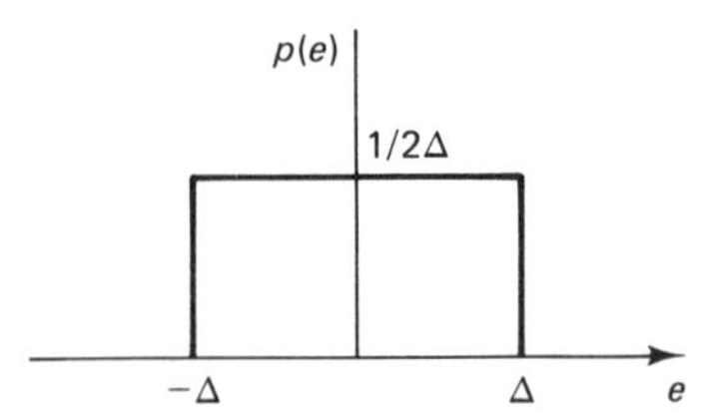

Figure 3.42 Quantization-error probability density function

never exceeds the step size, Δ. If we make the simplifying assumption that all signal amplitudes are equally likely, we conclude that the error is uniformly distributed over the range between $-\Delta$ and $+\Delta$, as shown in Fig. 3.42. The mean-square value of the quantization noise is given by

$$\text{mse} = \int_{-\Delta}^{\Delta} \frac{1}{2\Delta} e^2 \, de$$

$$= \frac{\Delta^2}{3} \tag{3-15}$$

In designing digital communication systems, a logical question is whether to use PCM or DM as the source encoding technique. As a designer, you would be concerned with many factors. Among these are complexity (i.e., cost), transmission bit rate (i.e., required system bandwidth), reliability, quantization noise, and the effects of transmission errors. It would be very nice if we could derive a simple formula that related SNR for PCM to that of DM. Unfortunately, many specific assumptions must be made to arrive at such a result. We shall present one particular approach to the comparison. The reader is referred to the references for other approaches. The bottom line is that under certain circumstances DM will provide the same SNR as PCM with a lower transmission bit rate. Under other circumstances the reverse is true. Adaptive delta modulation adds yet another parameter to complicate the analysis.

We start by trying to reduce the mean-square quantization error to a noise power at the output of the D/A converter. The demodulator normally includes a low-pass filter to smooth the staircase into a continuous curve. We must therefore somehow translate the quantization noise statistics into the power spectral density. This is not a simple analytical task, and it requires that a specific form be assumed for $s(t)$. We shall make several approximations in the following analysis.

Let us assume that the original $s(t)$ is a sawtooth type waveform. This is the simplest example of a waveform with amplitude that is uniformly distributed. The waveform, its quantized version, and the resulting quantization noise are shown in Fig. 3.43. The noise function is almost periodic with period T_s, the sampling period. It would be exactly periodic with a period equal to that of the sawtooth waveform if that period were an integral multiple of T_s. We are assuming that the step size and sampling period are chosen to avoid overload and, in this case, to give perfect symmetry. The power spectrum of $s_q(t)$ can be found precisely. It will be of the form $\sin^4 f/f^4$. The power spectrum of a triangle reaches its first zero at $f = 1/T_s$. The lobes

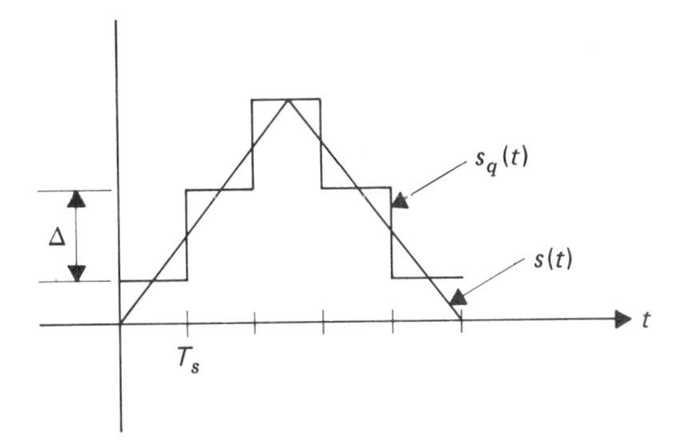

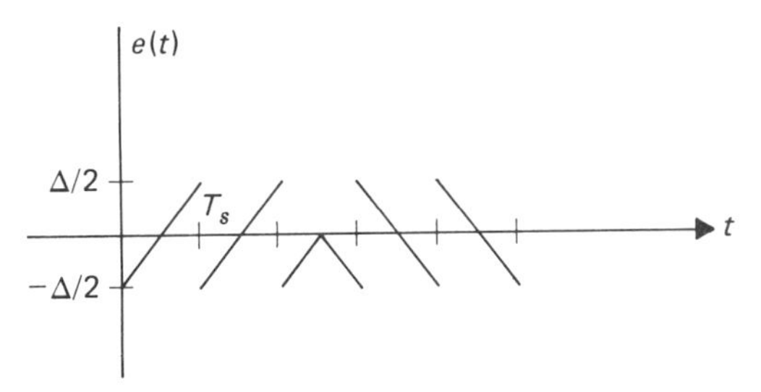

Figure 3.43 Analysis of quantization noise in delta modulation

beyond this point are attenuated as the fourth power of $1/f$. Therefore, there is little power outside of the main lobe, so we assume that all of the power is concentrated in a low-frequency band stretching to $f = 1/T_s$. Now, since we are assuming that sampling is taking place at a rate well above Nyquist, the cutoff frequency of $1/T_s$ can be assumed to be much larger than f_m. In fact the frequencies of interest are so far below the first zero of the power spectrum that we will assume the spectrum to be essentially flat over the range between 0 and f_m. The total power of the noise is the mean-square error we found earlier, $\Delta^2/3$. Therefore, if we assume a flat power spectral density, the power at the output of a low-pass filter with f_m cutoff is given by

$$N_q = \frac{\Delta^2}{3}\frac{f_m}{1/T_s} = \frac{\Delta^2}{3}\frac{f_m}{f_s}$$

where f_s is the number of samples per second.

Example 3-8

Given an audio waveform which is the sum of three sinusoids,

$$s(t) = 3\cos 500t + 4\cos 1000t + 4\cos 1500t$$

find the signal-to-quantization-noise ratio if this is coded using delta modulation.

Solution. We must first choose a step size and a sampling frequency for this waveform. The Nyquist rate is $T_s = \pi/1500$. Suppose for purposes of illustration we choose 32 times this rate—that is, $T_s = \pi/48{,}000$. At the fastest frequency, the function can change 4 volts in $\pi/3000$ second. In this amount of time, we have 16 sampling periods, so if a step size of $\frac{1}{4}$ volt is chosen, the staircase will overload only in extreme circumstances. The quantization noise power is given by

$$N_q = \frac{\Delta^2}{3} \frac{f_m}{f_s} = \frac{1}{48} \cdot \frac{1}{63} \approx 0.3 \text{ mW}$$

The signal power is $(3^2 + 4^2 + 4^2)/2$ or 20.5 watts. Finally, the signal-to-noise ratio is given by

$$\text{SNR} = \frac{20.5}{0.3 \times 10^{-3}} = 6.29 \times 10^4 \text{ or } 48 \text{ dB}$$

We now wish to compare the special-case result of Example 3-8 with that obtained using PCM encoding.

The signal-to-quantization-noise ratio for PCM with uniform quantization is given by Eq. (3-16):

$$\text{SNR} = \frac{12P_s}{\Delta S^2} \tag{3-16}$$

In this equation, P_s is the average signal power. If sampling is done at the Nyquist rate, the number of bits per second is

$$\text{BPS} = 2^{N+1}f_m$$

where N is the number of bits of quantization. The step size can then be written as

$$\Delta S = \frac{2V_{\max}}{2^N}$$

$$= \frac{4V_{\max}f_m}{\text{BPS}}$$

where $V_{\max}$ is the maximum amplitude of the signal. Combining these expressions yields a signal-to-noise ratio of

$$\text{SNR} = \frac{P_s \times 3 \text{ BPS}^2}{V_{\max}^2 \times 4f_m^2} \tag{3-17}$$

The equivalent result for delta modulation is found by starting with the expression for quantization noise power,

$$P_{N_q} = \text{VAR} \times \frac{f_m}{f_s}$$

where VAR is the variance of individual sample errors, f_s is the sampling frequency, and f_m is the maximum frequency of the signal. We shall assume that the slope of the staircase is greater than or equal to the maximum slope of $s(t)$ in order to avoid overloading. Thus, the step size and sampling rate are related by

$$\frac{\Delta}{T_s} > f'_{\max}$$

The maximum slope can be approximated by

$$f'_{\max} = V_{\max}\, 2\pi f_m$$

so

$$f_s > \frac{2\pi V_{\max} f_m}{\Delta}$$

and the noise power is bounded by

$$P_{N_q} < \frac{\text{VAR} \times \Delta}{2\pi V_{\max}}$$

Assuming a uniform error probability distribution, the variance is given by

$$\text{VAR} = \frac{\Delta^2}{3}$$

Finally,

$$P_{N_q} < \frac{V^2_{\max}\, 8\pi^2 f^3_m}{6\text{BPS}^3}$$

and the signal-to-noise ratio is bounded by

$$\text{SNR} > \frac{6P_s\text{BPS}^3}{8\pi^2 V^2_{\max} f^3_m}$$

Comparing this result to the result found for PCM in Eq. (3-17) yields

$$\frac{\text{SNR}_{\text{DM}}}{\text{SNR}_{\text{PCM}}} > \frac{\text{BPS}}{\pi^2 f_m}$$

Finally, with the bit rate given by

$$\text{BPS} = 2^{N+1} f_m$$

this becomes

$$\frac{\text{SNR}_{\text{DM}}}{\text{SNR}_{\text{PCM}}} > \frac{2^N}{\pi^2}$$

Thus, under the assumptions made, as the number of bits increases, the performance of delta modulation is superior to that of PCM for the same bit transmission rate.

It should be noted that the measure of performance, SNR, considers only *quantization noise.* In the next chapter, we will add the dimension of *transmission errors.* Therefore, one must exercise caution and not get too optimistic about the virtues of delta modulation versus PCM.

3.4 ALTERNATIVE FORMS OF MODULATION

In addition to PCM, DM, and ADM, there are numerous other methods for coding analog information into a digital format. The goal of each system is to send the information with maximum reliability and minimum bandwidth. We shall briefly discuss three of these methods in this section: *delta pulse code modulation* (DPCM), *differential PCM,* and *adaptive differential pulse code modulation* (ADPCM).

Delta modulation can be viewed in a way different from our previous approach. This new approach will point the way to variations. In delta modulation we approximate a continuous waveform with a staircase wave. At each sampling point, we develop an error term that is the difference between the signal and the staircase function. We quantize this error to develop a correction term, which is then added to the staircase function. In the case of basic DM, the quantization is done in units of 1 bit. In the case of delta PCM, we code the error into more than 1 bit of quantization and add this term to the previous staircase value, as shown in Fig. 3.44.

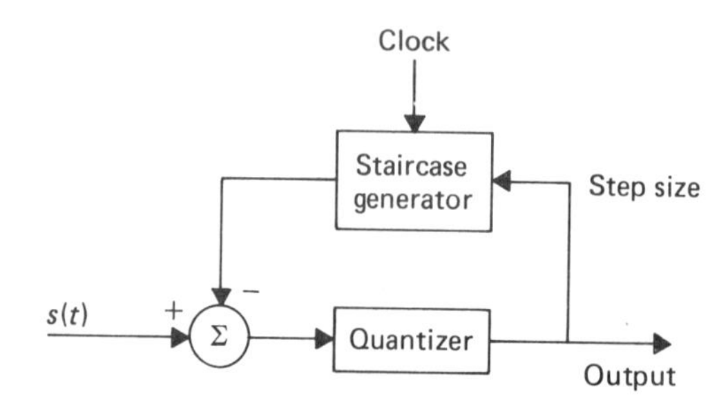

Figure 3.44 Delta PCM

Instead of the stair steps being of only one possible magnitude, they can now be one of two, four, eight, or any power of 2 sizes. At each sampling point, we must now send more than 1 bit of information, the various bits representing the PCM code for the error term. The advantage of DPCM over ordinary PCM is that, with proper choice of sampling interval, the error being quantized has a much smaller dynamic range than that of the original signal. For the same number of bits of quantization, we can therefore get much better resolution. The price we pay is complexity of the modulator. If the signal were always at its maximum frequency (the one that determines the sampling rate in PCM), DPCM would be essentially the same as PCM. However, since the signal frequencies are usually distributed over a range, adjacent samples are often correlated, and it is possible to get better perform-

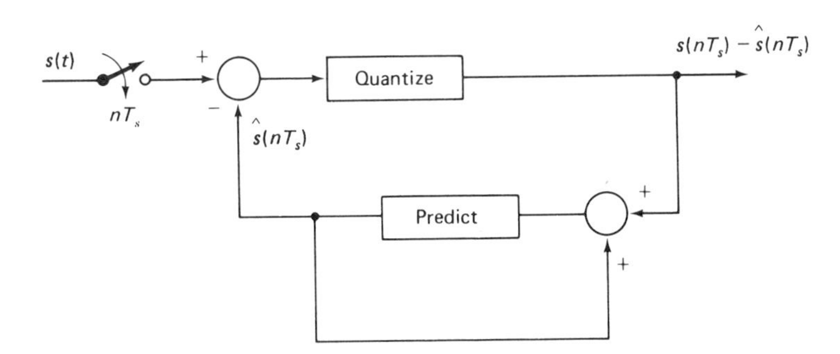

Figure 3.45 Differential PCM

ance from this system than from a PCM system with the same rate of bit transmission.

Differential PCM is another technique for sending information about changes in the samples rather than about the sample values themselves. The differential approach includes an additional step which is not part of delta PCM. The modulator does not send the difference between adjacent samples, but instead uses the difference between a sample and its *predicted* value. The prediction is made on the basis of previous samples. This is illustrated in Fig. 3.45. The symbol $\hat{s}(nT_s)$ is used to denote the predicted value of $s(nT_s)$.

The simplest form of prediction is linear prediction, where the estimate is a linear function of the previous sample values. Thus, using only one sample, we have

$$\hat{s}(nT_s) = As[(n-1)T_s] \tag{3-18}$$

where A is a constant. The predict block in Fig. 3.45 is therefore a multiplier of value A.

The problem is to choose A to make this prediction as *good* as possible. We get a measure of *goodness* by defining a prediction error as the difference between the sample and its estimate. Thus,

$$\begin{aligned} e(nT_s) &= s(nT_s) - s(nT_s) \\ &= s(nT_s) - As(n-1)T_s \end{aligned}$$

The mean-square value of this error is

$$\begin{aligned} \text{mse} &= E\{e^2(nT_s)\} \\ &= E\{s^2(nT_s) + A^2Es^2[(n-1)T_s] - 2AEs(nT_s)s[(n-1)T_s]\} \\ &= R(0)[(1+A^2) - 2AR(T_s)] \end{aligned}$$

where $R(t)$ is the autocorrelation of $f(t)$. This error can be minimized with respect to A by setting the derivative equal to zero:

$$\frac{d(\text{mse})}{dA} = 2AR(0) - 2R(T_s) = 0$$

or

$$E\{s[(n-1)T_s](s(nT_s) - As[(n-1)T_s])\} = 0 \tag{3-19}$$

thus yielding

$$A = \frac{R(T_s)}{R(0)}$$

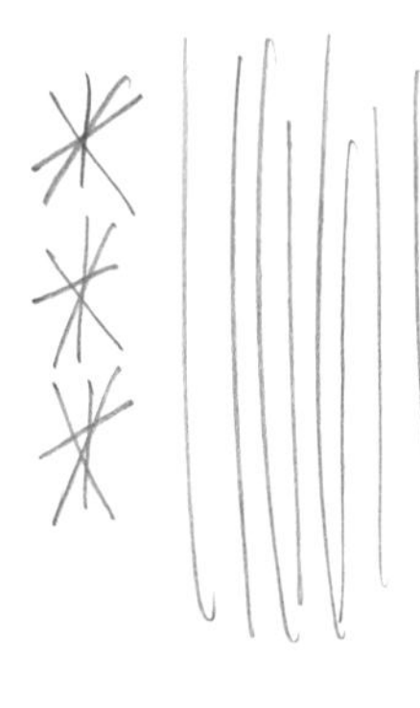

Equation (3-19) has an interesting intuitive interpretation. It states that the expected value of the product of the error with the measured sample is zero. That is, the error has no component in the direction of the observation. The two quantities are *orthogonal*. This makes sense, since, if the error *did* have a component in the direction of the observation, we could reduce that component to zero by readjusting the constant, A.

The predictor of Fig. 3.45 would take the most recent sample value (which it forms by adding the prediction to the difference term) and weight it by $R(T_s)/R(0)$. We assume that the input process has been observed sufficiently long to enable us to estimate its autocorrelation.

Example 3-9

Find the weights associated with a predictor operating on the two most recent samples, and evaluate its performance.

Solution. The prediction is given by

$$\hat{s}(nT_s) = As[(n-1)T_s] + Bs[(n-2)T_s]$$

where the object is to make the best choice of A and B. For this best choice, the error is orthogonal to the measured quantities. Thus,

$$E\{(s(nT_s) - As[(n-1)T_s] - Bs[(n-2)T_s])(s[(n-1)T_s])\} = 0$$

$$E\{(s(nT_s) - As[(n-1)T_s] - Bs[(n-2)T_s])(s[(n-2)T_s])\} = 0$$

Expanding these, we find that

$$R_s(T_s) - AR_s(0) - BR_s(T_s) = 0$$

$$R_s(2T_s) - AR_s(T_s) - BR_s(0) = 0$$

yielding for A and B:

$$A = \frac{R_s(T_s)[R_s(0) - R_s(2T_s)]}{R_s^2(0) - R_s^2(T_s)}$$

$$B = \frac{R_s(0)R_s(2T_s) - R_s^2(T_s)}{R_s^2(0) - R_s^2(T_s)}$$

The mean-square error using these values is

$$\begin{aligned}\text{mse} &= E\{[(s(nT_s) - \hat{s}(nT_s)]^2\} \\ &= E\{s^2(nT_s)\} - E\{\hat{s}(nT_s)s(nT_s)\} \\ &= R_s(0) - \frac{R_s(0)[R_s^2(T_s) + R_s^2(2T_s)] - 2R_s(2T_s)R_s^2(T_s)}{R_s^2(0) - R_s^2(T_s)}\end{aligned} \tag{3-20}$$

As a specific example, suppose that the autocorrelation function of $s(t)$ is as shown in Fig. 3.46, and that the sampling period, $T_s = 1$ s. Plugging these values into Eq. (3-20) yields

$$\text{mse} = 1.895$$

For comparison, if we had forgotten to measure $s[(n-1)T_s]$ and $s[(n-2)T_s]$, and simply guessed at 0 for the value, the mean-square error would have been 10.

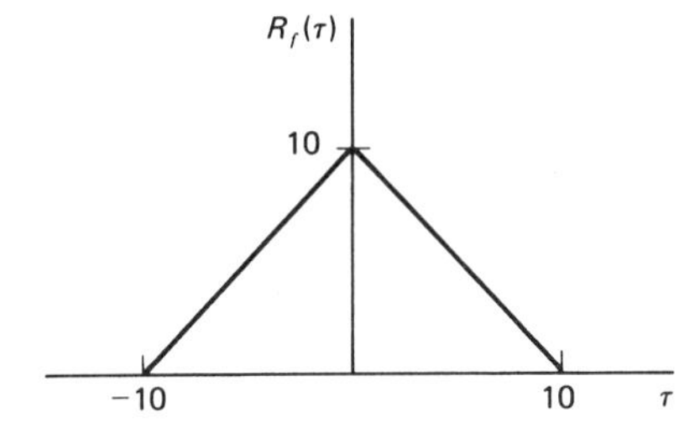

Figure 3.46 Autocorrelation function for Example 3-9

It can be further shown that since in Example 3-9,

$$\frac{R_s(2T_s)}{R_s(T)} \simeq \frac{R_s(T_s)}{R_s(0)}$$

the performance of the *single*-sample predictor would have almost equaled the performance of this *double*-sample predictor.

For a speech signal, it has been shown experimentally that a differential PCM system using the most recent sample in the predictor can save 1 bit per sample over PCM. That is, the differential PCM system can achieve the same error performance as the PCM system with 1 less bit. This frees the channel for other uses during the time that bit would be sent. The extra effort of differential PCM may therefore prove cost-effective.

In *adaptive DPCM,* the predictor coefficients do not remain constant for the entire transmission. For each group of a given length of samples, say n, we compute a covariance matrix, $[R_{ij}]$. We then use this matrix to solve for the predictor coefficients. The predictor coefficients are then changed for the next n samples. Since the predictor coefficients are no longer constant, there must be some way to assure that the receiver is using the same coefficients. The most common method for accomplishing this is to send the updated coefficients as *overhead,* usually multiplexed with the sample information.

We have presented the three most common alternate techniques for encoding analog information into a digital format. There are other possibilities, many of which represent perturbations of the basic techniques. For example, the three techniques described in this section can be coupled with companding to improve performance. In each case, we must weigh the performance improvement (i.e., decreased quantization noise) against the increasing complexity of the hardware.

3.5 VOICE TECHNIQUES

The various waveform coding techniques we have been discussing in this chapter can be applied to any bandlimited analog waveform. Why then do we include a separate section on voice techniques? The justification is that voice digitization is the most common form of digital communication, and in many ways it is the most highly developed. Many of the techniques that have been used for voice transmission are now being adapted to other forms of transmission.

We shall first discuss the typical voice waveform. We will then explore some of the various coding techniques, including linear predictive coding, adaptive predictive coding, time-assignment speech interpolation (TASI), and vocoders. We will briefly discuss speech synthesis and recognition, where developments are occurring at an accelerated pace; you are therefore referred to the technical literature for up-to-date information.

Typical speech waveforms possess frequencies from several tens of hertz to a high of the order of 10 kHz. In fact, high-quality sound reproduction systems often cover the band from 20 Hz to about 40 kHz. Although the ear does not respond to frequencies above perhaps 15 kHz (the upper limit decreases as one ages), the higher frequencies do contribute to an overall sensory appreciation of sound. FM broadcast radio uses an upper cutoff of 15 kHz, while AM radio uses only 5 kHz. The standard long-distance toll line cuts off at approximately 3 kHz, as do various forms of citizen radio. A typical average energy-vs.-frequency curve for certain human speakers would resemble that shown in Fig. 3.47. Note that the energy peaks at several hundred hertz, and that about 98% of the energy can be expected to lie below 3 kHz. The actual curve varies with age, sex, and county (i.e., the language being spoken), so this curve should be interpreted as an average.

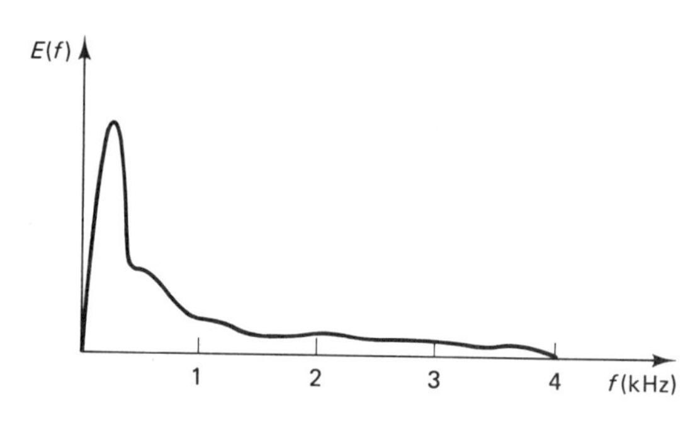

Figure 3.47 Energy spectral density for human speech

In evaluating digitization techniques for speech, we need a measure of success. The signal-to-quantization-noise ratio, a measure that we discussed earlier, is related to the mean-square quantization error. However, in speech transmission, subjective measures often take precedence over these mathematical measures. For example, word or sentence intelligibility and subjective qualities of the speech become important. Some people still cannot get used to computer-synthesized voice signals.

Voice digitization techniques fall into two distinct classes. The first treats the voice as an analog signal and attempts to code this to reproduce the time signal at the receiver. This is called *waveform digitization.* The second treats parameters of the speech signal, and the goal is to preserve the actual word information. Analysis of such systems involves breaking speech up into its basic components and waveform segments.

Waveform digitization is the subject of the earlier sections of this chapter. As an example, PCM can be used for speech waveforms, and the standard is the T1 system. This uses 8000 samples per second with 7-bit quantization of each sample value. Other common systems for speech waveform encoding are DPCM, ADPCM, DM, and ADM. Their advantages and disadvantages are discussed elsewhere in this text. The criteria for choosing a system include the signal-to-quantization-noise ratio and the complexity of the system. ADPCM has been shown to yield the highest SNR for a given bit rate. The other major candidate systems are ADM and companded PCM. At low bit rates, ADM has a higher SNR than PCM, but the situation reverses at higher bit rates. The trade-offs of system design are discussed in greater detail in Chapter 11.

Linear predictive coding is one technique for sending essential speech information using fewer bits per second. A typical system might only require 2.4 kbps to send a voice waveform, as compared to 56 kbps using PCM. The system is illustrated in Fig. 3.48. The analyzer is used to reduce the speech waveform to a minimum set of data. The waveform is first digitized using an A/D. The statistics of the waveform are calculated using autocorrelation analysis, and the essential frequency content is extracted using a pitch extractor. The pitch information, together with the parameters of a linear prediction filter (see Section 3.4, where we discuss linear prediction), are sent to the receiver, or synthesizer.

The synthesizer separates the pitch information and the filter parameters and simulates the proper filter to control the pitch generator. White noise forms the basis of reconstructing the intelligible speech signal.

Improvement in speech intelligibility is possible by using more sophisticated encoding schemes. The *adaptive predictive coding* (*APC*) scheme is a differential scheme which transmits the difference between the predicted and actual voice waveform samples, thus decreasing dynamic range and consequently decreasing quantization noise. The system also transmits amplitude, pitch, and filter coefficient information.

Speech information can be transmitted using bit rates below 1 kHz by employing *vocoders.* These are complex and expensive systems, and the synthesized

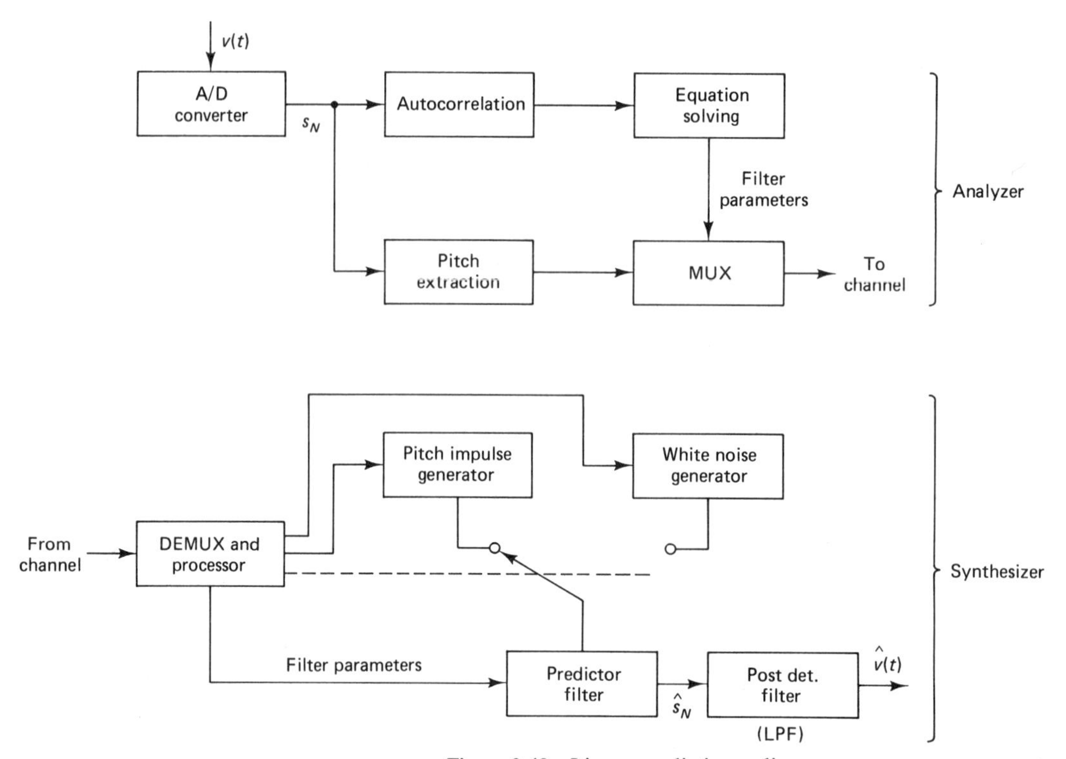

Figure 3.48 Linear predictive coding

speech has a computerlike synthetic quality. However, the speech is recognizable and intelligible. Vocoders extract essential voice information from the speech signal. They may calculate autocorrelations, *cepstrums* (spectrum of spectrum), or extract *formant* information. In one relatively simple implementation, the vocoder splits the speech into frequency bands and multiplexes this information. For example, the speech band can be divided into 300-Hz segments, with the information being extracted by the use of band-pass filters.

So far, we have been discussing encoding of a single voice signal. In telephone communications, we are interested in multiplexing a number of channels on a single line. We discuss time-division multiplexing elsewhere in this text. However, speech has some favorable properties that permit other forms of multiplexing. The average speaker transmits information less than 40% of the time. This is due to pauses between words and between sentences and pauses to listen to the other party who is speaking in a two-way conversation. *Time-assignment speech interpolation* (TASI) is a system which detects the presence of a speech waveform and shares the available bit rate among two or more users.

We can summarize the various techniques by comparing bit rate to send a single channel of voice information. This is shown pictorially in Fig. 3.49(a). Quality is difficult to measure, and depends upon SNR and various subjective

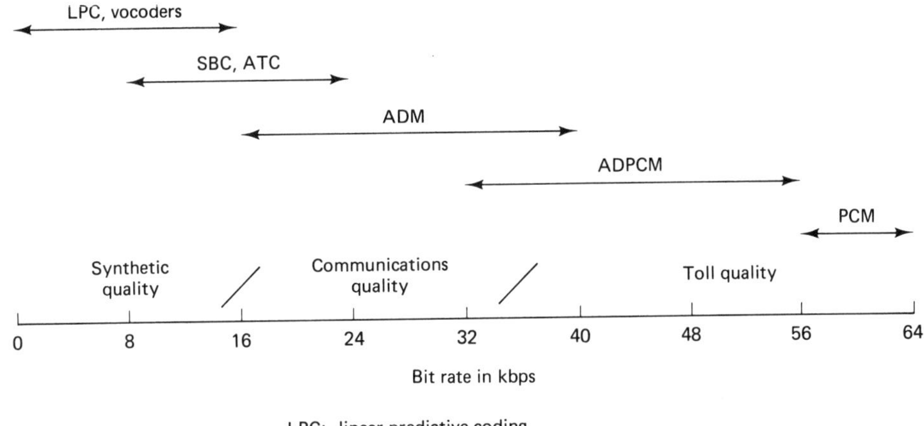

(a)

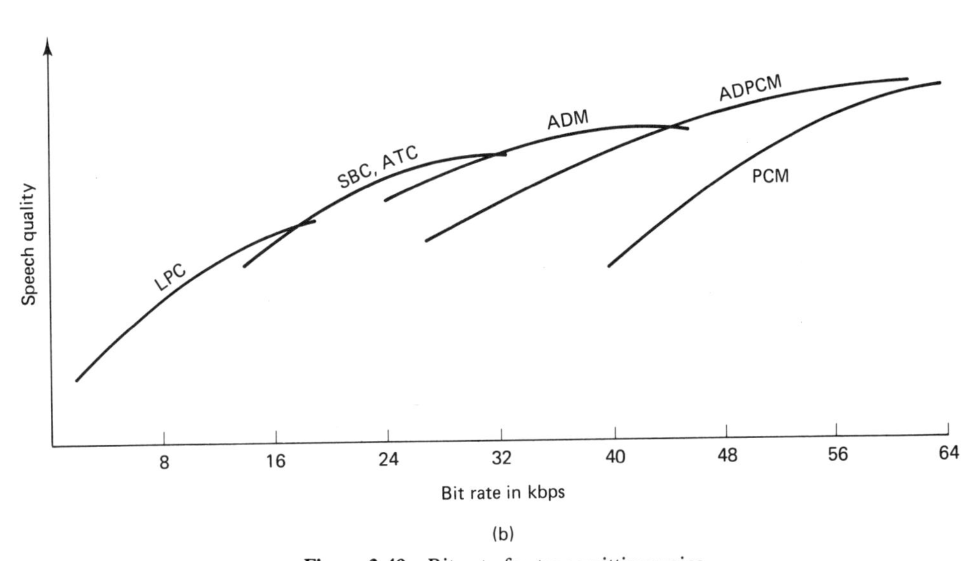

(b)

Figure 3.49 Bit rate for transmitting voice

criteria. We have combined various results that have appeared in the literature to present Fig. 3.49(b), which shows the speech quality as a function of bit rate. A third measure of comparison and design criterion is the cost of the systems. Vocoders are the most expensive, ADM systems the least expensive. However, as hardware becomes easier and easier to implement on VLSI, these differences can be expected to become less significant.

3.6 DIGITAL TELEVISION

Television is traditionally an analog medium. However, as the cost of digital electronics comes down, the advantages and tremendous versatility of digital circuitry are causing digital television to replace analog television. Uniform picture quality is possible over a wide area, with little degradation as distance from the transmitter increases. Sophisticated digital processing, including echo (ghost image) cancellation, can be employed to control errors. Encryption is possible to protect communication privacy (i.e., pay TV), and extra features (freeze frame, picture within a picture) become simple extensions of the digital processor.

The typical United States monochrome television signal is developed by scanning horizontally across a line and vertically across an entire frame. The resulting analog waveform has a maximum frequency of about 4 MHz. The picture contains approximately 211,000 picture elements (pixels), and the brightness (gray-scale level) must be transmitted 30 times each second (frame rate). If PCM is used, and each gray-scale level is quantized using 8 bits, we need to transmit over 50 Mbps. This is too high to be considered practical, so compression techniques are required. We discuss these in Section 4.4.1.2, so in the current section we will simply give an overview.

Interframe coding can be used to decrease the bit rate. Here, the difference signal between two successive frames is encoded and transmitted. Since the luminance of a particular pixel usually does not change much from frame to frame (i.e., over $\frac{1}{30}$ sec), the dynamic range of the differential signal is significantly decreased, and a lower bit rate is possible.

Additional bit-rate conservation is possible using data compression. This takes advantage of the fact that relatively large numbers of adjacent pixels often have the same level of luminance. Therefore, rather than sending each pixel independently, we can send areas of constant luminance by using parameters which define the area. This is discussed in the next chapter.

3.7 SUMMARY

This chapter deals with the important issue of source encoding—the process of changing an analog waveform into a digital signal.

We begin with a thorough examination of the operation which changes the analog continuous waveform into a list of numbers. We study several generic forms of sampler and examine the sources of error.

Section 3.2 presents pulse code modulation, which is a technique for changing a list of analog numbers into a sequence of 1's and 0's. We discuss modulators and demodulators and include some practical implementations. We then analyze quantization noise in both uniform and companded systems.

Delta modulation is presented in Section 3.3, where we include a discussion of adaptive systems such as the Song and Space Shuttle algorithms.

Section 3.4 presents alternative techniques for changing the analog samples into a digital sequence, including delta PCM, differential PCM, and adaptive differential PCM.

We conclude the chapter with discussions of voice and television applications. Within these two sections, we present the concepts of linear predictive coding, adaptive predictive coding, and time-assignment speech interpolation (TASI).

Now that we are adept at reducing analog continuous waveforms to digital sequences, the remainder of this text concentrates upon techniques for accurately transmitting these digital sequences.

PROBLEMS

3.1. A square wave, $s(t)$, with period $T = 1$ multiplies a time function $g(t)$, with $G(f)$ as shown below.

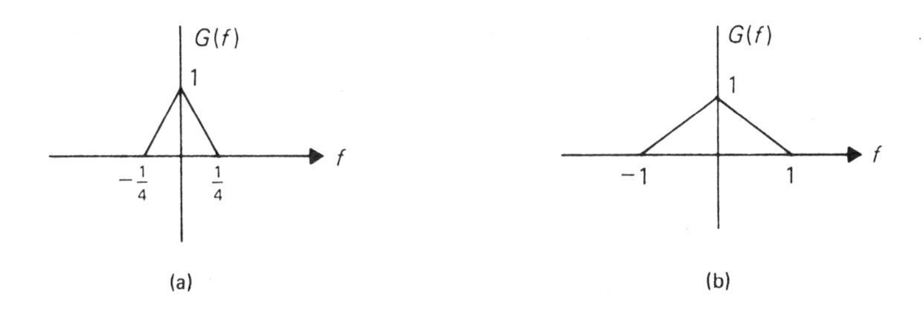

(a) Sketch the Fourier transform of $g_s(t) = g(t)s(t)$.
(b) Can $g(t)$ be recovered from $g_s(t)$?
(c) If $G(f)$ is as shown in sketch (b), can $g(t)$ be recovered from $g_s(t)$? Explain your answer.

3.2. You are given a function of time,

$$s(t) = \frac{\sin t}{t}$$

This function is sampled by being multiplied by $s_\delta(t)$ as shown below.

$$s_s(t) = s(t)s_\delta(t)$$

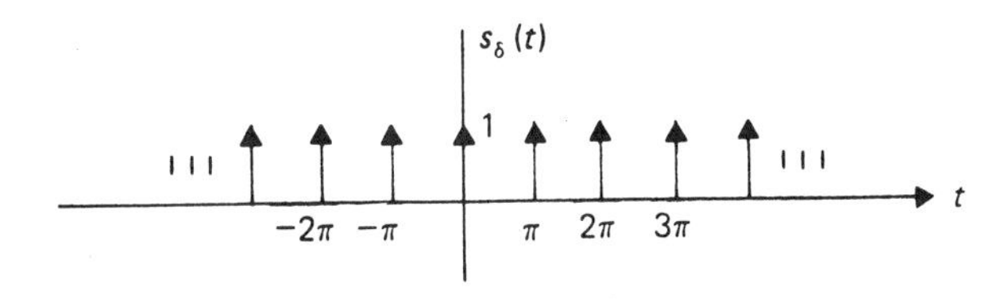

(a) What is the Fourier transform of the sampled function, $S_s(f)$?
(b) From your answer to part (a), what is the actual form of $s_s(t)$?
(c) Should your answer to part (b) have been a train of impulses? Did it turn out that way? Explain any apparent discrepancies.
(d) Describe a method of recovering $s(t)$ from $s_s(t)$ and show that it works in this specific case.

3.3. A signal, $s(t)$, with $S(f)$ as shown, is sampled by two different sampling functions, $s_{\delta 1}(t)$ and $s_{\delta 2}(t)$, where

$$s_{\delta 2}(t) = s_{\delta 1}\left(t - \frac{T}{2}\right) \text{ and } T = \frac{1}{2f_m}$$

Find the Fourier transform of the sampled waveforms,

$$s_{s1}(t) = s_{\delta 1}(t)s(t)$$
$$s_{s2}(t) = s_{\delta 2}(t)s(t)$$

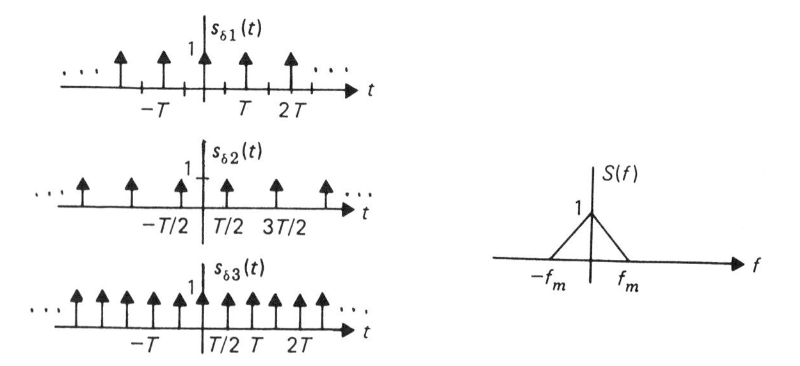

Now consider $s(t)$ sampled by $s_{\delta 3}(t)$, a train of impulses spaced $T/2$ apart. Note that this new sampling function is equal to $s_{\delta 1}(t) + s_{\delta 2}(t)$. Show that the transform of $s_{s3}(t)$ is equal to the sum of the transforms of $s_{s2}(t)$ and $s_{s1}(t)$. [*Hint:* Don't forget the phase of $s_{\delta 2}(t)$. You should find that when $S_{\delta 1}(f)$ is added to $S_{\delta 2}(f)$, every second impulse is canceled out.]

3.4. The function, $s(t) = \cos 2\pi t$, is sampled every $\frac{3}{4}$ second. Evaluate the aliasing error.

3.5. You are given a time function, $s(t)$:

$$s(t) = 5 \cos 2\pi \times 30t + 2 \cos 2\pi \times 55t$$

This is sampled at 100 samples per second. Evaluate the aliasing error.

3.6. Derive a block diagram of a 4-bit serial quantizer that would handle a voice signal with maximum frequency of 3 kHz and amplitude swing of ± 5 V. Illustrate the operations of the quantizer for sample values of -3 and $+2.6$ V.

3.7. Design a counting quantizer for 6-bit PCM and a voice signal as input. Assume the voice signal has a maximum frequency of 5 kHz. Choose all parameter values and justify your choices.

3.8. Find the output of the 3-bit ADC of Fig. 3.12 for the following voltage inputs:
(a) 0.327 V
(b) 0.631 V
(c) 0.751 V
(d) 1.000 V

3.9. Design a 5-bit serial quantizer which would operate upon a voice signal. The voice signal has maximum frequency of 5 kHz and an amplitude swing of ± 10 volts. Illustrate the operation of the quantizer for sample values of -2 and $+6.7$ V.

3.10. You are given a time signal $s(t) = \sin \pi t$. This signal is transmitted via 4-bit PCM. Find the PCM wave for the first 5 sec, and sketch the result.

3.11. A PCM wave is shown below, where voltages of $+1$ and -1 are used to send a 1 and a 0, respectively. Two-bit quantization has been used. Sketch a possible analog information signal, $s(t)$, that could have resulted in this PCM wave. Now assume that a pulse of $+1$ V represents a binary 0 and -1 V represents a 1. Sketch a possible $s(t)$.

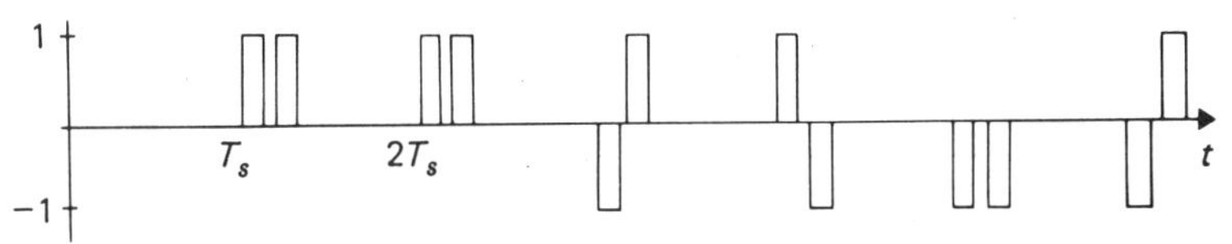

3.12. Two voice signals (3-kHz upper frequency) are transmitted using PCM with eight levels of quantization. How many pulses per second are required to be transmitted, and what is the minimum bandwidth of the channel?

3.13. The input to a delta modulator is $s(t) = 5t + 1$. The sampler operates at 10 samples/sec, and the step size is 1 V. Sketch the output of the delta modulator.

3.14. Delta modulation is used to transmit a voice signal with maximum frequency of 3 kHz. The sampling rate is set at 20 kHz. The maximum amplitude of the analog voice signal can be assumed to be 1 V. Discuss the choice of appropriate step size.

3.15. An adaptive delta modulator is to be used to transmit $s(t) = 5t + 1.1$. The sampler operates at 10 samples/sec. The step size depends upon the number of 1's in the four most recent samples. The step size is either 0.5 V, 1 V, or 1.5 V depending upon the number of 1's. That is, if the number of 1's is 0 or 4, the largest step size is used. If it is 1 or 3, the intermediate size is used. An equal number of 1's and 0's results in the smallest step size.

Sketch the staircase approximation to $s(t)$ for the first 3 sec and also sketch the transmitted digital waveform. Assume that all binary transmissions prior to $t = 0$ were 0.

3.16. Repeat Problem 3.15 using the Song algorithm.

3.17. Repeat Problem 3.15 using the Space Shuttle algorithm.

3.18. Find the mean-square quantization error if 3-bit PCM is used and the input signal is Gaussian distributed with quantization steps ranging over $\pm 3\sigma$, where σ^2 is the variance of the input.

3.19. White noise of power spectral density, $N_0/2$, forms the input to a predictor. The predictor is set to predict T seconds into the future based upon a single sample. Thus,

$$\hat{n}(t + T) = An(t)$$

(a) Find the optimum value for A.

(b) Now assume the noise is low-pass filtered prior to prediction. Find A and the mean-square error as a function of the filter cutoff, f_m.

3.20. A signal is triangularly distributed as shown below. Show that Eq. (3.4) will yield a uniform density for the quantization error.

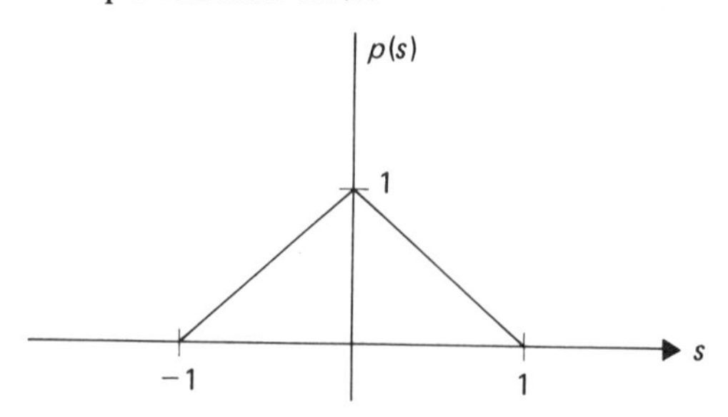

3.21. Design a PCM system that could be used to transmit the video portion of a black-and-white television signal. The video signal contains frequencies up to 4 MHz. A study of viewers has indicated that at least 70 gray-scale levels are required for a pleasing picture.

3.22. Find the mean-square quantization noise for the PCM modulation described in Problem 3.10.

3.23. A signal is Gaussian distributed with zero mean and variance σ^2. The signal is to be sent with 4-bit PCM, where the quantization steps are uniformly distributed over the range $\pm 3\sigma$. Find the mean-square quantization error.

3.24. (a) A signal is derived by passing a 1-kHz square wave through an ideal low-pass filter with cutoff at 5.1 kHz. The resulting signal is quantized using 6-bit PCM. Find the signal-to-quantization-noise ratio.

(b) Repeat the analysis assuming the μ-255 companding is used.

3.25. A signal is given by

$$s(t) = 10 \cos 100t + 17 \cos 500t$$

How many bits of quantization are required so that the signal-to-quantization-noise ratio is greater than 40 dB?

3.26. μ-255 companding is used in a quantization system. How many bits of quantization are required for the signal of Problem 3.25 in order to achieve a signal-to-quantization-noise ratio of 40 dB? Compare your answer to that of Problem 3.25.

3.27. Repeat Example 3-6 for 2 bits of quantization.

3.28. Repeat Example 3-6 for 4 bits of quantization.

3.29. Nonuniform quantization is to be used for 2-bit PCM where the signal is triangularly distributed as shown below.

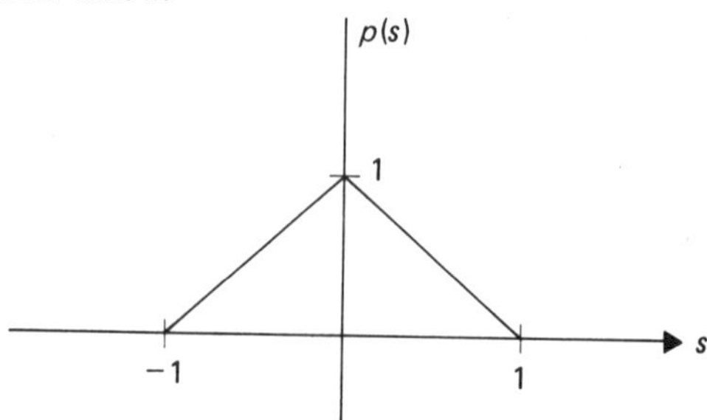

Equation (3.10) defines the quantization. Find the mean-square error and compare this to the mean-square error that would occur using uniform quantization. Also compare your answer to that obtained using μ-255 companding.

3.30. Find the mean-square quantization error when a signal, $s(t)$, with probability density as shown below, is

(a) uniformly quantized.

(b) compressed according to the formula

$$F(s) = \sqrt{|s|}\,\text{sgn}(s)$$

and then uniformly quantized. Assume 4-bit quantization is used.

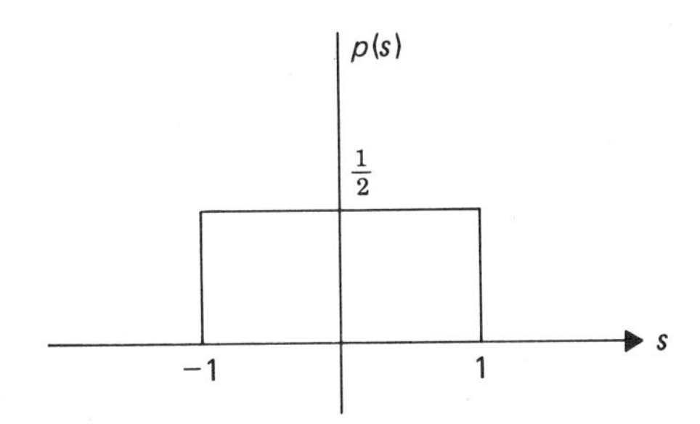

(c) compressed according to the μ-255 rule.

3.31. Repeat Problem 3.30 for 2-bit quantization. Then see if you can find a nonuniform quantizer that achieves lower mean-square error.

3.32. A compressor operates along a sinusoidal curve.

$$F(s) = V_{\max} \sin \frac{\pi s}{2V_{\max}}$$

as shown below.

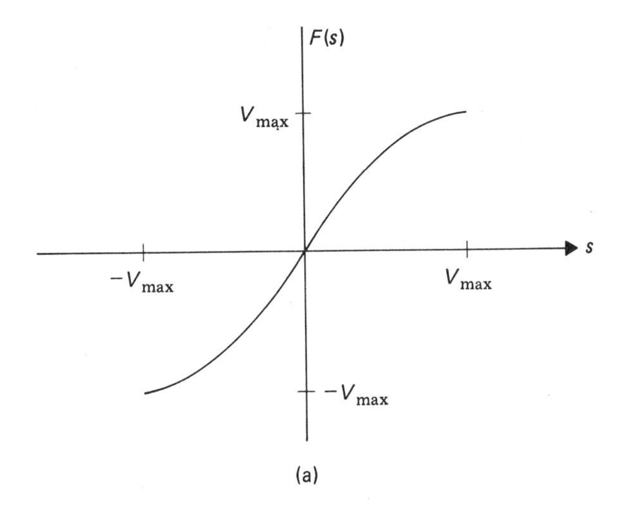

(a)

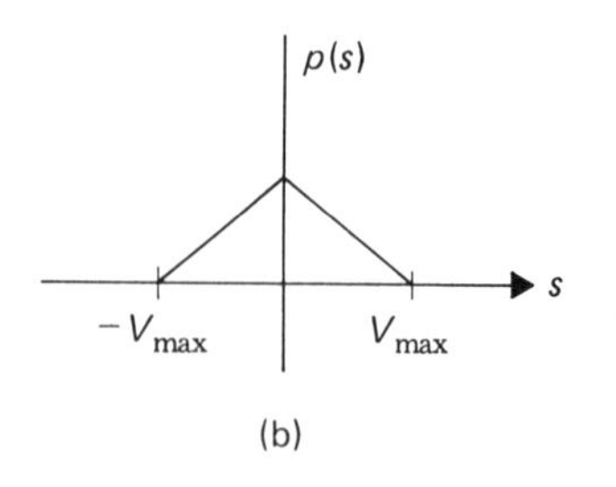

(b)

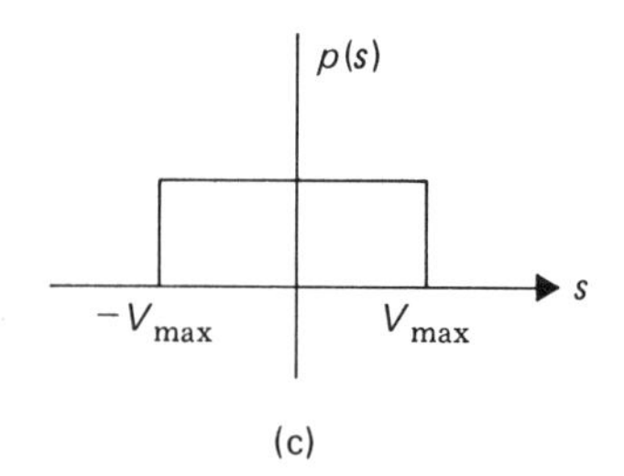

(c)

Find the improvement factor if
(a) the signal is a sinusoid.
(b) the signal is triangularly distributed as shown in the figure.
(c) the signal is uniformly distributed as shown in the figure.

3.33. A signal is derived by passing a 1-kHz square wave through a low-pass filter with cutoff at 5.1 kHz. Choose appropriate parameters for quantization by delta modulation and find the signal-to-quantization-noise ratio.

3.34. A signal to be transmitted is of the form

$$s(t) = 10 \cos 1000t + 5 \cos 1500t$$

This signal is to be quantized using delta modulation.
(a) Choose an appropriate sampling rate and step size.
(b) Using the values found in (a), find the signal-to-quantization-noise ratio.

3.35. A signal is given by

$$s(t) = 5t + 1$$

It is sampled at 10 times per second and quantized using 3-bit delta PCM. Sketch the output of the modulator.

3.36. Use linear prediction to approximate $s(t_0 + T)$ as a linear combination of $s(t_0)$ and its derivative, $s'(t_0)$. (Your answer will be in terms of the autocorrelation of the process.) Also find the mean-square error in this approximation.

3.37. Use linear prediction to estimate the value of $s(NT + \Delta)$ as a linear combination of $s(nT)$ for all n. Find the mean-square error in this approximation. Show that if the conditions of the sampling theorem are met, the mean-square error goes to zero.

3.38. A signal has a probability density given by

$$p(x) = \begin{cases} Ke^{-2|x|}, & -3 < x < 3 \\ 0, & \text{otherwise} \end{cases}$$

(a) Find the signal-to-quantization-noise ratio if 3 bits of uniform quantization are used.
(b) Find the signal-to-quantization-noise ratio if 3 bits of quantization are used for μ-255 companding.

3.39. You are given a signal, $s(t)$, with probability density, $p(s)$, as sketched below. This signal is to be converted from analog to digital using 5 bits of quantization.

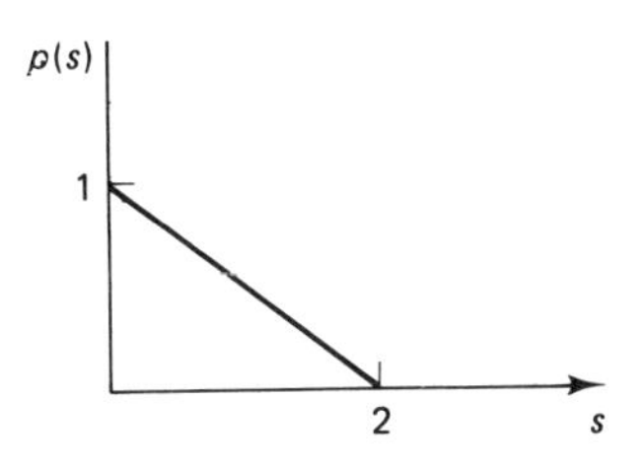

(a) Design a serial A/D converter for this signal.
(b) Find the signal-to-quantization-noise ratio at the output of this A/D converter.
(c) Suppose that the signal is now uniformly distributed between 0 and 2 (instead of following the density shown above). How many decibels of signal-to-quantization-noise ratio improvement would result?

4

Channel Encoding

Chapter 3 presented techniques for changing an analog signal into a sequence of 1's and 0's. We called this source encoding. Now that we have reduced the input to a data signal, we explore techniques for modifying that signal in order to achieve desirable results.

4.1 INFORMATION THEORY

There are a number of reasons for wanting to change the form of a data signal prior to transmission. In the case of English-language text, we start with a data signal having about 40 distinct messages (the letters of the alphabet, integers, and punctuation). We could transmit this using a signal alphabet consisting of 40 distinct voltage waveforms. This is rarely done, for the following reasons. First, the receiver would be complex, having to distinguish among 40 possible received signals. Second, with 40 possible received waveforms, the receiver would probably make numerous mistakes in distinguishing between messages. Third, since some signals occur much more often than others (e.g., those corresponding to "e" or "space"), the transmission is inefficient, as we prove shortly. Yet another reason might be that one wished privacy in transmission so that an unauthorized person could not listen in on the conversation.

A more likely approach to sending text is to first decide upon an alphabet of transmitted elements (perhaps this will be binary, with only two possible elements) and then to code the various possible messages into code words comprised of elements from that alphabet. Let us restate the major reasons for performing this coding operation: (1) to achieve higher transmission efficiency, (2) to simplify the transmitter and receiver design, (3) to decrease errors, and (4) to provide potential for privacy in transmission.

We concentrate upon two types of coding in this chapter: block and convolutional. Since entire textbooks are devoted to coding, one can approach the subject much more deeply than we intend to do herein. Our intent here is to illustrate various techniques of reducing data to binary information. Since we have previously learned (in Chapter 3) to reduce analog information to binary form, after the current discussion we will be ready to spend some time exploring techniques of sending a train of 1's and 0's.

4.2 MEASURE OF INFORMATION

The more *information* a message contains, the more work we are going to have to do to transmit it from one point to another. Some nontechnical examples should help to illustrate an intuitive concept of information before we give a rigorous definition.

Suppose you enjoy Mexican cuisine. You enter your favorite restaurant to order the following: one cheese enchilada, one beef taco, one order of refried beans, one order of rice, two corn tortillas, butter, a sprig of parsley, chopped green onions, and a glass of water. Instead of this entire list being written out longhand, the person taking the order probably communicates something of the form "*one number 2 combination*." Stop and think about this. Rather than attempting to decide whether to transmit the specific words via English spelling or by some form of shorthand, a much more basic approach has been taken. Let us look at another example.

A popular form of communication is the telegram. Have you ever called Western Union to send a wedding congratulations telegram? If so, you found that they try to sell you a standard wording such as "Heartiest congratulations upon the joining of two wonderful people. May your lives together be happy and productive. A gift is being mailed under separate cover." In the early days of telegraphy, the company noticed that many wedding telegrams sounded almost identical to that presented above. A primitive approach to transmitting this might be to send the message letter by letter. However, since the wording is so predictable, why not have a short code for this message? For example, let's call this particular telegram, "#79." The operator would then simply type the name of the sender and of the addressee and then #79. At the receiver, form #79 is pulled from the file and the address and signature are filled in. Think of the amount of transmission time saved!

As yet a third example, consider the (atypical, I hope) university lecture in which a professor stands before the class and proceeds to read directly from the textbook for two hours. A great deal of time and effort could be saved if the professor instead read the message, "pages 103 to 120." In fact, if the class expected this and knew they were past page 100, the professor could shorten the message to "0320." The class could then read these 18 pages on their own and get the same information. The professor could dismiss the class after 20 seconds and return to more scholarly endeavors.

I think that by now the point should be clear. A basic examination of the information content of a message has the potential of saving considerable effort in transmitting the message from one point to another.

4.2.1 Information and Entropy

The modest examples given at the start of this section indicate that, in an intuitive sense, some relatively long messages do not contain a great deal of information. Every detail of the restaurant order or the telegram is so commonplace that the person at the receiving end can almost guess what comes next.

The concept of information content of a particular message must now be formalized. After doing this, we shall find that the less information in a particular message, the quicker we can communicate the message.

The concept of information content is related to predictability. That is, the more likely a particular message, the less information is given by transmitting that message. For example, if today were Tuesday and you phoned someone to tell him or her that tomorrow would be Wednesday, you would certainly not be communicating any information. This is so because the probability of that particular message is "1".

The definition of information content should be such that it monotonically decreases with increasing probability and goes to zero for a probability of unity. Another property we would like this measure of information to have is that of *additivity.* If one were to communicate two (independent) messages in sequence, the total information content should be equal to the sum of the individual information contents of the two messages. Now if the two messages are independent, we know that the total probability of the composite message is the product of the two individual probabilities. Therefore, the definition of information must be such that when probabilities are multiplied together, the corresponding informations are added.

The logarithm satisfies these requirements. We thus *define* the information content of a message, x, as I_x, in Eq. (4-1):

$$I_x = \log \frac{1}{P_x} \tag{4-1}$$

In Eq. (4-1), P_x is the probability of occurrence of the message x. This definition satisfies the additivity requirement and the monotonicity requirement, and for $P_x = 1$, $I_x = 0$. Note that this is true regardless of the base chosen for the logarithm.

We usually use base 2 for the logarithm. To understand why, let us return to the restaurant example. Suppose that there were only two selections on the menu, and that past observation has shown that these two are equally probable (i.e., each is ordered half of the time). The probability of each message is therefore $\frac{1}{2}$, and if base 2 logarithms are used in Eq. (4-1), the information content of each message is

$$I = \log_2 (2) = 1$$

Thus, one unit of information is transmitted each time an order is placed.

Now let us think back to digital communication and decide upon an efficient way to transmit this order. Since there are only two possibilities, one binary digit would be used to send the order to the kitchen. A "0" could represent the first dinner and a "1" the second dinner on the menu.

Suppose that we now increase the number of items on the menu to four, with each having a probability of $\frac{1}{4}$. The information content of each message is now $\log_2 4$, or 2 units. If binary digits are used to transmit the order, 2 bits will be required for each message. The various dinners will be coded as 00, 01, 10, and 11. We can therefore conclude that if the various messages are equally probable, the information content of each message is exactly equal to the minimum number of bits required to send the message (provided this is an integer). This is the reason for commonly using base 2 logarithms, and in fact the unit of information is called the *bit of information.* Thus, in the last example, one would say that each menu order contains 2 bits of information.

When all the possible messages are equally likely, the information content of any single message is the same as that of any other message. In cases where the probabilities are not equal, the information content depends upon which particular message is being transmitted.

Entropy is defined as the average information per message. To calculate the entropy, we take the various information contents associated with the messages and weight each by the fraction of time we can expect that particular message to occur. This fraction is the probability of the message. Thus, given n messages, x_1 through x_n, the entropy is defined by

$$H = \sum_{i=1}^{n} P_{xi} I_{xi} = \sum_{i=1}^{n} P_{xi} \log\left(\frac{1}{P_{xi}}\right) \tag{4-2}$$

By convention, the letter H is used for entropy.

Example 4-1

A communication system consists of six possible messages with probabilities $\frac{1}{4}$, $\frac{1}{4}$, $\frac{1}{8}$, $\frac{1}{8}$, $\frac{1}{8}$ and $\frac{1}{8}$, respectively. Find the entropy.

Solution. The information content of the six messages is 2 bits, 2 bits, 3 bits, 3 bits, 3 bits, and 3 bits, respectively. The entropy is therefore given by

$$H = \frac{1}{4} \times 2 + \frac{1}{4} \times 2 + \frac{1}{8} \times 3 + \frac{1}{8} \times 3 + \frac{1}{8} \times 3 + \frac{1}{8} \times 3$$

$$= 2.5 \text{ bits/message}$$

It proves instructive to examine Eq. (4-2) for the binary message case. That is, we consider a communication scheme made up of two possible messages, x_1 and x_2. For this case,

$$P_{x2} = 1 - P_{x1}$$

and the entropy is given by

$$H_{\text{binary}} = P_{x1} \log\left(\frac{1}{P_{x1}}\right) + (1 - P_{x1}) \log\left(\frac{1}{1 - P_{x1}}\right)$$

This result is sketched as a function of P_{x1} in Fig. 4.1.

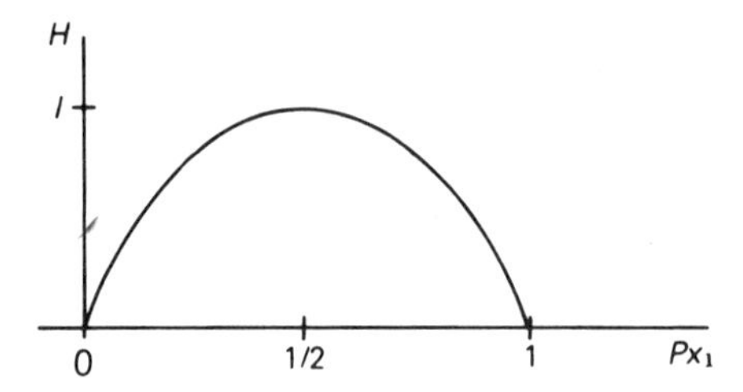

Figure 4.1 Entropy of binary communication system

Note that as either of the two messages becomes more likely, the entropy decreases. When either message has probability 1, the entropy goes to zero. This is reasonable, since, at these points, the outcome of the transmission is certain without our going to the trouble of sending any message. Thus, if $P_{x1} = 1$, we know message x_1 will be sent all the time (with probability 1). Alternatively, if $P_{x1} = 0$, we know that x_2 will be sent all the time. In these two cases, no information is transmitted by sending the message.

If we were able to show a three-dimensional plot for the three-message case, we would find a similar result with a peak in entropy for message probabilities of $\frac{1}{3}$.

The entropy function is symmetrical with a maximum for the case of equally likely messages. This is an important property that is used in later coding studies.

4.3 CHANNEL CAPACITY

In this section we investigate the rate at which information can be sent through a channel and the relationship of this rate to errors. Before we can do this, we must define the rate of information flow.

Assume that a source can send any one of a number of messages at a rate of r messages per second. For example, if the source is a Teletype machine, the rate is the number of symbols per second. Once we know the probability of each individual message, we can compute the entropy H in bits per message. If we now take the product of the entropy, H, with the message rate, r, we get the information rate in bits per second, denoted R. Thus,

$$R = rH \text{ bps}$$

For example, if the messages given in Example 4-1 were sent at 2 messages per second, the information rate would be 5 bps.

One can probably reason intuitively that, for a given communication system, as the information rate increases, the number of errors per second will also increase.

C. E. Shannon has shown that a given communication channel has a maximum rate of information, C, known as the *channel capacity*. If the information rate, R, is less than C, one can approach arbitrarily small error probabilities by intelligent coding techniques. This is true even in the presence of noise, a fact that probably contradicts your intuition.

The negation of Shannon's theorem is also true. That is, if the information rate R is greater than the channel capacity, C, errors cannot be avoided regardless of the coding technique employed.

We shall not prove this fundamental theorem here. We shall, however, discuss its application to several commonly encountered cases.

We consider the bandlimited channel operating in the presence of additive white Gaussian noise. In this case, the channel capacity is given by

$$C = B \log_2 (1 + S/N) \tag{4-3}$$

where C is the capacity in bits per second, B is the bandwidth of the channel in Hz, and S/N is the signal-to-noise ratio. Equation (4-3) makes intuitive sense. As the bandwidth of the channel increases, it should be possible to make faster changes in the information signal, thereby increasing the information rate. As S/N increases, one would expect to be able to increase the information rate while still preventing errors due to noise. Note that with no noise at all, the signal-to-noise ratio is infinity, and an infinite information rate would be possible regardless of the bandwidth.

Equation (4-3) might lead one to conclude that if the bandwidth approaches infinity, the capacity also approaches infinity. This last observation is not correct. Since the noise is assumed to be white, the wider the bandwidth, the more noise is admitted to the system (unless the noise is identically equal to zero). Thus, as B increases in Eq. (4-3), S/N decreases.

It is instructive to expand upon this last observation. That is, suppose that the total signal and noise power per hertz were fixed and we were trying to design the best possible channel. There is expense associated with increasing the system bandwidth, so we should attempt to find the maximum channel capacity. Suppose that the noise is white with power spectral density, $N_0/2$. Further assume that the signal power is fixed at a value S. The channel capacity is given from Eq. (4-3) as

$$C = B \log_2 \left(1 + \frac{S}{N_0 B}\right)$$

Note that B is in hertz and that N_0 is the power spectral density in watts per hertz. We now find the value that the channel capacity approaches as B goes to infinity:

$$\begin{aligned} C &= \lim_{B\to\infty} B \log_2 \left(1 + \frac{S}{N_0 B}\right) \\ &= \lim_{B\to\infty} \frac{S}{N_0} \log_2 \left(1 + \frac{S}{N_0 B}\right)^{N_0 B/S} \\ &= \frac{S}{N_0} \log_2 e = 1.44 \frac{S}{N_0} \end{aligned} \tag{4-4}$$

Equation (4-4) shows the maximum possible channel capacity as a function of signal power and noise spectral density. In an actual system design, the channel capacity will be compared to this figure and a decision will be made whether further increase in bandwidth is worth the expense.

Suppose now that you were asked to design a binary communication system and you cranked the bandwidth of your communication channel, the maximum signal power, and the noise spectral density into Eq. (4-4). You come up with a maximum information rate, C. You plug in the sampling rate and the number of bits of quantization, and you arrive at a certain information transmission rate, R bps. Aha! you observe, R is less than C, so Shannon tells you that if you do enough work encoding this binary data train, you can achieve arbitrarily low probability of error. But you have already coded the original signal into a train of binary digits. What do you do next?

The next section discusses answers to this question. As applied to the example above, most procedures would entail grouping combinations of bits and calling the result a new word. For example, we can group by twos to yield four possible messages, 00, 01, 10, and 11. Each of these four possible messages can be transmitted using some code word. By so doing, the probability of error can be reduced.

As an example, if you wished to transmit this textbook to a class of students, you could read each letter aloud as, for example,

r-e-a-d-e-a-c-h-l-e-t-t-e-r-a-l-o-u-d

Alternatively, you could group letters together and send the groups as a code word from an acceptable dictionary. For the example above, you send

read-each-letter-aloud

Suppose that in reading the individual letters, you slur the "d" sound and one student receives it as "v". That student receives the erroneous message,

r-e-a-v-e-a-c-h-l-e-t-t-e-r-a-l-o-u-d

However, in the word coding, the student would receive

reav-each-letter-aloud

and since "reav" is not an acceptable code word, the student could correct the error to interpret the received message as

read-each-letter-aloud

and thus make no error. The essence of coding to achieve arbitrarily small errors amounts to grouping into longer and longer code words. The longer the code words, the more different the dictionary entries will be, and the less likely it will be that individual errors cannot be corrected.

4.4 CODING

We now have a feel for the concept of entropy and are ready to see how this concept helps us design a transmission system with desirable properties. We begin with a general statement of the problem.

Given M possible messages, we wish to convert these into M possible code words. The code words can be selected to achieve objectives such as *efficiency, error correction,* or *security.* We discuss codes to achieve efficiency in the next section under the heading *entropy coding.* Such codes attempt to send the information using the minimum number of bits (or symbols). After that discussion, we consider error detection and correction, illustrating these concepts with both block and convolutional codes. We delay discussion of coding for secure communications until Chapter 12 of this text.

We need first to establish some basic concepts. Even in the absence of additive noise, codes must be carefully designed to avoid decoding errors. The first potential problem relates to the concept of *unique decipherability.* As an example, suppose that there were four possible messages to be transmitted, and these were coded into binary numbers as follows:

$$M_1 = 1, \qquad M_2 = 10, \qquad M_3 = 01, \qquad M_4 = 101$$

Suppose now that you are sitting at the receiver and receive the sequence 101. You would not know if this constituted M_4 or either of the paired message sets, M_2M_1 or M_1M_3. Therefore, this choice of code words yields a code that is not uniquely decipherable. Some thought would convince you that a code is uniquely decipherable if no code word forms the starting sequence (known as the *prefix*) of any other code word. Thus, for example, the following four-message code is uniquely decipherable:

$$M_1 = 1, \qquad M_2 = 01, \qquad M_3 = 001, \qquad M_4 = 0001$$

The prefix restriction property is sufficient but not necessary for unique decipherability. For example, the code

$$M_1 = 1, \quad M_2 = 10, \quad M_3 = 100, \quad M_4 = 1000$$

is uniquely decipherable, even though each code word is the prefix of every other code word to its right. The major difference between this and the first example in this section is that no code word can be formed as a combination of other code words. There is, however, one problem with this code. Again suppose that you are sitting at the receiver, and you receive 10. Until you look at the next two received bits, you do not know whether you are receiving the message M_2, M_3, or M_4. We say that this code is uniquely decipherable but not *instantaneous.*

Example 4-2

Which of the following codes are uniquely decipherable? For those that are uniquely decipherable, determine whether they are instantaneous.
(a) 0, 01, 001, 0011, 101
(b) 110, 111, 101, 01
(c) 0, 01, 011, 0110111

Solution. (a) This is not uniquely decipherable, since the first and last words when sent in sequence, 0101, could be interpreted as 01, 01—that is, two transmissions of the second word.
(b) This is uniquely decipherable since all but one word starts with a "1" and has length 3. If the start of a 3-bit sequence is not "1", we know it is the only 2-bit word. This code is also instantaneous, since no code word is the prefix of another word.
(c) This is uniquely decipherable, since all words start with a single zero, there is no repeated zero in any word, and no word is a combination of other words. It is not instantaneous, since each of the first three words is a prefix of at least one other word.

4.4.1 Entropy Coding

It is of interest to find uniquely decipherable codes of minimum length. This will allow maximum transmission rates through the channel. Examination of the codes presented so far in this section shows that different messages are coded into words of different length. When talking about the length of a code, we therefore must refer to the *average* length of the code words. This average is computed by taking the probabilities of each message into account. It is clearly advantageous to assign the shorter code words to the most probable messages. The *Morse code* follows this rule by assigning the shortest code to the letter e.

A fundamental theorem exists in noiseless coding theory. The theorem states that:

> For binary-coding alphabets, the average code word length is greater than, or equal to, the entropy.

This is intuitively satisfying, and it gives a whole new interpretation to the concept of average information, or entropy.

Denoting this average word length by $\bar{n}$, the theorem then states

$$\bar{n} \geqslant H$$

To prove this important theorem, we start with the property of the logarithm given in Eq. (4-5), where ln is the natural logarithm.

$$\log_2 (x) \leqslant \frac{x-1}{\ln 2} \tag{4-5}$$

This is so because the log is convex and its slope at $x = 1$ is $1/\ln 2$. The two sides of the inequality are sketched in Fig. 4.2.

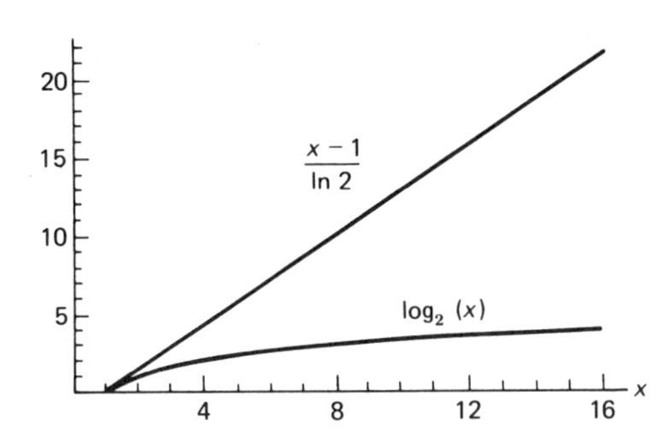

Figure 4.2 Comparison of two sides of Eq. (4-5)

We define

$$q_i = \frac{2^{-n_i}}{\sum_{i=1}^{M} 2^{-n_i}}$$

and

$$x = \frac{q_i}{p_i}$$

where n_i is the length of code word i, p_i the probability of message i, and M the number of messages.

From Eq. (4-5) we have

$$\log \frac{q_i}{p_i} \leqslant \frac{q_i - p_i}{p_i \ln 2}$$

$$\sum_{i=1}^{M} p_i \log \frac{q_i}{p_i} \leqslant \sum_{i=1}^{M} \frac{q_i - p_i}{\ln 2} = 0$$

The last equality (= 0) is true because the p_i sum to 1 and the q_i also sum to 1.

Finally,

$$\sum_{i=1}^{M} p_i \log q_i \leqslant \sum_{i=1}^{M} p_i \log p_i$$

and plugging in for q_i, we have

$$-\sum_{i=1}^{M} p_i n_i \leqslant \sum_{i=1}^{M} p_i \log p_i \sum_{i=1}^{M} 2^{-n_i}$$
$$\leqslant \sum_{i=1}^{M} p_i \log p_i \qquad (4\text{-}6)$$

The last result is true because

$$\sum_{i=1}^{M} 2^{-n_i} \leqslant 1$$

for distinct code words.

The left side of Eq. (4-6) is the average code word length, and the right side is the entropy. We have thus proven the theorem. Equality obtains only if the length of each code word is such that $2^{-n} = p_i$. This, of course, requires that the probabilities of every message be inverse powers of 2. Note that had we considered a system other than binary, the theorem would be modified as follows:

$$\bar{n} \geqslant \frac{H}{\log L}$$

where L is the number of symbols in the code alphabet.

Example 4-3

Find the minimum average length of a code with four messages with probabilities $\frac{1}{8}$, $\frac{1}{8}$, $\frac{1}{4}$, and $\frac{1}{2}$, respectively.

Solution. The entropy is given by

$$\tfrac{1}{8} \times 3 + \tfrac{1}{8} \times 3 + \tfrac{1}{4} \times 2 + \tfrac{1}{2} \times 1 = 1.75 \text{ bits}$$

which is also the minimum average length for this code. We note that one possible optimum code is

$$M_1 = 000, \qquad M_2 = 001, \qquad M_3 = 01, \qquad M_4 = 1$$

This code is uniquely decipherable, and the average length is 1.75 bits.

4.4.1.1 Variable-length codes. If the various messages to be transmitted do not have equal probabilities, efficiencies are possible by allowing the code words to have unequal length. For example, suppose that there are four messages with probabilities $\frac{1}{8}$, $\frac{1}{8}$, $\frac{1}{4}$, and $\frac{1}{2}$, respectively. One way to code these into binary words is to use 00, 01, 10, and 11 to send the four possible messages, with a resulting average length of 2 bits. If we instead use 111, 110, 10, and 0 (which is uniquely decodable and instantaneous—convince yourself of this!), the average length is given by

$$\tfrac{1}{8} \times 3 + \tfrac{1}{8} \times 3 + \tfrac{1}{4} \times 2 + \tfrac{1}{2} \times 1 = 1.75$$

In effect, we are coding the more probable messages into shorter code words. In this particular case, the average word length matches the entropy, so we cannot possibly find a code with a smaller average length. Our purpose in this section is to illustrate techniques for finding the best possible code.

One way to derive variable-length codes is to start with constant-length codes and expand subgroups. For example, starting with a 1-bit code, we have two code words, 0 and 1. We can expand this to five code words by taking the "1" and expanding this to 100, 101, 110 and 111, yielding the five code words

0

100

101

110

111

Another way is to start with the 2-bit code 00, 01, 10, 11 and expand any one of these four words into two words. If 01 is chosen for expansion, we get the five-word code

00

010

011

10

11

The question now is: Of the many ways to do this expansion, which results in the code of minimum average length?

We present two techniques for finding efficient variable-length codes. These techniques lead to the Huffman and to the Shannon-Fano codes.

Huffman codes provide an organized technique for finding the best possible variable-length code for a given set of messages. We present the procedure using a specific example.

Suppose that we wish to code five words, s_1, s_2, s_3, s_4, and s_5 with probabilities $\frac{1}{16}$, $\frac{1}{8}$, $\frac{1}{4}$, $\frac{1}{16}$, and $\frac{1}{2}$, respectively. The Huffman procedure can be accomplished in four steps.

Step 1. Arrange the messages in order of decreasing probability. If there are equal probabilities, choose any of the various possibilities.

Word	Probability
s_5	$\frac{1}{2}$
s_3	$\frac{1}{4}$
s_2	$\frac{1}{8}$
s_1	$\frac{1}{16}$
s_4	$\frac{1}{16}$

Step 2. Combine the bottom two entries to form a new entry with probability that is the sum of the original probabilities. If necessary, reorder the list so that probabilities are still in descending order.

Word	Prob.	Prob.
s_5	$\frac{1}{2}$	$\frac{1}{2}$
s_3	$\frac{1}{4}$	$\frac{1}{4}$
s_2	$\frac{1}{8}$	$\frac{1}{8}$
s_1	$\frac{1}{16}$ } →	$\frac{1}{8}$
s_4	$\frac{1}{16}$ }	

Note that the bottom entry in the right-hand column is a combination of s_1 and s_4.

Step 3. Continue combining in pairs until only two entries remain.

Word	Prob.	Prob.	Prob.	Prob.
s_5	$\frac{1}{2}$	$\frac{1}{2}$	$\frac{1}{2}$	$\frac{1}{2}$
s_3	$\frac{1}{4}$	$\frac{1}{4}$	$\frac{1}{4}$ } →	$\frac{1}{2}$
s_2	$\frac{1}{8}$	$\frac{1}{8}$ } →	$\frac{1}{4}$ }	
s_1	$\frac{1}{16}$ } →	$\frac{1}{8}$ }		
s_4	$\frac{1}{16}$ }			

Step 4. Assign code words by starting at right with the most significant bit. Move to the left and assign another bit if a split occurred. The assigned bits are un-

derlined in the chart below.

Word	Prob.	Prob.	Prob.	Prob.
s_5	$\frac{1}{2}$	$\frac{1}{2}$	$\frac{1}{2}$	$\frac{1}{2}$ 0
s_3	$\frac{1}{4}$	$\frac{1}{4}$	$\frac{1}{4}$ 10	$\frac{1}{2}$ 1
s_2	$\frac{1}{8}$	$\frac{1}{8}$ 110	$\frac{1}{4}$ 11	
s_1	$\frac{1}{16}$ 1110	$\frac{1}{8}$ 111		
s_4	$\frac{1}{16}$ 1111			

Finally, the code words are given by

$$s_1 \rightarrow 1110$$
$$s_2 \rightarrow 110$$
$$s_3 \rightarrow 10$$
$$s_4 \rightarrow 1111$$
$$s_5 \rightarrow 0$$

Note that at each assignment point, two possible assignments are possible. In addition, when there are three or more equal lowest probabilities, the choice for combination is arbitrary.

The average length is given by

$$\bar{L} = 4 \times \tfrac{1}{16} + 3 \times \tfrac{1}{8} + 2 \times \tfrac{1}{4} + 4 \times \tfrac{1}{16} + 1 \times \tfrac{1}{2} = \tfrac{15}{8}$$

If block coding had been used, the length would have been 3. The entropy of the code is given by

$$H = \tfrac{2}{16} \log(16) + \tfrac{1}{8} \log(8) + \tfrac{1}{4} \log(4) + \tfrac{1}{2} \log(2)$$
$$= \tfrac{15}{8} \text{ bits}$$

the same as the average length of this Huffman code. Thus, the Huffman procedure gave us a code with maximum efficiency. This was possible since all probabilities were powers of $\frac{1}{2}$.

Example 4-4

Find the Huffman code for the following seven messages with probabilities as indicated:

S_1	S_2	S_3	S_4	S_5	S_6	S_7
0.05	0.15	0.2	0.05	0.15	0.3	0.1

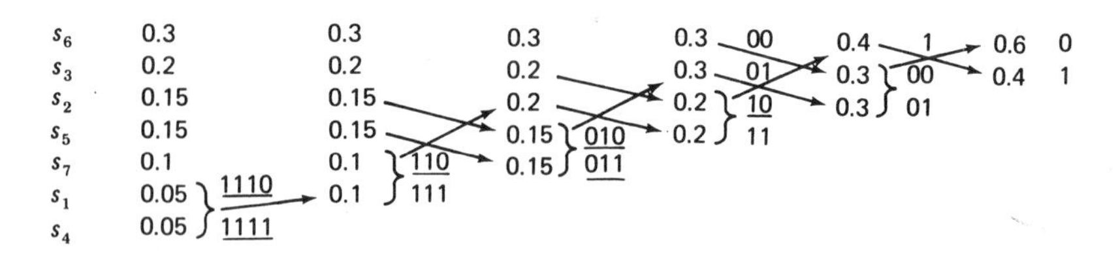

Figure 4.3 Development of Huffman code for Example 4-4

Solution. The steps are illustrated by the chart in Fig. 4.3. Note that at the third, fourth, and fifth steps, a reordering was necessary to place the probabilities in descending order. The final code is

$$S_1 \to 1110$$
$$S_2 \to 010$$
$$S_3 \to 10$$
$$S_4 \to 1111$$
$$S_5 \to 011$$
$$S_6 \to 00$$
$$S_7 \to 110$$

The average length of the code is

$$\bar{L} = 4(0.05 + 0.05) + 3(0.15 + 0.15 + 0.1) + 2(0.2 + 0.3)$$

$$= 2.6 \text{ bits}$$

The entropy is given by

$$H = 0.3 \log\left(\frac{1}{0.3}\right) + 0.2 \log\left(\frac{1}{0.2}\right) + 0.3 \log\left(\frac{1}{0.15}\right)$$
$$+ 0.1 \log\left(\frac{1}{0.1}\right) + 0.1 \log\left(\frac{1}{0.05}\right)$$
$$= 2.57 \text{ bits}$$

We should note that the Huffman codes are formulated to minimize the average length. The resulting codes do not necessarily possess error-detection and correction potential. Looking at the result of Example 4-4, the resulting code could not even *detect* a single bit error. You should convince yourself (by exhaustive testing of the 21 possibilities) that *every* possible 1-bit error will cause a decoding error.

One disadvantage of the Huffman code is that we cannot start assigning code words until the entire combination process is completed. That is, every one of the columns must be developed before the first code word can be assigned. The coding process is often performed by a special-purpose microcomputer. This last observation indicates that considerable storage may be required.

The *Shannon-Fano* code is similar to the Huffman, a major difference being that the operations are performed in a forward, rather than backward, direction. Thus, the storage requirements are considerably relaxed and the code is easier to implement. While it often leads to average lengths that are the same as those of the Huffman code, the results of Shannon-Fano are not always as good as those of Huffman.

We again illustrate the technique with an example—the same example we used earlier for the Huffman code.

Step 1. Arrange the messages in order of decreasing probability. If there are equal probabilities, choose any of the various possibilities.

Word	Probability
s_5	$\frac{1}{2}$
s_3	$\frac{1}{4}$
s_2	$\frac{1}{8}$
s_1	$\frac{1}{16}$
s_4	$\frac{1}{16}$

Step 2. Partition the messages into the most equiprobable subsets. That is, we start at the top or bottom and divide the group into two sets. We find the total probability of the upper set and the total probability of the lower set. We choose the dividing line that results in the closest two probabilities. In this case, the dividing line falls below the first entry, resulting in exactly $\frac{1}{2}$ for the probability of the entries above and below the line. This is illustrated below:

Word	Probability
s_5	$\frac{1}{2}$
---	---
s_3	$\frac{1}{4}$
s_2	$\frac{1}{8}$
s_1	$\frac{1}{16}$
s_4	$\frac{1}{16}$

We now assign a 0 to all members of one set, and a 1 to all members of the other (the choice is arbitrary). Suppose we choose 0 for the top set and 1 for the bottom. If a set contains only one entry, the process for that set is concluded. Thus, the code word used to send s_5 is 0, and we need look no longer at that message. We now concentrate upon the other set and repeat the subdivision process. Thus, after one more subdivision, we have

Word	Probability	
s_3	$\frac{1}{4}$	Code word 10
s_2	$\frac{1}{8}$	Code word 11
s_1	$\frac{1}{16}$	
s_4	$\frac{1}{16}$	

Note that once again, the subdivision worked out perfectly, as the probability both above and below the line is exactly $\frac{1}{4}$. We have added a second bit to the code words, using a 0 for that bit above the line and a 1 below. Since there is only one entry in the top set, we are finished, and the code for s_3 is 10. Continuing the subdivision with the bottom set, we have

Word	Probability	
s_2	$\frac{1}{8}$	Code word 110
s_1	$\frac{1}{16}$	Code word 111
s_4	$\frac{1}{16}$	

Finally, we subdivide the bottom set to get

Word	Probability	
s_1	$\frac{1}{16}$	Code word 1110
s_4	$\frac{1}{16}$	Code word 1111

The resulting code words are

s_1	1110
s_2	110
s_3	10
s_4	1111
s_5	0

For this example, the result turned out to be exactly the same as that using Huffman coding.

Example 4-5

Find the Shannon-Fano code for the following seven messages with probabilities as indicated:

s_1	s_2	s_3	s_4	s_5	s_6	s_7
0.05	0.15	0.2	0.05	0.15	0.3	0.1

Solution. We order these and subdivide as follows. The code words are underlined.

Message	Probability				
s_6	0.3	0.5 0	0.3 <u>00</u>		
s_3	0.2		0.2 <u>01</u>		
s_2	0.15	0.5 1	0.3 10	0.15 <u>100</u>	
s_5	0.15			0.15 <u>101</u>	
s_7	0.1		0.2 11	0.1 <u>110</u>	
s_1	0.05			0.1 111	0.05 <u>1110</u>
s_4	0.05				0.05 <u>1111</u>

The resulting code is

s_1	1110
s_2	100
s_3	01
s_4	1111
s_5	101
s_6	00
s_7	110

Once again, the Shannon-Fano technique has resulted in a code of the same average length (2.6 bits—see Example 4-4) as that which results using the Huffman code.

We have illustrated two techniques for reducing a set of messages to a binary code with excellent efficiency. In all of this, we assume that the messages are given, and that they cannot be combined prior to coding. In fact, greater efficiencies are often possible with message combinations. We illustrate this for an example with two messages. Suppose the messages have probability

s_1	0.9
s_2	0.1

The entropy is given by

$$H = -0.9 \log 0.9 - 0.1 \log 0.1 = 0.47 \text{ bits}$$

We would therefore hope to arrive at a code with average length close to this value. However, using either the Huffman or Shannon-Fano technique results in assigning a 0 to one of the words and a 1 to the other word. The average length is then 1 bit per message. This is more than twice the minimum given by the entropy. How can we possibly improve upon the situation?

Suppose we combine the messages in pairs. We then have four possible two-message sets. Assuming independence of the messages, the possible sets and resulting probabilities are:

s_1s_1	0.81
s_1s_2	0.09
s_2s_1	0.09
s_2s_2	0.01

Using the Shannon-Fano method, we assign the code words as follows:

s_1s_1	0.81	0
s_1s_2	0.09	10
s_2s_1	0.09	110
s_2s_2	0.01	111

The average word length is given by

$$\bar{L} = 1 \times 0.81 + 2 \times 0.09 + 3 \times 0.10 = 1.29 \text{ bits}$$

Now since each combined message represents two of the original messages, we divide this number by 2 to find that 0.645 bits are being used to send each of the original messages.

Suppose now that we combine three messages at a time to get the following message probabilities and code words:

$s_1s_1s_1$	0.729	0
$s_1s_1s_2$	0.081	100
$s_1s_2s_1$	0.081	101
$s_1s_2s_2$	0.009	11100
$s_2s_1s_1$	0.081	110
$s_2s_1s_2$	0.009	11101
$s_2s_2s_1$	0.009	11110
$s_2s_2s_2$	0.001	11111

The average length of the codes is 1.598 bits, so the average length per original message is

$$\bar{L} = \frac{1.598}{3} = 0.533 \text{ bits}$$

Note that as we combine more and more messages, the average length approaches the entropy. This average length will equal the entropy if the probabilities are all inverse powers of 2. As more and more messages are combined, the probabilities approach such powers more and more closely.

4.4.1.2 Data compression. *Data compression* is a term applied to a wide variety of techniques for reducing the number of bits required to send a given message. Entropy coding is one form of data compression.

The success of data-compression techniques depends upon the properties of the message. For example, entropy coding becomes most effective when the probabilities of the various messages are far from being equal. The other techniques we shall briefly describe depend upon sequential properties of the messages. That is, they depend upon symbols occurring in a predictable order.

As an example, consider the encoding of a television picture. In the United States, a TV picture contains 426 visible picture elements (pixels) in each horizontal line. If we talk about black-and-white TV, we need to send the brightness (luminance) of each pixel. Suppose we decide to transmit 7 bits of information. That is, we quantize the luminance into 2^7, or 128 different levels. This would represent a high quality of resolution. We therefore need 7×426, or 2982 bits to transmit the information in each line using a PCM scheme. Now a standard TV picture often contains a sequence of adjacent pixels with the same luminance. That is, as we trace across a horizontal line, we may encounter as many as several hundred pixels with the same brightness (suppose there is one figure in the middle of the screen and the background is uniform, or suppose we are sending text on a uniform background). In such cases, we can use a data-compression technique known as *run-length coding* to reduce the number of bits required to send the signal. Instead of sending the luminance of every single pixel, we send the starting position and luminance of the first of a number of pixels with the same brightness. In order to send the position, we need 9 bits of information, since $2^9 = 512$ and there are 426 different positions. Thus, we need 9 bits for the location and 7 bits for the luminance, for a total of 16 bits. For runs of three or more identical brightnesses, we save bits by using this technique. For example, if 10 adjacent bits have the same luminance, we need $10 \times 7 = 70$ bits to send these individually, but only 16 bits to send them using run-length coding. This concept can lead to even greater savings if extended to two dimensions. Thus, we would have to specify only the luminance and corner coordinates of any rectangle of constant luminance.

One disadvantage of run-length coding is that the data signal occurs at a nonuniform rate. That is, without such coding, bits are being sent at a constant rate. However, with the coding, large uniform brightness areas result in lower data rates. The system therefore requires a buffer. An additional shortcoming is that errors propagate, since the system has memory. A bit error in a system which uses PCM to send individual pixel information results in a luminance error for that pixel only. However, if run-length coding is used, a bit error can affect the brightness of the entire line segment (or rectangle).

Prediction can be used in other forms of data compression. If the future values of data can be predicted from the present and past values, there is no need to send all the data. It may be sufficient to send the current data values plus some key parameters to aid in the prediction. This can be likened to sending a particular curve. We can send the value of the ordinate at every value of the abscissa, or we can send key parameters. If the curve is a straight line, we need only send the slope and intercept. If it is a parabola, other parameters are sufficient. We can even send the various coefficients in a series expansion if this results in a savings of transmission time.

4.4.2 Linear Block Encoding

In linear block encoding, constant-length groupings of the message are coded into constant-length groupings of the code bits. In a basic binary communication system, only two symbols are generated by the source. Coding begins by combining these symbols into groups. One possible approach is to start with a binary communication system and group bits to form any desired number of possible messages. For example, by combining groups of 3 bits, we can form eight possible message words: 000, 001, 010, 011, 100, 101, 110, and 111. Each of these eight possible message words can now be coded into one of eight different code words. The code words need not necessarily be of the same length as the message words. In fact, to achieve error-correction capabilities, *redundancy* must be introduced by making the code words longer than the message words.

In this section, we examine error-control capability for randomly distributed errors. That is, we assume that the actual bits which reverse during transmission are randomly distributed within the message. This is not true for *burst errors*, where there is a high probability of bit errors occurring among a sequence of adjacent bits. Techniques for decreasing the effects of burst errors can be found in the references.

Distance between code words. We now turn our attention to the *noisy coding problem* and a discussion of *error-correcting codes*. We restrict our attention here to codes whose words all have the same length. We also consider only binary codes.

To analyze the error-correction capabilities of various codes, we first define the concept of distance between two binary code words. The distance between two equal-length binary code words is defined as the number of bit positions in which the two words differ. For example, the distance between 000 and 111 is 3; the distance between 010 and 011 is 1. The distance between any one code word and the word formed by changing 1 bit is "1."

Suppose now that we transmit one of the eight possible 3-bit words. Assume that the channel is noisy and that one bit position is incorrectly received. Since every possible 3-bit combination is used as a message, the received 3-bit combination will be identical to one of the code words. For example, if 101 is transmitted and an error occurs in the third bit, 100 is received. There is no way for the receiver to know that 100 was not the transmitted word. An error is therefore going to occur.

Suppose now that the dictionary of code words is such that the distance between any two words is at least 2. The following eight code words have this property:

0000, 0011, 0101, 0110, 1001, 1010, 1100, 1111

The reader should verify that the minimum distance between code words is 2.

Now suppose that we transmit one of these eight words and that a single bit error occurs during transmission. Since the distance between the received word and the transmitted word is 1, the received word cannot be identical to any of the dictionary words. For example, suppose that 0101 is transmitted and an error occurs in the third bit. 0111 is received. This is not one of the eight acceptable words, and the system can tell that an error was made.

It is instructive to think of distance in terms of a multidimensional space. In the previous example of 3-bit code words, we can plot this in three dimensions. Each code word is a point at the corner of a unit cube, as shown in Fig. 4.4. Starting at each corner, if a single bit error is made, we move along one of the edges to another corner a distance of 1 unit away.

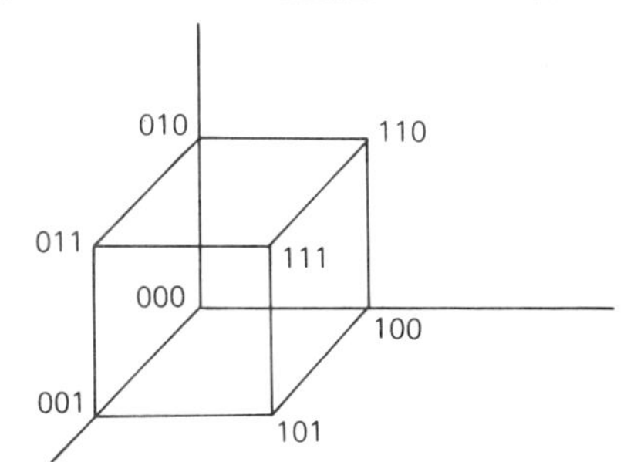

Figure 4.4 Distance relationships for 3-bit code

In the 4-bit code-word case, we need an abstract figure with eight points representing the four-dimensional code words. This requires a figure in four-dimensional space. An error of one bit position moves us away from the code word along a particular direction corresponding to the bit in error. Since the distance between code words is 2, at least 2 bits must be changed to go from one code word to another. We can draw four-dimensional spheres of radius 1 around each code-word point. Figure 4.5 shows three representative code-word points and their associated spheres. A 1-bit error will take us to some point on the sphere. Such a point on the sphere can be reached by changing one bit in each of four possible code words. Using the 4-bit example presented before as an illustration, assume that 0111 is received. This could have resulted from 1111 being transmitted and an error occurring in the first bit position. It could also result from 0011, 0101, or 0110 with errors occurring in the second, third, or fourth bit positions, respectively. Thus, a single or triple error can be detected but not corrected.

Suppose now that the minimum distance between code words is increased to 3 units. We then see that if a single error is made, the received word will be distance 1

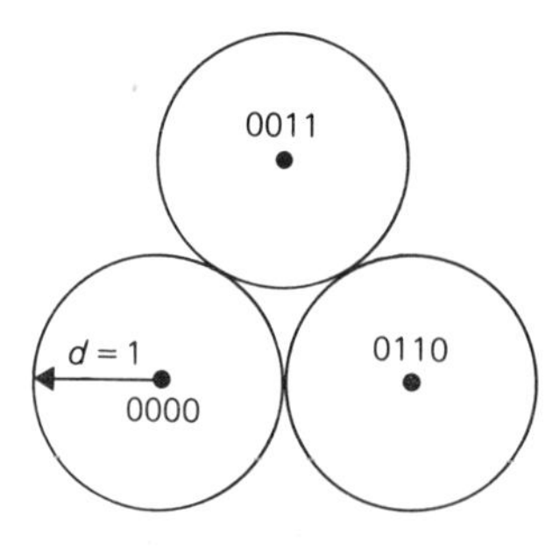

Figure 4.5 Distance for 4-bit code

from the correct code word and at least 2 units of distance from every other code word. An intelligent receiver can therefore correct the single-bit error.

As an example, suppose that the system consisted of three code words, 01111, 10011, and 01000. The minimum distance between code words is 3 units. If, for example, 01111 were transmitted and the second bit were in error, 00111 would be received. The distance from this received word to the second code word is 2, and to the third it is 4. We could therefore easily decide that this should be decoded as the first code word, 01111. The message received is decoded into the closest dictionary word. The same technique is used in conversation.

The student should become convinced that, in general, if the minimum distance between code words is $D_{\min}$, errors involving up to $D_{\min} - 1$ bits can be detected. If $D_{\min}$ is an even number, errors up to $D_{\min}/2 - 1$ can be corrected. If $D_{\min}$ is odd, errors up to $D_{\min}/2 - \frac{1}{2}$ can be corrected. Note carefully the difference between error detection and error correction. For example, if $D_{\min}$ were 6, up to two bit errors can be corrected or up to five bit errors can be detected. One must decide in advance whether error correction or detection is the desired mode of operation. For example, if error correction is being performed in the $D_{\min} = 6$ case, an error in five bit positions would place us only 1 unit away from another code word, and the receiver could erroneously decode the received message as that code word. If error detection is being performed, any received word that is not identical to a code word will be rejected.

Since distance is so critical to error detection and correction, the obvious next question is: How do we increase distance between code words? The simplest way to increase distance is by repetition.

Suppose that we again have eight message words to transmit: 000, 001, 010, 011, 100, 101, 110, and 111. Let us repeat each of these 3-bit groupings to yield the code words 000000, 001001, 010010, 011011, 100100, 101101, 110110, and 111111. The minimum distance between code words is 2 units, and we can therefore detect (but not correct) single errors. This repetition process is not a very efficient method of achieving a distance of 2. Earlier in this section we presented eight code words with a minimum distance of 2 using only 4-bit words. It would be a waste of transmission time to use 6-bit words if 4-bit words can accomplish the same error detection. In fact, the longer word would also have a greater probability of a single-bit error during transmission.

Code length. Viewing Fig. 4.4 and its generalized version in Fig. 4.5, it should not be surprising to find that given the length, n, of code words (and therefore the dimension of the space), there is a limit to how many code words exist with a given minimum distance. This is analogous to a supervisor's job in assigning floor space to engineers in a nonprogressive company that uses large rooms. Given that the assignment is in two dimensions (i.e., every engineer will keep his or her feet on the floor), and given a minimum spacing between desks, there is a maximum number of engineers that can be put into the room.

A straightforward theorem attributed to Hamming states that the number of code words, M, of length n with capability of correcting E errors (i.e., minimum distance greater than $2E$) is bounded by the following:

$$M \leqslant \frac{2^n}{\sum_{i=0}^{E} \binom{n}{i}}$$

where

$$\binom{n}{i} = \frac{n!}{i!(n-i)!}$$

The proof relates to a term called *sphere packing* (the reason for this name will become obvious). Viewing Fig. 4.5, we see that the question can be restated as follows: Given that the radius of each sphere is E and that the spheres do not intersect (otherwise, E errors could not be corrected), how many spheres can be packed into the space?

With code length equal to n, there are 2^n possible binary code words. We first ask how many of the 2^n possible code words can be in any one sphere. Given that the code word at the center of the ith sphere is designated c_i, all code words with distances of up to E from c_i are in the sphere. There are n words with distance 1, $\binom{n}{2}$ with distance 2, and so on up to $\binom{n}{E}$ with distance E. Thus, the total number of code words within one sphere is

$$\sum_{i=0}^{E} \binom{n}{i}$$

Now if there are M spheres, the total number of code words within all of these spheres is

$$M \sum_{i=0}^{E} \binom{n}{i}$$

The total number of possible (distinct) code words of length n is 2^n, so the total number within the spheres cannot exceed this number. We are using the fact that the spheres cannot intersect if we are to be able to correct E errors. Therefore,

$$M \sum_{i=0}^{E} \binom{n}{i} \leqslant 2^n$$

and the theorem is proven.

Algebraic codes. We have already seen several possible codes that achieve some level of error identification. We now investigate an organized technique for formulating code words and a complementary organized technique for recovering the original word and identifying errors at the receiver. The class of codes we consider next is known as *algebraic codes*.

The one-error-detecting code presented early in this section was a simple form of algebraic code known as a *single-parity-bit check code*. We repeat the message and code words for that example in the following table:

Message	Code Word
000	0000
001	0011
010	0101
011	0110
100	1001
101	1010
110	1100
111	1111

Note that the coding process can be described as an addition of a fourth bit to the message word such that the total number of 1's in any code word is even. The receiver simply checks to see that the total number of 1's is even and, if so, ignores the fourth bit in each code word. If the number of 1's is odd, the receiver rejects the word, since a bit error is present. This rejection is known as an *erasure* and would normally require that the word be retransmitted. We thus can detect but not correct one bit error. Note that if two bit errors occur, the parity will remain even and the receiver will incorrectly decode the word. The probability of two bit errors is often small enough to ignore.

Example 4-6

Three-bit message words are encoded by adding a fourth parity-check bit. The coded words are transmitted in a manner which results in a bit-error probability of 5×10^{-3}. Find the probability of an undetected error at the receiver.

Solution. An undetected error will occur in a message if a particular 4-bit code word experiences either two or four bit errors. If either one or three bit errors occur, the parity test will fail, and the receiver will detect an error. The bit error rate is 5×10^{-3}. The probability of any particular 2 bits being in error is therefore 25×10^{-6} (we assume independence). Since there are six possible combinations of 2 bits in any 4-bit word (i.e., first and second bit, first and third bit, first and fourth bit, second and third bit, second and fourth bit, and third and fourth bit), the probability of two bit errors is

$6 \times (5 \times 10^{-3})^2 \times (1 - 5 \times 10^{-3})^2$. The probability of four bit errors is 625×10^{-12}. The probability of either two or four bit errors is therefore approximately 1.49×10^{-4}. In this system, we can expect about three incorrectly decoded messages in every 20,000 transmitted messages.

In a similar manner, we can calculate the probability of a detected error (i.e., 1- or 3-bit errors) at the receiver. The student should verify that this probability is approximately 2×10^{-2}. We can therefore expect about three detected errors out of every 100 transmitted messages. Putting these two results together, we see that in any group of 10,000 transmitted messages, we can expect about three messages to be incorrectly decoded, about 280 messages to result in parity errors (erasures), and the remaining 9717 messages to be correctly decoded.

We now generalize this type of coding. Suppose that the message words consist of m bits (in the example, above $m = 3$). We therefore have a set of up to 2^m distinct message words. We consider code words that add n parity bits to the m message bits to end up with code words of length $m + n$ bits. Thus, if a_i is an original message bit and c_i is a parity check bit, each code word will be of the form

$$a_1 a_2 a_3 \cdots a_m c_1 c_2 c_3 \cdots c_n$$

Note that of the 2^{m+n} possible code words, only 2^m are used.

Each check bit is chosen to achieve even parity when combined with specific message bits. For example, c_1 can be chosen to yield an even number of 1's when combined with a_1, a_3, a_4, and a_6; c_2 can be chosen to yield an even number of 1's when combined with a_1, a_2, and a_5; and so on. Some elementary Boolean algebra and matrix skills permit us to express this relationship in a general format. The check bits are chosen in such a way as to satisfy

$$[H]\bar{T} = 0 \tag{4-7}$$

In Eq. (4-7), $\bar{T}$ is the $m + n$ column vector

$$\bar{T} = \begin{bmatrix} a_1 \\ a_2 \\ a_3 \\ \cdot \\ \cdot \\ \cdot \\ a_m \\ c_1 \\ c_2 \\ \cdot \\ \cdot \\ \cdot \\ c_n \end{bmatrix}$$

and $[H]$ is the $n \times (n + m)$ matrix

$$[H] = \begin{bmatrix} h_{11} & h_{12} & \cdots & h_{1m} & 1 & 0 & 0 & 0 & \cdots & 0 \\ h_{21} & h_{22} & \cdots & h_{2m} & 0 & 1 & 0 & 0 & \cdots & 0 \\ \cdot & \cdot & \cdots & \cdot & \cdot & \cdot & \cdot & \cdot & \cdots & 0 \\ h_{n1} & h_{n2} & \cdots & h_{nm} & 0 & 0 & 0 & 0 & \cdots & 1 \end{bmatrix}$$

Note that the $n \times n$ matrix formed by partitioning the right part of $[H]$ is an identity matrix. $h_{11}, h_{12}, \ldots, h_{nm}$ are each either equal to 0 or 1. The sums in Eq. (4-7) are taken using modulo-2 addition. Thus, if the number of 1's is even, the sum is zero. To appreciate Eq. (4-7) more fully, we expand the first two rows:

$$h_{11}a_1 + h_{12}a_2 + \cdots + h_{1m}a_m + c_1 = 0$$

$$h_{21}a_1 + h_{22}a_2 + \cdots + h_{2m}a_m + c_2 = 0$$

Example 4-7

Given the algebraic code with 4-bit message words and 3 parity-check bits, $[H]$ is defined as

$$[H] = \begin{bmatrix} 1 & 1 & 0 & 1 & 1 & 0 & 0 \\ 1 & 0 & 1 & 1 & 0 & 1 & 0 \\ 0 & 1 & 1 & 1 & 0 & 0 & 1 \end{bmatrix}$$

Find the code words corresponding to the 16 possible message words.

Solution. The three equations corresponding to the matrix equation $[H]\bar{T} = 0$ are

$$a_1 + a_2 \qquad + a_4 + c_1 = 0$$

$$a_1 \qquad + a_3 + a_4 + c_2 = 0$$

$$a_2 + a_3 + a_4 + c_3 = 0$$

Thus, c_1 is chosen to achieve even parity when combined with the first, second, and fourth bits; c_2 is chosen to achieve even parity when combined with the first, third, and fourth bits; c_3 achieves even parity when combined with the second, third, and fourth message bits. The following code words result:

Message Word	Code Word
0000	0000000
0001	0001111
0010	0010011
0011	0011100
0100	0100101
0101	0101010
0110	0110110
0111	0111001

Message Word	Code Word
1000	1000110
1001	1001001
1010	1010101
1011	1011010
1100	1100011
1101	1101100
1110	1110000
1111	1111111

We now transmit the code words. The receiver forms the product of the received word with the matrix [H], and if this product is not equal to zero, the receiver knows that at least one error was made in transmission. That is, we know that $[H]\bar{T} = 0$ for all acceptable code words. Therefore, if the product of [H] with the received vector is not zero, we know that word received is not one of the acceptable code words. Of course, one would hope that this elaborate system will do more than simple error detection. We were able to accomplish error detection with only 1 parity bit.

The value of algebraic coding arises in multiple-error detection and in error correction. Suppose that the transmitted vector is denoted as $\bar{T}$. We define an error vector $\bar{E}$ which contains a "1" in each bit position in which an error occurs. The received vector is therefore of the form

$$\bar{R} = \bar{T} + \bar{E}$$

where, again, the addition is modulo-2. Error correction is possible if the vector $\bar{E}$ can be isolated at the receiver. If we multiply the received vector by the matrix [H], we have

$$[H]\bar{R} = [H](\bar{T} + \bar{E}) = [H]\bar{T} + [H]\bar{E} = [H]\bar{E}$$

In order to arrive at the final term in this equation we used the fact that $[H]\bar{T} = 0$, since $\bar{T}$ is an acceptable code word. The product $[H]\bar{E}$ is defined as $\bar{S}$, the *syndrome*. The dictionary defines syndrome as "a number of symptoms occurring together and characterizing a specific disease or condition." In fact, the error syndrome characterizes the specific bit error. Let us first assume that a single bit error occurs. $\bar{E}$ would then be a vector composed of a single "1," and the remaining positions would be zero. If we take the product of this with [H] to form the syndrome, the result is a vector that is identical to one column of [H], that column being the one corresponding to the bit position in error.

Example 4-8

Given the [H] defined in Example 4-7, a message word 1010 is transmitted. A bit error occurs in the fourth bit position. Find the syndrome.

Solution. The transmitted code word is 1010101, and the received word is 1011101. We form the product $[H]\bar{R}$ to get

$$[H]\bar{R} = \begin{bmatrix} 1 & 1 & 0 & 1 & 1 & 0 & 0 \\ 1 & 0 & 1 & 1 & 0 & 1 & 0 \\ 0 & 1 & 1 & 1 & 0 & 0 & 1 \end{bmatrix} \begin{bmatrix} 1 \\ 0 \\ 1 \\ 1 \\ 1 \\ 0 \\ 1 \end{bmatrix} = \begin{bmatrix} 1 \\ 1 \\ 1 \end{bmatrix}$$

We note that the result is equal to the fourth column of $[H]$, thus identifying an error as having occurred in the fourth bit position.

In order to correct a single error using this technique, it is necessary that $[H]$ contain no duplicate columns. It is also necessary that no column of $[H]$ be composed of all 0's, since if the syndrome were a vector composed of all 0's, we would not know whether we were seeing that column or the result of no errors.

If an error of more than one bit error occurs, the syndrome will be equal to the sum of the corresponding columns of $[H]$. If, for example, errors occurred in the third and fourth bits of the example above, the syndrome would be

$$\bar{S} = \begin{bmatrix} 1 \\ 0 \\ 0 \end{bmatrix}$$

This would be incorrectly interpreted as a single bit error in the fifth bit position. For the present, we restrict our discussion of algebraic decoders to the single-error-correction case. However, note that the syndrome can correct two errors, provided that no two columns sum to either zero or to another column or to the sum of two other columns.

Algebraic codes are not difficult to implement. Calculation of the various vector products is equivalent to summing combinations of bit positions. This is accomplished by entering the codes into shift registers, tapping off at the appropriate positions, and feeding these outputs into a summation device. Those readers familiar with digital processing know that modulo-2 addition of two inputs is performed in the EXCLUSIVE OR logic block.

Figure 4.6 shows the implementation of the vector product of (1011010) and a received word. This represents the second of three such multiplications required for the receiver in Example 4-8.

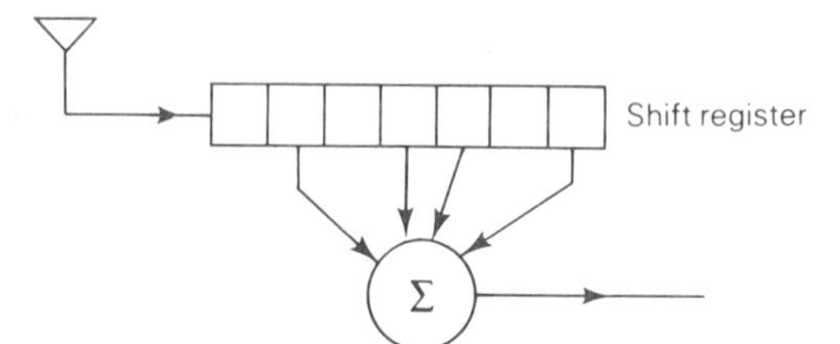

Figure 4.6 Shift-register implementation of algebraic code

As each bit enters a storage block in Fig. 4.6, the previous contents of that block are shifted one position to the right. For example, if the shift register contains the binary number 1011001 and a binary zero enters the left-hand block, the new contents of the register will be 0101100.

We have seen that a single error can be located, and therefore corrected, by using the syndrome. Correction of more than one error would involve finding code words with minimum distances greater than 4 units. If we are successful in devising such codes, the basic decoding technique involves finding the distance between a received word and each of the acceptable code words. The received word is then decoded into the code word which is the minimum distance from the received word (i.e., the closest word).

4.4.3 Convolutional Coding

The improvement in error performance for block encoding is realized when redundancy is added. That is, parity bits are appended to the message in order to increase the distance between code words, thereby providing for error detection and/or correction. In order to increase error-correction capability, it is necessary to increase the amount of redundancy. This increased redundancy means that additional transmitted energy must be expended in order to provide for error detection and correction.

An alternative to block encoding is convolutional coding. In this type of code, we no longer consider individual blocks of bits as code words. Instead, a continuing stream of information bits is operated upon to form the coded message. The source generates a continuing message sequence of 1's and 0's, and the transmitted sequence is generated from this source sequence. The transmitted sequence need not be any longer than the message sequence. That is, the technique does not add redundancy bits. Instead, it achieves error-correction capability by introducing memory into the system. Message bits are combined in a particular manner in order to generate each transmitted bit. The convolutional code is therefore capable of providing error correction without requiring extra transmission energy to send additional redundancy bits.

The technique of generating the transmitted sequence is to *convolve* the source sequence with a fixed binary sequence. Thus, a particular transmitted bit, t_n, is generated from a combination of source bits, $s_n, s_{n-1}, s_{n-2}, \ldots, s_{n-k}$, according to the convolution equation,

$$t_n = \sum_{k=-\infty}^{n} s_k h_{n-k} \tag{4-8}$$

The h's in Eq. (4-8) are either 1 or 0, and the addition is modulo-2. The equation can be implemented with a shift register and a modulo-2 adder.

Figure 4.7 shows a general implementation of Eq. (4-8). The switches in the figure are closed if the associated h in Eq. (4-8) is 1, and open if it is 0.

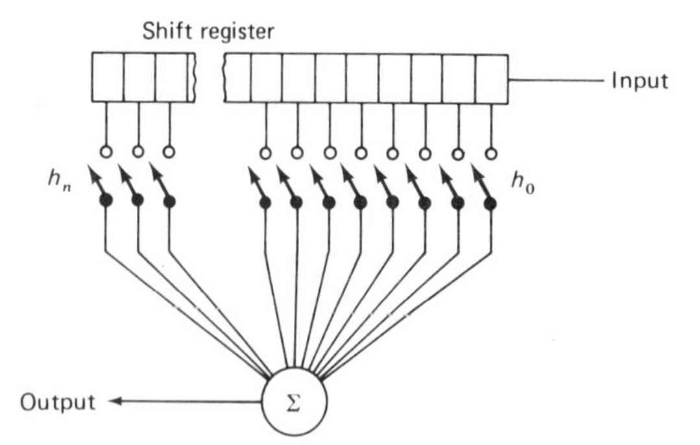

Figure 4.7 Shift-register implementation of digital convolution

In the application of convolutional coding we often transmit more than 1 bit for each input bit. Thus, referring to Fig. 4.7, we may shift in 1 input bit, set the switches to correspond to a particular set of h's, and generate the first output bit. Then, before feeding in another input bit, we reset the switches to correspond to a second set of h's and transmit a second bit. If 2 output bits are transmitted for each input bit, the code is referred to as a *rate-$\frac{1}{2}$ convolutional code.* In sending a rate-$\frac{1}{2}$ convolutional code, one often chooses the first bit of each transmitted pair to be identical to the information (input) sequence. This would result from setting $h_0 = 1$ and $h_n = 0$ for $n \neq 0$. In such cases, the code is known as a *systematic code.*

Example 4-9

As an example of a rate-$\frac{1}{2}$ nonsystematic convolutional code, consider a 3-bit shift register with the first set of h's given by 101 and the second set by 011. This is illustrated in Fig. 4.8. If the input sequence is 11010010000 . . . (the most recent bit is on the right), find the output sequence.

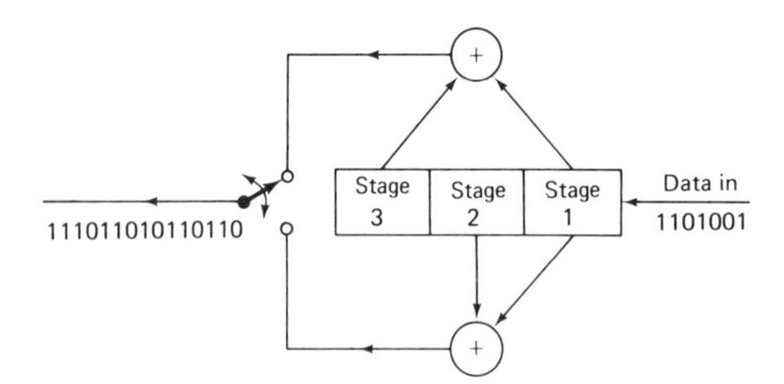

Figure 4.8 Convolutional code generator for Example 4-9

Solution. The output sequence would be

1 1 1 0 1 1 0 1 0 1 1 0 1 1 0 1 1 0 0 0 0 0 0 . .

To arrive at this output sequence, we have assumed that the system is initialized with a string of zeros prior to receiving the first bit of the input sequence and that the last input bit is followed by a string of zeros. We find the first, third, fifth, and so on, bits in the output sequence by adding the current input bit to the input bit two periods in the past (corresponding to an h of 101). To find the second, fourth, sixth, and so on, output bits, we add the current bit to the immediately preceding bit (corresponding to an h of 110).

Keeping track of the input bits, shift register storage bits, and convolutions has a way of raising one's blood pressure. (The solution to Example 4-9 went through three iterations, and will probably have to be corrected in the next printing of this text.) It is desirable to find alternative, and perhaps more graphic, ways of presenting the coding operation. One such method is the *tree,* which indicates the output sequence for *every* possible input sequence. It would appear that with sequences of length N, as N increases the tree will spread its branches geometrically. That is, with each unit increase in N, the number of input combinations doubles. However, we are saved by the observation that the output depends only upon a limited number of input bits, that number being equal to the number of storage elements in the generating shift register. The tree therefore becomes repetitive after this number of levels of branches.

Figure 4.9 shows the tree for the encoder of Example 4-9. Starting at point A, the first input bit determines whether branching takes place to B or C. If the input bit is a 0, the upward path is taken, arriving at point B. The corresponding output is 00, which is indicated on that branch. If the input is a 1, a transition to point C occurs and the output is 11. Once at either B or C, inputs of 0 or 1 cause transitions to either D, E, F, or G. The process continues. For example, assuming an input of 1101 (the start of the input of Example 4-9), we trace a path from A to C, G, J, and M with output 11101101 (thankfully agreeing with the example). Note that at each junction, the two paths yield complementary outputs. That is, if an input of 1 at a particular point yields an output of 10, then a 0 input gives 01. This is so because, for this particular example, each of the 2 output bits is formed as a sum that includes the most recent bit. Therefore, if all previous bits remain the same while the most recent bit changes from 0 to 1, each output would be expected to change.

The tree would not be of much use if it kept spreading its branches by a factor of 2 for each movement of one level to the right. However, it can be observed that after three levels of splitting, the top half of the tree is identical to the bottom half. Figure 4.9 takes this into account by having two points labeled H and two labeled I, J, and K. Once arriving at any one of these points, the output for a given succeeding input sequence will not depend upon which of the two equivalent points forms the starting point. Similarly, at the next level there is another split. Therefore, once the system settles down, there are only four significant *states.* We will clarify this concept in a moment.

Because of the combinations of points in the tree, it is often more desirable to simplify the structure to arrive at which is known as a *trellis.* The trellis corresponding to the tree of Fig. 4.9 is presented as Fig. 4.10. The points have been labeled to correspond to those in the tree, and outputs are again indicated on each branch. At each point, the trellis splits into two paths, the upper representing a 0 input, and the

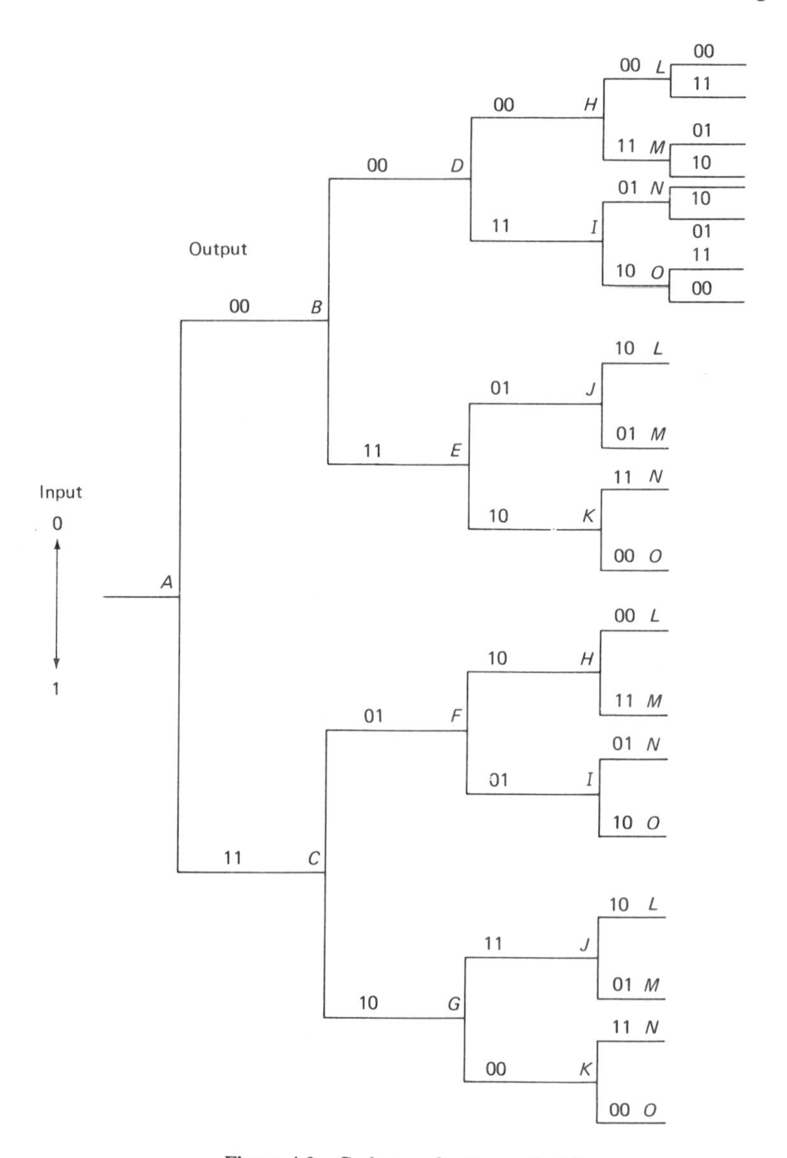

Figure 4.9 Code tree for Example 4-9

lower, a 1. You should study Fig. 4.10 until you are convinced that it is equivalent to Fig. 4.9. If you do this in less than 5 minutes, you are kidding yourself.

We now make the very important observation that all points in a horizontal row of the trellis exhibit equivalent behavior. For example, once at any one of *A*, *B*, *D*, *H*, or *L*, a subsequent input sequence will produce the same output regardless of which of these five points we start at. The same is true of the combinations *CEIM*,

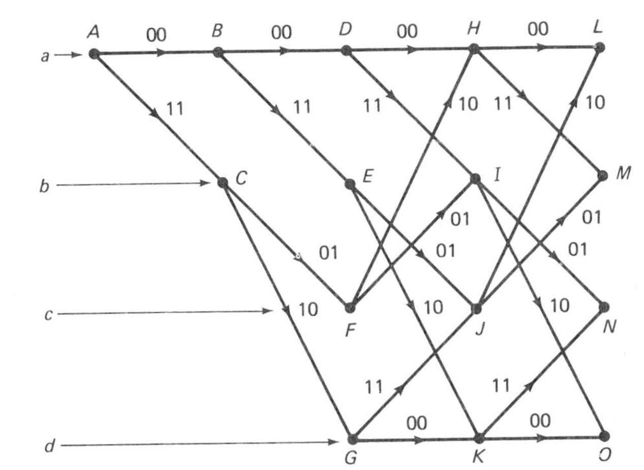

Figure 4.10 Trellis for Example 4-9

FJN, and *GKO*. Note that although we stopped the trellis at *LMNO*, we could have continued ad infinitum, and the structure would have repeated over and over again. It is therefore possible to consider this system as one possessing four states. The states represent the system memory, and the fact that there are four states results from the observation that the two previous input bits contain all the relevant history that affects the output due to a particular input bit. There are four possible combinations of these two *history* or *memory* bits.

We can convey the same information as is in the trellis by drawing a *state-transition diagram.* The four states are indicated as nodes, a, b, c, and d, corresponding to the four horizontal levels in the trellis. The state-transition diagram is shown in Fig. 4.11. Thus, state a is equivalent to points A, B, D, H, and L in the trellis, state b is equivalent to C, E, I, and M, and so forth.

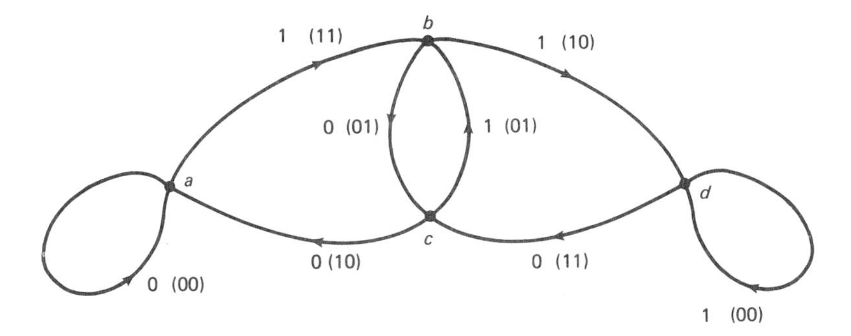

Figure 4.11 State-transition diagram for Example 4-9

Once at a state, there are two possible input bits. For each of these two inputs, there are two corresponding output bits. The transitions are indicated with directed arrows, with the input bit indicated beside the arrow. The corresponding output bits are in parentheses adjacent to the input bit. For example, starting at state b, an input of 1 causes a transition to state d with output 10. Note that an input of 0 while in state a, or 1 while in state d, does not change the state, and results in an output of 00. The student should redo Example 4-9 using the state-transition diagram to (hopefully) arrive at the same output sequence. Note that it should be assumed that the system starts in state a, since all prior inputs were zero. If for some reason all prior inputs had been 1, the steady state would be state d.

Example 4-10

Sketch the tree, trellis, and state-transition diagram for the convolutional code generated by the shift register shown in Fig. 4.12. This is a rate-$\frac{1}{2}$ code.

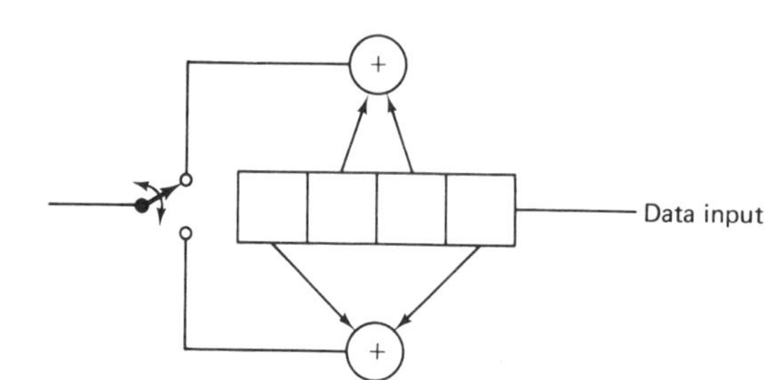

Figure 4.12 Convolutional code generator for Example 4-10

Solution. The tree is found by shifting the various inputs into the shift register and solving for the outputs of the two summing devices. The resulting tree is drawn in Fig. 4.13. The trellis is found from the tree by recognizing that the output depends upon the current input bit and the 3 bits immediately prior to this input. There are thus eight unique states in this system. The trellis is shown in Fig. 4.14. Each horizontal line in the trellis defines a state. It can be seen that the behavior leaving any point along the same line is repeated at every other point. The state-transition diagram can then be derived from the trellis. This is shown in Fig. 4.15.

Let us not lose sight of the coding problem. Hopefully, the convolutional code, in addition to being relatively easy to implement in a bit-by-bit format, also possesses error-correcting and -detecting capability. It is therefore necessary to examine the *distance* between acceptable code words. If two input sequences differ only in the last (most recent) bit, the code words will differ only in the most recent K bits, where K is the reciprocal of the rate of the code. That is, if two input sequences are identical except for the most recent bit, the coded output sequences will be identical until the coder arrives at that last input bit. If a rate-$\frac{1}{2}$ code is used, 2 bits are generated from that last input bit, and those 2 bits will be complementary (if the most recent bit is used in all combinations, as in Example 4-8). This result is not surprising, since the output coded sequence is only a function of the most recent input bit and the previous state of the system. To be fair to convolutional codes, we

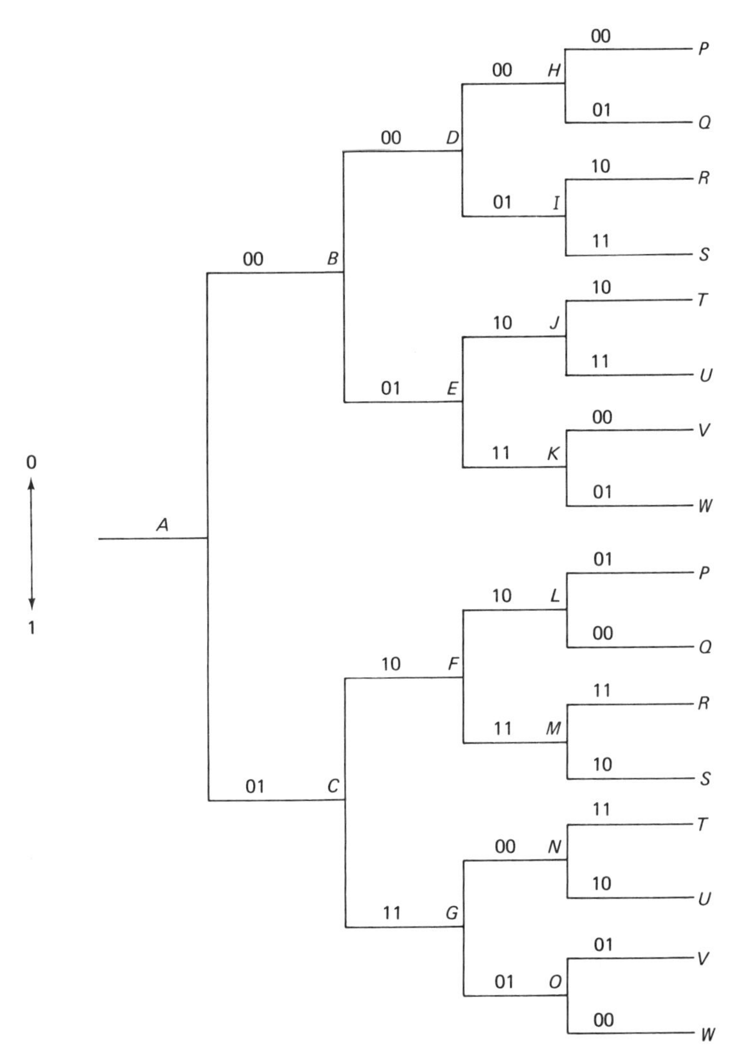

Figure 4.13 Code tree for Example 4-10

usually modify the distance concept slightly by comparing code words whose input sequences have different first (oldest) bits. For example, considering 3-bit inputs in the code displayed in Fig. 4.9, we make the initial assumption that the first input bit differs among the two code words. Therefore, to find the minimum distance between acceptable code words, we compare code words that start with states *AB* with those starting with states *AC*. This clearly gives a head start of a distance of 2. We could examine all eight possible output sequences (each 6 bits long), and find distances between all pairs, taking one from the top half of the tree and one from the bottom half. We would thus make 16 comparisons and would find a minimum dis-

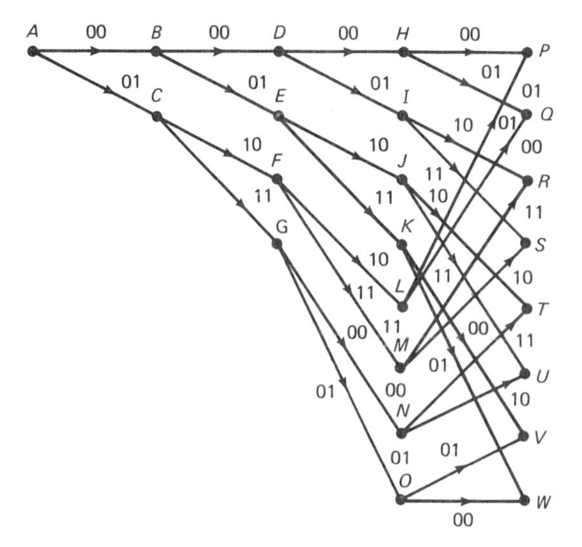

Figure 4.14 Trellis for Example 4-10

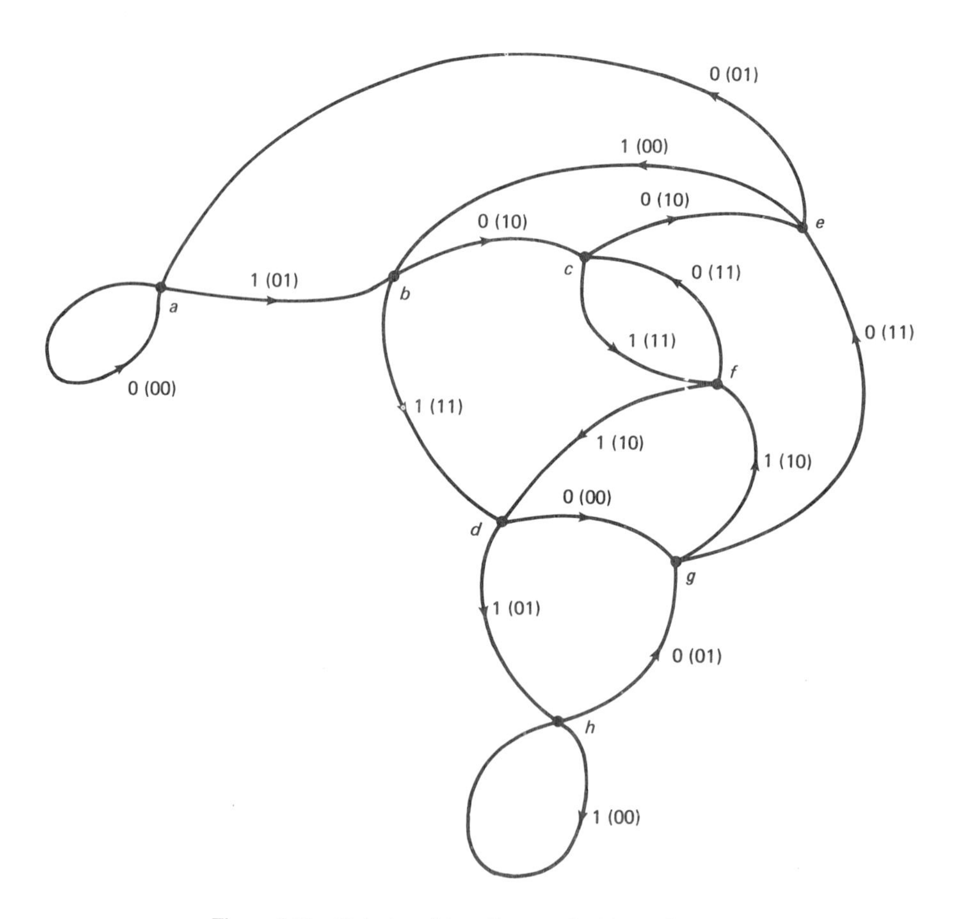

Figure 4.15 State-transition diagram for Example 4-10

tance of 3. It can be shown that this is the same as examining the four paths starting with *AC* and finding the *weight* (number of 1's) of each of these four output sequences. The minimum distance between words is equal to this minimum weight (3 for the path *ACGK*). Therefore, the code presented can correct one error and detect two errors, provided that the first-bit input constraint is observed.

We should note that the analysis above assumed that the input sequences were only 3 bits in length. In contrast to block encoding, convolutional coding deals with the input bit by bit, so the total message length must be considered as the input length. This will normally be considerably greater than 3 bits.

Decoding. Convolutional codes can be decoded in a manner not unlike that used for algebraic codes. A *syndrome* is computed at the receiver and used to operate upon the received sequence to recover the message sequence. In that manner, we can treat codes with rates below 1 as if parity bits were being added. The student is referred to the references for details. We shall concentrate upon an elegant and simple decoding technique known as the *Viterbi algorithm.* In essence, this algorithm is equivalent to comparing the received sequence to all possible transmitted sequences and choosing the sequence that is closest (minimum distance).

The algorithm is easily implemented and relates directly to the trellis presented earlier in this section. Referring back to Example 4-9 and the trellis of Fig. 4.10, suppose that the received sequence were 10101010. To decode this, we start the algorithm by redrawing the trellis, indicating the distance between each pair of received bits and the corresponding bits on the trellis. This results in the trellis of Fig. 4.16. As an example, the 1 on path *AB* results from comparing the first two received bits, 10, to 00 as indicated in Fig. 4.10. These two differ in one bit position. We now examine various paths from the start of the trellis and add the distances. These cumulative distances are indicated in brackets adjacent to each node. Thus, to reach point *E*, we go through a total distance of 2. At the third and subsequent levels, there are two possible paths leading into each node, generating two (sometimes equal) cumulative distances. Although both of these are written in brackets (the first entry corresponds to the upper path), at this point we reject the path with the greatest distance. Thus, if there are two input sequences that get us to the same intermediate point (state), we decode as the sequence with the minimum resulting distance. As an example, although point *I* can be reached by a path with distance 3 (*ABDI*) from the received word and another path with distance 5 (*ACFI*), we reject the greater distance path (*ACFI*) and continue from this point maintaining the distance 3 path. Note if two paths have equal distance, either one can be chosen. It can be shown that the probability of decoding error will be the same for either of these equal-distance paths. This path-rejection step is significant, since without it we would be calculating distance for every possible transmitted word. The number of words we would have to check distances for increases geometrically with the length of the sequence. Because of the path-rejection techniques applied at each level of received bit combinations, we must carry forth only four cumulative distances to

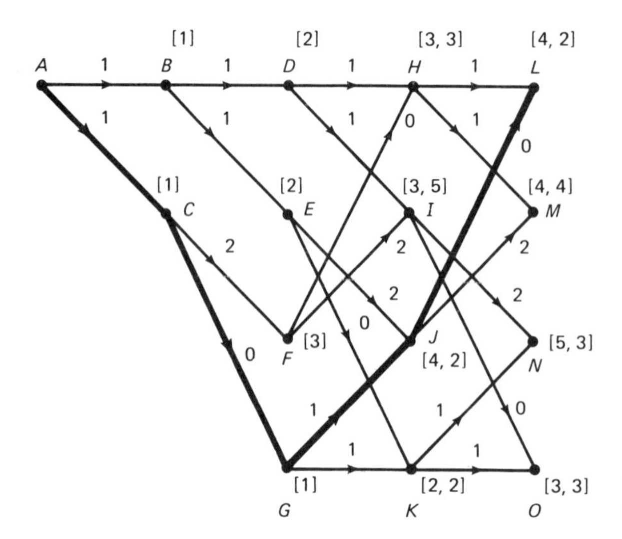

Figure 4.16 The Viterbi algorithm for convolutional decoding

the next level. Scanning the output, we find one path with distance 2, and this one is chosen. This corresponds to the path *ACGJL* through the trellis and is indicated by a heavy line in Fig. 4.16. That path is generated with an input 1100, which is the result of the decoding operation. As a check, we note that an input of 1100 would be coded into 11101110, which has distance 2 from the received word.

4.4.4 Cyclic Coding

A *cyclic code* is a *parity-check code* with the property that a *cyclic shift* of any code word results in another acceptable code word. A cyclic shift is a shift of each bit by one position, where the end bit is folded around to the other end. Thus, one possible cyclic shift of 1011011 would be 0110111.

Cyclic codes can be formed by starting with parity-check matrices that have cyclicly related rows. For example, the following three matrices would generate cyclic codes.

$$[H_1] = \begin{bmatrix} 000101 \\ 001010 \\ 010100 \\ 101000 \end{bmatrix}, \quad [H_2] = \begin{bmatrix} 110 \\ 101 \\ 011 \\ 110 \end{bmatrix}, \quad [H_3] = \begin{bmatrix} 110110 \\ 101101 \\ 011011 \\ 110110 \end{bmatrix}$$

Note that $[H_3]$ is formed by repeating $[H_2]$ twice.

Example 4-11

Find the code words associated with each of the three matrices presented above.

Solution. $[H_1]$ represents a code with 4 parity bits and 2 message bits, thus having four possible code words. These words are

000000, 111111, 101010, 010101

Note that a cyclic shift of any word results in an acceptable code word.

Although it would appear that $[H_2]$ defines a code with 4 bits, the code words can only have total length 3. The matrix is degenerate, as the fourth row is a duplicate of the first and the third row can be derived by adding the first two rows. Thus, writing the parity equations results in only two independent equations, and the code has 2 check bits and 1 information bit. The two code words are

000 and 111

Finally, the code associated with $[H_3]$ has 2 check bits (same reasoning as $[H_2]$) and 4 message bits. The 16 possible codes words are

000000	111111	
110110	011011	101101
110001	111000	011100
001110	000111	110001
101010	010101	
100100	010010	001001

A cyclic shift of any of these 16 words results in one of the code words.

Analysis of cyclic codes is facilitated using some results from linear algebra. We define the *characteristic polynomial* of a square matrix, $[H]$, as

$$\psi(\lambda) = |\,[H] - \lambda[I]\,| \tag{4-9}$$

$[I]$ is the identity matrix, and the right side of this equation is the determinant.

Example 4-12

Find the characteristic polynomial of the matrix

$$[H] = \begin{bmatrix} 0100 \\ 0010 \\ 0001 \\ 1010 \end{bmatrix}$$

Solution. Equation (4-9) becomes

$$|\psi(\lambda)| = \begin{vmatrix} -\lambda & 1 & 0 & 0 \\ 0 & -\lambda & 1 & 0 \\ 0 & 0 & -\lambda & 1 \\ 1 & 0 & 1 & -\lambda \end{vmatrix} = \lambda^4 + \lambda^2 + 1$$

where all the arithmetic is modulo-2.

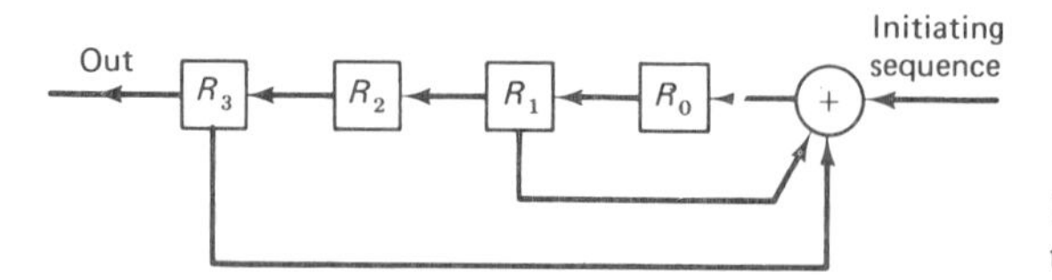

Figure 4.17 Shift-register implementation of cyclic code

A shift-register implementation of the code can be derived from the characteristic polynomial. Powers of λ indicate orders of shift involved. Thus, the polynomial of Example 4-12 can be implemented using the shift register of Fig. 4.17. Each block in this figure is a storage cell which holds a binary number that it receives at its input (right side) for one period. It then transfers that number to the left. For example, if we start with 0101 stored in the storage cells, the output of the summing device is 0, and at the next clock time, the storage cells contain 0010. This yields a summer output of 0, yielding 0001 as the next set of states. We continue with states 1000, 0100, 1010, and 0101 and thus develop a cyclic code.

We now take a few moments to present key results from *linear algebra.* These will prove important in understanding characteristic polynomial properties as they relate to the PN code. We will also find these concepts useful later during a discussion of cryptography in Chapter 12. Finally, the student who approaches the technical literature in the area of coding cannot avoid linear algebraic concepts. We begin with some key definitions.

A *commutative group* is a set of elements combined with an operation we shall denote as $\times$, which satisfies the following five rules:

1. *Closure:* If a and b are in the group, then $a \times b$ is also in the group.
2. *Identity, I:* If a is in the group, then

$$a \times I = I \times a = a$$

3. *Inverse:* For every element in the group a, there exists another element, a^{-1}, such that

$$a \times a^{-1} = I$$

4. *Associative law:* For every set of three elements, a, b, and c,

$$(a \times b) \times c = a \times (b \times c)$$

5. *Commutative law:* For any two elements,

$$a \times b = b \times a$$

A *field* is a set of at least two elements and two operations (call them + and $\times$) such that

1. The set forms a commutative group under one of the operations.

2. The set forms a commutative group under the second operation except that no inverse exists.
3. The set is distributive as follows:

$$a \times (b + c) = (a \times b) + (a \times c)$$

Note that the binary numbers form a field, as we can check as follows:

1. Closure is obeyed under both multiplication and addition, since if any two binary numbers are added or multiplied, the result is a binary number. That is,

$$1 + 1 = 0, \qquad 1 + 0 = 0 + 1 = 1$$

$$0 + 0 = 0, \qquad 0 \times 1 = 1 \times 0 = 0$$

$$1 \times 1 = 1, \qquad 0 \times 0 = 0$$

2. The identity element for addition is 0 and for multiplication is 1.
3. The inverse of 1 under addition is 1, since

$$1 + 1 = 0$$

The inverse of 0 under addition is 0, since

$$0 + 0 = 0$$

0 has no inverse under multiplication, since there is no element which can multiply 0 to yield 1.
4. The associative law holds for both operations, as an exhaustive test will show.

A finite field consisting of q elements is known as a *Galois field* (after the French mathematician Galois) and is designated GF(q). The binary field with the two operations of multiplication and addition is therefore known as GF(2).

We now define *division* of polynomials as follows:

$a(x)$ and $b(x)$ are polynomials with coefficients drawn from a field, F. There exist a $q(x)$ and an $r(x)$ such that

$$a(x) = q(x)b(x) + r(x)$$

$q(x)$ is the *quotient* and $r(x)$ the *remainder.* The degree of $r(x)$ is less than the degree of $b(x)$.

We now define operations *modulo a polynomial* much as standard modulo arithmetic is defined. For example,

10 modulo 4

is the remainder when 10 is divided by 4, or 2. $a(x)$ *modulo* $p(x)$ is the remainder, $r(x)$, that results when $a(x)$ is divided by $p(x)$.

Example 4-13

Find $x^3 + 1$ modulo $x^2 + 1$ in GF(2).

Solution. We divide $x^3 + 1$ by $x^2 + 1$ to get

$$\begin{array}{r} x \\ x^2 + 1 \overline{)x^3 + 1} \\ \underline{x^3 + x} \\ 1 - x \end{array}$$

so the remainder is $1 - x$, but since the coefficient -1 is not acceptable in GF(2), we add 2 to this coefficient to get $1 + x$ as the remainder. As a check, we see that

$$x(x^2 + 1) + (x + 1) = x^3 + 1$$

in GF(2), which checks.

We need just one more concept and then we are finished. $a(x)$ is *reducible* (factorable) *over* GF(q) if there exist a $b(x)$ and a $c(x)$ each of degree greater than or equal to 1, such that

$$a(x) = b(x)c(x)$$

If such a factoring does not exist, we say that $a(x)$ is *irreducible over GF(q)*.

The coding we are about to discuss is related to these polynomial rules if we draw a one-to-one correspondence between code vectors and polynomials in t. Thus, given the code word

$$(c_0, c_1, c_2, \ldots, c_{n-1})$$

we express this as the polynomial

$$c(x) = c_{n-1}x^{n-1} + c_{n-2}x^{n-2} + \cdots + c_1x + c_0$$

This polynomial is then used to denote the n-bit code word. In a similar manner, the information block of k bits is expressed as the data polynomial

$$d(x) = d_{k-1}x^{k-1} + d_{k-2}x^{k-2} + \cdots + d_1x + d_0$$

The code words are related to the uncoded words by the generating matrix. The rows of the generating matrix can be related to a *generating polynomial.*

$$g(x) = x^{n-k} + g_{n-k-1}x^{n-1-1} + g_{n-k-2}x^{n-k-2} + \cdots + 1$$

The code words are then found by multiplying the polynomial representing the uncoded word by the generating polynomial. That is,

$$c(x) = d(x)g(x)$$

Note that all operations are performed modulo-2.

If the generating polynomial is a factor of the polynomial

$$f(x) = x^n + 1$$

the resulting code will be cyclic.

Example 4-14

Generate a (7, 4) cyclic code.

Solution. Since the length of the code words is 7 bits, we start with the polynomial

$$f(x) = x^7 + 1$$

This can be factored into

$$f(x) = (x + 1)(x^3 + x + 1)(x^3 + x^2 + 1)$$

Any of the factors can be used to form a cyclic code. However, since the length of the message blocks is specified as 4, we need a polynomial of order $n - k = 3$. We can choose either of these two factors. Suppose we choose $(x^3 + x + 1)$. The code words are then generated by multiplication. For example, the coded word associated with 1001 is found by taking the product,

$$(x^3 + 1)(x^3 + x + 1) = x^6 + x^4 + x + 1$$

Note we have used the fact that in modulo-2 arithmetic, $x^3 + x^3 = 0$. The code word is then given by 1010011. In a similar manner, the code is developed:

Message	Code Word
0000	0000000
0001	0001011
0010	0010110
0100	0101100
1000	1011000
0111	0110001
1110	1100010
1011	1000101
0011	0011101
0110	0111010
1100	1110100
1111	1101001
1001	1010011
0101	0100111
1010	1001110
1101	1111111

Pseudonoise. A particular class of characteristic polynomials yields a set of cyclic codes with very desirable distance properties. These are known as *maximal irreducible polynomials.*

The code resulting from the irreducible polynomials is known as the *PN code*. PN is an abbreviation for pseudonoise. We shall see that the binary sequence, generated according to the rules for this code, bears a resemblance to white noise (i.e., the autocorrelation is similar to an impulse). We illustrate the code for two representative irreducible polynomials

$$\lambda^3 + \lambda + 1$$

$$\lambda^4 + \lambda^3 + 1$$

The corresponding shift-register implementations are shown in Fig. 4.18. Suppose that the system of Fig. 4.18 is initialized with any 3-bit number other than 000. For example, choose 010. With this starting state, the output would be

$$01001110100111\ldots$$

Since the generator operates upon the 3 most recent bits, once the generating sequence (010) repeats in the example above, the entire remaining sequence will also repeat. The above code is therefore periodic with a period of 7 bits.

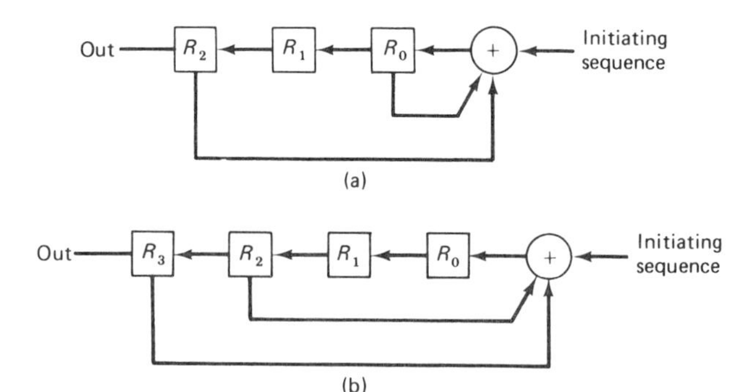

Figure 4.18 Shift-register implementation of PN code

We now examine the first 9 bits of the output sequence shown above. If we examine these in groups of 3 adjacent bits, we see that every possible 3-bit combination occurs, with the exception of 000. That is, the first 3 bits are 010: the second, third, and fourth bits are 100; the third, fourth, and fifth bits are 001; and so on. Therefore, each of the seven possible 7-bit code sequences is a cyclic shift of every other sequence. The PN code is *cyclic*.

The code word composed of all 0's is usually not used. Excluding this, the seven possible 7-bit PN code words are

$$\begin{array}{lll} 0111001, & 1110010, & 1100101 \\ 1001011, & 0010111, & 0101110 \\ 1011100 & & \end{array}$$

The PN code words have some very useful uniform properties. Observe that each of the words contains exactly four 1's. Further observe that the distance between any pair of code words is 4 units. We now find the autocorrelation of a PN sequence.

Autocorrelation is defined as

$$R(i) = n - D(c_m, c_{m-i})$$

where n is the length of the code, c_m is a code word, and c_{m-i} is the code word formed by a cyclic shift of i units. D is the distance function. The autocorrelation is therefore the number of bits in common between a code word and its shifted version. In the example above, suppose we wished to evaluate the autocorrelation at a shift of 2 units. Since a shift of 2 units results in another code word, we are simply finding the number of bit positions in which two of the words are the same. This is equal to 3. At a shift of zero, the autocorrelation is equal to the number of bits in the code word, or 7. Thus, the autocorrelation is equal to 7 for a shift of zero and to 3 for all other values of shift. This is shown in Fig. 4.19. Observe that the autocorrelation of the PN sequence resembles a digital impulse. This is the reason the PN type of code is known as pseudonoise. In applications where white noise is required, the PN sequence is often used. Because of the impulse type autocorrelation, this sequence proves useful in timing applications.

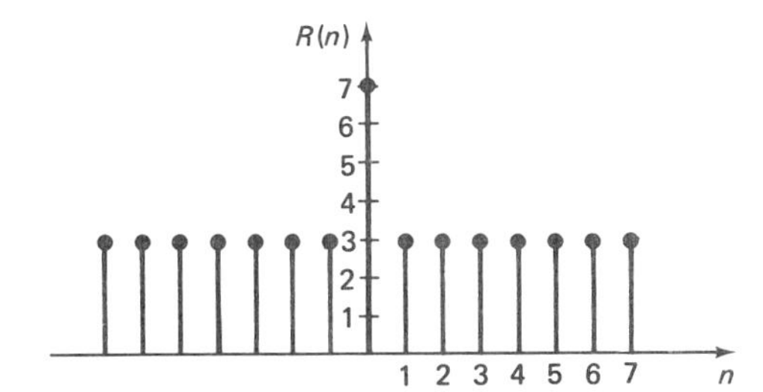

Figure 4.19 Autocorrelation of PN code

In general, for a generator length of n bits, the PN code-word length will be $2^n - 1$. For $n = 3$, the length is $2^3 - 1$, or 7 bits. The distance between code words will always be 2^{n-1}. The difference between the number of zeros and ones in every code word will be one.

Example 4-14

Find the PN code for the shift register shown in Fig. 4.20.

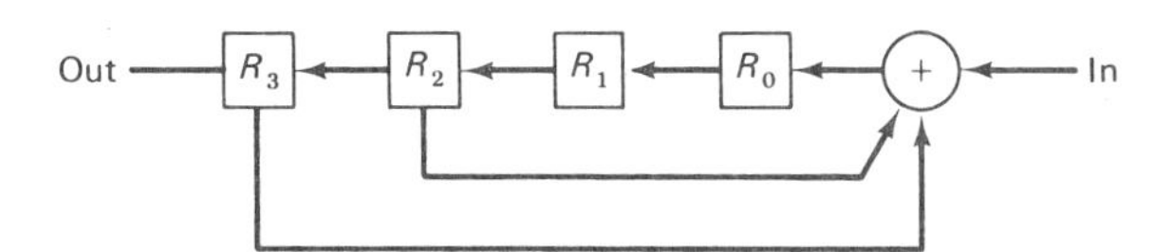

Figure 4.20 Code generator for Example 4-14

Solution. For the 4 bits of storage shown in Fig. 4.20, we can start with any 4-bit number except 0000. Using 0110, we get an output

$$0110101111000100110\ldots$$

This has period $2^4 - 1 = 15$ and distance between code words of $2^3 = 8$.

4.5 THROUGHPUT

Suppose a block code is used for error detection in a noisy environment. We add parity bits to allow detection of any number of bit errors up to a certain limit. One of three things can happen each time a code word is sent.

1. The code can contain no errors and be accepted (CA).
2. The code can contain one or more errors and be rejected (RJ).
3. The code can contain errors, but the errors go undetected (e.g., an even number of errors will not be detected using a single parity check bit). We refer to this as a *false alarm* (FA).

A particular code may be very good at detecting errors, yet provide for very poor communication. That is, if at least one bit error is made during every transmission, and the code rejects every transmission, it is accomplishing what it has been designed to do. Nonetheless, we are not communicating any information.

Throughput is defined as the probability of code acceptance. It is the probability of no errors in the received symbol word. Thus, if the probability of bit error is given by P_e, and the transmitted words consist of n bits, the throughput is given by

$$P_{CA} = (1 - P_e)^n$$

Throughput can be increased by reducing the probability of bit error or by decreasing n. We investigate techniques of reducing the bit error rate in the second part of this text, where we examine methods of sending binary information. Reduction of the bit length, n, has the undesirable effect of increasing the probability of false alarm, since it becomes more likely for one acceptable word to be changed into another acceptable word as the length of the words decreases (i.e., the distance between code words decreases).

In many applications the approach is to choose an acceptable false-alarm rate and, subject to this constraint, design the code for maximum P_{CA}. These are the type of trade-off design considerations we consider later in this text.

4.6 CRITERIA FOR CODE SELECTION

We shall briefly examine several criteria for code selection, including coding gain, error-detection and -correction capability, and throughput.

When we study techniques for sending binary information, we will see that the probability of bit error is dependent upon the signal-to-noise ratio. As the signal-to-noise ratio is raised, the probability of bit error, P_e, will decrease. Error-control coding is an alternative to increasing the signal-to-noise ratio. That is, by providing for correction of a number of bit errors, we can achieve the same overall performance as would obtain with increases in the signal-to-noise ratio (without error-control coding). The *coding gain* is defined as the reduction of signal-to-noise ratio permitted by the error-control coding. Thus, for example, if the same overall correct message transmission can occur using coding and one-half of the signal-to-noise ratio, the coding gain for that code is 3 dB. Thus, coding can be thought of as either providing a lower probability of error for a given signal-to-noise ratio or permitting a lower signal-to-noise ratio yet achieving the same probability of error.

If the important criterion for code selection is throughput, then it is important for the probability of rejection to be made small. We therefore wish to maximize the probability of code acceptance, given by

$$P_{CA} = (1 - P_e)^n$$

If the number of parity bits (code redundancy) is increased, the throughput will decrease. This is so because, in a practical system, increasing redundancy implies less time to send each bit. For example, if we must send 10^4 words per second, and each word is 10 bits long, we have 10 microseconds to send each bit. On the other hand, if each word is increased to 20 bits, we have only 5 microseconds to send each bit. This decreases the energy per bit (nominally by a factor of 2), which decreases the signal-to-noise ratio, thereby increasing the probability of bit error and decreasing the probability of code acceptance. Obviously, code design requires some significant trade-off decisions.

On the other hand, if the criterion is to decrease the probability of false alarm, we need to increase the distance between code words. This is accomplished by having high redundancy (i.e., many check bits).

Code performance can also be evaluated against the Shannon limit. Recall that, assuming noise is white, the channel capacity is given by

$$C = B \log_2\left(1 + \frac{S}{N_0 B}\right)$$

with an upper limit of

$$C_{max} = \frac{1.44S}{N_0} \tag{4-10}$$

The signal power per bit is given by the energy per bit, E_b, divided by the time per bit, T_b:

$$S = \frac{E_b}{T_b}$$

But if we communicate at the channel capacity, C, then

$$T_b = \frac{1}{C}$$

Thus,

$$S = E_b C$$

and, plugging this into Eq. (4-10), we have

$$C = \frac{1.44 E_b C}{N_0}$$

or

$$\frac{E_b}{N_0} = \frac{1}{1.44} = -1.6 \text{ dB}$$

Thus, if the signal-to-noise ratio (bit energy divided by N_0) is above -1.6 dB, we can theoretically communicate up to the channel capacity. This provides an additional criterion for code selection.

Block codes are excellent for error detection but typically provide relatively low coding gains when used for error correction. This is so because too high a percentage of bits is used for parity.

On the other hand, convolutional codes provide relatively high coding gains. They are very useful for error correction. They are useless for error detection. This is so because *something* comes out of the decoder regardless of the number of errors in the received sequence.

The fact that block codes excel for error detection while convolutional codes excel for error correction has led to a hybrid application known as *concatenated codes.* This is a combination of two coding techniques into an "inner" and an "outer" code. A message can be encoded using block codes, which leads to excellent error detection. The resulting block code can then be encoded using convolutional coding, providing for excellent error correction.

4.7 SUMMARY

This chapter explores channel encoding, which is the process of modifying the binary signal from the source encoder in order to achieve certain desirable features. Among these features are efficient use of the channel and increased immunity toward noise and channel errors.

We begin the chapter with a study of classical information theory and by defining the measure of information. This leads naturally to a study of Shannon's important channel-capacity theorem. The results of this theorem form an ideal bound against which practical systems are compared.

Section 4.4 begins a study of coding, and the basic measures of entropy, efficiency, error correction, and decipherability are introduced. Section 4.4.1 explores entropy coding, where we attempt to devise codes with rates approaching the entropy. Huffman and Shannon-Fano coding are presented. Block encoding is then introduced in Section 4.4.2 as a method of detecting and correcting bit errors. Section 4.4.3 presents convolutional coding and compares this to block encoding. The tree, trellis, and state-transition diagram are all used to describe convolutional codes. The section ends with a discussion of the Viterbi algorithm for maximum-likelihood decoding of convolutional codes.

Section 4.4.4 explores cyclic codes and presents some results from linear algebra. These results are then extended to explore the important class of cyclic codes known as pseudonoise (PN).

The concept of throughput is presented in Section 4.5. The final section, 4.6, compares the various coding techniques and establishes criteria for selecting the proper technique for a given application. Coding gain is introduced as a significant measure of code performance.

PROBLEMS

4.1. A communication system consists of three possible messages. The probability of message A is p, and the probability of message B is also p. Plot the entropy as a function of p.

4.2. A communication system consists of four possible messages. $P(A) = P(B)$ and $P(C) = P(D)$. Plot the entropy as a function of $P(A)$.

4.3. The probability of rain on any particular day in Las Vegas, Nevada, is 0.01. Suppose that a weather forecaster in that city decides to save effort by predicting no rain every day of the year. What is the average information content of each forecast (in bits per day)? Make the (simplistic) assumption that the information content of an incorrect prediction is zero. Repeat your calculations if the forecaster now decides to predict rain every day.

4.4. A TV picture contains 211,000 picture elements. Suppose that this can be considered as a digital signal with eight brightness levels for each element. Assume that each brightness level is equally probable.

(a) What is the information content of a picture?

(b) What is the information transmission rate (bits per second) if 30 separate pictures (frames) are transmitted per second?

(c) Assume that the English language consists of about 50,000 equiprobable words (a highly unrealistic assumption). What is the information content of a typical 1000-word message?

(d) Is a picture worth a thousand words?

4.5. The list below shows the probability of each alphabet letter in standard English text. (Note this would be different in technical text—for example, the letter X appears in many equations.)

A	0.081	B	0.016	C	0.032
D	0.037	E	0.124	F	0.023
G	0.016	H	0.051	I	0.072
J	0.001	K	0.005	L	0.040
M	0.022	N	0.072	O	0.079
P	0.023	Q	0.002	R	0.060
S	0.066	T	0.096	U	0.031
V	0.009	W	0.020	X	0.002
Y	0.019	Z	0.001		

(a) Find the entropy associated with sending a single letter.
(b) What is the information content of the three-letter word THE? Assume that letters are independent of each other. (This is a highly unrealistic assumption. For example, the probability of the letter U given that the previous letter is Q is close to unity.)
(c) What is the information content of the three-letter word JUT?
(d) You are told that the probability of occurrence of the word THE is 0.027. That is, out of every 1000 words, on the average there will be 27 occurrences of the word THE. Using this result, what is the information content of the word THE? What does this imply about the letter-independence assumption in part (b)?

4.6. Find the channel capacity, C, if the $S/N = 6$ dB and a standard broadcast AM channel is used (5-kHz bandwidth). Repeat for a broadcast TV channel (6-MHz bandwidth) and compare your answers.

4.7. A voice-grade telephone line has a bandwidth of 3 kHz. If you wished to transmit at a rate of 5 kbps, what signal-to-noise ratio is required?

4.8. Which of the following are uniquely decipherable codes? Of these, which are instantaneous?
(a) 010, 0110, 1100, 0001, 00011, 00110.
(b) 0, 010, 01, 10.
(c) 0, 100, 101, 11.

4.9. Find the minimum average length of a code with five messages having probabilities
(a) $\frac{1}{8}, \frac{1}{8}, \frac{1}{8}, \frac{1}{8}, \frac{1}{2}$.
(b) 0.2, 0.2, 0.3, 0.15, 0.15.

4.10. Find the minimum distance for the following code consisting of four code words:

0111001, 1100101, 0010111, 1011100

How many bit errors can be detected? How many bit errors can be corrected?

4.11. Find the minimum distance for the following code consisting of four code words:

0111001, 1100101, 0010111, 0101001

How many by errors can be detected? How many bit errors can be corrected?

4.12. Four-bit mesage words are encoded by adding a fifth parity bit. if the bit-error rate is 10^{-4}, find the probability of a word error.

4.13. In a particular binary communication system, the probability of a 0 is 0.95. Bits are independent of each other. Since a 0 is much more likely than a 1, you decide to design a

run-length coding scheme to combine multiple consecutive zeros into a single code word.

Propose a scheme that would compress a run of zeros into a code indicating the length of the run. Be sure that your system is uniquely decipherable. Find the average bit rate for your system and find the percentage improvement when this rate is compared to the original bit transmission rate.

4.14. An 8-bit word is formed from 4 information bits and 4 parity bits with the parity bits given by

$$c_1 = a_2 + a_3 + a_4$$

$$c_2 = a_1 + a_2 + a_3$$

$$c_3 = a_1 + a_2 + a_4$$

$$c_4 = a_1 + a_3 + a_4$$

(a) Find the generating matrix, $[H]$.
(b) How many errors can be detected?
(c) How many errors can be corrected?
(d) Demonstrate the decoding process for any single message.

4.15. A 10-bit code (2^{10} separate messages) has an eleventh parity bit added. How many of the possible 11-bit words are not used?

4.16. Find the Huffman code for five messages with probabilities as given below. Compare the average code-word length to the entropy in each case.
(a) $\frac{1}{8}, \frac{1}{8}, \frac{1}{8}, \frac{1}{8}, \frac{1}{2}$.
(b) 0.2, 0.2, 0.3, 0.15, 0.15
(c) 0.1, 0.2, 0.3, 0.2, 0.2

4.17. Find the Huffman code for messages with the following probabilities:
(a) 0.2, 0.2, 0.2, 0.2, 0.2
(b) 0.1, 0.15, 0.2, 0.2, 0.25, 0.05, 0.025, 0.025
(c) 0.4, 0.2, 0.2, 0.1, 0.1

4.18. Repeat Problems 4.16 and 4.17 using Shannon-Fano coding.

4.19. **(a)** Find the shift-register implementation of a convolutional code with impulse response

$$\{h_n\} = \{1 \quad 1 \quad 0 \quad 1 \quad 0 \quad 1\}$$

(b) Find the output code word if the information input message is

$$\{1 \quad 1 \quad 1 \quad 0 \quad 1 \quad 0 \quad 1 \quad 1 \quad 0\}$$

4.20. A rate-$\frac{1}{2}$ convolutional code is constructed with constraint length 4. The first set of h' given by 10000 and the second set by 01011. Find the output sequence if the input quence is given by

$$\{1 \quad 0 \quad 1 \quad 1 \quad 0 \quad 1 \quad 0 \quad 1 \quad 1 \quad 0\}$$

4.21. Sketch the tree, trellis, and state-transition diagram for the rate-$\frac{1}{2}$ codes generated shift registers shown on page 206. Also find the minimum distance of the res code words.

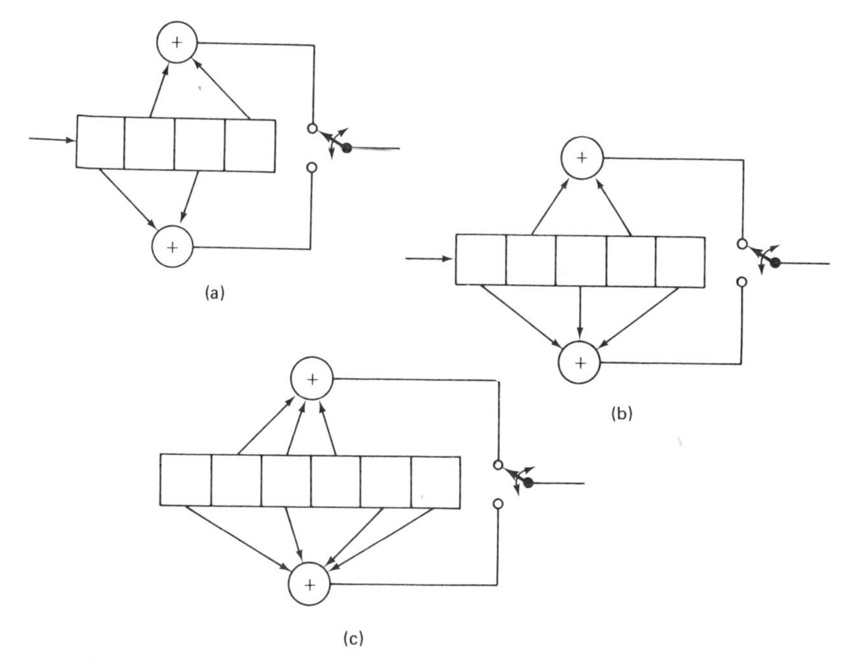

4.22. For the coders of Problem 4.21, decode the following received sequence

$$1\ 0\ 1\ 0\ 1\ 0\ 1\ 0\ 1\ 0 \ldots$$

4.23. Find the characteristic polynomial of the following matrices. If it is not possible, discuss the reasons.

$$\begin{bmatrix} 1 & 0 & 0 & 0 \\ 0 & 1 & 0 & 0 \\ 0 & 1 & 1 & 1 \\ 0 & 0 & 1 & 1 \end{bmatrix}, \quad \begin{bmatrix} 1 & 1 & 1 & 1 \\ 1 & 0 & 1 & 0 \\ 1 & 1 & 0 & 0 \\ 0 & 0 & 0 & 1 \end{bmatrix}, \quad \begin{bmatrix} 1 & 0 & 0 & 0 \\ 0 & 1 & 0 & 1 \\ 0 & 0 & 1 & 0 \\ 1 & 1 & 1 & 0 \end{bmatrix}$$

4.24. Find $x^4 + x + 1 \bmod x + 1$ in GF(2).

4.25. A code is specified by the polynomial

$$\lambda^5 + \lambda + 1$$

(a) Is this a cyclic code?
(b) Is this a PN code?
(c) If not, can you factor the polynomial?

4.26. For the PN generator polynomial

$$\lambda^3 + \lambda + 1$$

find

(a) the PN code for an initiating sequence of 010.
(b) the length of the code words.

(c) the autocorrelation of the code words.
(d) the number of errors that can be detected.
(e) the number of errors that can be corrected.

4.27. Generate a (7, 4) cyclic code which is different from the one presented in Example 4-14.

5

Data Transmission and Reception

5.1 SIGNAL CHARACTERISTICS

In this section we present the various common time-domain signal formats used in digital communication.

One of the simplest proposals for transmitting a sequence of 1's and 0's is to transmit a 1-volt signal for a digital one and a 0-volt signal for a digital zero. This is known as *unipolar* transmission, since the signal deflects from zero in only one direction. A slight modification of this natural choice is often made. In fact, we consider sending a constant of $+V$ volts for a digital one and $-V$ volts for a digital zero. This is known as *bipolar* transmission. V is chosen on the basis of available power and other physical considerations. We use $+V$ and $-V$ rather than $+V$ and 0 since, as we shall see later, it is advantageous to make the two signals as different as possible. Figure 5.1 shows a signal waveform that might be used to send the digital sequence 1 0 0 1 0.

The specific signaling format illustrated is NRZ, or *non-return to zero.* This particular type of transmission is known as NRZ-L, where the voltage level corresponds to the logic level. Alternatively, we could have represented a logic 1 with $-V$ volts and a logic 0 with $+V$ volts.

Figure 5.2 illustrates NRZ and RZ (return to zero) signaling for the sequence 1 1 0 0 0 1 0 1. Note that in RZ, the pulse returns to the "0" state before reaching the end of the sampling interval. Thus, the pulses are not as wide as in the NRZ case, and this change increases the bandwidth, as we shall see in the next section.

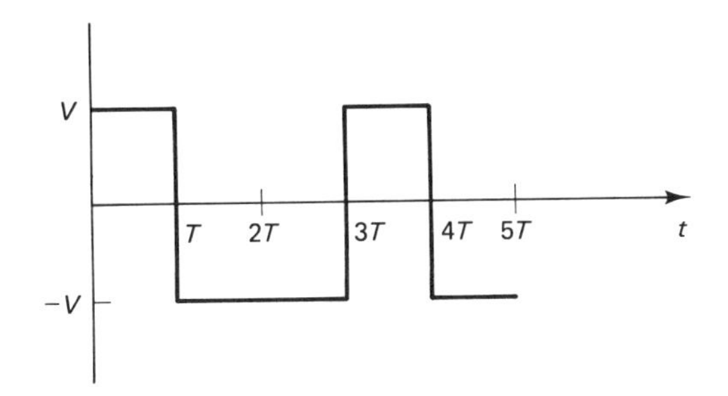

Figure 5.1 Bipolar NRZ signal format

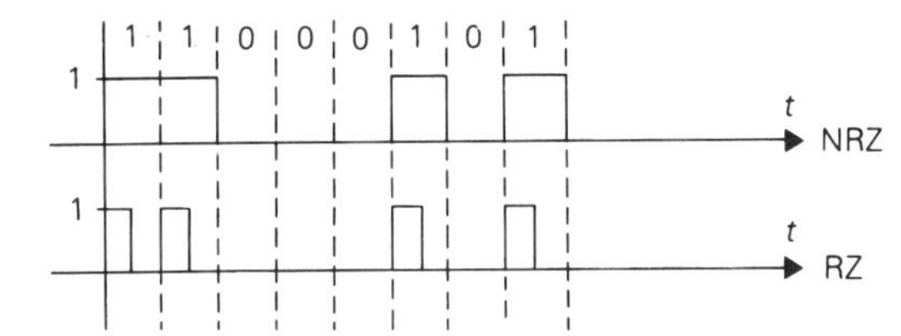

Figure 5.2 Unipolar NRZ and RZ signal formats

There are two particular problems associated with NRZ-L transmissions. First, when the data are *static* (i.e., no change from bit interval to bit interval), there is no transition in the transmitted waveform. This can cause serious timing problems when we try to establish bit synchronization. That is, we will see in Chapter 6 that establishing the proper local clock signals requires transitions in the received waveform. The second problem occurs with data inversion. If the levels are reversed during transmission (i.e., $+V$ is interpreted as $-V$ at the receiver), the entire data train will be inverted and every bit will be in error. Inversion can occur in several ways. We shall see later that the most common form of inversion occurs when the information is being sent by varying phase angles. A time delay corresponding to a 180-degree phase shift results in a data inversion. Additionally, some systems consist of numerous electronic devices (e.g., op-amps), each of which inverts the signal. It may sometimes prove difficult to keep track of the number of inversions, or indeed, different signals may encounter different numbers of inverters.

For these reasons we often choose *differential* forms of encoding. In such techniques, the data are represented as changes in levels rather than by the particular signal level. We shall investigate the NRZ-M and NRZ-S systems. The terminology of M and S stems from MARK and SPACE, a carryover from the telegraph days.

In the NRZ-M system, a data "1" is represented by a change in level between two consecutive bit times while a data "0" is represented by no change. Figure 5.3 illustrates the NRZ-M waveform for the NRZ-L signal of Fig. 5.1. Note that we have started the NRZ-M signal at $+V$ volts. We could have started the signal at $-V$ volts.

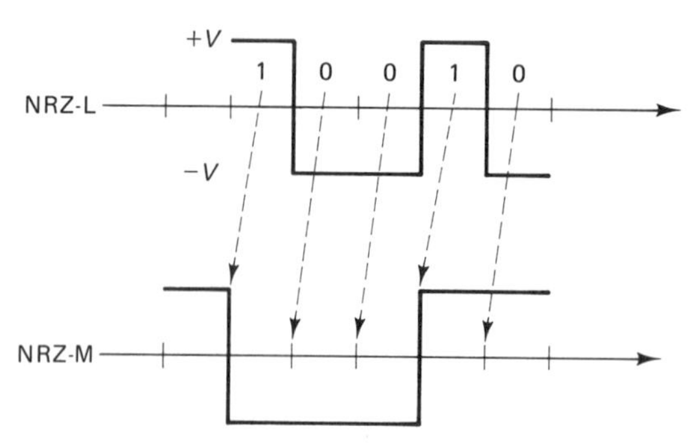

Figure 5.3 NRZ-M differential signal format

We can implement an NRZ-M encoder with the use of an exclusive OR gate and a time delay. This is shown in Fig. 5.4. We start with the NRZ-L signal, and exclusive-OR this with the delayed NRZ-M output. Thus, if the input bit is a 1, the output represents an inversion of the previous output. That is, if we take the exclusive-OR operation of something with 1, the result is the inverse of that something. Thus,

$$1 \oplus 1 = 0$$

$$1 \oplus 0 = 1$$

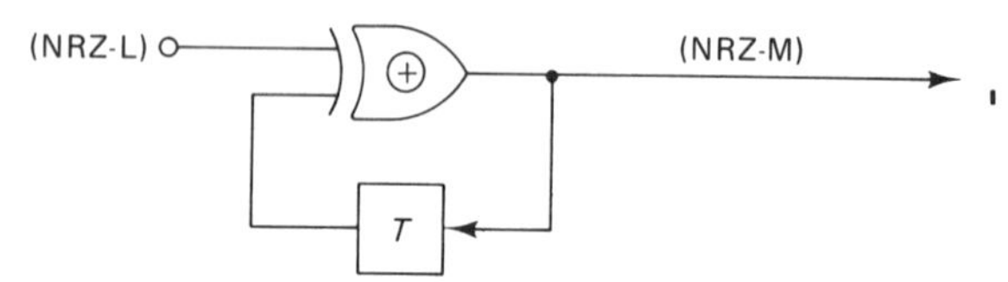

Figure 5.4 NRZ-M encoder

The decoder from NRZ-M simply compares the NRZ-M signal to a delayed version of itself. If the two are identical, we know that a 0 is being sent, and if the two are different, a 1 is being sent. This describes the operation of an exclusive-OR gate, so the decoder can be implemented as in Fig. 5.5.

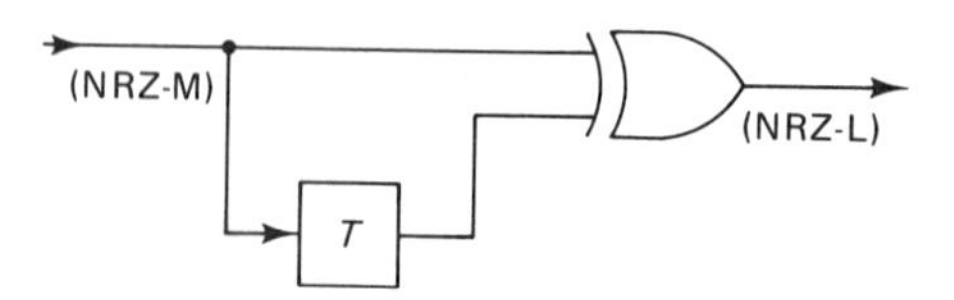

Figure 5.5 NRZ-M decoder

The NRZ-S system is almost the same as the NRZ-M, except that the two outputs are reversed. A data 1 is represented by no change in levels between two consecutive bit times, while a data 0 is represented by a change. Figure 5.6 shows the NRZ-L, NRZ-M, and NRZ-S representations for the same data signal we have been illustrating. The encoders and decoders for NRZ-S are the same as those for

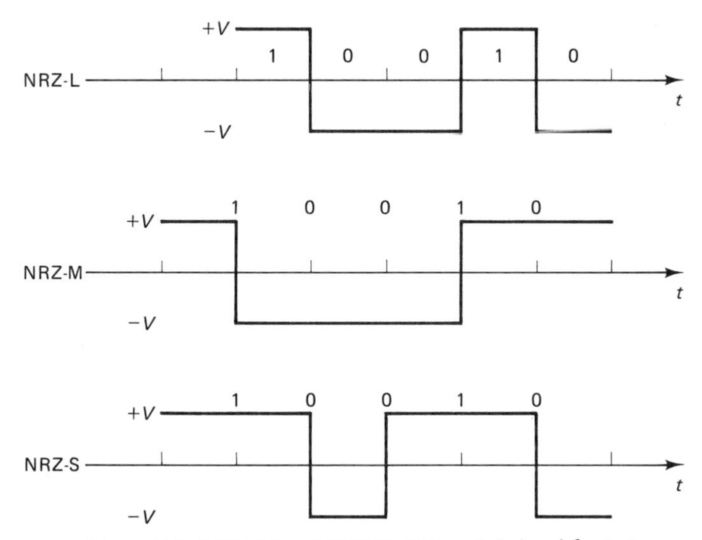

Figure 5.6 NRZ-M and NRZ-S differential signal formats

NRZ-M except for the addition of an inverter. Derivation of these systems is left to the problems at the back of this chapter.

While the NRZ-S and NRZ-M differential systems have solved the problem of waveform inversion, they have not solved the problem of loss of timing information. For example, in the NRZ-M system, a train of data 0's will result in a transmitted waveform without transitions. Likewise, in the NRZ-S system, a train of data 1's will result in a loss of bit timing information.

The *biphase,* or *Manchester,* formats overcome this static data and timing problem. Figure 5.7 shows the NRZ-L and biphase-L waveforms for the same data train we have been examining. Also shown on the figure are the basic waveforms used to send a 1 and a 0. Note that a 1 is sent by transmitting $-V$ volts for the first half of the bit interval and $+V$ volts for the second half. A 0 is sent with the inverse signal. Thus, a transition always occurs at the midpoint of each bit interval.

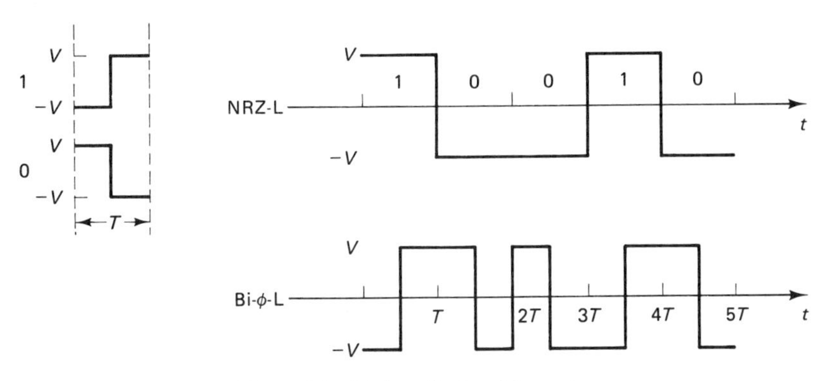

Figure 5.7 Biphase-L signal format

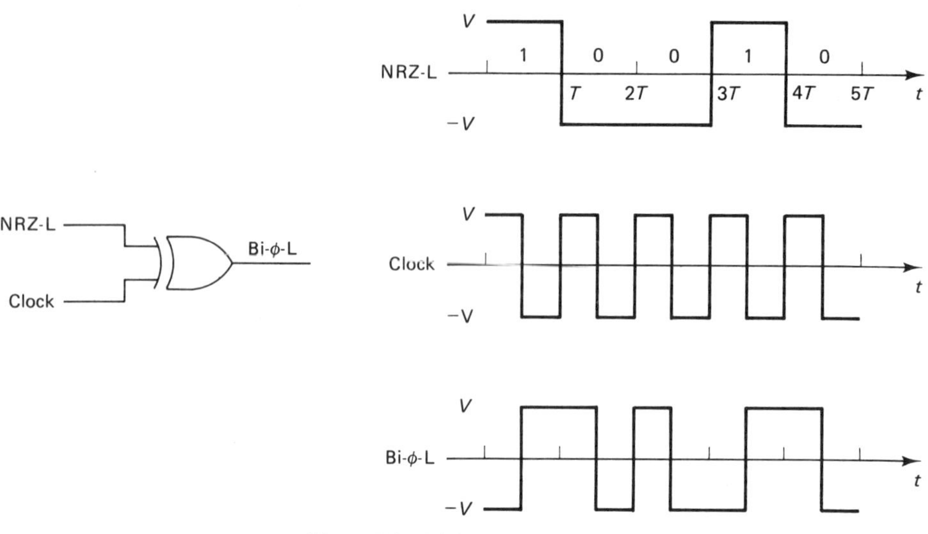

Figure 5.8 Biphase-L encoder

Figure 5.8 shows an encoder for biphase-L. We start with NRZ-L and feed both this and a double-frequency clock signal into an exclusive-OR gate. When the data logic level is 0, the output of the gate is the same as the clock. When the data logic level is 1, the output of the gate is the inverse of the clock. Alternatively biphase-L can be sent with the reverse of these two signals. We have thus solved the timing problem, but signal inversion during transmission or reception will still result in bit reversal, causing unacceptable bit errors.

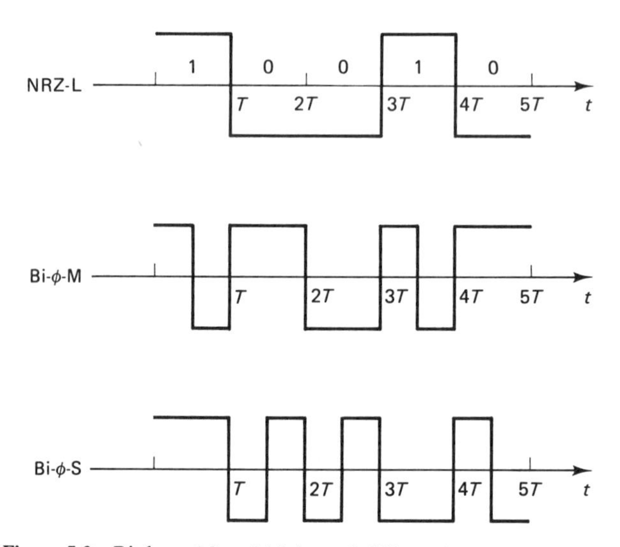

Figure 5.9 Biphase-M and biphase-S differential signal formats

We can combine differential encoding with biphase encoding to solve both problems. The biphase-M and biphase-S formats are differential. A transition occurs at the *beginning* of every bit period, unlike the biphase-L where transitions occur in the middle of the intervals. In the biphase-M system, a data 1 is represented by an additional transition in the midperiod, while a 0 is represented by no midperiod transition. In the biphase-S system, the data 1 results in no midperiod transition, while the data 0 results in a midperiod transition. This is illustrated in Fig. 5.9.

The encoder for biphase-S is shown in Fig. 5.10. We start with NRZ-M, delay it by one-half of the bit interval, and exclusive-OR this with the double-frequency clock. The result is to invert every other segment of the NRZ-M signal. If we wanted biphase-M instead of biphase-S, we would start with NRZ-S.

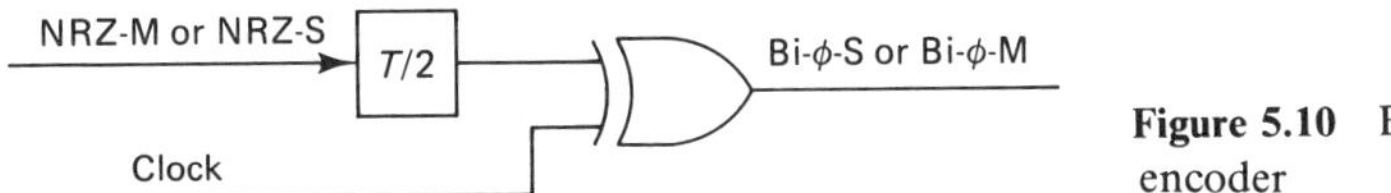

Figure 5.10 Biphase-S and biphase-M encoder

5.2 DATA RANDOMNESS AND SIGNAL FREQUENCY CHARACTERISTICS

In this section, we shall examine the frequency characteristics of the various data formats.

We start by examining NRZ format signals. It is important that we know the frequency characteristics of these signals, both to determine required channel characteristics (e.g., bandwidth, distortion) and to know the effects of additive noise. If the original data signal is periodic, the resulting NRZ signal will be a periodic function of time. Its frequency spectrum is therefore composed of discrete lines at the fundamental frequency and harmonic multiples of this frequency. For example, if we examine the NRZ-L signal with a periodic alternating train of 0's and 1's as input, the result is a square wave of period $2T$, where T is the bit period. The power spectral density is then as shown in Fig. 5.11, where the envelope is a $(\sin^2 f)/f^2$ type function. The total power of this signal is V^2, since when the signal is squared, the result is a DC signal at V^2 volts. This total power is equal to the integral under the power spectral density curve (in this case, the sum of the areas of the impulses).

If the data now consist of random sequences of 1's and 0's, the power spectral density is continuous and can be shown to be equal to

$$S(f) = V^2 T \frac{\sin^2(\pi f T)}{(\pi f T)^2} \tag{5-1}$$

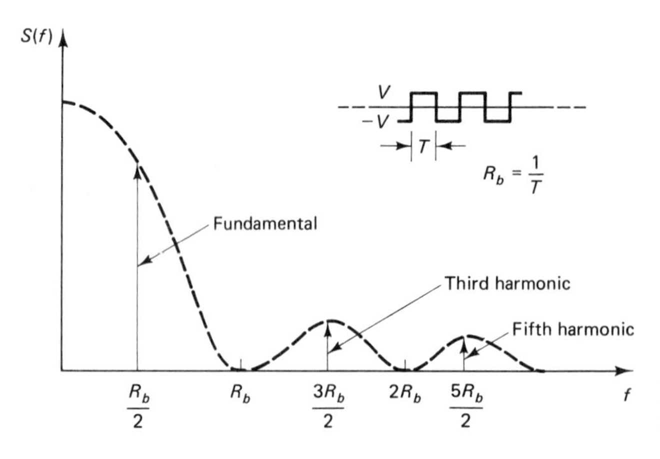

Figure 5.11 Power spectral density of periodic NRZ signal

This is sketched in Fig. 5.12. Note that approximately 91% of the total area (power) is contained in frequencies below $1/T$, the bit rate.

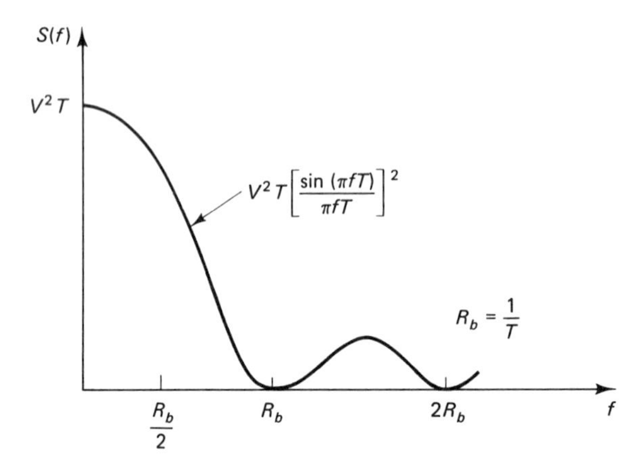

Figure 5.12 Power spectral density of random NRZ signal

We now examine the frequency content of a biphase signal. If the sequence is all 1's or all 0's in the biphase-L format, the resulting signal will be a square wave of period T. This has a power spectral density as shown in Fig. 5.13. Note that this is the same shape as that of Fig. 5.11, except that the frequency axis has been scaled by a factor of 2. The channel must therefore pass twice the bandwidth for biphase as it does for NRZ-L.

The power spectral density of biphase for random data must be calculated by considering all possible sequences. For example, it is possible to have a resulting waveform segment which resembles either the single- or double-frequency components for NRZ-L and biphase worst-case periodic signals. Also note that there is no DC component, since the signal always spends half of its time at $+V$ and half at

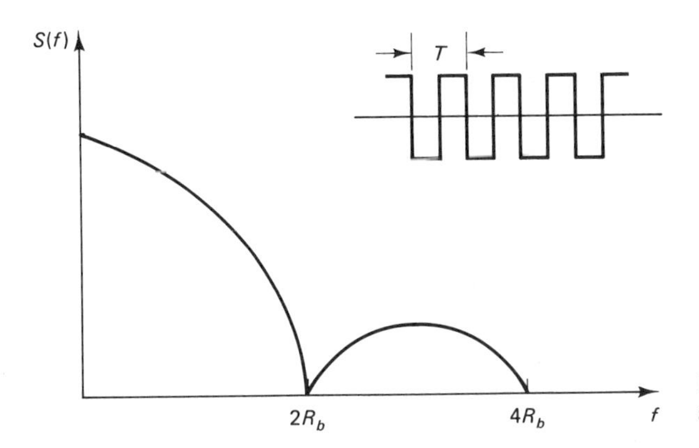

Figure 5.13 Power spectral density of static data biphase-L signal

$-V$. The power spectral density is given by

$$S(f) = \frac{V^2T}{2}\left[\frac{\sin^2(\pi fT/2)}{\pi fT/2}\right]^2 \tag{5-2}$$

This is sketched in Fig. 5.14. Note that 65% of the power lies in frequencies below $1/T$, and 86% lies in frequencies below $2/T$ Hz.

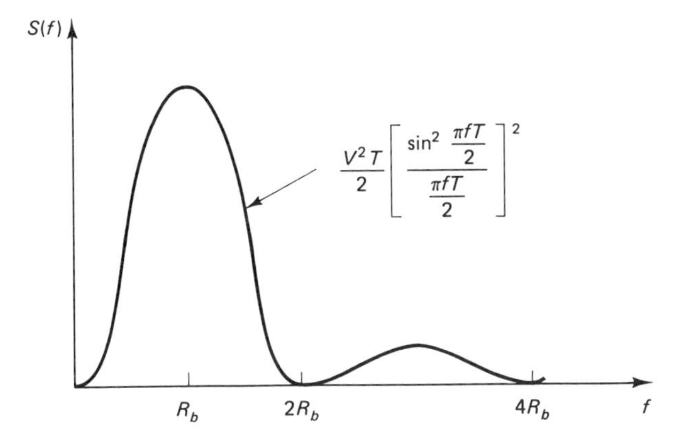

Figure 5.14 Power spectral density of random data biphase-L signal

5.3 MULTIPATH

When a signal is transmitted through a channel, it often experiences reflections. This occurs whether the transmission is confined to a wire or is in space. If the wave reflects from some interface or object and then reaches the receiver, it experiences a longer path than the direct path from the transmitter to the receiver. Thus, the reflected signal is delayed relative to the direct-path transmission. This results in the received waveform being a summation of the transmitted signal and various weighted time shifts of that signal. This phenomenon is known as *multipath,* or *echo*. If the transmission takes place only over two major paths (one reflection and one direct), we refer to this as *specular multipath*. On the other hand, if there are multiple reflections with differing delays, the multipath is known as *diffuse*. It is easier to reduce the effects of specular multipath using filters (known as equalizers) than it is to reduce the effects of diffuse multipath.

Multipath leads to *intersymbol interference*, since the delayed version of the waveform will extend into the next sampling interval. The multipath effect is well known in television, where it manifests itself as ghost images. It is particularly disturbing at radio frequencies where reflections from the ionosphere occur. In fact, without extra work, data rates above about 100 bits per second are not possible.

Many techniques have been devised for decreasing the effects of this distortion. For example, upon reflection, the polarization of a circularly polarized signal reverses direction. This is the reason that polarized sun glasses are effective against reflections. Thus, with redesign of antennas, circularly polarized signals can be made relatively insensitive to multipath.

A mathematical technique for analyzing multipath is the *cepstrum*. The cepstrum is related to the Fourier transform of the received signal. Looking at the specular multipath case, the received waveform can be written as

$$r(t) = s(t) + as(t - T) \tag{5-3}$$

where $s(t)$ is the direct-path received signal. The Fourier transform of this received waveform is

$$R(f) = S(f)[1 + ae^{-j2\pi fT}] \tag{5-4}$$

The second term in this expression represents a complex sinusoidal term. The power spectrum of the received waveform is

$$G_R(f) = G_S(f)[1 + a^2 + 2a \cos 2\pi fT] \tag{5-5}$$

where $G_S(f)$ is the power spectrum of the direct-path signal, $s(t)$. To find the power spectrum, we take the log of the expression above. This changes the product to a sum and isolates the cosine term:

$$\log [G_R(f)] = \log [G_S(f)] + \log (1 + a^2 + 2a \cos 2\pi fT) \tag{5-6}$$

The second term in Eq. (5-6) is periodic with period $1/T$ (note this is a function of frequency, so the period has units of hertz). Therefore, the transform of this log, where f is the independent variable, will have an impulse at T. Thus, with this new tool, we can identify the time shifts of the original signal and thereby reduce multipath by subtracting such a term from the received signal. The cepstrum has received a great deal of attention in radar and sonar studies, because the time delay of reflections yields ranging information.

Multipath is one of the many driving forces behind the transition from analog to digital processing. For example, we mentioned that multipath causes ghost images in television. These effects are time-dependent, and in fact may occur when aircraft or land vehicles are moving near the receiving antenna. Ghost images can also occur in cable systems if proper attention is not paid to line terminations. Digital television can use simple, microprocessor controlled circuitry to adaptively cancel multipath effects, thereby virtually eliminating the associated problems.

5.4 MATCHED FILTERS

The requirements placed upon a digital (data) receiver are radically different from those for an analog receiver. The *analog* receiver must reconstruct a waveform that is as close as possible to the information signal present at the transmitter. The *data* receiver is not the least bit concerned with reconstructing the information waveform that existed at the transmitter. Instead, it is content to decide at each sampling point which of the possible symbols in the transmission coding alphabet is being received. This opens up processing possibilities that could not be considered in the analog system. Stated another way, the receiver can distort the signal as much as it wishes if such distortion helps to make a decision as to which symbol is being communicated.

The actual mechanism of making a decision at the receiver is usually one of processing the incoming signal in some prescribed manner, and then viewing a sample value of the output of this processor. This sample value is then manipulated in some manner to yield a decision. The processing prior to sampling is such that the sample value is most likely to yield the correct decision. This involves decreasing the effects of additive noise.

The *matched filter* is one technique for processing the received signal. The criterion for designing the filter is that the output at a sample time T is to have *maximum signal-to-noise ratio*. We designate the input to the filter as $s(t) + n(t)$, where $s(t)$ is the signal and $n(t)$ is the noise. The filter output is $s_0(t) + n_0(t)$, where $s_0(t)$ is the signal portion and $n_0(t)$ is the noise portion of the output. This is illustrated in Fig. 5.15. The filter is chosen to maximize $s_0^2(T)/n_0^2(T)$. Since the denominator of this expression is usually random, we use the average of the random variable. The output signal-to-noise ratio is given by

$$\rho = \frac{s_0^2(T)}{\overline{n_0^2(T)}} = \frac{\left|\int_{-\infty}^{\infty} S(f)H(f)e^{j2\pi fT}\,df\right|^2}{\int_{-\infty}^{\infty}|H(f)|^2 G_n(f)\,df} \tag{5-7}$$

The numerator in Eq. (5-7) is the square of the inverse Fourier transform of the product of the input transform with the system function. Thus it is the square of the deterministic time sample, $s_0(T)$. The $G_n(f)$ in the denominator is the power spectral

$s(t) + n(t)$ → $H(f)$ → $s_o(T) + n_o(T)$

Figure 5.15 Signal plus noise input to linear system

density of the input noise. Thus the denominator integrand is the power spectral density of output noise.

We wish to choose $H(f)$ to maximize the expression of Eq. (5-7). The choice is simplified if we apply *Schwartz's inequality* to the numerator. Schwartz's inequality states that for all functions, $f(x)$ and $g(x)$,

$$\left| \int f(x)g(x)\,dx \right|^2 \leqslant \int |f^2(x)|\,dx \int |g^2(x)|\,dx$$

Textbook writers enjoy giving an elegant proof of this theorem, and although it breaks the continuity, I shall not break with tradition. We start by noting that for real functions,

$$\int [f(x) - Tg(x)]^2\,dx \geqslant 0$$

for all $f(x)$, $g(x)$, and T. This is true because the integrand, being a square, is nonnegative. Expanding this, we find

$$T^2 \int g^2(x)\,dx - 2T \int f(x)g(x)\,dx + \int f^2(x)\,dx \geqslant 0 \qquad (5\text{-}8)$$

The left side of Eq. (5-8) is quadratic in T. Since the value is constrained never to go negative, the quadratic cannot have distinct real roots. Therefore, its discriminant must be nonpositive. Thus,

$$4\left[\int f(x)g(x)\,dx\right]^2 - 4\int f^2(x)\,dx \int g^2(x)\,dx \leqslant 0$$

and the inequality is established for real functions. The extension to complex functions is left as an exercise at the end of this chapter.

We now wish to apply the inequality to Eq. (5-7). We modify the numerator of the equation in a seemingly unmotivated way that will factor out a term identical to the denominator:

$$\left| \int_{-\infty}^{\infty} S(f)H(f)e^{j2\pi fT}\,df \right|^2 = \left| \int_{-\infty}^{\infty} \frac{S(f)}{\sqrt{G_n(f)}} H(f)\sqrt{G_n(f)}\, e^{j2\pi fT}\,df \right|^2$$

The square-root operation is unambiguous, owing to the fact that $G_n(f)$ can never be negative, since it is a power spectrum. Now applying Schwartz's inequality, we find

$$\rho \leqslant \frac{\int_{-\infty}^{\infty} |H(f)|^2 G_n(f)\,df \int_{-\infty}^{\infty} |S(f)|^2/G_n(f)\,df}{\int_{-\infty}^{\infty} |H(f)|^2 G_n(f)\,df}$$

$$\leqslant \int_{-\infty}^{\infty} \frac{|S(f)|^2}{G_n(f)}\,df$$

An upper bound has therefore been established on the signal-to-noise ratio at the output of the filter. If by some hocus-pocus we can guess at an $H(f)$ which yields this maximum, we need look no further. Starting at Eq. (5-7) makes the guess almost obvious. By observation, if

$$H(f) = Ce^{-j2\pi fT} \frac{S^*(f)}{G_n(f)} \tag{5-9}$$

is plugged into Eq. (5-7), the maximum signal-to-noise ratio is achieved. C is an arbitrary constant and factors out, since it appears in both the numerator and denominator of Eq. (5-7).

Equation (5-9) gives the system function of the optimum filter. In order to find the impulse response, the inverse Fourier Transform of $H(f)$ is evaluated. Let us assume that $n(t)$ is white noise, so $G_n(f)$ is a constant, $N_0/2$. The filter impulse response then becomes

$$h(t) = \tfrac{1}{2} CN_0 s(T - t) \tag{5-10}$$

This inverse transform is found by recognizing that the transform of $s(-t)$ is the complex conjugate, $S^*(f)$. The term $e^{-j2\pi fT}$ represents a time shift. A filter with this impulse response is called the *matched filter.* Note that this equation might represent a nonrealizable filter.

Example 5-1

Find the impulse response of the matched filter for the two time functions shown in Fig. 5.16.

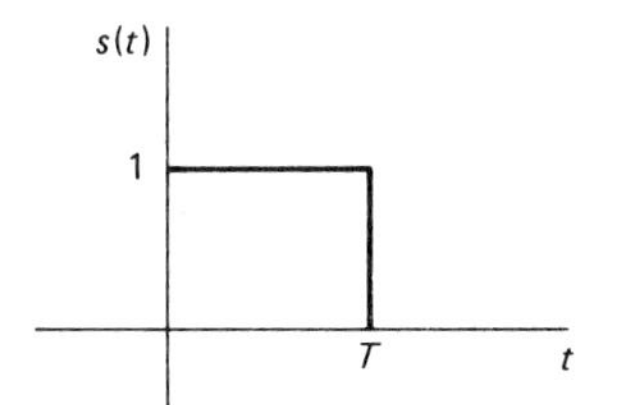

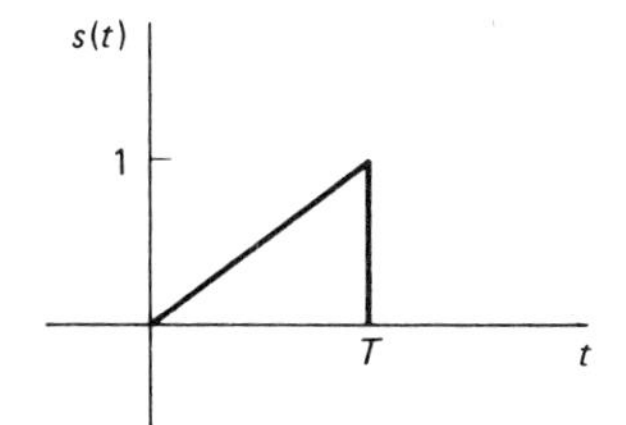

Figure 5.16 Time functions for Example 5-1

Solution. The $h(t)$ is derived directly from Eq. (5-10). The result is sketched in Fig. 5.17.

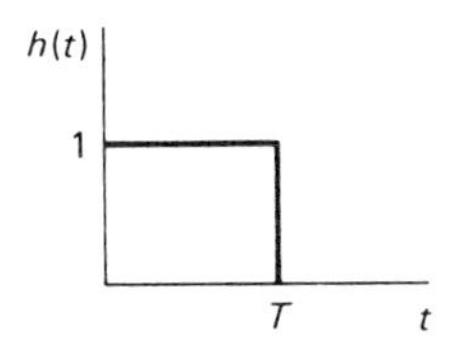

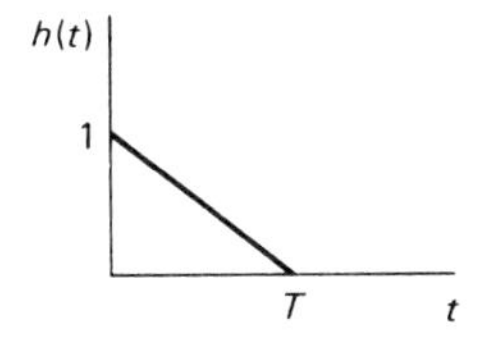

Figure 5.17 Impulse response for Example 5-1

The actual time function at the output of the matched filter can be found by convolving the input time function with the impulse response. Therefore,

$$s_o(t) + n_o(t) = [s(t) + n(t)] * h(t)$$
$$= \int_0^T [s(\tau) + n(\tau)]s(T - t + \tau)\, d\tau$$

and at time T,

$$s_o(T) + n_o(T) = \int_0^T [s(\tau) + n(\tau)]s(\tau)\, d\tau \qquad (5\text{-}11)$$

Thus, the matched filter is equivalent to the system of Fig. 5.18. The operation being performed by the system in Fig. 5.18 is called *correlation* (i.e., multiply two time functions together and integrate the product). For that reason, the matched filter is sometimes referred to as a *correlator.* In a generalized sense, the filter is finding the projection of the input signal in the direction of the information, $s(t)$. Since the system is aligned in the $s(t)$ direction, the output signal-to-noise ratio is thereby maximized.

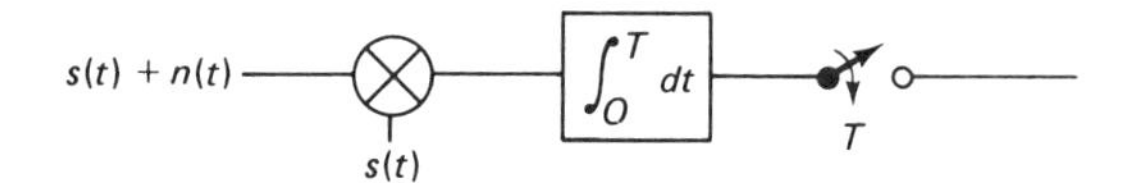

Figure 5.18 Signal correlator

Example 5-2

Find the output signal-to-noise ratio of a matched filter where the signal is

$$s(t) = A \qquad \text{for } 0 \leqslant t \leqslant T$$

in white noise with power spectral density $N_0/2$.

Solution. The matched filter simply multiplies the received waveform by a constant, A, and then integrates this product between 0 and T. We note that the multiplying constant could be any value without affecting the signal-to-noise ratio, since it multiplies both the signal and the noise waveform. The output signal is

$$s_o(T) = \int_0^T s^2(t)\, dt = A^2T$$

The output noise is

$$n_o(T) = A\int_0^T n(t)\, dt$$

and the noise power is the mean-square value of this expression. That is, we take the expected value of the square of the integral (note that the square of an integral is *not* the integral of the square—look at Schwartz's inequality).

$$E\left\{A^2 \iint_0^T n(t)n(\tau)\, dt\, d\tau\right\} = \frac{A^2N_0}{2}\int_0^T dt$$
$$= \frac{A^2N_0T}{2}$$

The signal power is simply the square of the signal value, or A^4T^2. The signal-to-noise ratio is therefore given by

$$\frac{S}{N} = \frac{A^4T^2}{A^2N_0/2} = \frac{2A^2T}{N_0}$$

5.5 DECISION THEORY

When a data signal arrives at the receiver, during each sampling interval the receiver must decide which message was transmitted. An entire theory has been developed to indicate the "best" choice the receiver can make. In this section we develop the basic concepts of this *decision theory*. These concepts will be developed independent of particular signal waveforms.

A *hypothesis* is defined as a statement of a possible condition. Thus, in a binary system, there are two possible hypotheses: "0" is being sent and "1" is being sent. We observe some waveform, and based upon this observation, choose between the two hypotheses. The choice is made to optimize something.

Let us start with some terminology. H_0 and H_1 are the hypotheses that a zero or one is being sent, respectively. $P(H_0)$ and $P(H_1)$ are the *a priori probabilities* of the two hypotheses. By using the term *a priori* we imply that there is some way of determining these probabilities before the fact—that is, before the experiment is performed. Thus, in a particular code, if 1's and 0's are sent an equal fraction of the time, the two a priori probabilities are $\frac{1}{2}$ each.

Let y be the observation variable. We can then form two conditional probabilities:

$$P(H_0 \mid y) \qquad \text{and} \qquad P(H_1 \mid y)$$

The first is the probability that hypothesis zero is true given the observed variable, and the second is the probability that hypothesis one is true given the observation. A natural decision rule is to choose the hypothesis that has the maximum conditional probability associated with it. The decision rule can be stated: Choose hypothesis zero if

$$P(H_0 \mid y) > P(H_1 \mid y)$$

or equivalently, if

$$\frac{P(H_0 \mid y)}{P(H_1 \mid y)} > 1 \tag{5-12}$$

Otherwise, choose H_1. This is really quite obvious. As an example, suppose that you measure the relative humidity to be 70% at 7 P.M. and you want to decide whether or not it will rain tomorrow. If you had sufficient information and were able to determine that the probability of its raining tomorrow given that the humidity at 7 P.M. is 70% is 0.7, and the probability of its not raining given this humidity is 0.3, you would of course guess that it will rain. These conditional probabilities are known as *a posteriori probabilities,* since they are calculated after the fact (experiment). The decision criterion of Eq. (5-12) is called the *maximum a posteriori criterion* (*MAP*).

It is sometimes computationally difficult to find the conditional probabilities used above. A more natural approach is to find the conditional probabilities in the other direction. These probabilities are $P(y|H_0)$ and $P(y|H_1)$. This becomes apparent in a data communication system where the conditional probabilities might be the probability density of the matched filter output assuming that a 0 is transmitted, and the density assuming that a 1 is transmitted. Note that if y is a continuous variable rather than discrete, we replace the capital P with a small p, denoting probability density, and introduce a new notation, $p_0(y)$ and $p_1(y)$, for the densities of the observation variable, y, under the two hypothesis conditions. The maximum a posteriori criterion can be modified to this format by noting that

$$P(H_0|y) = \frac{p_0(y)P(H_0)}{p(y)}$$

and

$$P(H_1|y) = \frac{p_1(y)P(H_1)}{p(y)}$$

Substituting this into Eq. (5-12), we arrive at the criterion to choose H_1 if

$$\lambda(y) = \frac{p_1(y)}{p_0(y)} > \frac{P(H_0)}{P(H_1)} \tag{5-13}$$

The ratio of conditional probabilities on the left side of Eq. (5-13) is called the *likelihood ratio,* $\lambda(y)$, and will prove of critical importance to us. The numerator and denominator of Eq. (5-13), that is, $p_0(y)$ and $p_1(y)$, are known as *likelihood functions.* If the a priori probabilities are equal, Eq. (5-13) indicates that we choose the hypothesis with the greatest likelihood. This is known as *maximum-likelihood detection.*

Example 5-3

Design a maximum a posteriori detector for the two conditional probability (likelihood) functions shown in Fig. 5.19. Assume that the a priori probabilities of the two hypotheses are equal.

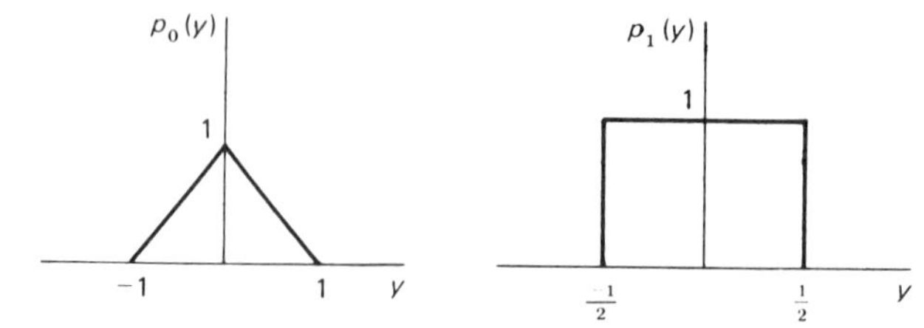

Figure 5.19 Probability functions for Example 5-3

Solution. The maximum a posteriori criterion tells us to choose H_1 if

$$\frac{p_1(y)}{p_0(y)} > \frac{P(H_0)}{P(H_1)} = 1$$

By inspection of Fig. 5.19, we see that for $|y| < \frac{1}{2}$, the likelihood ratio is greater than 1, so we choose H_1. For $\frac{1}{2} < |y| < 1$, the ratio is zero, so we choose H_0.

For $y \geqslant 1$, the likelihood ratio is indeterminate, but we are saved by the observation that the probability of measuring a value of y greater than 1 is zero. The only point left is the origin, where the likelihood ratio is 1. Since we have defined the criterion with strict inequality required, a measurement of $y = 0$ should lead to a choice of H_0. Actually, since the probability of measuring exactly 0 (or any exact value) is zero, that is of no consequence.

In summary, the detector measures y. If $|y| > \frac{1}{2}$ or $y = 0$, the detector decides in favor of H_0. If $|y| < \frac{1}{2}$, the detector decides in favor of H_1.

5.5.1 Bayes Criterion

So far, we have applied intuition to the decision problem. It is time to formalize things. The first question to be asked is: What is meant by *optimum*? It is rumored that Wiener was once asked how he knew that a system he described at a seminar was optimum. He stared the questioner in the eyes, shrugged his shoulders, and stated "I did the best I could!" Although this response may appear flippant, in fact Wiener was responding that the system was indeed optimum under the constraint of his abilities.

Well, there are many ways to define optimum. For example, we could seek to maximize the probability of a correct decision. Or we can minimize the probability of a certain kind of error. These might seem equivalent, but a simple example will illustrate the difference. Suppose that you were a weather forecaster in the Napa Valley in California, and you were charged with the responsibility of predicting frost so that wine growers could run their automatic sprinklers (or smudge pots, if they are not ecologically responsible winemakers). You would probably want to minimize the probability of saying there would be no frost when indeed there was a frost. You would even be willing to settle for a reasonable number of *false alarms,* since these only cost the producers water as opposed to destroying wine grapes. Therefore, although your criterion may not be to maximize the probability of correct decision, it could nonetheless be optimum under the circumstances.

Bayes approached the criterion problem by defining *cost* functions. The result of this analysis can be applied to almost all other optimization criteria, as we shall soon see.

There are four possible combinations of hypotheses and decisions in a binary situation. We assign a cost to each of these four. C_{ij} is the cost associated with choosing hypothesis H_i when in fact H_j is true. It might seem strange assigning a cost to a correct decision (C_{00} and C_{11}) and, indeed, this is included to make the C_{ij} into a matrix. In many cases, C_{00} and C_{11} are zero. If, however, the rewards for the two correct decisions are not equal, one might sometimes want to take this into account by having unequal costs.

We define an average cost, called the *risk,* as the weighted average of the four cost functions. To do this, we introduce one more piece of notation, D_i. This represents a decision for hypothesis H_i. Using this, the average cost is

$$\begin{aligned}\bar{C} &= C_{00}P(D_0, H_0) + C_{01}P(D_0, H_1) + C_{10}P(D_1, H_0) \\ &\quad + C_{11}P(D_1, H_1) \\ &= C_{00}P(H_0)P(D_0 \mid H_0) + C_{01}P(H_1)P(D_0 \mid H_1) \\ &\quad + C_{10}P(H_0)P(D_1 \mid H_0) + C_{11}P(H_1)P(D_1 \mid H_1)\end{aligned} \tag{5-14}$$

We now recognize that

$$P(H_0) + P(H_1) = 1$$

and

$$P(D_1 \mid H_i) + P(D_0 \mid H_i) = 1$$

Equation (5-14) can be reduced to

$$\begin{aligned}\bar{C} &= P(H_0)C_{10} + [1 - P(H_0)]C_{11} \\ &\quad - P(H_0)(C_{10} - C_{00})P(D_0 \mid H_0) \\ &\quad + [1 - P(H_0)](C_{01} - C_{11})P(D_0 \mid H_1)\end{aligned} \tag{5-15}$$

The objective is to devise a decision rule that minimizes $\bar{C}$ in Eq. (5-15). The first two terms in this are independent of the decision rule, so we confine our attention to the last two terms. Note that

$$P(D_i \mid H_j) = \int_{R_i} p_j(y)\, dy$$

where R_i is the set of observation values, y, which lead to a decision in favor of H_i. Stated another way, the objective of designing a decision rule is to split the range of observations, y, into two regions, R_0 and R_1. If the observation falls in R_0, we decide H_0. Using this notation, Eq. (5-15) can be rewritten as follows, where the first two terms have been indicated as a constant, K:

$$\bar{C} = K + \int_{R_0} \{[1 - P(H_0)](C_{01} - C_{11})p_1(y) - P(H_0)(C_{10} - C_{00})p_0(y)\}\, dy \tag{5-16}$$

The problem now becomes one of choosing R_0 to minimize the cost. Everything else in Eq. (5-16) is assumed known.

As y varies over all possible values, the integrand in Eq. (5-16) can be positive or negative. $\bar{C}$ is minimized if R_0 is chosen as that combination of values for which the integrand is negative. The decision rule thus becomes: Choose H_0 if

$$\frac{p_1(y)}{p_0(y)} \lessgtr \frac{P(H_0)(C_{10} - C_{00})}{[1 - P(H_0)](C_{01} - C_{11})} \tag{5-17}$$

H_1 is chosen if the inequality is reversed. We note that if the cost of a correct decision is zero and the cost of an incorrect decision is 1, Eq. (5-17) is identical to Eq. (5-13). That is, the Bayes criterion and the maximum a posteriori criterion are identical.

The threshold against which the likelihood ratio is compared will be given the symbol λ_0. Thus, if

$$\lambda(y) = \frac{p_1(y)}{p_0(y)} > \lambda_0$$

we decide H_1. Note that λ_0 increases if $P(H_0)$ increases and/or C_{10} increases. This is intuitively satisfying, since increasing the a priori probability of H_0 and/or increasing the cost of incorrectly choosing H_1 should bias the decision rule against choosing H_1. Increasing λ_0 does just this.

Example 5-4

Design a Bayes detector for a priori probabilities

$$P(H_1) = \tfrac{2}{3}, \qquad P(H_0) = \tfrac{1}{3}$$

and costs

$$C_{00} = C_{11} = 0, \qquad C_{01} = 1, \qquad C_{10} = 3$$

The likelihood functions are both Gaussian, unit variance, with means of 0 and 1 for hypotheses 0 and 1, respectively.

$$p_0(y) = \frac{1}{\sqrt{2\pi}} e^{-y^2/2}$$

$$p_1(y) = \frac{1}{\sqrt{2\pi}} e^{-(y-1)^2/2}$$

Solution. The Bayes detector compares the likelihood ratio to

$$\frac{P(H_0)(C_{10} - C_{00})}{\{1 - P(H_0)\}(C_{01} - C_{11})} = \frac{\frac{1}{3} \times 3}{\frac{2}{3} \times 1} = \frac{3}{2}$$

If

$$\lambda(y) = \frac{p_1(y)}{p_0(y)} > \frac{3}{2}$$

choose H_1. Plugging in for $p_0(y)$ and $p_1(y)$, we have

$$\lambda(y) = \frac{(1/\sqrt{2\pi})e^{-(y-1)^2/2}}{(1/\sqrt{2\pi})e^{-y^2/2}} = \exp\left(\frac{2y - 1}{2}\right)$$

The test is then specified by: Choose H_1 if

$$\exp\left(\frac{2y-1}{2}\right) > \frac{3}{2}$$

Taking logs of both sides, this becomes: Choose H_1 if

$$y > \frac{1}{2} + \ln\left(\frac{3}{2}\right) = 0.905$$

5.5.2 Minimum-Error Criterion

The Bayes cost is not terribly different from the probability of error. If the cost of a correct decision is zero and the cost of either error is 1, the average cost in Eq. (5-14) becomes

$$\bar{C} = P(D_1, H_0) + P(D_0, H_1) \tag{5-18}$$

Since the two events included in Eq. (5-18) are disjoint and represent all ways in which an error can be made, this average cost is truly the probability of error. Thus, with this particular choice of cost function, the Bayes criterion becomes a minimum-error-probability criterion. The rule is therefore to choose H_1 if

$$\lambda(y) = \frac{p_1(y)}{p_0(y)} > \frac{P(H_0)}{1 - P(H_0)}$$

Note that the detector again is equivalent to the maximum a posteriori detector.

Example 5-5

Design a minimum error detector for equal a priori probabilities and likelihood functions,

$$p_1(y) = \frac{1}{\sqrt{2\pi(\frac{1}{2})}} e^{-y^2/(1/2)}$$

$$p_0(y) = \begin{cases} \frac{1}{2} & \text{for } |y| < 1 \\ 0 & \text{otherwise} \end{cases}$$

Solution. The likelihood ratio is given by

$$\lambda(y) = \begin{cases} \dfrac{2}{\sqrt{2\pi(\frac{1}{2})}} e^{-y^2/(1/2)}, & |y| < 1 \\ \infty, & |y| > 1 \end{cases}$$

The minimum-error criterion with equal a priori probabilities is to choose H_1 if $\lambda(y) > 1$.

If $|y| > 1$, the likelihood ratio is infinity, so clearly H_1 is chosen. If $|y| < 1$, choose H_1 if

$$\frac{2}{\sqrt{2\pi(\frac{1}{2})}} e^{-y^2/(1/2)} > 1$$

$$y^2 < -\frac{1}{2} \ln\left(\frac{\sqrt{2\pi}}{4}\right)$$

$$|y| < 0.483$$

5.5.3 Minimax Criterion

Design of the *Bayes detector* requires that we be able to assign meaningful *costs* and that we have a way of finding, or assuming, *a priori probabilities* for the hypotheses. The *minimum-error detector* (or equivalently the maximum a posteriori) requires that we know the *a priori probabilities,* but we need *not* concern ourselves with assigning costs. What happens if we have a reliable way of assigning costs but have no idea of the a priori probabilities? A method known as *minimax* proves optimum for this scenario.

Let us begin by assuming that you are foolhardy enough to take a wild guess at the a priori probabilities. Since the costs are assumed given, you then design a Bayes detector. The average cost of this detection scheme is given by Eq. (5-19), which is a repeat of Eq. (5-14):

$$\begin{aligned}\bar{C} &= C_{00}P(H_0)P(D_0 \mid H_0) + C_{01}P(H_1)P(D_0 \mid H_1) \\ &\quad + C_{10}P(H_0)P(D_1 \mid H_0) + C_{11}P(H_1)P(D_1 \mid H_1)\end{aligned} \tag{5-19}$$

Once the receiver is designed, the conditional probabilities, $P(D_i \mid H_j)$, are fixed. That is, these four probabilities depend only upon the decision regions and not upon the a priori probabilities. If $P(H_0)$ now deviates from the value used in the Bayes design, $\bar{C}$ deviates from its optimum minimum value. In fact, since $P(H_0)$ appears as a linear term on the right side of Eq. (5-19), $\bar{C}$ will vary linearly with $P(H_0)$. This is illustrated in Fig. 5-20. As $P(H_0)$ varies from the value that was guessed, the average cost follows the straight line shown. Thus, the cost might be smaller or larger than what was expected. It is significant to compare the cost with that achievable using a properly designed Bayes detector. Figure 5.21 repeats Fig. 5.20 with a typical Bayes cost curve superimposed. The meaning of the Bayes curve is as follows. At every value of $P(H_0)$, the optimum Bayes detector is designed for the given costs. The straight line intercepts the Bayes cost curve at the point corresponding to the guessed a priori probability. The straight line must be tangent to the Bayes curve, since the cost along the straight line could never be less than that of the perfectly designed Bayes detector. This is true since, by design, the Bayes detector minimizes cost for the given a priori probability. It would be inconsistent if a lower cost could be achieved by guessing. It can be shown that the Bayes error curve is convex. The "guess at $P(H_0)$" detector is therefore sensitive to changes in $P(H_0)$ and may yield a cost which is significantly greater than that expected. *Except,* what if we

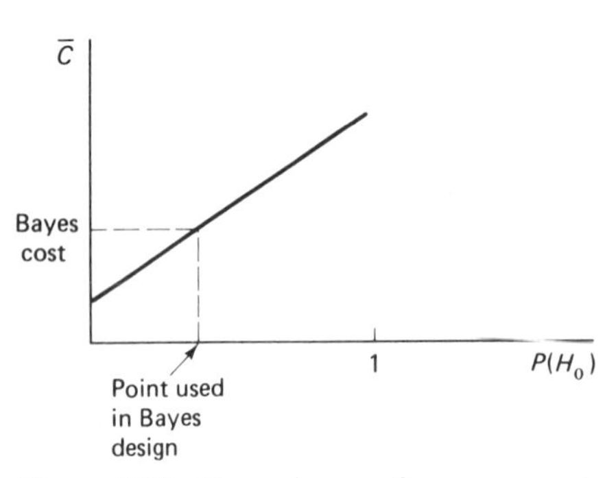

Figure 5.20 Dependence of average cost upon a priori probability

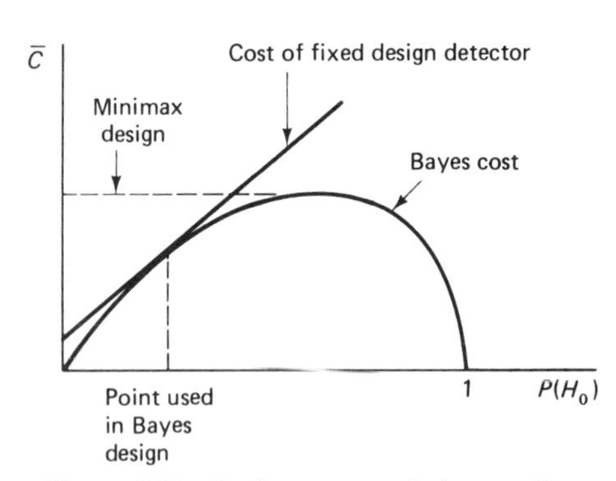

Figure 5.21 Performance relative to Bayes detector

designed for the peak of the Bayes curve? The tangent would then be a horizontal line, and the cost for the detector would be insensitive to changes in $P(H_0)$. Thus, although we might be lucky and get smaller costs with the design of Fig. 5.20, a more conservative approach is to accept the peak cost on the Bayes curve, and design for that peak. This procedure is known as *minimax,* since we are minimizing the cost for the maximum choice of $P(H_0)$. To design the minimax detector, we let $P(H_0)$ be a variable and design the Bayes detector. The average cost is then calculated, and $P(H_0)$ is chosen to maximize this cost.

The minimax design can often be simplified by working with the observation that the slope of the line tangent to the maximum is zero. Thus, differentiating Eq. (5-19) with respect to $P(H_0)$ and setting the derivative equal to zero, we have

$$C_{10}P(D_1 \mid H_0) + C_{00}P(D_0 \mid H_0) = C_{01}P(D_0 \mid H_1) + C_{11}P(D_1 \mid H_1) \tag{5-20}$$

The conditional probabilities in Eq. (5-20) are a function of the choice of decision region. We therefore choose the decision regions to satisfy this equation.

Example 5-6

Design a minimax receiver for the cost matrix

$$C = \begin{bmatrix} 0 & 2 \\ 1 & 0 \end{bmatrix}$$

and the likelihood functions shown in Fig. 5.22. Also guess at a priori probabilities,

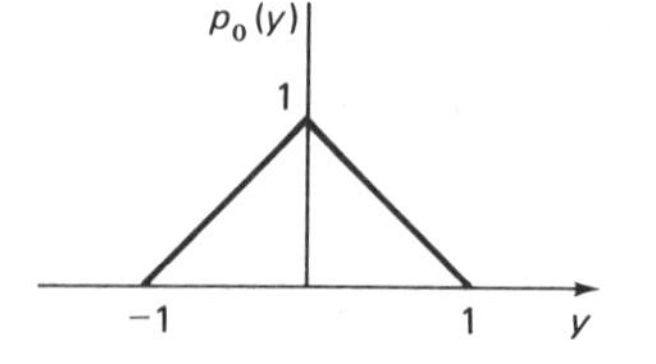

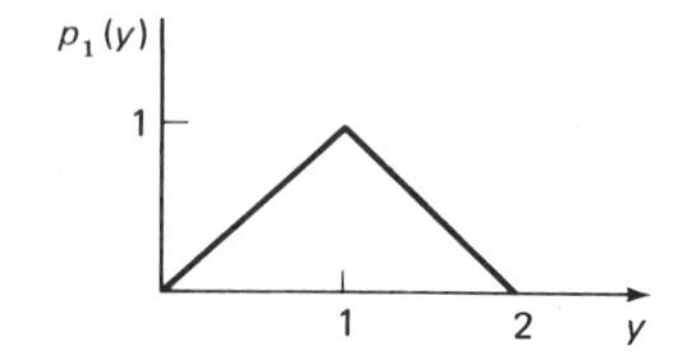

Figure 5.22 Likelihood functions for Example 5-6

design a Bayes detector for this guess, and compare performance of the two detectors.

Solution. We apply Eq. (5-20) to get

$$P(D_1 | H_0) = 2P(D_0 | H_1) \tag{5-21}$$

Assuming that the threshold is somewhere between 0 and 1, we proceed to calculate the conditional error probabilities for insertion in Eq. (5-21). Denoting the threshold as y_0, we have

$$P(D_1 | H_0) = \frac{(1 - y_0)^2}{2}$$

$$P(D_0 | H_1) = \frac{y_0^2}{2}$$

Therefore, Eq. (5-21) becomes

$$\frac{(1 - y_0)^2}{2} = y_0^2, \qquad y_0 = 0.414 \quad \text{or} \quad -2.1414$$

We reject the negative value and choose 0.414 for the threshold. Using this value, we can calculate the probability of the two types of error:

$$P(D_1 | H_0) = \frac{(1 - y_0)^2}{2} = 0.172$$

$$P(D_0 | H_1) = \frac{y_0^2}{2} = 0.207$$

If we now design a Bayes detector for a guess of equal a priori probabilities, the criterion is to choose H_1 if

$$\lambda(y) = \frac{p_1(y)}{p_0(y)} > \frac{1}{2}$$

Solving this for y between 0 and 1, we have

$$\lambda(y) = \frac{1}{1 - y} > \frac{1}{2}$$

$$y > \frac{1}{2} - \frac{1}{2}y$$

$$y > \frac{1}{3} = y_0$$

Using this threshold value, we find the error probabilities:

$$P(D_1 | H_0) = \frac{(1 - \frac{1}{3})^2}{2} = 0.222$$

$$P(D_0 | H_1) = \frac{(\frac{1}{3})^2}{2} = 0.056$$

and the probability of error,

$$P_e = (0.222 + 0.056)/2 = 0.139$$

Comparing this to the minimax result, we see the probability of deciding H_0 when H_1 is true is lower for the Bayes, but the probability of deciding H_1 when H_0 is true is higher. In order to assess the dangers of the Bayes-guess technique, suppose that the a priori probabilities were not equal but that $P(H_0) = 0.95$ and $P(H_1) = 0.05$. Then the Bayes detector yields an error probability of

$$P_e = 0.95(0.222) + 0.05(0.056) = 0.2$$

with an average cost of

$$\bar{C} = 0.95(0.222)(1) + 0.05(0.056)(2) = 0.217$$

The minimax detector gives

$$P_e = 0.95(0.172) + 0.05(0.207) = 0.183$$

$$\bar{C} = 0.95(0.172)(1) + 0.05(0.207)(2) = 0.184$$

for the same a priori probabilities. This represents a considerable improvement.

5.5.4 Neyman-Pearson Criterion

We started with the Bayes criterion where it was assumed that all costs and a priori probabilities were known. We then developed the minimum-error criterion for those cases where costs are not known, yet a priori probabilities are known. The situation was then reversed and the minimax criterion was developed for the case when costs are known but a priori probabilities are unknown. There is only one scenario left, that of maximum ignorance. Suppose that there is no meaningful way to assume a priori probabilities, nor can cost functions be assigned in any reasonable way. We then throw up our hands and change the method of approach.

There are two ways of making an error. One can decide H_1 when in fact H_0 is true, or decide H_0 when H_1 is true. The first of these is known as an *error of the first kind,* or *false alarm.* The second is, quite logically, an *error of the second kind,* or *miss.* The names evolved from the radar scenario where H_1 is the hypothesis that a target is present and H_0 indicates that no target is present. It should be intuitively clear that as a design is changed to lower the probability of one type of error, the probability of the other type will increase. Thus, returning to our weather-predicting example, the forecaster can avoid the possibility of failing to predict a frost if a frost is forecast every single day. This makes the probability of one type of error equal to 0, but the other probability of error is 1.

In recognition of the various trade-offs, and with some sensitivity to the radar problem (radar at the time being in the early stages of development), the following criterion was devised. A maximum acceptable false-alarm probability is assumed given. A system is then designed to maximize $P(D_1 | H_1)$, subject to the false-alarm probability's not exceeding the given level. Since $P(D_1 | H_1) = 1 - P(D_0 | H_1)$, this criterion is equivalent to minimizing the probability of an error of the second kind subject to the constraint that the probability of an error of the first kind does not exceed some given value.

Looking back in history to the motivating radar problem is helpful. The radar people knew that as they designed a radar system to reduce the probability of missing a detection of a real target, they were simultaneously increasing the probability of a false alarm (detection of a target that was not there). In the absence of any meaningful way to assign costs and a priori probabilities, they determined a partial type of costs due to false alarms. Assuming that the response to a false alarm was deployment of aircraft (a missile today), they determined how often they could afford a false alarm, and they designed the system to achieve exactly this level of performance. In that way, they knew they were reducing the probability of a miss as far as possible, subject to the practical constraint.

Although this may appear quite simple, some additional work is needed before we can design the detector. If you were simply told to design the detector with a specified false-alarm rate, you probably would not know where to start. That is, without knowing the general form of the detector, the design is impossible. Specifying a detector design is equivalent to specifying the regions for deciding H_0 and H_1. In many cases, these regions are disconnected and are specified by more than one dividing point. Even in the simple case of Example 5-3, we found a disconnected region which required that two dividing points be specified. In attempting to find the location of the dividing points from the given false-alarm rate, one would find only one equation in a number of unknowns. The exact number of unknowns is not even known until the problem is solved, so we are really in a dilemma.

It is necessary to reduce the problem to a single unknown, since only a single condition is given. We accomplish this by showing that the likelihood ratio is still the critical parameter.

Restating the problem, we must minimize the probability of a miss given the constraint that the probability of false alarm is equal to P_{FA}. When a minimization problem is stated with constraints, the approach that is often used is that of LaGrange multipliers. We then must minimize

$$E = P_M + \lambda_0 P_{FA}$$

where λ_0 is the arbitrary LaGrange multiplier (our use of the symbol for likelihood ratio will be justified in a moment). Rather than reinvent the wheel, we note that minimization of E is really a special case of the Bayes detector where $\bar{C}$ is minimized. Looking at the general expression for Bayes risk in Eq. (5-19), we see that with the following choice of costs, the Bayes risk is exactly the same as E:

$$C_{00} = C_{11} = 0, \qquad C_{01} = \frac{1}{P(H_1)}, \qquad C_{10} = \frac{\lambda_0}{P(H_0)}$$

We can thus minimize E with a Bayes detector. Plugging the cost values above into Eq. (5-17), we have the following detector design: Choose H_1 if

$$\frac{p_1(y)}{p_0(y)} > \lambda_0$$

Note that λ_0 was the arbitrary LaGrange multiplier, so one might fairly ask what has been accomplished. In fact, we now know that the form of the receiver is a likelihood-ratio detector where the threshold, λ_0, is chosen to achieve the given false-alarm probability. We have therefore reduced the problem to a single equation in a single unknown, λ_0.

Example 5-7

Design a Neyman–Pearson detector for the following likelihood functions,

$$p_1(y) = \frac{y}{4}\exp\left[-\frac{1}{8}(y^2 + 2)\right] I_0\left(\frac{y}{2}\right), \qquad y > 0$$

$$p_0(y) = \frac{y}{4}e^{-y^2/8}, \qquad y > 0$$

and a maximum false-alarm rate of

$$P_{FA} = 10^{-4}$$

Solution. The first likelihood function is the *Ricean density,* where I_0 is a modified Bessel function. The second function is a *Rayleigh density.* These two densities will result when certain types of detectors are applied to data communication, as we shall see in Chapter 8. Fortunately, the Neyman–Pearson detector can be designed without using the first density function, since the detector depends only upon false-alarm probabilities.

It is first necessary to determine the general form of the detector. We shall assume (and see why later) that the detector compares the measured parameter to a threshold, y_0, and chooses H_1 if the measurement exceeds that threshold. Some caution is advised at this point, since not all problems yield a simple threshold. Even the simple densities of Example 5-5 yielded a situation different from the assumption above. In any case, if H_1 is decided for measurements greater than y_0, the probability of false alarm is given by

$$P_{FA} = \int_{y_0}^{\infty} \frac{y}{4} e^{-y^2/8}\, dy$$

This function should be set equal to the given probability, 10^{-4}. The function is easy to integrate, yielding

$$-e^{-y^2/8}\Big|_{y_0}^{\infty} = e^{-y_0^2/8} = 10^{-4}$$

$$y_0^2 = -8 \ln(10^{-4}) = 8.58$$

$$y_0 = 2.93$$

The Neyman–Pearson detector measures y and chooses H_1 if y is greater than 2.93.

5.5.5 Receiver Operating Characteristic

In the following chapters, we use the concepts of decision theory to design data receivers. These receivers will perform some type of signal processing, set a threshold, and then decide which of the possible message symbols is being transmitted. The previous sections should have convinced you that as the detector design is varied (i.e., the threshold is moved), one type of error increases while the other type decreases.

We normally design the detector in accordance with some set of criteria, and this will determine the value of the threshold. However, in certain applications it will prove useful to examine the receiver performance as a function of the threshold setting. This can be done by plotting either the probability of false alarm as a function of threshold setting, or the probability of miss as a function of this threshold. Yet another approach is to plot the probability of miss as a function of the probability of false alarm. That is, we choose a value of threshold and calculate the two error probabilities. This yields one point on the curve. The threshold is then varied to find other points on the curve.

The *receiver operating characteristic* (*ROC*) is a plot of "1 minus the probability of miss" as a function of the probability of a false alarm. When the probability of miss is subtracted from 1, we get the probability of correctly choosing hypothesis 1, $P(D_1 \mid H_1)$, sometimes known as the probability of detection. (The use of this term is another carryover from the radar problem. The reader is cautioned that this probability is *not* the same as the probability of correct decision.)

Note that since the threshold becomes a variable in finding ROC curves, it is not necessary to give any consideration to which detection criterion is being used.

Example 5-8

Given the likelihood functions

$$p_0(x) = \frac{1}{\sqrt{2\pi}} e^{-x^2/2}$$

$$p_1(x) = \frac{1}{\sqrt{2\pi}} e^{-(x-m)^2/2}$$

develop the ROC curves for various choices of m.

Solution. We first comment that in our later work, variation of m will be analogous to changes in the signal power, or equivalently, in the signal-to-noise ratio. Therefore, as we develop the various ROC curves, each will apply for a particular signal-to-noise ratio.

If m is positive, this becomes a simple one-threshold problem. The receiver compares x to x_0, and if $x > x_0$, it chooses H_1. The two probabilities to be plotted are therefore given by

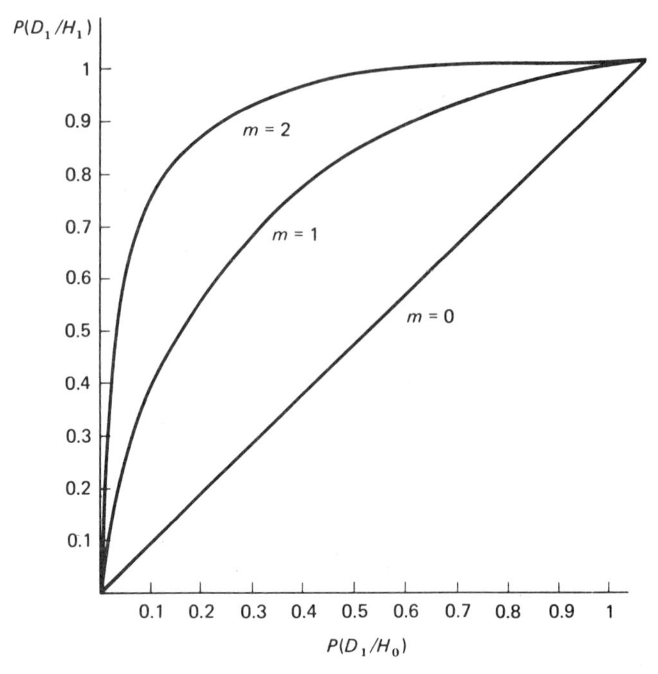

Figure 5.23 ROC curves for Example 5-8

$$P(D_1 \,|\, H_0) = \frac{1}{\sqrt{2\pi}} \int_{x_c}^{\infty} e^{-x^2/2}\, dx$$

$$= \frac{1}{2} \operatorname{erfc}\left(\frac{x_0}{\sqrt{2}}\right)$$

$$P(D_1 \,|\, H_1) = \frac{1}{\sqrt{2\pi}} \int_{x_c}^{\infty} e^{-(x-m)^2/2}\, dx$$

$$= \frac{1}{2} \operatorname{erfc}\left(\frac{x_0 - m}{\sqrt{2}}\right)$$

Figure 5.23 illustrates $P(D_1 \,|\, H_1)$ vs. $P(D_1 \,|\, H_0)$ for $m = 0$, 1, and 2.

5.5.6 Concluding Remarks

The study of decision theory normally now extends the observation space to cover vectors and then continuous-time functions. That is, in the development above, y is assumed to be a single random variable, perhaps a sample of a time function. If a detector is designed to base a decision upon a single observation of a received time function, it is reasonable to think that performance can be improved by making additional observations (e.g., more than one sample). In the limit, we might continuously observe a time function, $y(t)$, in order to make a decision.

Since we have already discussed matched filters and accepted a single sample of their output as a reasonable observation upon which to base a binary decision, it is not necessary to go any further in detection theory for our particular applications.

5.6 MULTIPLEXING

Multiplexing is the sending of more than one signal at a time through the communication channel. There are two broad classes of multiplexing, one being the dual of the other. These are *frequency-division multiplexing* (*FDM*) and *time-division multiplexing* (*TDM*).

If a transmitted digital signal occupies a narrow band of frequencies, other signals can be transmitted at the same time, provided that their frequencies do not overlap. Frequencies can be shifted at will using modulation (heterodyning). This

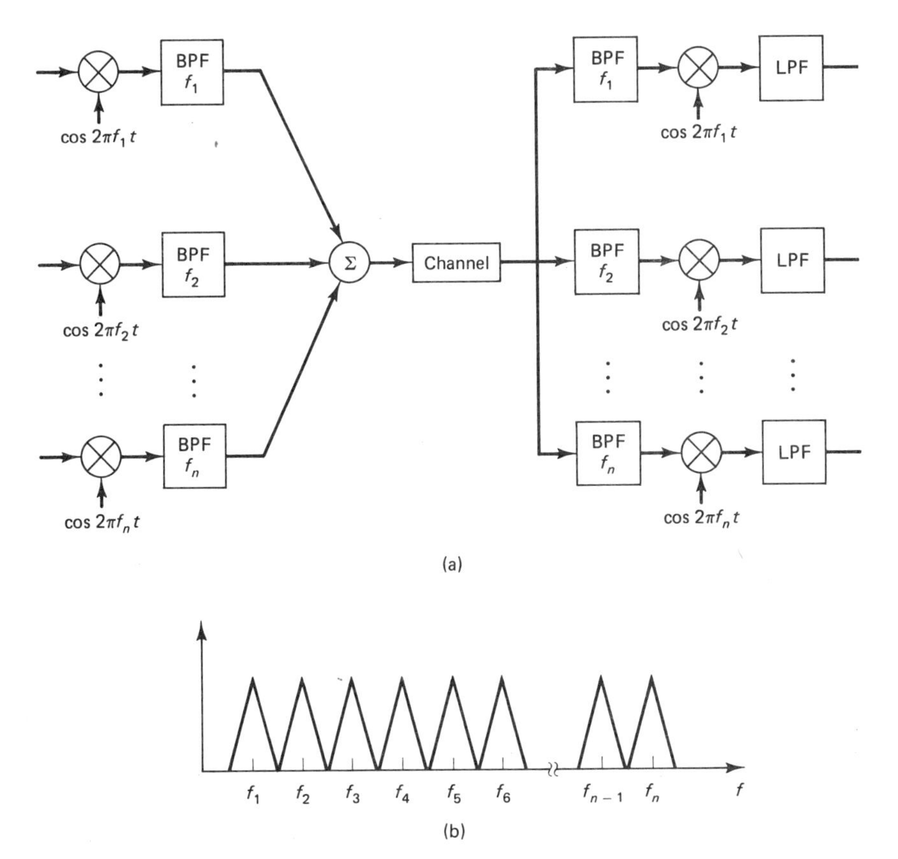

Figure 5.24 Frequency-division multiplexing

is the technique used in all analog modulation schemes. Figure 5.24(a) illustrates the multiplexing and demultiplexing operation. The filters in the transmitter are required to suppress "out-of-band" signal energy. The various frequency assignments are illustrated in Fig. 5.24(b).

In most forms of digital communication, we have the luxury of considering a second form of multiplexing. Since the information is discrete in time, the transmission scheme can provide rest periods between transmissions. For example, in a PCM transmission scheme, we may have to transmit 6 bits every sampling period (6-bit quantization). If these 6 binary digits can be sent in a lot less than the time between samples, we have additional time wherein the channel can be used to send other signals. In fact, in most practical situations, there is enough time to send many signals using the same channel. This interleaving of symbols is known as *time-division multiplexing.* Figure 5.25 illustrates this for a 3-bit PCM signal and multiplexing of four signals. At the receiver, a commutator is required to sort out the various signals. The interleaving, multiplexing, and sort operation is analogous to automobiles entering and leaving a highway using entrances and exits.

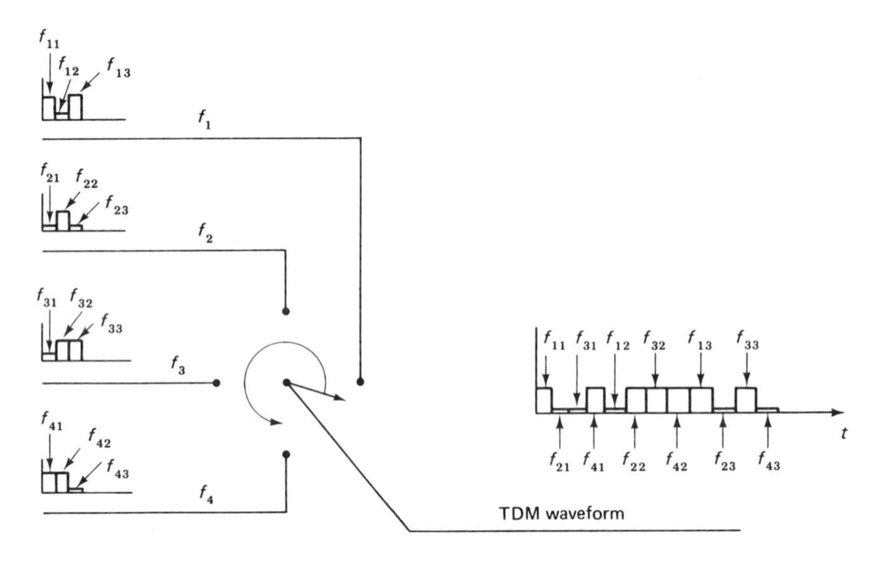

Figure 5.25 Time-division multiplexing

In TDM, synchronization and other timing considerations take on a new meaning. If the commutator in the receiver is not synchronized to that in the transmitter, messages can be interchanged. This could be most annoying in a telephone conversation, if you were talking to one person yet the response was coming from another person. In a data communication system, loss of such synchronization could prove fatal.

5.6.1 T1 Carrier System

In the early 1960s, the need developed to provide capability for commercial and public digital communications. This need arose first with voice channels. The Bell System already had an elaborate carrier system in operation, with cables that could handle bandwidths up to several megahertz. The major advantages of digital communication were well known. That is, greater noise immunity is possible. As a bonus, advances in electronics have made digital equipment less expensive than analog. In addition, the signalling information required to control telephone switching operations could be economically transmitted in digital form.

The *Bell System T1 digital carrier communication system* was formulated to be compatible with existing carrier communication systems. The existing equipment was designed primarily for interoffice trunks within an exchange, and therefore was suited for relatively short distances of about 15 to 65 kilometers.

The system develops a 1.544-megabit/second pulsed digital signal for transmission down the cable. The signal is developed by multiplexing 24 voice channels. Each channel is first sampled at a rate of 8000 samples/second. The samples are then quantized into 127 discrete levels—63 positive, 63 negative, and zero. Thus, 7 bits are required to send each sample value. The coding for each channel is illustrated in Fig. 5.26. An eighth bit is added to each encoded sample for *signalling* information. The 24 channels are then time-division multiplexed as shown in Fig. 5.27. The interleaved order of channels is due to the method of combining, which was designed to use existing multiplex equipment. The quantized samples from the 24 channels form what is known as a frame, and since sampling is done at a rate of 8 kHz, each frame occupies the reciprocal of this, or 125 microseconds.

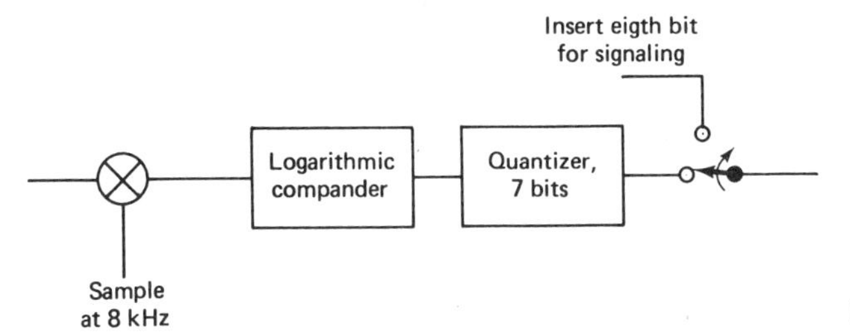

Figure 5.26 Coding for T1 system

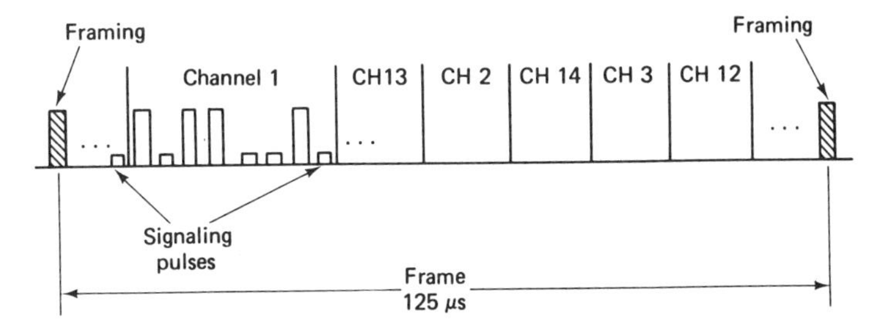

Figure 5.27 Time-division multiplexing for T1 system

There are 192 information and signalling bits in each frame (24 × 8), and a 193rd bit is added for *frame synchronization*. This framing signal is a fixed pattern of alternating 1's and 0's in every 193rd pulse position. Such an alternating pattern will rarely be found in any other position for more than two or three consecutive frames, so synchronization is fairly easy to obtain. Lack of frame synchronization would lead to unacceptable crosstalk between adjacent channels.

With 193 bits per frame and 8000 frames per second, the product yields a transmission rate of 1,544,000 bits per second.

We see that the signalling information for each channel can contain 8000 bits/second of information. If this is not sufficient, as is the case in some foreign dialing, the least significant bit of the information can be used to double the number of bits of signalling information. This, of course, lowers the quantization resolution from 127 to 63 levels.

The T1 carrier system can be used to send data signals as well as PCM, since the basic system is simply a data communication system transmitting 1.544 Mbps. In reality, it is not quite that simple. It is necessary to maintain compatibility with the voiced-oriented T1 system. This means developing the various timing signals to control the receiver clocks. The framing pulses must be sent as with the voice system. The data are therefore usually transmitted in blocks of 8-bit characters.

The TDM system described above assumes that the various signals are being sampled at the same rate, and that each sample is being coded into the same number of bits. That is, the bit rate for each individual channel is identical to that of every other channel. In real systems this might not be the case. As an example, we could be multiplexing a television signal with a voice signal. The bit rate for the television signal is orders of magnitude greater than that of the voice signal. Indeed, even if the individual data rates are nominally the same, small variations occur due to jitter.

The problem of slightly varying rates can be handled with the use of buffers which store a number of bits prior to transmission. Thus, even though the bit rate at the input to the buffer may be varying slightly, the output bit rate can be held constant. The size of the buffer depends upon the magnitude of the variations. If the buffer is too small, it may run out of storage space during increases in the input bit rate, and information bits may be lost. The opposite problem occurs if the input data rate decreases sufficiently that the buffer "runs out of bits." That is, at a time when the buffer should send a bit to the multiplexer, it may be empty. In such cases, *bit stuffing* can be used, where non-information-carrying bits are inserted into the output sequence. Of course, these must be deleted at the receiver, so they must be readily identifiable.

One way to give the receiver the necessary information to allow it to ignore the stuffing bits is to place these bits at regular intervals, and provide a form of identification to tell the receiver whether stuffing is present or not. For example, in the T1 transmission system, each frame consists of 192 information bits plus one frame sync bit. Suppose we now multiplexed several T1 systems together, and occasionally added a stuffing bit at the end of a particular frame. Additional control bits are added which instruct the receiver whether or not a stuffing bit has been added at the end of a particular frame. If the receiver observes the control bits, it knows whether to ignore the stuffing bit (i.e., the 194th bit of a single T1 channel).

Figure 5.28 shows a typical T-carrier structure which is used in North America. Various voice and data inputs are combined into a single 1.544-Mbps T1 channel. This could comprise of 24 multiplexed PCM voice channels, or a combination of voice and data. Four T1 channels are multiplexed into a single T2 signal, with a bit rate of 6.3 Mbps. Note that four times 1.544 is 6.176 Mbps, so the T2 structure contains additional control bits (some used to identify the presence of stuffing bits). The multiplexing continues through various hierarchies, with the T3 structure being a combination of seven T2 structures, and the T4 comprising six T3 groupings. Pictures can be transmitted at the T2 level, but television requires the T3 (43.8 Mbps) level. The *cross connectors* in the figure are not multiplexers. They simply indicate that various types of sources can be fed into the multiplexer.

Without the use of buffers, multiplexers can be synchronous or *asynchronous*.

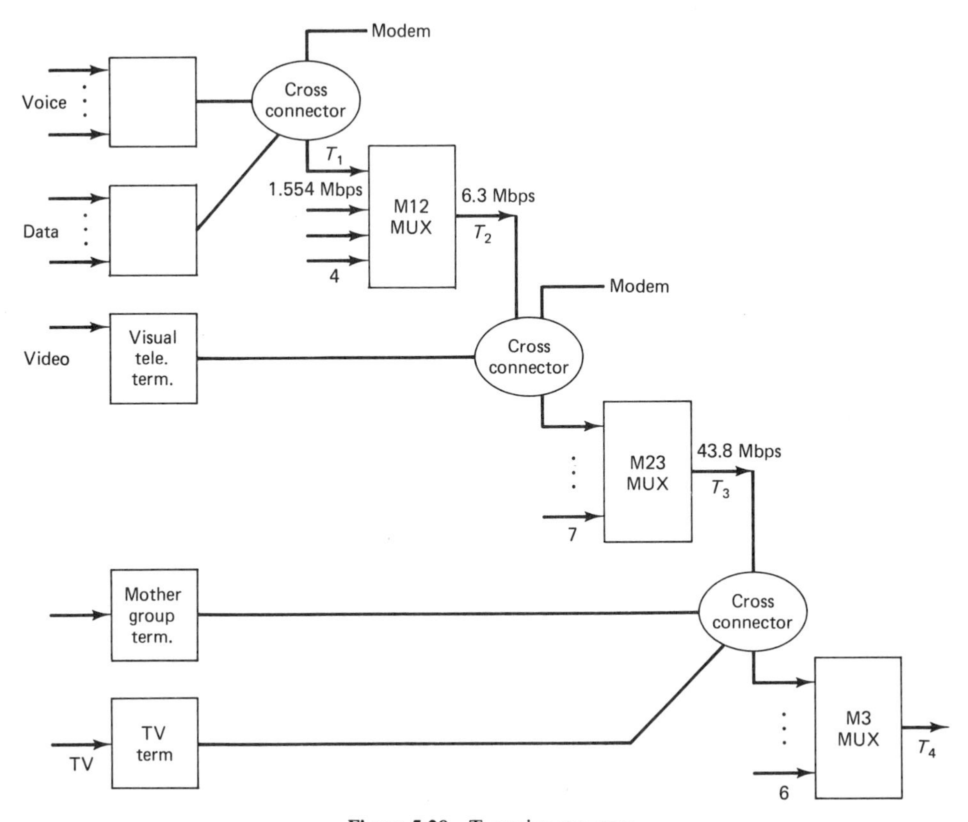

Figure 5.28 T-carrier structure

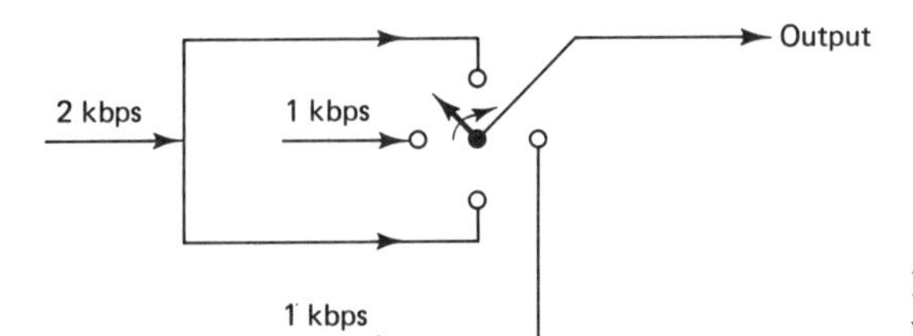

Figure 5.29 Multiplexing of signals with different frequencies

In synchronous multiplexing, the individual channel clocks are derived from a common source. In order to muliplex sources with differing data rates, it is necessary to find a common factor, much as one finds a lowest common denominator when adding fractions together. As a simple example, if you were to multiplex three sources, two producing data at 1 kbps and the other at 2 kbps, you might decide to run your system clock at 4 kHz. Then, every other clock pulse, you would receive a bit from the 2-kbps channel, say on the 1st, 3rd, 5th, 7th, etc. pulses of the system clock. Then, every fourth clock pulse, you could receive a bit from one of the 1-kbps channels, say on the 2nd, 6th, 10th, 14th, etc. pulses of the system clock. Finally, on the 4th, 8th, 12th, 16th, etc. pulses you would receive a bit from the second 1-kbps channel. This is illustrated in Fig. 5.29. Finding the lowest system clock rate becomes a computational problem.

Alternatively, you could use asynchronous multiplexing, where buffers are used to allow individual channel clocks to be independent. This is illustrated in Fig. 5.30.

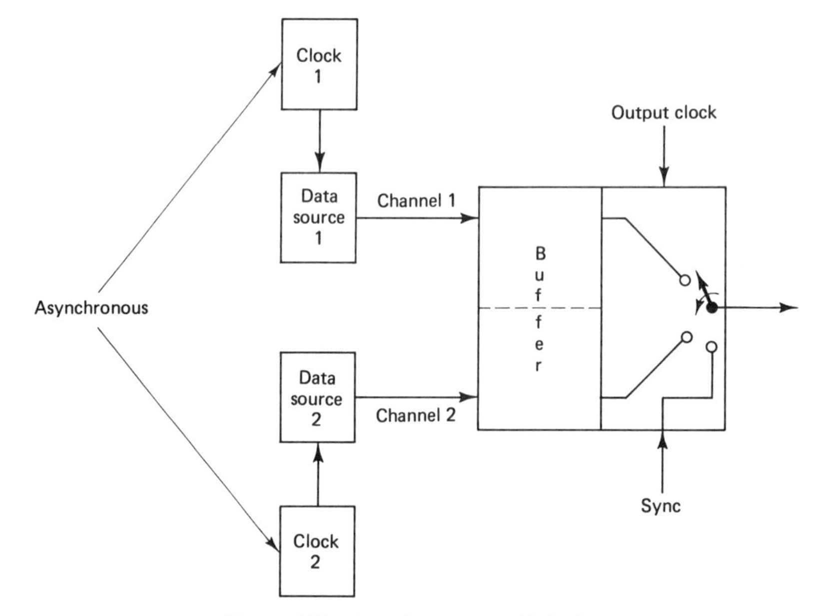

Figure 5.30 Asynchronous multiplexing

5.7 TIME-DIVISION MULTIPLE ACCESS

Time-division multiplexing shares the channel among various users by bit-interleaving the components of the mesasge. Alternatively, *time-division multiple access* (*TDMA*) is a technique for sharing a communication channel among a number of users by assigning each transmitter a time slot. The time slots are chosen to be nonoverlapping when the corresponding signals arrive at the receiver. Since the distance from the various transmitters to the receiver is not necessarily the same, the time slots must be carefully chosen with this in mind. That is, although signals may be nonoverlapping at the time of transmission, if two signals take different amounts of time to reach a receiver, they may overlap upon reception. This consideration is not too critical with relatively short transmission paths. Small differences in transmission time can be accounted for by inserting dead bands between transmissions. This does, however, become a serious consideration in satellite communication systems, where the distances, and therefore propagation times, are greater.

5.8 INTRODUCTION TO PROTOCOLS

Protocols are rules that provide for the orderly exchange of data between users. Although protocols are discussed in considerable detail in Section 13.5, we introduce them here in order to put the overall transmission and reception process into perspective.

Just as there are diplomatic protocols to keep government moving, there must be protocols in almost everything else that involves interaction among communicators. Let us take the classroom as an example. We examine the specific situation of a student asking a question. The student perhaps raises a hand to indicate the presence of a message. The teacher acknowledges the raised hand either by voice or by a nod of the head. This might be called *initialization* of the communication link. If the teacher is not ready for the question, the student's raised hand may be ignored, or the teacher may request that the student wait. The initialization also serves the function of establishing the fact that the student has a message ready to send to the teacher. The student then asks a question. The teacher says: "Do you mean that you want to know why the equation is . . . ?" This part of the intercourse is a type of error-control function. The student responds that the question was understood. The teacher answers the question. The student, being polite, says "Thank you" and the teacher, being equally polite, says "You are welcome," whereupon the teacher turns back to the blackboard, thus finalizing the interaction.

Now think about a data communication interaction, say between a terminal and a remote computer. Much of what was taken for granted in the example above now becomes a serious system-design consideration. We start by considering protocols for a single transmitter and receiver. The protocols for a time-sharing type of situation are more complex.

There are three modes in which data communication can take place. These are known as *simplex, half duplex,* and *full duplex.* In the simplex mode, information flow is only in one direction from the transmitter to the receiver. An example of simplex transmission is television (if data are sent), or airline flight monitors at airports. The half-duplex channel can accommodate transmission in both directions, but only one at a time. This is the most common form of data communication in use. The full duplex allows information to travel in both directions at the same time. Full-duplex operation normally requires two separate channels (e.g., four wires), or multiplexing of the two directions.

Protocols are divided into various levels, depending upon their function. The first, or lowest, level specifies hardware interface parameters for getting individual bits of data from the terminal onto the line. This level of protocol encompasses standards such as the Electronic Industries RS-232-C, which is widely used (although newer standards are replacing it as data rates increase and demands upon the systems become greater). The low levels also include elementary controls to assure that the line is ready to accept data. These are known as *handshaking* sequences, where the source issues a "request to send" signal, and if the channel is ready to accept the data, it replies with a "clear to send."

Line protocols represent a higher level and add sophistication such as error detection. Thus, the line protocol normally operates on blocks of data and concentrates upon getting these blocks through the channel correctly. In the context of these protocols, the channel is a single data link between two terminals. Since messages often traverse more than one link, a higher-level protocol is required to oversee the overall process from start to finish. These are known as *end-to-end protocols,* and we shall briefly concentrate upon them as applied to half duplex, since many of the components are common to other levels. This type of transmission is also the most prevalent.

A message may consist of any number of blocks of data. The source commonly sends a single block and waits for an acknowledgment from the receiver before transmitting the next block. The receiver performs error detection (see Chapter 4) and sends an ACK (acknowledge) if all is proper. Otherwise, it sends a NAK (negative acknowledge), and the source retransmits the block. This procedure leads to delay, of course, due to propagation delays in the channel and the actual time required to send the acknowledgment. The longer the data blocks, the more efficient the system is, but longer blocks have higher probabilities of error.

In reality, more control signals than those described above are needed. The source must be able to ask the receiver for an ACK or NAK if none has been received. Such an inquiry is known as a *reply request* (ENQ). Let us carry this a bit further by looking at two particular scenarios. First assume that a block has been properly received and an ACK sent back, but the acknowledgment never reaches the transmitter. A reply request causes the receiver to retransmit the ACK signal, and all is well. But now suppose that the transmitted block never reached the receiver. The reply request would cause the receiver to repeat its most recent acknowledgment for the previous block, and we shall assume that was an ACK. The transmitter thinks all is well, and continues without repeating the lost block of data.

A relatively simple way to avoid single dropped-block errors is to use a system similar to parity-check coding. The receiver alternates between two different acknowledgment signals, call them ACK-0 and ACK-1. If a reply request results in an ACK that is the same as the most recent one, the source knows that a block has been lost.

To round out this end-to-end protocol, control messages must be added to start or end an interchange and to permit delays if the receiver is not ready to receive the next block of data. In fact, one type of protocol for binary synchronous communications has 15 control characters, marking such things as start of text, end of text, start and end of blocks and intermediate blocks, start of heading, temporary text delay, and others. Some of these were presented as part of the ASCII code in Section 1.3. The interested reader should consult the references for more detail.

Multipoint communication involves more than two locations. In such cases, one location is the *control* and the others are *tributary* stations. The control *polls* the other stations to see if any are ready to send messages. It then selects a station, notifies it to start transmitting, and the other protocols are similar to those discussed earlier.

The protocols discussed so far are considered as *block-oriented* protocols, as they operate upon characters of data. There is a class of newer protocols which are *bit-oriented.* These combine information into a frame with a start and stop (called *flags*) and some control functions. These protocols are faster, since they do not have to constantly perform the back-and-forth handshaking operations.

5.9 SUMMARY

This chapter presents an assortment of topics related to the physical transmission of signals. It deals with the environment within which digital transmission occurs, and as such, contains information that is critical to the operation of communication systems.

The chapter begins with discussion of the various signal formats, including RZ, NRZ, and biphase. The characteristics, advantages and disadvantages of each scheme are explored.

Section 5.2 examines the various data formats with respect to frequency characteristics. We derive expressions for the power spectral density of digital transmissions under both periodic and random data assumptions.

Section 5.3 introduces the concept of multipath, or echo. We discuss the effects of this transmission phenomenon upon intersymbol interference and bit errors. The analysis technique known as cepstrum is introduced.

The extremely important topic of matched filters is examined in Section 5.4. The filter characteristics are derived, and properties are explored in a number of examples.

Decision theory forms the mathematical basis for detection of digital signals within noise. The theory develops criteria for deciding which of the various digital signals (e.g., 0 or 1) is being transmitted during any particular interval of time. The choice of criteria depends upon how much information is known about the transmission system and about the form of signal expected at the receiver. We derive expressions for the maximum a posteriori, maximum likelihood, minimum error, Bayes, minimax, and Neyman-Pearson detectors. The section concludes with a discussion of receiver operating characteristics (ROCs).

The next two sections discuss multiplexing techniques which are used to share a channel among various information sources. The T1 carrier system is presented as an example of a time-division multiplexed system.

The chapter concludes with a brief introduction to protocols. This topic is explored in detail later in the text within the context of computer communication. For the present, we concentrate upon the essentials of setting up a digital message exchange.

PROBLEMS

5.1. A 1-kHz sine wave is sampled 5000 times each second and encoded using 3-bit PCM (you may choose 2-bit plus sign or 3-bit encoding).
 (a) Sketch the waveform for the first 2 msec assuming that the NRZ-L format is used.
 (b) Repeat part (a) for the NRZ-M differential format.
 (c) Repeat part (a) for the biphase-L format.

5.2. The signal shown below is sampled every second and encoded using 4-bit PCM.
 (a) Sketch the waveform for the first 5 seconds assuming that NRZ-L is used.
 (b) Sketch the waveform for the first 5 seconds if the biphase-M format is used.
 (c) Sketch the block diagram of a system that would convert the NRZ-L signal to a biphase-M signal.
 (d) Sketch the block diagram of a system that would convert the biphase-M signal to an NRZ-L signal.

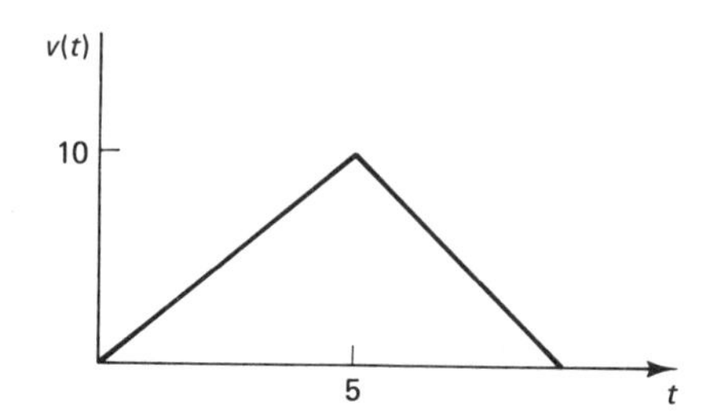

5.3. Prove that about 91% of the total power of a random NRZ-L signal is contained in frequencies below the bit rate. [*Hint:* Integrate the expression of Eq. (5-1).]

5.4. Prove that about 65% of the total power of a random biphase-L signal is contained in frequencies below the bit rate, and that about 86% of the power is contained in frequencies below twice the bit rate. [*Hint:* Integrate the expression of Eq. (5-2).]

5.5. Find the cepstrum of a periodic square wave with fundamental frequency of 1 kHz.

5.6. A channel consists of a direct path and a reflected path, where the transmission time for the reflected path is 1 μs greater than that of the direct path. Assuming that the input is a pure sinusoid, find the channel response as a function of frequency.

5.7. **(a)** Find the matched filter for the signal

$$s(t) = \begin{cases} 10 \cos 2\pi \times 1000t, & 0 \leqslant t \leqslant 0.01 \\ 0, & \text{otherwise} \end{cases}$$

in white noise with power spectral density $N_0/2 = 1$.

(b) What is the output signal-to-noise ratio?

5.8. **(a)** Find the matched filter for the signal

$$s(t) = \begin{cases} 5 \cos 2\pi \times 1000t, & 0 \leqslant t \leqslant T \\ 0, & \text{otherwise} \end{cases}$$

in white Gaussian noise with $N_0 = 10^{-4}$ W/Hz and $T = 5 \times 10^{-5}$ s.

(b) Find the output signal-to-noise ratio at T.

(c) As T approaches infinity, find the limiting value of the signal-to-noise ratio.

5.9. Under hypothesis 0, the received variable, x, has density $p_0(x)$ as shown below, and under hypothesis 1, the density is $p_1(x)$. $P(H_0) = \frac{1}{4}$ and the cost of an incorrect decision is 2. The cost of a correct decision is 0.

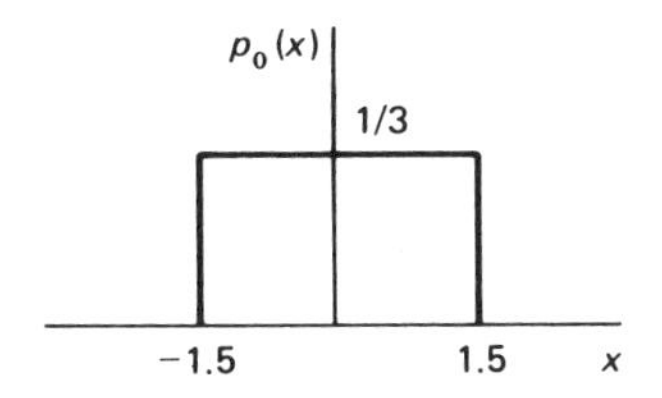

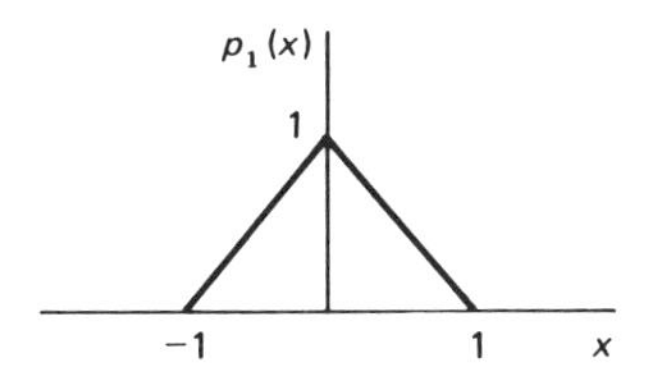

(a) Design a Bayes detector.

(b) Design a minimum-error detector.

(c) If Neyman–Pearson detection is used, and $P_{FA} = 0.1$, what is the form of the detector?

5.10. A received variable, x, is distributed as follows under the two hypotheses:

$$p_1(x) = \frac{1}{\sqrt{2\pi}} e^{-x^2/2}$$

$$p_0(x) = 2e^{-2x} \qquad \text{for } x \geqslant 0$$

The a priori probabilities are $P(H_0) = \frac{1}{4}$; $P(H_1) = \frac{3}{4}$. The cost matrix is

$$C = \begin{bmatrix} 0 & 1 \\ 1 & 0 \end{bmatrix}$$

(a) Design a Bayes detector and find the resulting average cost.
(b) Design a minimum-error detector and find the resulting probability of error.
(c) Design a Neyman–Pearson detector for a false-alarm probability of 0.1. Find the probability of miss for this detector.

5.11. Design a minimax detector for a binary communication system with likelihood functions as shown below and cost matrix

$$C = \begin{bmatrix} 0 & 2 \\ 1 & 0 \end{bmatrix}$$

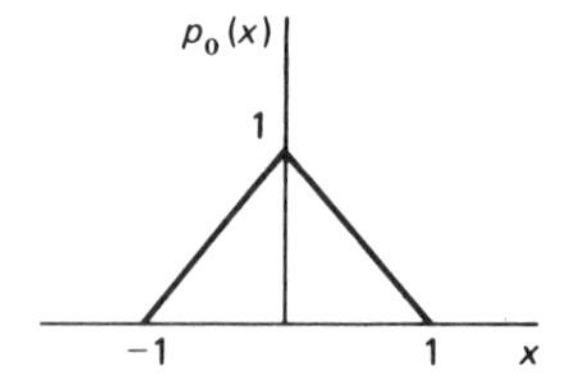

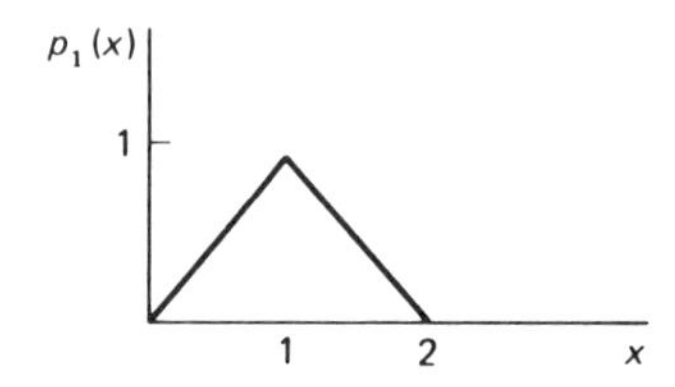

(a) What is the detector criterion in terms of the observation?
(b) Find the resulting probability of false alarm and probability of miss.
(c) Find the average cost assuming equal a priori probabilities.
(d) Find the average cost for a Bayes detector assuming equal a priori probabilities, and compare this to your answer for part (c).

5.12. A choice is to be made between hypotheses H_0 and H_1 on the basis of a single measurement of a quantity, x. Under hypothesis $H_0, x = n$, and under $H_1, x = s + n$. Both s and n are nonnegative independent random variables with density functions

$$p(n) = be^{-bn}U(n)$$

$$p(s) = ce^{-cs}U(s)$$

Assume that c is larger than b. You are given neither costs nor a priori probabilities. Design a detector for a maximum false-alarm probability of 0.1, and find the probability of detection, $P(D_1 | H_1)$, for this detector.

5.13. A binary communication channel is characterized by the error probabilities

$$P(r_1 | s_0) = 0.3, \qquad P(r_0 | s_1) = 0.4$$

where the symbol s_i denotes sending message i and r_i denotes receiving message i.

The a priori probabilities of sending a 1 or 0 are

$$P(s_1) = 0.6, \qquad P(s_0) = 0.4$$

(a) Design a Neyman–Pearson detector using a false-alarm probability of 0.1. (Bear in mind that the only possible received values in the binary channel are 0 and 1.)

(b) Design a minimum-error-probability detector and find the minimum error.

5.14. A random variable, x, has the following probability density functions under the two hypotheses:

$$p_0(x) = \begin{cases} 1 - |x|, & |x| < 1 \\ 0, & \text{otherwise} \end{cases}$$

$$p_1(x) = \begin{cases} \frac{1}{2} - \frac{1}{4}|x|, & |x| < 2 \\ 0, & \text{otherwise} \end{cases}$$

Choosing H_0 when H_1 is true costs twice as much as choosing H_1 when H_0 is true. Correct choices cost nothing. Find the Bayes test under various conditions of a priori probabilities of the two hyotheses.

5.15. You are given the conditional densities of a received variable, y, as shown below. The a priori probability of H_0 is $P(H_0) = 0.9$. $C_{11} = C_{00} = 0$, $C_{01} = 5$, and $C_{10} = 25$.

(a) What is the Bayes decision rule?

(b) What is the average cost?

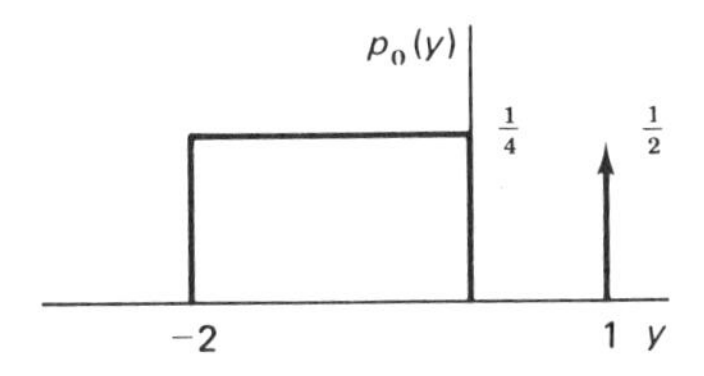

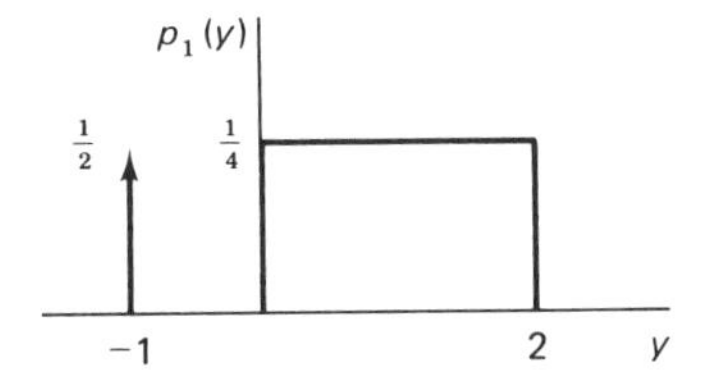

5.16. You are given the conditional densities of the received variable, y, as shown below.

(a) $P(H_0) = \frac{1}{4}$, $C_{00} = C_{11} = 0$, $C_{01} = 2$, and $C_{10} = 5$. Design a Bayes detector and find the minimum average cost.

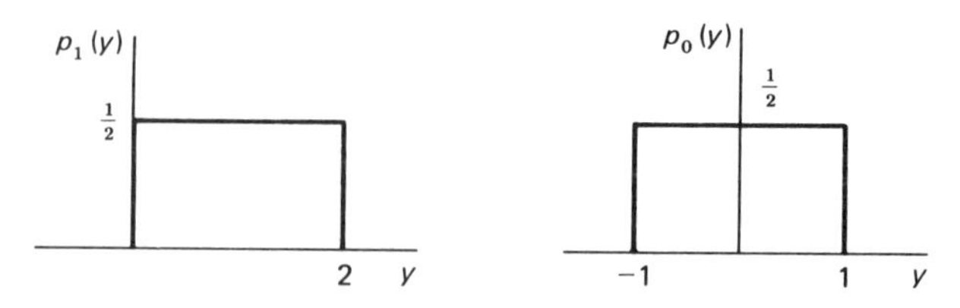

(b) Design a minimum-error detector for the a priori probabilities of part (a), and find the error.

(c) Design a Neyman–Pearson detector for a false-alarm probability of 0.1.

5.17. Design a detector for the following conditional probabilities. Assume that $\lambda_0 = 2$. Also find $P(D_1 | H_1)$.

$$p_1(y) = \frac{1}{\sqrt{2\pi}\sigma} \exp\left[-\frac{(y-1)^2}{2\sigma^2}\right]$$

$$p_0(y) = \begin{cases} \frac{1}{2}, & 0 \leqslant y \leqslant 2 \\ 0, & \text{otherwise} \end{cases}$$

5.18. You are told that there is a correlation between relative humidity and temperature. You wish to decide whether the relative humidity is above or below 50% based upon one measurement of the temperature. The joint probabilities are given in the following table.

	Temperature (°F)						
Humidity	20	30	40	50	60	70	80
10	0.01	0.01	0.01	0.01	0.01	0.01	0.01
20	0.02	0.02	0.02	0.02	0.02	0.02	0.02
30	0.01	0.01	0.01	0.01	0.01	0.01	0.01
40	0.02	0.02	0.02	0.02	0.02	0.02	0.02
50	0	0	0	0	0	0	0
60	0.03	0.03	0.03	0.03	0.02	0.02	0.02
70	0.02	0.02	0.02	0.02	0.03	0.03	0.03
80	0.01	0.01	0.01	0.01	0.01	0.01	0.01
90	0.02	0.02	0.02	0.02	0.02	0.03	0.03

You are given the a priori probabilities:

$$P(\text{humidity} < 50\%) = \tfrac{1}{4}$$

$$P(\text{humidity} > 50\%) = \tfrac{3}{4}$$

Design a minimum-error test to make this decision.

5.19. A choice is to be made between hypothesis H_0 and H_1 on the basis of a single measurement of a quantity x. Under hypothesis 0, $x = n$, and under hypothesis 1, $x = s + n$. Both s and n are independent Gaussian random variables with mean 1 and variance 1. Using the Neyman–Pearson criterion, find the detector rule to yield a false-alarm rate of 0.1. Using this detector, what is the probability of correctly detecting H_1? That is, find $P(D_1 \mid H_1)$.

5.20. A random variable x is distributed according to a Cauchy distribution,

$$p(x) = \frac{m}{\pi(m^2 + x^2)}$$

The parameter m can take on either of two values, m_0 and m_1, where m_0 is less than m_1. Design a detector to decide between the two hypotheses on the basis of a single measurement of x. Use the Neyman–Pearson criterion with a false-alarm probability of 0.15.

5.21. A random variable has likelihood functions as given in Problem 5.11. Plot the ROC curve. Now assume that $p_1(x)$ is shifted to the left by 0.5 (i.e., change the mean from 1 to 0.5). Plot the ROC curve, and compare to the previous curve.

5.22. You are given the signal shown below.

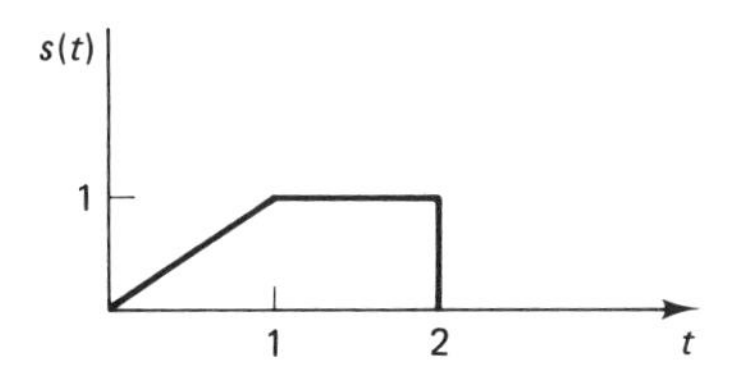

(a) Design a matched filter [i.e., find $h(t)$] for this signal.

(b) Find the signal-to-noise-ratio at the output of the matched filter assuming that the additive noise is white with a power of 1 watt/Hz.

(c) Write an expression for the transfer function, $H(f)$, of this filter.

(d) Suppose that the signal-plus-noise now forms the input to a low-pass filter with a cutoff frequency of 2 Hz. What is the degradation in output SNR when compared to that of the matched filter?

6

Timing

This chapter considers the important issue of timing. Timing is the operation that significantly distinguishes a digital communication system from an analog system. Before a receiver can even begin to decide which of the various symbols it is receiving (e.g., 0's or 1's in a binary system), it must establish symbol timing. That is, it must synchronize a local clock with the timing of the received symbols. Section 6.1 discusses techniques for accomplishing this symbol synchronization.

Once the symbol time is established and the receiver decoder is making decisions regarding which symbols are being received, it is necessary to establish other time frames, including character/word and block/message synchronization. That is, it is not sufficient to decode a long string of symbols unless these symbols can be associated with the correct message. This is particularly important in time-division multiplex transmissions, where the symbols must be decommutated. Frame synchronization is considered in Section 6.2.

Messages often begin with a particular code word which is readily recognizable at the receiver. Section 6.3 considers criteria for such code selection.

Section 6.4 presents several examples of solutions to timing problems as applied to contemporary systems.

6.1 SYMBOL SYNCHRONIZATION

Since the received waveform is usually in the form of an electrical signal extending over all time, it is important to be able to chop the time axis into segments corresponding to the signal segments for each symbol.

Data transmission can be either *synchronous,* in which symbols are transmitted at a regular periodic rate, or *asynchronous,* where spacing between words or message segments is no longer regular. Asynchronous transmission is often given the descriptive name *start/stop*. Asynchronous communication requires that symbol synchronization be established at the start of each message segment or code word. This requires a great deal of *overhead.* In the synchronous mode, symbol timing can be established at the very beginning of the transmission, and only minor adjustments are needed thereafter.

The problem of bit, or symbol, synchronization is greatly simplified if a periodic component exists in the incoming symbol sequence. For example, in the return-to-zero (RZ) signaling system, a transition occurs at the midpoint of every interval in which a 1 is being sent. We can therefore decompose the waveform into a sum of a periodic square wave plus a random pulse signal. This is shown in Fig. 6.1, where a representative binary sequence is being sent.

Because the RZ signal contains a periodic component, the power spectrum consists of a continuous portion (due to the random signal component) and a series of discrete spectral lines at the bit rate and its harmonics. A very narrow band-pass filter tuned to the vicinity of the bit rate can extract the clock information. Alternatively, feedback control systems can be used to "lock on" to the periodic component. The simplest such system is the phase-locked loop.

The *phase-locked loop* (*PLL*) is a feedback system where the error term is the

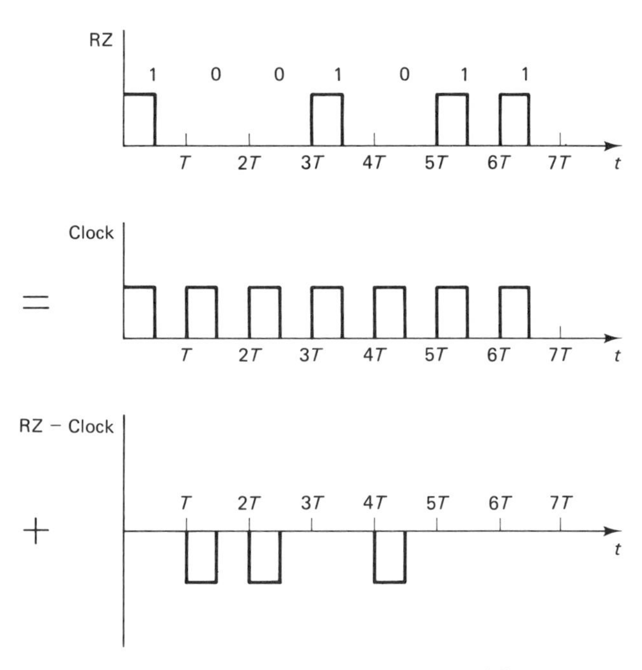

Figure 6.1 Periodic component in RZ signal format

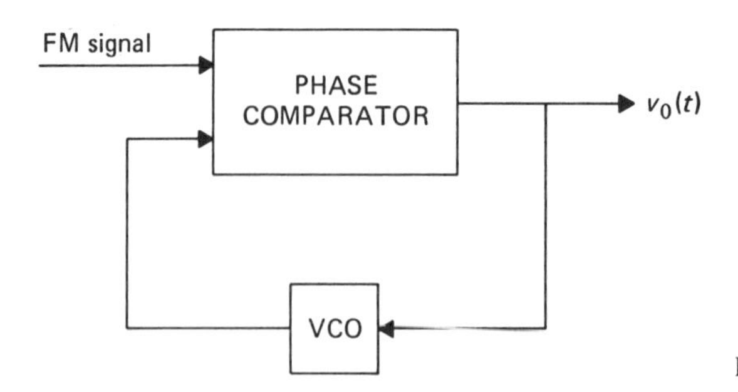

Figure 6.2 Phase-locked loop

phase difference between the input signal and a reference sinusoid. The PLL incorporates a *voltage-controlled oscillator* (*VCO*) in the feedback loop. A VCO is an oscillator whose output frequency is a function of its input voltage. A typical loop configuration is shown in Fig. 6.2. The loop compares the phase of the input signal to that of the signal at the output of the VCO. If the phase difference between these two signals is anything other than zero, the output frequency of the VCO is adjusted in a manner which forces this difference down to zero.

The output of the *phase detector*, $v_o(t)$, forms the input to the VCO. If, for example, the input frequency is higher than the VCO frequency, the phase difference between the two will increase linearly with time, causing the VCO frequency to increase until it matches that of the input. Thus, the VCO output attempts to follow the input frequency. Since the frequency of the VCO output is proportional to the voltage at its input, it is this input voltage which also tries to follow the frequency of the loop input signal.

The simplest method for accomplishing phase comparison consists of a product operation followed by a low-pass filter, as shown in Fig. 6.3. Suppose that the two inputs to the comparator are $\cos(2\pi f_c t + \theta_1)$ and $\cos(2\pi f_c t + \theta_2)$. The output of the low-pass filter will then be $0.5\cos(\theta_1 - \theta_2)$. This output will be a maximum if the two phases are equal. It is preferable to have minimum rather than maximum output when phases are equal, since feedback loops normally drive a particular parameter toward zero. To achieve a minimum output at the desired condition, one of the inputs is changed from cosine to sine. The output of the low-pass filter will then be $0.5\sin(\theta_1 - \theta_2)$. This output will be zero if the phases are equal. That is the desired lock condition.

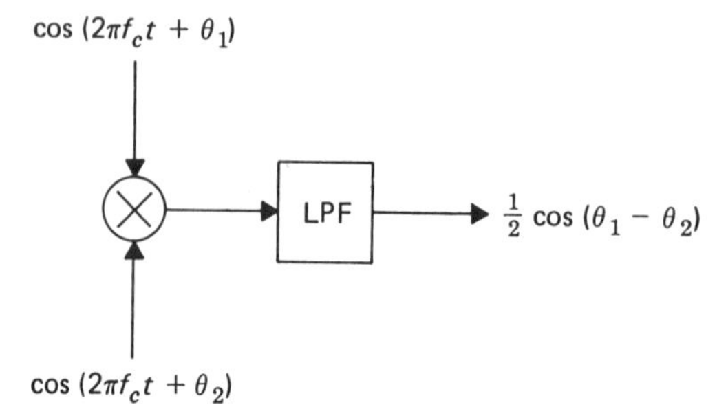

Figure 6.3 Phase comparator

In many cases, it is desirable to have a comparator output which varies *linearly* with the phase difference rather than sinusoidally (i.e., a *linear* phase detector). If the input cosine and sine signals are severely clipped, they can be thought of as square waveforms. A representative situation is illustrated in Fig. 6.4. The product of the two square waves is now averaged by the low-pass filter. This average is proportional to the fraction of time that the square waves are equal. Therefore, the output amplitude is linearly related to the phase difference.

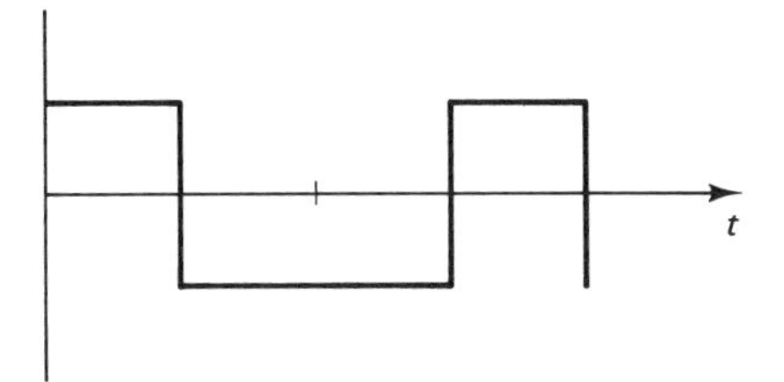

Figure 6.4 Phase comparator with linear relationship

It is important to know loop transient behavior both to examine acquisition time and to know how the loop behaves in the presence of noise. We wish to make a linear approximation to the loop in order to develop a differential equation for its operation. This will lead to simple results for the transient response. To begin this analysis, we redraw the loop as shown in Fig. 6.5, using a multiplier for the phase detector. It is assumed that the loop filter will pass only the low-frequency output of the multiplier.

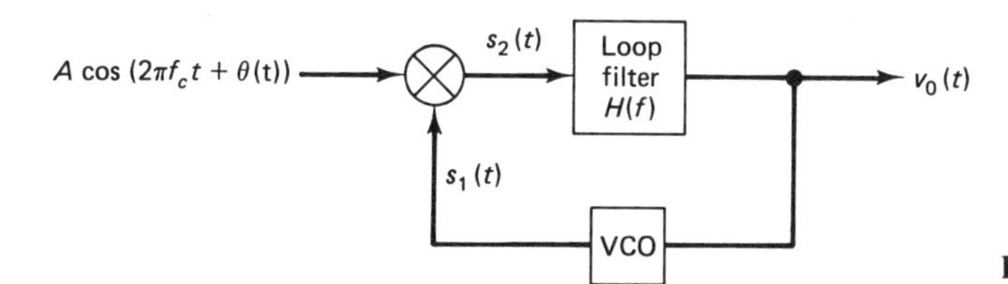

Figure 6.5 Modified phase-locked loop

We assume that the VCO has an output whose amplitude is B and whose frequency is given by

$$f(t) = f_0 + k_0 v_o(t)$$

That is, with a zero input to the VCO, the oscillation frequency is f_0. The frequency varies linearly with the VCO input voltage. Thus, the VCO output $[s_1(t)]$ is given by

$$s_1(t) = B \sin 2\pi [f_0 t + k_0 \int v_o(t)\, dt]$$

The term k_0 is the constant relating the VCO output frequency to the input amplitude. The signal at the output of the multiplier, $s_2(t)$, results from taking the product of two sinusoids. This product can be expanded into one term at the sum and one term at the difference between the two frequencies. We focus attention upon the difference frequency term, since the higher frequency will normally not propagate through the loop filter. The term of interest is given by

$$s_2(t)_{lf} = 0.05AB \sin\left[\int \theta(t)\,dt - 2\pi k_0 \int v_o(t)\,dt\right] \tag{9-6}$$

We will now convert the variables from time functions to phase factors. We define

$$\theta_o(t) = 2\pi k_0 \int v_o(t)\,dt$$

The phase-locked loop can then be redrawn as in Fig. 6.6, where the input is now the phase of the input waveform.

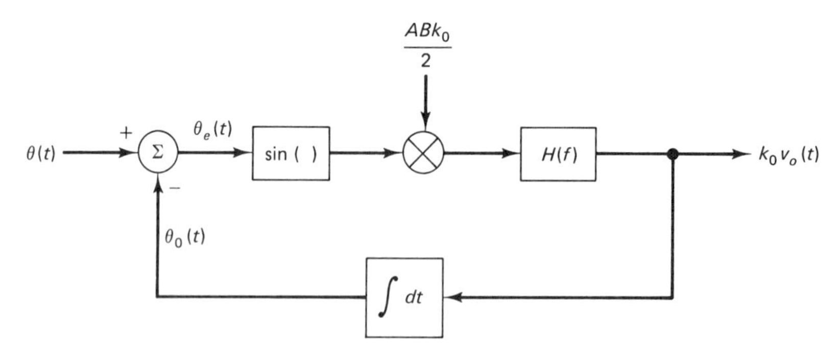

Figure 6.6 Phase-locked loop with angular variables

The *phase error* is

$$\theta_e(t) = \theta(t) - \theta_o(t)$$

When the loop is in the lock condition, $\theta_e(t) = 0$. We can linearize this model by eliminating the sin ( ) block in Fig. 6.6. This is justified by assuming that the loop is close to the lock condition, and thus $\theta_e(t)$ is small. Indeed, had we used the linear phase detector of Fig. 6.4, this block would not have appeared in the first place. The resulting loop is shown in Fig. 6.7.

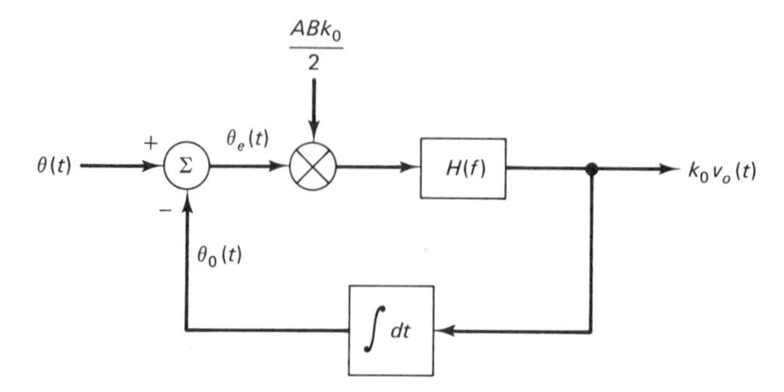

Figure 6.7 Simplified version of phase-locked loop

If the loop filter is an ideal low-pass filter, the PLL is known as *first-order*. The loop equations then become

$$v_o(t) = 0.5AB[\theta(t) - \theta_o(t)]$$
$$= 0.5AB\left[\theta(t) - k_o 2\pi \int v_o(t)\,dt\right] \tag{6-1}$$

When Eq. (6-1) is differentiated, we obtain

$$\frac{dv_o}{dt} = 0.5AB\frac{d\theta(t)}{dt} - \pi ABk_0 v_o(t)$$

The derivative, $(d\theta/dt)/2\pi$, can be considered as the input to the system, since this is the frequency variation. Thus, denoting this input signal as $\Delta f(t)$, the system differential equation becomes

$$\frac{dv_o}{dt} + \pi ABk_0 v_o(t) = \pi AB\,\Delta f(t) \tag{6-2}$$

The overall system function is the ratio of output to input Fourier transform and is given by

$$H_L(f) = \frac{V_o(f)}{\Delta F(f)} = \frac{1}{\dfrac{j2f}{AB} + k_0} \tag{6-3}$$

The magnitude of this system function is shown in Fig. 6.8. The half-power frequency is $f = k_o AB/2$. This gives a measure of the range of frequency variation over which the loop can respond properly.

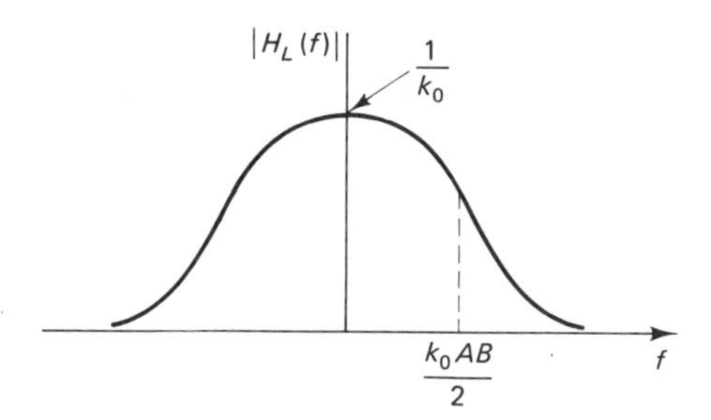

Figure 6.8 Phase-locked loop magnitude response

An indicator of how quickly the loop can respond to changes in $f(t)$ is given by the step response. We solve Eq. (6-2) for a unit step input function to obtain:

$$a(t) = \frac{1}{k_0}[1 - \exp(-\pi ABk_0 t)]u(t) \tag{6-4}$$

We now briefly analyze the PLL operation in the presence of additive noise (additional details can be found in the references). We start by assuming that the input to the loop is the sum of the original input and narrowband noise. Thus, we replace the input in Fig. 6.2 with

$$A \cos [2\pi f_0 t + \theta(t)] + x(t) \cos 2\pi f_0 t - y(t) \sin 2\pi f_0 t$$

This can be expanded using trigonometric identities to yield a sinusoid at a frequency of f_0 with a time varying phase angle given by

$$\tan^{-1}\left[\frac{A \sin \theta(t) + y(t)}{A \cos \theta(t) + x(t)}\right]$$

The phase detector compares the phase of this signal with that of the signal at the output of the VCO. If this is done, and we assume a first-order loop, a high signal-to-noise ratio, and slow angular variations at the output of the VCO, we can modify the block diagram of Fig. 6.7 as shown in Fig. 6.9. This linearized model can be used to find the variance of the phase angle at the output of the VCO. Since the noise is introduced as a linear term in Fig. 6.9, it can be shown that the phase variance is inversely proportional to the input signal to noise ratio. The details are left to the problems at the back of this chapter.

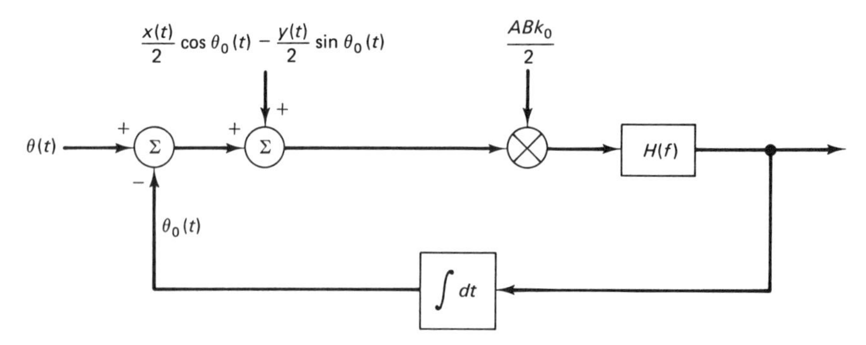

Figure 6.9 Linearized phase-locked loop model

Example 6-1

A phase-locked loop has the following parameters:

$$A = B = 6; \; k_0 = 10 \text{ Hz/V}$$

The loop is locked with $f_0 = 10$ kHz. The input abruptly changes to a frequency of 10,010 Hz. Plot the output frequency as a function of time.

Solution. The frequency input is a step function, so we can use Eq. (6-4). The time constant of the exponential is given by

$$\frac{1}{\pi A B k_0} = 8.84 \times 10^{-4} = 0.88 \text{ msec}$$

The output frequency is sketched in Fig. 6.10.

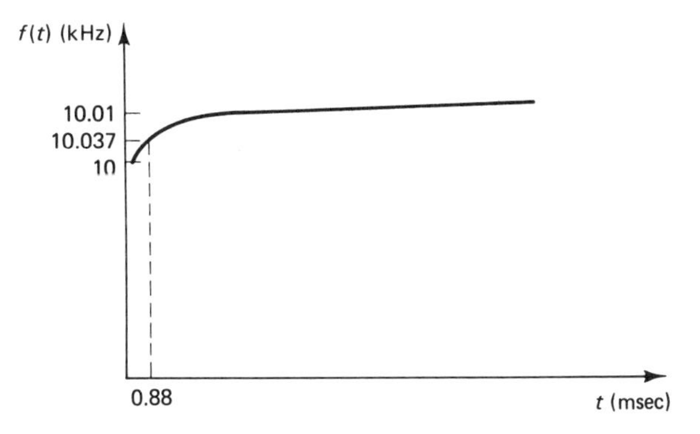

Figure 6.10 Output frequency for Example 6-1

Example 6-2

The input to a phase-locked loop abruptly changes frequency back and forth from 9.9 kHz to 10.1 kHz. The rate of this oscillation in frequency is 100 Hz. $A = B = 6$ and $f_0 = 10$ kHz. Find k_0 for proper loop tracking of this varying frequency.

Solution. The larger the value of k_0, the shorter the transient response time of the loop, so the faster it will respond to the input frequency changes. However, as k_0 increases, the loop transfer function widens and higher-frequency terms are passed through the loop low-pass filter.

The input frequency variation has a fundamental of 100 Hz, and the first nonzero harmonic (since the variation is a square wave) is at 300 Hz. Therefore, if we set the half-power frequency of the loop transfer function to 100 Hz, the harmonic will be severely attenuated. We thus set

$$\frac{k_0AB}{2} = 100 \text{ Hz}$$

and

$$k_0 = 5.56 \text{ Hz/V}$$

The RZ system can be used in those cases where simplicity is important and signal-to-noise ratio is high. The system is not very power-efficient, since no signal is being sent half the time. We therefore suffer a "factor-of-2" or 3-dB performance degradation when compared to NRZ or biphase signaling. However, with NRZ or biphase transmission and random data, the received signal does *not* contain a periodic component at the bit rate. We must therefore resort to more sophisticated bit-synchronization approaches.

We shall explore three of these techniques—the maximum a posteriori, the early-late gate, and the digital data transition tracking loop.

In the *maximum a posteriori* (*MAP*) approach to bit synchronization, we observe the received signal over a finite-length interval and decode the signal using an assumed locally generated clock. The decoded signal is then correlated with a stored replica of the known symbol sequence, and the clock is adjusted to achieve maximum correlation. Note that perfect correlation is usually not obtainable, owing to bit errors which occur in transmission. Further note that this scheme requires part of the transmitted signal to be dedicated to this function. That is, the receiver must know what it is looking for, so signal information cannot be transmitted during this synchronization period. MAP synchronization can be performed in serial or parallel, but either technique entails problems. In the serial mode, adjustments are made in the local clock, and the next group of received symbols is correlated with the stored sequence. This requires a long acquisition time, as the known sequence must be transmitted more than once. Alternatively, the receiver could try different clock adjustments simultaneously, using parallel processing. This can become quite expensive in terms of hardware.

Because of the these shortcomings, practical approaches have been devised which approximate the MAP synchronizer. Most of these techniques use adaptive control loops which lock on to the desired timing by minimizing an error term.

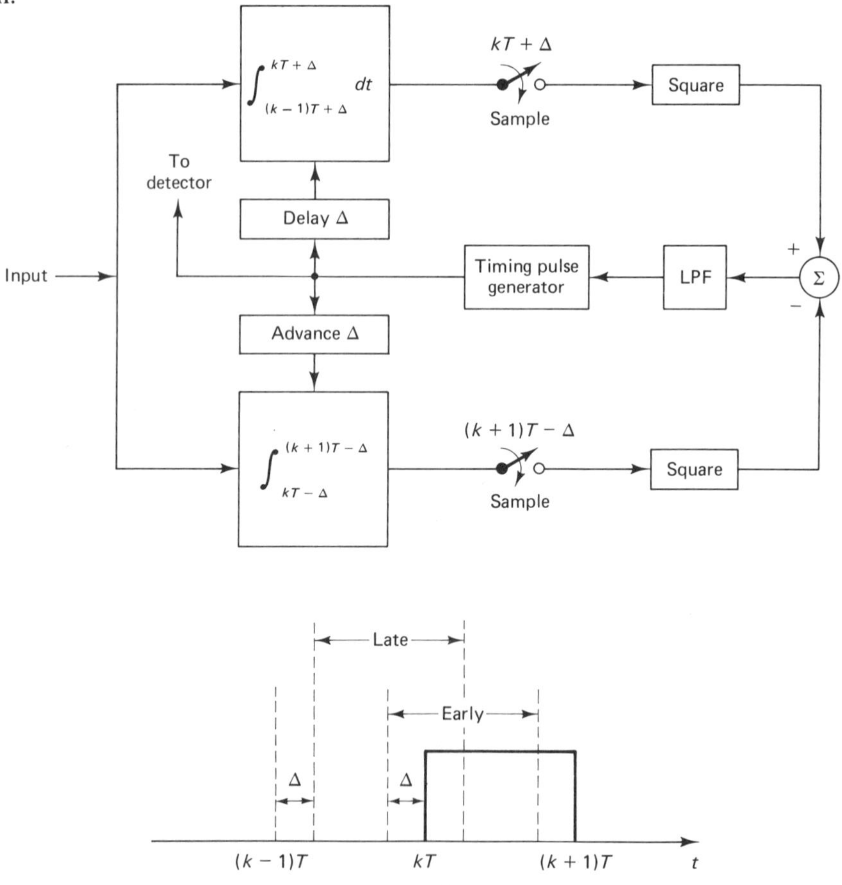

Figure 6.11 Early-late gate

The *early-late gate* is one such approach. A block diagram of the system is shown in Fig. 6.11. The error signal is defined as the difference between the square of the area under the early gate and the square of the area under the late gate. This error goes to zero when the early and late gates coincide. In this case, the correct timing is at the midpoint of the gate periods. When the error is nonzero, Δ changes in a direction to cause the error to go to zero. Note that when the signal is zero, both the early and late gates have zero output, and no adjustments occur.

One implementation of the *digital data transition tracking loop* (DTTL) is shown in Fig. 6.12. When this loop is in lock, the upper portion integrates the input pulse over periods corresponding to the bit period, while the lower portion integrates over intervals spanning transitions. This is best understood by examining typical waveforms when the loop is in lock.

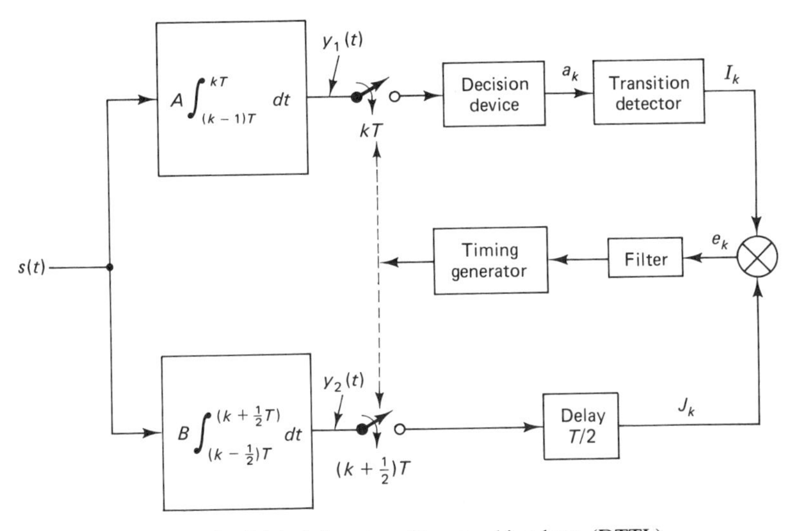

Figure 6.12 Digital data transition tracking loop (DTTL)

Figure 6.13 illustrates a typical set of waveforms. Figure 6.13(a) shows a representative NRZ-L baseband waveform. Figures 6.13(b) and (c) show the output waveforms for the in-phase and quadrature integrators. Note that the quadrature integrator dumps (sets to zero) at the midpoints of the intervals.

The remainder of the system compares integrator outputs from one interval to the next. Figure 6.13(d) shows the output of the decision device, which is simply a sampler and hard limiter. It tests the output of the in-phase integrator to see if it is positive or negative. We indicate a positive output with a $+1$ pulse, and a negative output with a -1 pulse. If two adjacent pulses are the same, it indicates that no transition occurs in the original waveform. On the other hand, if two pulses are different, a transition occurs. The transition detector compares adjacent outputs and creates the I_k waveform of Fig. 6.13(e). In the case where no transition occurs, we

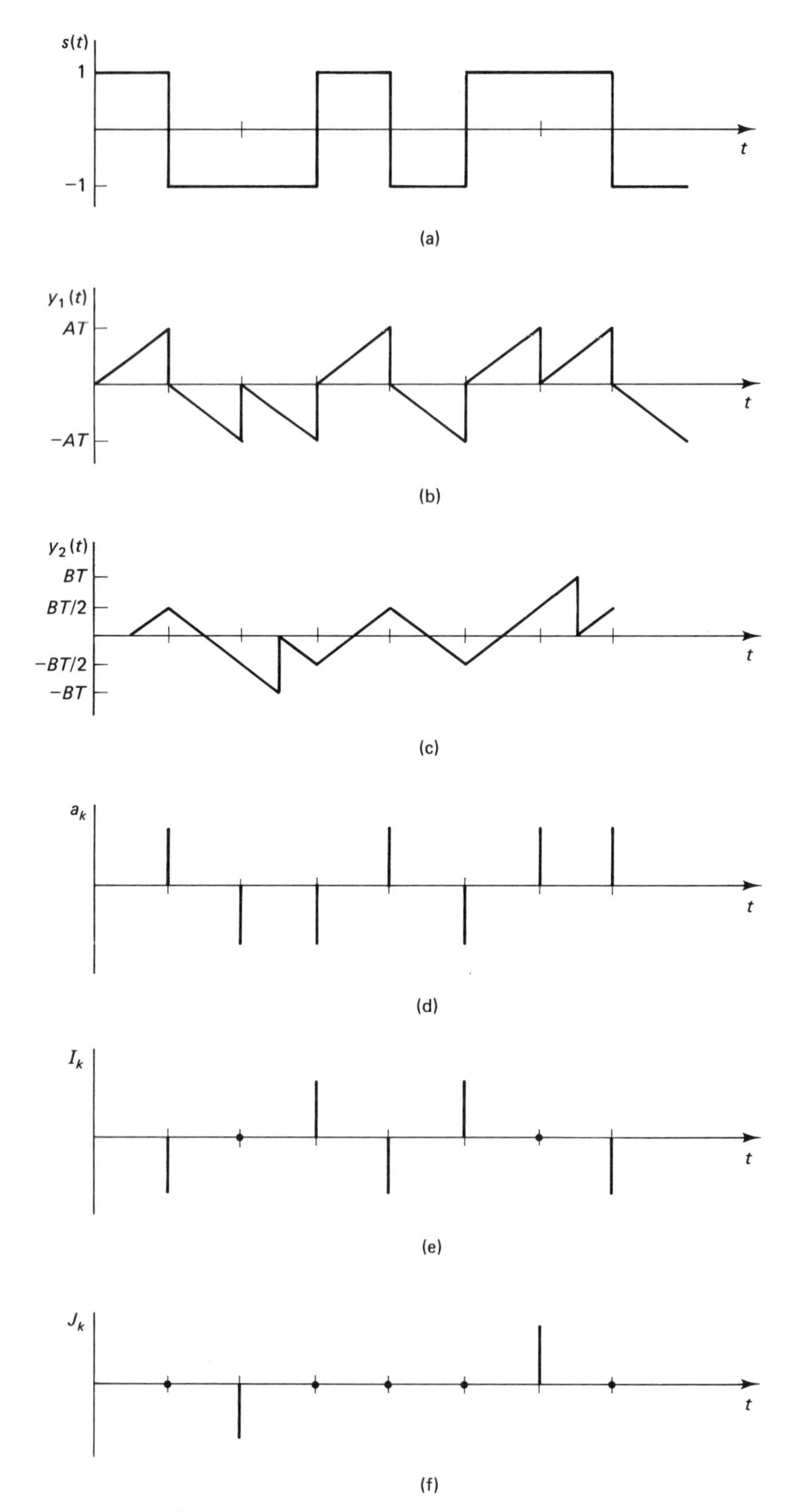

Figure 6.13 Typical waveforms for the DTTL

6.3 CODES FOR SYNCHRONIZATION

The previous section discussed the process of inserting a distinctive symbol sequence to mark the beginning of each frame. This sequence is known as a *marker* or *prefix*. Let us assume that this sequence is m bits long. The receiver then correlates each m-bit received sequence with a stored version of the known prefix. If the correlation exceeds a specified threshold, the search mode is exited.

We desire a prefix sequence for which there is a minimum probability of falsely establishing frame synchronization at the wrong location.

For example, suppose we choose the 7-bit prefix sequence 0001101. The received sequence will be of the form

$$\ldots x\,x\,x\,x\,x\,x\,0\,0\,0\,1\,1\,0\,1\,x\,x\,x\,x\,x \ldots$$

where "x" is used to denote a portion of the information message. We will consider the x's to be randomly distributed with equal probability of each being 0 or a 1. The *correlation* between the stored sequence and the portion of the received sequence is the difference between the number of bit agreements and the number of bit disagreements. Thus, the correlation ranges from $+7$, when all bits are identical in the two sequences, to -7, when all bits are different. Thus, when synchronization is achieved, in the absence of bit errors, the correlation is 7. Suppose we compare the two sequences and that the timing is incorrect by one bit period. We then compare

$$x\,0\,0\,0\,1\,1\,0$$

with

$$0\,0\,0\,1\,1\,0\,1$$

The correlation will be either 1 or -1, depending upon whether x is 0 or 1, respectively. In the other direction, we compare

$$0\,0\,1\,1\,0\,1\,x$$

with

$$0\,0\,0\,1\,1\,0\,1$$

Again, the correlation will be either 1 or -1, depending upon whether x is 0 or 1, respectively.

It can be seen that, in designing the prefix sequence, we wish to obtain minimum correlation of the sequence with truncated shifted versions of itself (contrast this with the PN sequence, where the shifted versions are not truncated—they are cyclic).

The truncated correlation, for a shift of k positions, is given by

$$C_k = \sum_{i=1}^{m-k} x_i x_{i+k}$$

where x_i is either $+1$ or -1, and m is the length of the prefix.

Among the most promising codes for use in synchronization are the *Barker code,* for which the correlation is limited to a magnitude of 1 for $k \neq 0$. Barker codes have been discovered for lengths of 3, 7, 11, and 13. Thus, for example, with a length of 13, the correlation for a zero shift is 13, while for any nonzero shift, the correlation is bounded in magnitude by 1. Unfortunately, Barker codes do not exist for lengths greater than 13. The known Barker codes are given by

1 1 0

1 1 1 0 0 1 0

1 1 1 0 0 0 1 0 0 1 0

1 1 1 1 1 0 0 1 1 0 1 0 1

In a very noisy environment, we may require prefixes longer than 13 bits in length. *Neuman-Hofman* codes are designed to minimize the maximum value of cross correlation between the prefix pattern and the incoming data stream. As an example, the Neuman-Hofman code of length 24 is

0 0 0 0 0 1 1 1 0 0 1 1 1 0 1 0 1 0 1 1 0 1 1 0

The correlation, with zero shift, is 24. The maximum magnitude correlation for other shifts occurs at a shift of 10 and results in a correlation of -4. The reason we are concerned with negative correlations (i.e., magnitude) is that these codes are often used in differential systems where bit reversals may occur. Thus, rather than look for the maximum positive correlation to declare frame synchronization, we look at the magnitude of the correlation.

6.4 DESIGN EXAMPLE

Suppose you are asked to design a system for military applications which must transmit 1 million information bits per second. The information is arranged in frames of 1000 bits each. The environment is extremely noisy, and the application is highly sensitive. You must design the frame synchronization system to have a probability of falsely locking at the wrong point of less than 10^{-6}. Frame sync must be acquired within 5 frames with a probability greater than 99.9%.

We are at quite an early point in our study of digital communications to tackle a design problem of this magnitude. However, we shall use it to establish parameters and motivate the further studies.

A very important quantity that the problem has not specified is the bit error rate. The problem states that the environment is very noisy, so we could expect bit error rates of the order of 10^{-2} or more. As the bit transmission rate increases, less time is available to send each bit, so the energy per bit decreases and the pulse spreading (intersymbol interference) increases. Thus, as the transmission rate increases, the bit error rate also increases. This is an important observation, since lengthening of the frame synchronization prefix must increase the transmission rate. The problem specifies that each frame contains 1000 information bits, and the information must be transmitted at 1 Mbps. Thus, the bit transmission rate will be $(1000 + N)/1000$ times 1 Mbps.

The actual relationship between bit error rate and transmission speed depends upon the noise power, the bandwidth of the channel, channel factors such as multipath, the transmission scheme, and the detector design. Since we need this information to solve the problem, and many of these topics are not covered until later in this text, let us assume that the bit error rate is related to the transmission rate by the graph of Figure 6.17. The key values are summarized in the following table:

Transmission rate (Mbps)	Bit error rate ($\times 10^{-2}$)
1	1
1.001	1.0005
1.002	1.0011
1.003	1.0018
1.004	1.0026
1.005	1.0035
1.006	1.0045
1.007	1.0056
1.008	1.0068
1.009	1.0081
1.01	1.0095

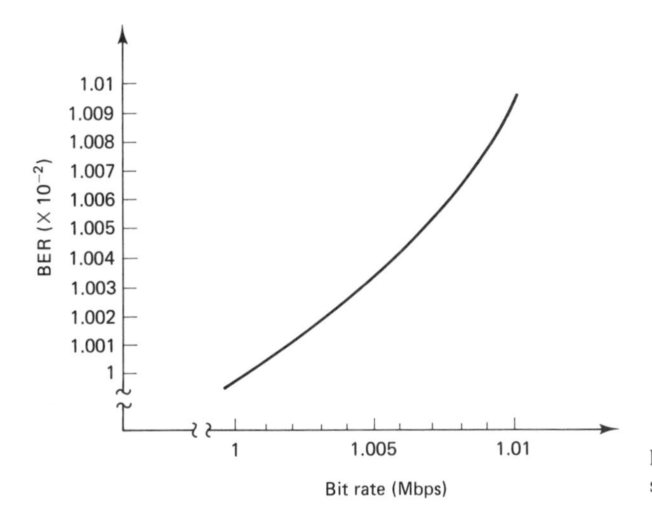

Figure 6.17 Bit error rate vs. transmission rate for design example

We are now ready to try various framing techniques. In order to start the analysis, let's assume that a single framing bit is added to each frame, and that the receiver looks for that bit over a single frame. This is clearly not acceptable, since the probability of false lock would be 50%—that is the probability that any information bit would be the same as the transmitted frame sync bit. Even if the receiver scans over 5 adjacent frames (as permitted in the specifications), the probability of the information's having any specified 5-bit pattern is 2^{-5} or 3.1×10^{-2}. The specifications call for a false-lock probability of less than 10^{-6}. Clearly, we must go to longer frame sync sequences.

We will start by concentrating upon the probability of acquiring synchronization in order to gain facility with the probability calculations.

Suppose we decide to try a Barker code of length 3—that is, 110. Suppose further that we set the threshold at the correlation of 3. That is, we look for perfect agreement between the received prefix and the locally generated replica. The probability of locking in a single frame will then be the probability that no bit errors occurred in the prefix, or

$$\Pr(\text{lock}) = (1 - \text{BER})^3 = (1 - 0.010018)^3 = 97\%$$

This does not meet the specifications, so there is no need to even examine the probability of falsely locking on the wrong point. We must investigate either lowering the threshold or viewing more than a single frame at a time.

Suppose we decide to lower the correlation threshold to 2. Then the probability of locking in a single frame is still the probability of no bit errors. This is true since a single bit error will reduce the correlation by 2. Thus, the probability of locking in a single frame still does not meet the specifications. We will not even waste time trying to lower the required correlation threshold to 1, since it should be clear that the probability of false locking under this condition would be unacceptable.

The alternative is to increase the number of frames over which we look for agreement. Suppose we examine the received sequence over two frames, and look for a total correlation of at least 4. We can then permit a single bit error over the 6 bits, so the probability of acquisition becomes

$$(1 - \text{BER})^6 + 6 \times \text{BER} \times (1 - \text{BER}^5) = 99.85\%$$

This still does not meet the specifications. If we wish to permit 2 bit errors over the two frames (i.e., look for a total correlation of at least 2), we will meet this specification but not be close to the false-lock specification. We must look over at least three frames. Suppose we do this and require a total correlation of at least 5. That is, we are examining 9 bits and are permitting up to 2 bit errors. The probability of acquisition is then

$$(1 - \text{BER})^9 + 9 \times \text{BER} \times (1 - \text{BER})^8 + 36 \times \text{BER}^2 \times (1 - \text{BER})^7 = 99.99\%$$

We have therefore met the first specification and must now examine the probability of false lock. Unfortunately, this is where we get into trouble. The probability of falsely locking is the probability of getting at least 7 agreements between the synchronization sequence (three 3-bit transmissions over three frames) and the data. If the data are considered to be random, this probability of false lock is

$$(0.5)^9 + 9 \times (0.5)^8(0.5)^1 + 36 \times (0.5)^7(0.5)^2 = 46 \times (0.5)^9$$
$$= 8.98 \times 10^{-2}$$

which clearly does not meet the specification. Even if we were to look over the full five frames permitted by the specifications and require perfect agreement (doing so could not achieve the required probability of acquisition), the probability of false alarm would be

$$(0.5)^{15} = 3 \times 10^{-5}$$

This does not meet the specification of 10^{-6} maximum probability of false lock. Obviously, we need a longer synchronization sequence.

Suppose we go to the 7-bit Barker sequence. Let us approach the analysis by examining acquisition probability as a function of required correlation. Note that the bit error rate increases to 1.0056×10^2, since we must transmit at a faster rate. If we use the full five frames, we are transmitting 35 synchronization bits. The probability of errors among these 35 bits is found from the binomial distribution and is given as follows:

$$\text{Pr (0 errors among 35 bits)} = (1 - \text{BER})^{35} = 0.702$$

$$\text{Pr (1 error among 35 bits)} = 35 \times (\text{BER})^1(1 - \text{BER})^{34} = 0.2496$$

$$\text{Pr (2 errors in 35 bits)} = \binom{35}{2}(\text{BER})^2(1 - \text{BER})^{33} = 8.62 \times 10^{-2}$$

$$\text{Pr (3 errors in 35 bits)} = \binom{35}{3}(\text{BER})^3(1 - \text{BER})^{32} = 4.82 \times 10^{-3}$$

$$\text{Pr (4 errors in 35 bits)} = \binom{35}{4}(\text{BER})^4(1 - \text{BER})^{32} = 3.914 \times 10^{-4}$$

$$\text{Pr (5 errors in 35 bits)} = \binom{35}{5}(\text{BER})^5(1 - \text{BER})^{31} = 2.465 \times 10^{-5}$$

Since the specifications require a 99.9% probability of locking within five frames, we see that permitting up to three errors among the 35 synchronization bits would meet this specification. That is, if three errors are permitted, the probability of NOT locking is the probability of four or more errors, which is approximately 4×10^{-4}. Thus, the probability of acquisition is approximately 99.96%, which meets the specification.

We require a correlation of at least 29 [$35 - 2 \times 3$)] to declare that frame synchronization is achieved. It is now necessary to check that the probability of false lock is within specifications. The probability of achieving agreement between 32 of the 35 bits, assuming random data, is given by

$$\Pr(\text{false lock}) = 1 \times (0.5)^{35} + 35 \times (0.5)^{34}(0.5)^{1} + \binom{35}{2}(0.5)^{33}(0.5)^{2} + \binom{35}{3}(0.5)^{32}(0.5)^{3} = 2.09 \times 10^{-7}$$

This meets the specification, so we have been successful at configuring one possible frame synchronization scheme.

We have ignored other design trade-off considerations, including the entire issue of hardware and implementation. We will be ready to tackle these issues after learning about specified transmission techniques.

6.5 SUMMARY

Timing considerations are extremely important in digital communication. The first operation required of a digital detector is to establish a local clock which is synchronized to the symbol transmission rate. This operation is required before the detector can start making decisions regarding the particular digital sequence being transmitted. The operation, known as symbol synchronization, is discussed in Section 6.1. The operation of the phase-locked loop is analyzed for both steady-state and transient response. This loop is used for transmitted carrier communications. In the case of suppressed carrier, we use the MAP approach and several control-loop approximations to this approach. These include the early-late gate and the data transition tracking loop (DTTL). We also briefly examine non-linear clock recovery systems.

Once symbol synchronization has been established, the next step is to acquire frame synchronization. This operation partitions the received symbols into frames representing specific messages. The particular approach of using a prefix is discussed in Section 6.2. We explore the PN sequence as a potential choice for the prefix.

We define the various modes of frame synchronization as search, check, and lock.

Section 6.3 examines particular binary codes with properties suited to use as prefixes. The Barker and Neuman-Hofman codes are presented as having desirable properties for use as the prefix.

The final section of the chapter discusses the various trade-off considerations required in designing a synchronization system. This is done in the context of designing a military system for use in a very noisy environment. The specifications for the system are extremely rigid. In particular, the choices of the length and type of prefix are fully explored.

PROBLEMS

6.1. A first-order phase-locked loop has the following parameter values:

$$A = B = 1, \qquad k_0 = 5 \text{ Hz/V}, \qquad f_0 = 10 \text{ kHz}$$

Plot the output frequency of the PLL as a function of time if a sinusoidal input of frequency 10,010 Hz is applied at time zero.

6.2. Repeat Problem 6.1 if the input is changed from a sinusoid to a square wave.

6.3. The clock in a certain system has jitter. Its frequency is given by

$$f(t) = 10^4 + 100 \cos (2\pi \times 300t)$$

Design a PLL that will track this within 10 Hz.

6.4. In the system of Fig. 6.9, the additive white Gaussian noise has a power of 10^{-5} watts/Hz. The loop has parameters

$$A = B = 1, \qquad k_0 = 5 \text{ Hz/V}$$

The noise-free loop input is a sinuosid of frequency 10 kHz. Find the loop output and the variance of the frequency variation of this output.

6.5. The input to a PLL is a pulse train of very narrow pulses at a frequency of 10 kHz. The pulse frequency varies slightly by a maximum amount of 5 Hz. This variation occurs at a maximum frequency of 100 Hz. That is, the time it takes for the frequency to deviate by the maximum amount of 5 Hz away from 10 kHz is no shorter than 100 msec. Design a first-order PLL to accurately track the input frequency.

6.6. Suppose that in a maximum a posteriori bit-synchronization scheme, a known message of 50 bits in length is sent in order to provide for symbol synchronization. The probability of correct reception is as shown below, where the abscissa is the amount of timing mismatch. That is, when the receiver clock is perfectly synchronized, there is a 0.99 probability of correct reception. With a mismatch of one-half of the bit period, the probability of correct detection drops to 0.5. The system is considered synchronized if there are 45 or more agreements between the transmitted and detected waveform. Find the probability that the system locks to within 1% of correct synchronization.

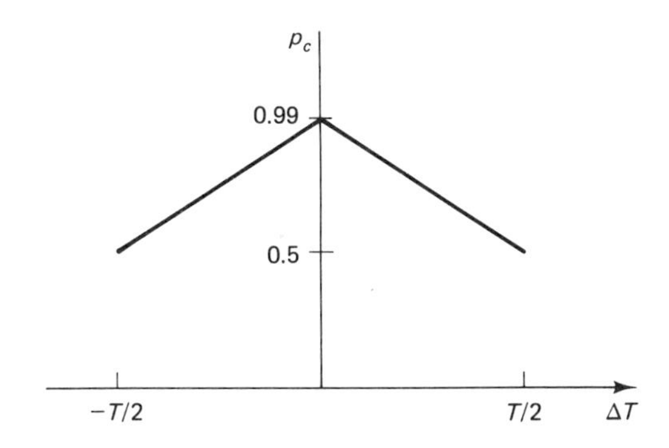

6.7. Assume that the early-late gate of Fig. 6.11 is in perfect lock with a baseband signal at the input. Sketch the waveform at the various points within the block diagram.

6.8. The DTTL is shown in Fig. 6.12. Repeat the waveform sketches of Fig. 6.13, assuming that the system is one-fourth of a bit period away from the lock position.

6.9. The system of Fig. 6.14 is used to recover clock information with a biphase-L coded input signal. Sketch the waveform at each point in this system if the input is a biphase-L signal representing the binary train 00101110.

6.10. Find the probability of false word synchronization using a 15-bit-length PN code as a preamble.

6.11. A frame-synchronization scheme is to be designed with a search mode, a check mode, and a lock mode. Each frame contains 250 bits of information and a preamble of length N bits. The frame must be transmitted within 1 sec. The error rate is given by

$$\text{BER} = 0.5 - 0.5e^{-f/2500}$$

where f is the number of bits per second being transmitted.

(a) If $N = 10$ and frame synchronization must be acquired within 10 frames (with 99% probability), find the threshold to be used in the search mode, T_S.

(b) Once the system is in lock, we wish it to remain in this condition with probability 99.9%. Find the threshold value, T_L.

(c) Repeat parts (a) and (b) if N is raised to 100 bits.

6.12. Calculate the correlation for the Barker code of length 7.

6.13. Calculate the correlation for the Neuman-Hofman code of length 24.

6.14. Repeat the design example of Section 6.4 if the environment is no longer noisy. Assume that the bit error rate is 0.1% of the value given by Fig. 6.17.

7

Baseband Systems

In analog communication, the baseband signal is the unaltered analog information signal. For example, in voice communication, the basic audio waveform would be considered to be the baseband signal. The analog baseband signal is often used without modification. This is particularly true for wire communication. For example, the audio signal in the wire between an amplifier and the speaker in an analog sound system is usually baseband. When the communication channel is space, the baseband signal must be modified prior to transmission. This is true for both digital and analog communication.

A digital communication system starts with a list of numbers. These numbers must be converted to waveforms for transmission. This chapter focuses on the criteria for selection of signal waveshapes.

The list of numbers must somehow be converted into an electrical signal. The system is known as *baseband* if the conversion is done in such a way as to result in a low-frequency electrical signal (extending approximately to DC or zero frequency).

The study of baseband systems establishes some benchmarks that will prove useful in evaluating carrier systems. In addition, the baseband signal will often form the starting point for developing nonbaseband signals.

This chapter begins with a general discussion of baseband signals, including coding and decoding techniques. We then examine intersymbol interference and methods of reducing the effects of channel distortion. The performance of the baseband system is discussed in Section 7.4. The matched filter detector is analyzed in detail in Section 7.5. The discussion in that section is sufficiently general to permit

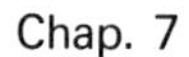

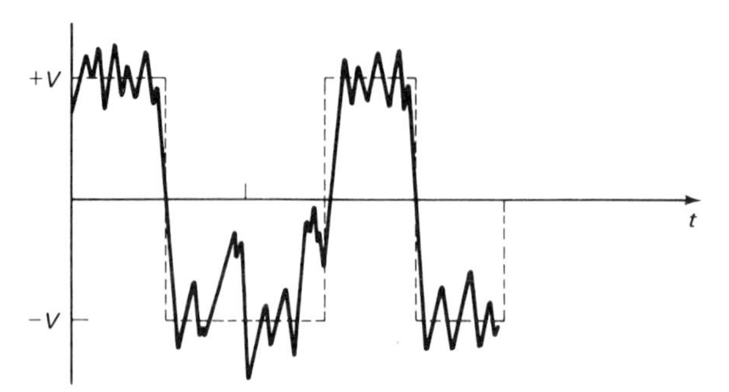

Figure 7.1 Example of noisy baseband signal

the results to be applied directly to the binary systems presented in the three chapters that follow.

The next section introduces M-ary communication, where the number of symbols in the message alphabet is increased beyond two.

Section 7.7 discusses techniques for communicating at rates higher than the basic channel would seem to permit. The resulting intersymbol interference is controlled in a manner that permits elimination of its detrimental effects. We conclude with a discussion of the threshold effect and with a study of system design.

7.1 INTRODUCTION

A baseband digital system is a system that transmits low-frequency signals in order to communicate digital information. The baseband signal design presented in this chapter is a digital version of analog pulse-amplitude modulation (PAM).

A simple proposal for transmitting a sequence of 1's and 0's might be to transmit a 1-V signal for a digital 1 and a 0-V signal for a digital zero. This is known as *unipolar* transmission, since the signal waveform deflects from zero in only one direction. A slight modification of this natural choice is usually made. In fact, we consider sending a constant of $+V$ volts for a digital 1 and $-V$ volts for a digital 0. This is called *bipolar* transmission. V is chosen on the basis of available power and other physical considerations. We use $+V$ and $-V$ rather than $+V$ and zero because, as we shall see shortly, there are advantages to making the two signals as different as possible.

Figure 7.1 shows a signal waveform that might be used to send the digital sequence 10010. The dashed curve is the signal, and the solid curve is an example of what this signal might look like at the receiver once it is corrupted by additive noise in the channel.

Channel distortion is not being taken into account in this sketch. To do so would involve time delay and a rounding and spreading of the pulses. The specific coding scheme illustrated is the NRZ, or nonreturn to zero.

The receiver must determine whether the unperturbed signal is $+V$ or $-V$ within each sampling interval. Methods for making this decision are described later in this chapter.

Baseband transmission is used primarily where the communication channel is a coaxial cable. With other types of communication, much information would be lost, since the majority of noncable channels do not pass low frequencies.

To assess the distortion effects of channels upon the signal, we need some idea of the frequency spectrum of the waveform. We presented this briefly in Chapter 5 but will repeat some of those results here along with an abbreviated derivation. We do so because we will be generalizing the pulse shapes later in this chapter. That is, we will not always be using square-shaped pulses.

The unperturbed waveform can be considered as the output of the idealized system illustrated in Fig. 7.2. This system starts with a train of ideal impulses with strengths of $+1$ and -1, depending upon the data bit to be transmitted. This train of impulses then forms the input to a filter whose impulse response is a pulse of height V and width T. The output of the filter is the required baseband signal.

The Fourier transform of the output is found by multiplying the Fourier transform of the impulse train with the transfer function of the filter:

$$\begin{aligned} Y(f) &= S_\delta(f)H(f) \\ &= S_\delta(f)Ve^{-j\pi fT}\frac{\sin \pi fT}{\pi f} \end{aligned}$$

The expression for the impulse train can be written as

$$s_\delta(t) = \sum_n a_n\delta(t - nT)$$

where the a_n are either $+1$ or -1, depending upon the corresponding information bit. The Fourier transform of a single term in this sum is given by

$$\mathscr{F}[a_m\delta(t - mT)] = a_m e^{-j2\pi fmT}$$

Therefore, the Fourier transform of the impulse train is

$$S_\delta(f) = \sum_n a_n e^{-j2\pi fnT} \tag{7-1}$$

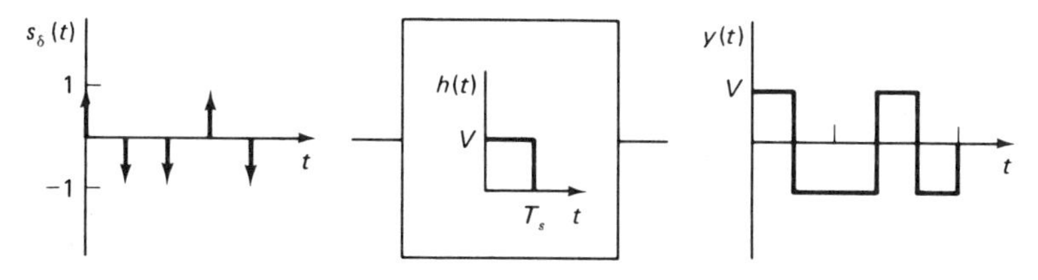

Figure 7.2 System to generate baseband signal

If the a_n are random and zero-mean, then the expected value of $S_\delta(f)$ is zero except at the points $f = k/T$. At these points in frequency, we are summing an infinity of zero-mean variables. More importantly, we note that the transform is periodic with basic period of $f_0 = 1/T$.

Evaluation of $S_\delta(f)$ is simple for some periodic $s_\delta(t)$ examples. Suppose first that $a_n = 1$ for all n. We see that when $f = k/T$, all the terms in the summation of Eq. (7-1) are unity and the sum is infinity. This is not surprising, since the Fourier transform of any periodic time function is a train of impulses at harmonics of the fundamental frequency. The time function, its transform, and the transform of the output, $Y(f)$, are illustrated in Fig. 7.3. The result may appear surprising until you observe that the baseband signal corresponding to this choice of a_n is a constant.

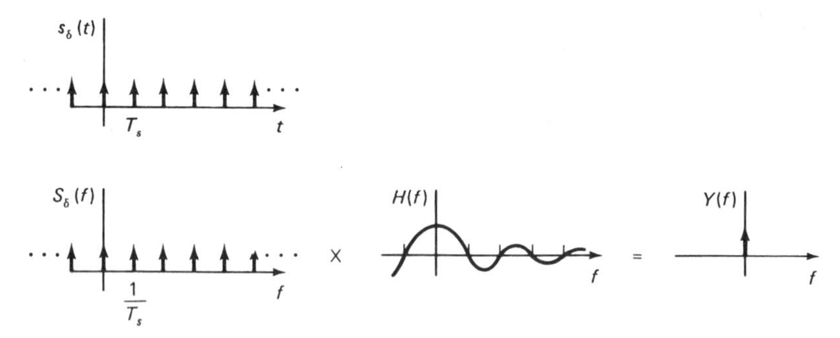

Figure 7.3 Evaluation of baseband Fourier transform for static data

If we now change the sequence to an alternating train of 1's and -1's—that is,

$$a_n = (-1)^n$$

we have the result of Fig. 7.4. Note that the impulses follow an envelope of $(\sin f)/f$ shape, so we can define a bandwidth. We usually choose the distance out to the first zero of the envelope to specify bandwidth, but the exact definition depends upon the specific application. The envelope observation is fairly general, since the transform of the impulse train is periodic. That is, although a different sequence would lead to a different transform, the transform will always repeat at intervals of $1/T$, and therefore periods beyond the $1/T$ basic period will be reduced according to the $(\sin f)/f$ multiplication.

We now formalize this result by considering the case of random bipolar NRZ-L data. The power spectral density of a time function, $s(t)$, is given by taking the ratio of energy to time. We do this over a limited time interval and then take the limit as the interval length becomes infinite. Setting the time-interval length to ΔT, the power spectral density is given by

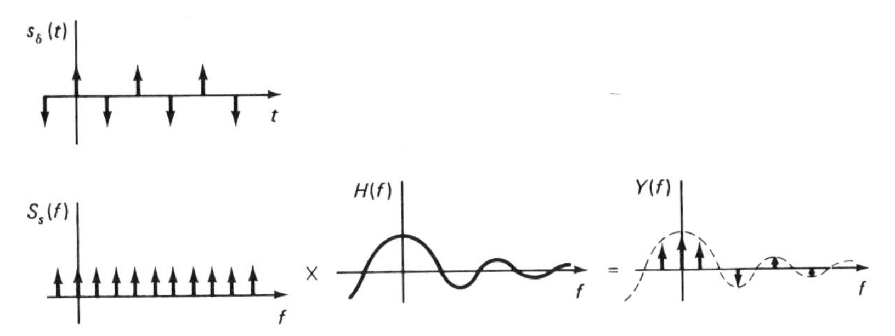

Figure 7.4 Frequency spectrum of periodic baseband signal for periodic data

$$G(f) = \left|\left[\lim_{\Delta T\to\infty} \frac{1}{\Delta T}\int_{-\Delta T/2}^{\Delta T/2} s(t)e^{-j2\pi ft}\,dt\right]^2\right| \tag{7-2}$$

The bracket in Eq. (7-2) is the Fourier transform of the truncated signal. Since $s(t)$ is NRZ bipolar, it is $+V$ or $-V$ within each interval. As ΔT increases, we keep adding identical terms, since the squaring operation makes the contribution of each period identical. We can therefore perform the operation over one period to get

$$G(f) = \frac{1}{T}\left|\int_{-T/2}^{T/2} Ve^{-j2\pi ft}\,dt\right|^2 = V^2T\left(\frac{\sin \pi fT}{\pi fT}\right)^2 \tag{7-3}$$

This power spectral density is sketched in Fig. 7.5, which is a repeat of Fig. 5.11.

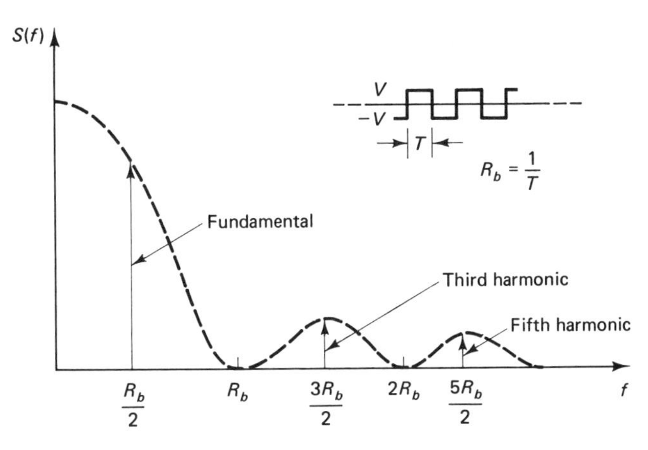

Figure 7.5 Frequency spectrum of random NRZ-L baseband signal

Suppose now that the data are sent using a biphase code. The power spectral density is then given by

$$G(f) = \lim_{\Delta T \to \infty} \frac{1}{\Delta T} \left| \int_{-\Delta T/2}^{0} \pm Ve^{-j2\pi ft}\, dt + \int_{0}^{\Delta T/2} \mp Ve^{-j2\pi ft}\, dt \right|^2$$

$$= \lim_{\Delta T \to \infty} V^2 \frac{1}{\Delta T} \left| \frac{-2 + e^{j\pi f \Delta T} + e^{-j\pi f \Delta T}}{j2\pi f} \right|^2$$

$$= \lim_{\Delta T \to \infty} \frac{V^2}{\Delta T} \left| \frac{1 - \cos \pi f \Delta T}{\pi f} \right|^2$$

$$= \lim_{\Delta T \to \infty} \frac{4V^2}{\Delta T} \left| \frac{\sin^2(\pi f \Delta T/2)}{\pi f} \right|^2$$

Once again, as ΔT goes to infinity, we add equal components for each sampling period. Thus, the power spectral density is given by

$$G(f) = V^2 T \left[\frac{\sin^2(\pi f T/2)}{\pi f T/2} \right]^2 \tag{7-4}$$

This expression is plotted as Fig. 7.6, which is a repeat of Fig. 5.14.

The channel distorts the waveform and causes time dispersion. This in turn, causes intersymbol interference.

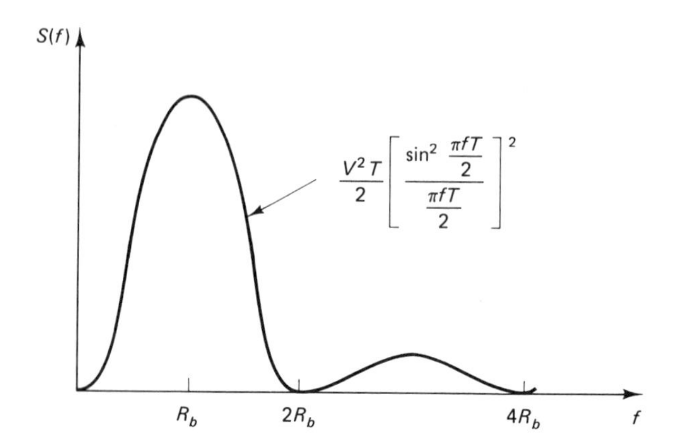

Figure 7.6 Frequency spectrum of random biphase baseband signal

7.1.1 Coding and Decoding

Baseband *coders* are systems whose input is a data signal and output is a baseband time signal. The form of the coder depends upon the form of the data signal. For example, if the data signal is already in the form of an electrical signal with two possible values, generation of the baseband signal may simply involve shifting the levels and, perhaps, holding the values throughout the sampling interval. The coder may also include a filter to shape the signal prior to transmission through the channel.

Decoding is a more complex operation. Viewing the baseband signal of Fig. 7.1, we might start by simply sampling the signal at the midpoint of each interval. These sample values could then be compared to zero—the signal is decoded as a 1 if the sample is positive and 0 if the sample is negative. A receiver operating on this principle is illustrated in Fig. 7.7.

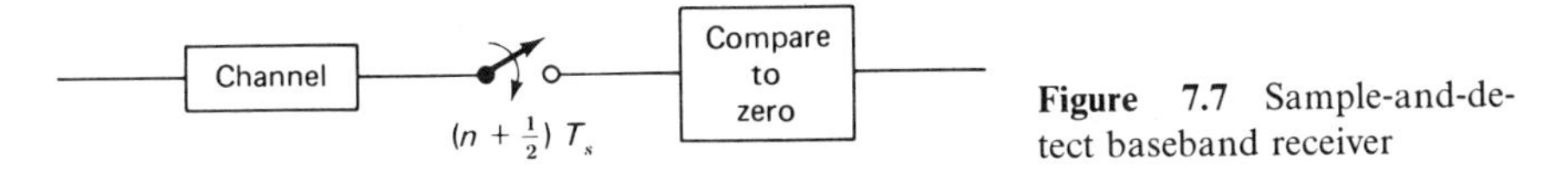

Figure 7.7 Sample-and-detect baseband receiver

The sample value is a random variable because of the presence of noise. The additive noise can be modeled as filtered white noise, where the filtering is done by the channel characteristic. The noise would have zero mean and a variance σ^2. The sample would therefore be a random variable with mean value of $\pm V$, depending upon which binary digit is being sent in that interval. The variance of the sample would be the variance of the noise, which is approximately equal to the bandwidth of the channel multiplied by the power per hertz of the white noise.

One might suggest modifying the receiver of Fig. 7.7 by adding a filter at the input. This filter reduces the noise and could possibly be chosen so that it does not change the mid-interval value of the signal. A low-pass filter might be chosen. Such a filter has a response to the signal pulse of the form $(\sin t)/t$. In fact, though, we justified earlier that the matched filter is the best type of filtering to maximize the signal-to-noise ratio at the input to the decision device.

Let us start by ignoring the channel characteristics. The filter would then be matched to a perfect square pulse of $+V$ volts in order to look for the presence of a data 1 in the interval. The matched filter would have impulse response $h(t) = V$ for $0 < t < T$. Convolving the received signal with this is equivalent to correlating the signal with the voltage pulse. Therefore, the receiver would be as shown in Fig. 7.8.

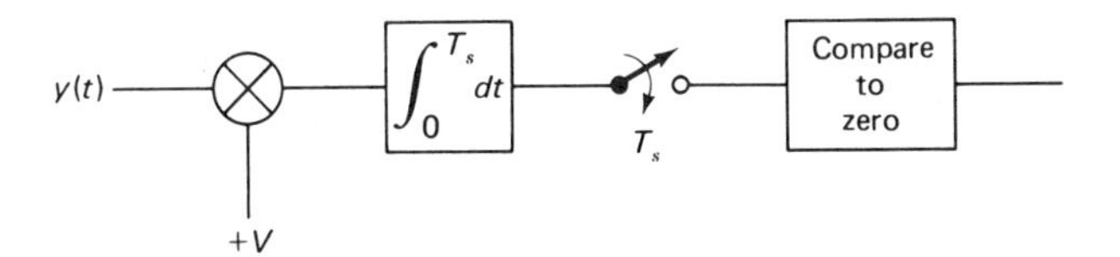

Figure 7.8 Matched filter detector for baseband

A second receiver, matched to the $-V$-volt pulse, would be needed to look for the digital 0. However, since the output of that second receiver would simply be the negative of the output of the system shown in Fig. 7.8, we can simplify the overall design. Rather than build two receivers and compare the outputs to see which is larger, we need simply stick with the single receiver illustrated and compare its output to zero. Thus, if the output is positive, we decode as a 1. If the output is negative, we decode as a 0. The system is then reset to examine the next time interval. The reset operation is known as *dumping,* and the system is known as an *integrate-and-*

dump circuit. A realization using an operational amplifier is shown in Fig. 7.9. The operational amplifier with capacitive feedback is an integrator. The switch labeled nT_s^+ closes instantaneously just after the end of the interval, thereby discharging the capacitor for the start of the next interval.

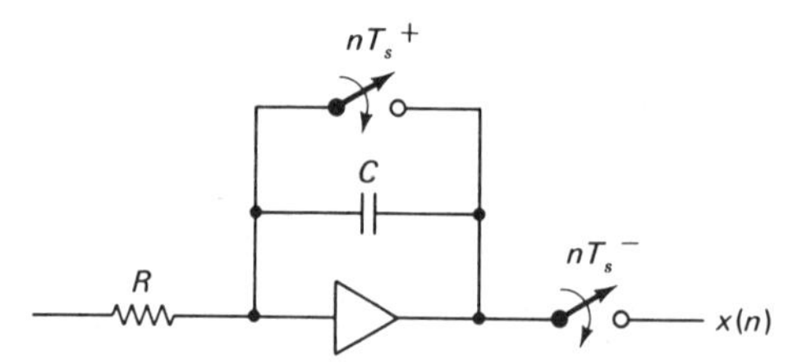

Figure 7.9 Integrate-and-dump detector for baseband

If the multiplication by V in Fig. 7.8 were replaced with multiplication by any constant, the performance would not be affected, since the output is being compared to zero. The error probabilities are also not affected, since both the signal and noise components are multiplied by the same constant. Thus, with a choice of $V = 1$, the multiplier can be completely removed.

7.2 INTERSYMBOL INTERFERENCE

Before performing a mathematical analysis of intersymbol interference, we shall begin with a pictorial approach in order to introduce the concept of *eye patterns.* Suppose that binary information is transmitted using a pulse-type waveform. A 1-V pulse is used to send a binary 1 and a 0-V pulse for a binary 0 (unipolar). When this waveform goes through the system, it is distorted. Among other effects, any sharp corners of the wave are rounded, since the system cannot pass infinite frequency. Therefore, the values in previous sampling intervals affect the value within the present interval. If, for example, we send a long string of 1's, we would expect the channel output to eventually settle to being a constant, 1. Similarly, if we send a long string of 0's, the output should eventually settle toward 0. If we alternate 1's and 0's, the output might resemble a sine wave, depending upon the frequency cutoff of the channel.

Therefore, if we examine a single interval in which a binary 1 is being transmitted, the output waveform within that interval will depend upon the particular sequence that preceded the interval in question. If we now plot all possible waveforms within the interval, including those for a 1 and those for a 0 in the interval, we get a pattern that resembles a picture of an eye.

Figure 7.10 shows some representative transmitted waveforms and the resulting receiver waveform. The partial eye pattern is sketched. The eye pattern is therefore the superposition of many waveforms within one sampling interval, the components of this composite waveform being the signals due to all possible preceding data strings. The number of individual waveforms contributing to the eye pattern depends upon the memory of the system. For example, if the system tran-

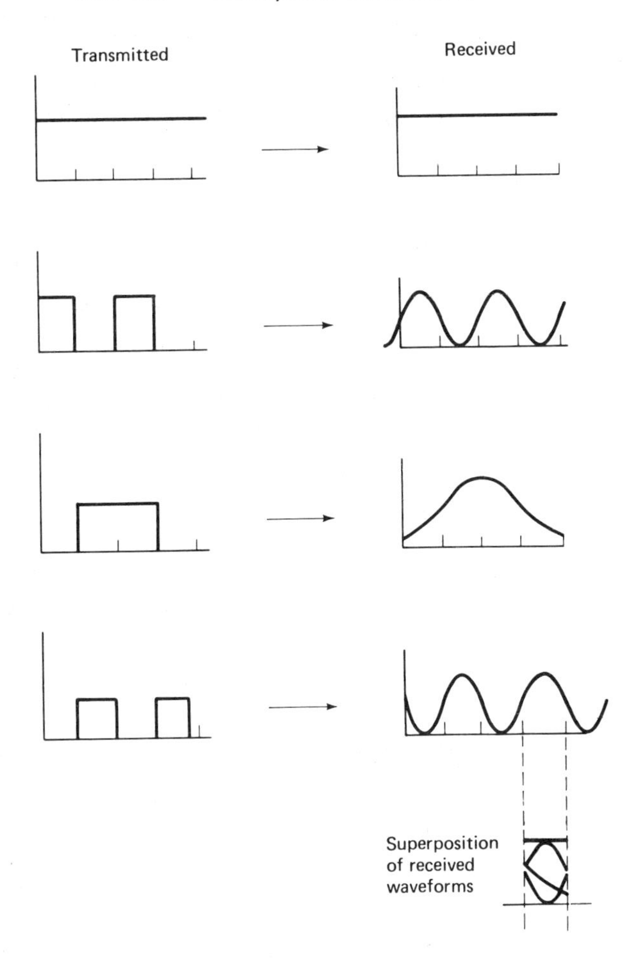

Figure 7.10 Generation of eye pattern

sient response extends over six sampling intervals, the particular pattern of six most recent bits determines the waveform within the interval. There are thus 2^6 or 64 waveforms making up the eye pattern.

If a random binary pulsed waveform forms the input to a system and the output is observed on an oscilloscope synchronized to the sampling interval, the eye pattern will be displayed on the screen. The eye opening (blank space in the center) is extremely significant. Suppose that you were at the receiver and wished to decide whether a 1 or 0 was being transmitted. One possible approach would be to sample the received waveform in the middle of each interval and to compare the sample value with the midpoint between a 0 and a 1. A sample greater than the midpoint would lead to a decision that a 1 was being sent. Similarly, a sample less than the midpoint would cause you to decide that a zero was being sent. As long as all components of the eye due to a 1 in the interval are separated from those due to a 0, this decision scheme will reproduce the transmitted signal with no errors (we are, of course, assuming no additive noise). If, however, the eye closes, errors will be made in receiving certain sequences. Note that for the distortionless channel, the eye consists of two horizontal lines, one at the top and one at the bottom of the interval.

Now that we have a qualitative idea of the problems associated with intersymbol interference, we shall take a more analytic approach toward the analysis. Let us assume bipolar baseband signals where $s_T(t)$ is the transmitted waveform used to send a binary 1 and $-s_T(t)$ is used to send a binary 0. In the simple case discussed in the previous section, $s_T(t)$ was a pulse of height $+V$ volts and duration T. Since the transmitted pulse is time-limited to the sampling interval, no intersymbol interference results at the receiver unless distortion occurred during transmission. We will assume $s(t)$ to be the actual pulse shape incident upon the receiver circuitry when a single bit of information is sent.

The received signal, $y(t)$, is a superposition of the individual pulse signals and can be written as a summation of the form

$$y(t) = \sum_{n=-\infty}^{K} a_n s(t - nT) + n(t) \tag{7-5}$$

In Eq. (7-5), $n(t)$ is the additive channel noise, and the a_n are either $+1$ or -1, depending upon whether a binary 1 or 0 was being sent. In the format shown, n represents the number of intervals from time zero. For example, a_2 would be $+1$ if the binary digit at time $2T$ was a binary 1. K is the most recent sampling point. That is, K is the largest integer less than t/T. Suppose that we examine Eq. (7-5) at particular sampling points, $t = mT + T/2$. These sampling instants occur at the midpoint of each interval. This can, of course, be easily modified for any point within the interval. The sample values at the receiver are then of the form

$$y\left(mT + \frac{T}{2}\right) = \sum_{n=-\infty}^{m} a_n s\left[\frac{T}{2} + (m - n)T\right] + n\left(mT + \frac{T}{2}\right) \tag{7-6}$$

For simplicity, we shall adopt the following notation to change these time samples into a discrete numerical sequence:

$$t_i = iT + \frac{T}{2}$$

$$x_i = x\left(iT + \frac{T}{2}\right)$$

Thus, x_i is the sample value at the middle of the ith interval. Using this notation, Eq. (7-6) becomes

$$y_m = \sum_{n=-\infty}^{m} a_n s_{m-n} + n_m \tag{7-7}$$

We shall ignore the additive channel-noise term in the remainder of this section, as this topic will be dealt with separately. The intersymbol interference is represented by the summation term in Eq. (7-7), where the $n = m$ term in the sum is the desired signal component, and the other terms represent the interference from past binary information.

We rewrite Eq. (7-7) as

$$y_m = a_m s_o + \sum_{n=-\infty}^{m-1} a_n s_{m-n}$$

$$= a_m s_o + I_m$$

where I_m is the intersymbol interference term. This term depends upon the specific shape of the signal, $s(t)$, and upon the particular sequence of binary digits transmitted. We will first examine a few parameters of I_m and then will talk in general terms about the distribution of this interference term.

The peak value of the interference can be found by assuming that the particular sequence is such as to make all the terms in the summation positive. Therefore,

$$I_m(\text{peak}) = \sum_{n=-\infty}^{m-1} |s_{m-n}| \tag{7-8}$$

Given $s(t)$, this expression is simple to compute.

We now look at the mean-square value of the interference. The square of the summation is first expanded, and then the expected-value operation is moved inside the summation to encompass the only random term:

$$\overline{I^2} = E\left\{\left[\sum_{n=-\infty}^{m-1} a_n s_{m-n}\right]^2\right\}$$

$$= \sum_{j=-\infty}^{m-1} \sum_{i=-\infty}^{m-1} E\{a_i a_j\} s_{m-i} s_{m-j}$$

Finally, we assume that the a_i are independent of each other and that both values, $+1$ and -1, are equally likely. In terms of the binary data signal, this implies that 1's and 0's are transmitted with equal probability, and adjacent bits are independent of each other. The expected value of the a_i is then zero, and, owing to the independence, the cross product cancels out. Thus, the only nonzero terms in the summation occur when $i = j$, yielding

$$\overline{I_m^2} = \sum_{n=-\infty}^{m-1} s_{min}^2 \tag{7-9}$$

Example 7-1

Find the peak and mean-square intersymbol interference for a signal resulting from sending ideal impulse samples through a channel with a triangular passband characteristic as shown in Fig. 7.11.

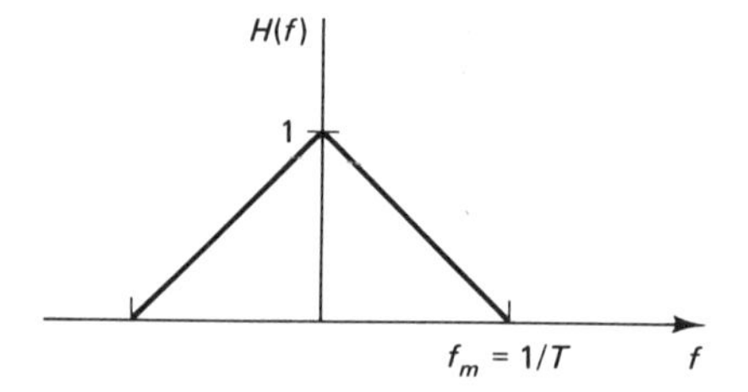

Figure 7.11 Channel characteristic for Example 7-1

Solution. We first find the impulse response of the channel.

$$h(t) = \frac{T}{\pi^2 t^2} \sin^2\left(\frac{\pi t}{T}\right)$$

From this result, we find the sample values

$$s_0 = \frac{4}{\pi^2 T}$$

$$s_1 = \frac{4}{\pi^2 3^2 T}$$

$$\vdots$$

$$s_n = \frac{4}{\pi^2 (2n + 1)^2 T}$$

We can immediately plug these results into Eq. (7-8) to get

$$I_m(\text{peak}) = \sum_{n=-\infty}^{m-1} |s_{m-n}|$$

$$= \frac{4}{\pi^2 T} \sum_{n=-\infty}^{m-1} \frac{1}{[2(m - n) + 1]^2}$$

and from Eq. (7-9),

$$\overline{I_m^2} = \sum_{n=-\infty}^{m-1} s_{m-n}^2$$

$$= \left(\frac{4}{\pi^2 T}\right)^2 \sum_{n=-\infty}^{m-1} \frac{1}{[2(n - m)]^4}$$

Now that we have done the easy part, we turn our attention to the more general characteristics of the interference. Ideally, we would like to find a *probability distribution* of the received samples, y_m. Starting with a simple case, if only two terms in the sum of Eq. (7-7) are significant, the received signal sample can be written as

$$y_m = a_m s_0 + a_{m-1} s_1$$

This expression can take on one of four possible values, these being

$$\pm(s_0 + s_1)$$
$$\pm(s_0 - s_1)$$

The probability of each of these four values is $\frac{1}{4}$. If the sum is now extended to include three terms, eight possible values occur:

$$\pm(s_0 + s_1 + s_2)$$
$$\pm(s_0 + s_1 - s_2)$$
$$\pm(s_0 - s_1 + s_2)$$
$$\pm(s_0 - s_1 - s_2)$$

Each of these has probability $\frac{1}{8}$, corresponding to one possible sequence of 3-bit length.

In the limit, as more and more terms in the sum are included, we would expect the probability distribution to approach a continuous distribution. This is so because the s_{m-n} will approach zero as n approaches minus infinity [i.e., the $s(t)$ is approaching zero as t approaches infinity for any real system]. Thus, the actual distribution of y_m fills in the space between

$$s_0 - \sum_{-\infty}^{m-1} |s_{m-n}|$$

and

$$s_0 + \sum_{-\infty}^{m-1} |s_{m-n}| \tag{7-10}$$

The actual distribution is difficult to evaluate. We could treat the problem as a convolution of an infinity of distributions, each of which is a pair of impulses at + and − the corresponding sample value of $s(t)$. If, for example, $s(t)$ were a ramp, the final distribution would be uniform, but unfortunately, the summation in Eq. (7-8) would not converge for this choice of $s(t)$.

Finding the distribution of the received samples is exciting, but it does not represent the bottom line in the analysis of the data system. In fact, what we really want to know is the effect of the intersymbol interference on detection capability; that is, what is the probability of making a detection error?

We take the simple symmetrical case wherein the receiver compares the samples, y_m, to zero and decides that a binary one was sent if the sample is positive, and a binary zero is sent if the sample is negative. The probability of a bit error assuming that a 1 is sent is, therefore,

$$P_e = \Pr\{y_m < 0\} \qquad \text{given that } a_m = 1$$

In the absence of noise, this is the probability that the intersymbol interference is more negative than $-s_0$. We sketch the first steps in a general approach to finding this probability, and some specific examples are considered in the problems at the back of the chapter. The approach we present is extremely awkward and is almost trial-and-error. Hopefully, the motivated student may see a way to approach this in a cleaner manner: it could make a good project. Of course, if the complete probability distribution of y_m had been evaluated as described previously, the probability of error could then be found by integrating this function over the range of values specified in Eq. (7-10).

Let us first arrange the sample value magnitudes, s_m for $m > 0$, in order. That is, reorder them so that the absolute-value sequence monotonically decreases (for many examples, no reordering is required to accomplish this). We then form a dichotomy of the sequence in the following manner. Determine the dividing point wherein the sum of all values to the left of this point is greater than the sum to the right plus the zero value. We then reason that if the binary sequence is such that each member of the left-hand set contributes negatively to the sum, then regardless of the sign of the terms to the right, the total sum will be more negative than s_0, thus leading to a bit error. If the number of terms in the left segment is called N_1, then 2^{N-N_1} sequences are included in the set described, where N is the total number of samples. This is true since the N_1 bits are fixed, and the remaining bits can each be one of two values. If these were the only ways in which the inequality condition could be met, the probability of error would be

$$\frac{2^{N-N_1}}{2^N} = 2^{-N_1}$$

We now see whether there are additional sequences by carefully relaxing the condition that all terms in the left group have negative contributions. We start with the rightmost bit in this group and examine whether a change in sign could be compensated for by changes to the right. We note that if the sequence converges rapidly enough such that the sum of all terms to the right of a particular point does not exceed the single term at that point, our work is finished.

Example 7-2

Given the following sample values:

n	0	1	2	3	4	5
$s(n)$	1	−2	0.5	0.4	−0.21	−1.3

find the probability of an error caused by intersymbol interference.

Solution. We first arrange the samples in order, using the magnitudes and eliminating $s(0)$.

$$2 \quad 1.3 \quad 0.7 \quad 0.4 \quad 0.21$$

We now divide this after the second entry, since, with this division, the sum on the left is 3.3 and that on the right is 1.31, the difference being greater than $s(0)$.

$$2 \quad 1.3 \,|\, 0.7 \quad 0.4 \quad 0.21$$

First note that had the division been moved one position left, the left-hand sum would be 2 and the right-hand sum would be 2.61. That would be unacceptable, as the left sum must exceed the right by at least $s(0)$. The reasoning is that, with the division shown, if the 2 and 1.3 terms oppose $s(0)$, there is no combination of the three remaining terms that could avoid an error. Getting specific, since this is probably cloudy, suppose that $a_0 = 1$, $a_1 = 1$, and $a_5 = 1$. The sum of the corresponding three terms is $1 - 2 - 1.3$, or -2.3. Regardless of the sign of the remaining three terms, the overall sum can never go positive, thus resulting in a bit error.

Returning to the dichotomy, we see three terms on the right side, and these can take on any values. This results in 2^3, or 8, possible sequences that will cause errors.

We now see if there are any sequences besides these which can cause errors. If the "2" moves across the dividing line, there is no way it can be compensated for. But if the "1.3" moves as follows:

$$2 \,|\, 1.3 \quad 0.7 \quad 0.4 \quad 0.21$$

we can compensate in one of two ways,

$$2 \quad 0.7 \quad 0.21 \,|\, 1.3 \quad 0.4$$

or

$$2 \quad 0.7 \quad 0.4 \,|\, 1.3 \quad 0.21$$

That is, with all terms to the left of the division subtracting from (opposing) s_0, an error will occur regardless of the sign of terms to the right of the division. The first of these adds two possible sequences (not four, since two of these were included in the original set) and the second adds one more possibility (since two were included in the original set of eight, and one was in the previous set of two).

There are thus 11 sequences out of the $2^5 = 32$ possibilities that will cause error, for an error probability of

$$P_e = \tfrac{11}{32}$$

This is a rather high probability, but the example is somewhat unrealistic, with $s(1)$ and $s(5)$ both larger in magnitude than $s(0)$.

7.2.1 What Is the Best Pulse Shape?

Suppose that a waveform, $s(t)$, is used to transmit a binary "1" and that $-s(t)$ is used to send "0." The question then arises as to the best choice of $s(t)$. We start with a natural choice, the square-transform case. That is, we assume that the transmitted pulses are generated using an impulse fed through an ideal low-pass filter (the channel). This is shown in Fig. 7.12. Note that the idealized $+V$-volt pulse is not used, since this pulse could not possibly pass through a bandlimited channel. It has frequency components reaching to infinity.

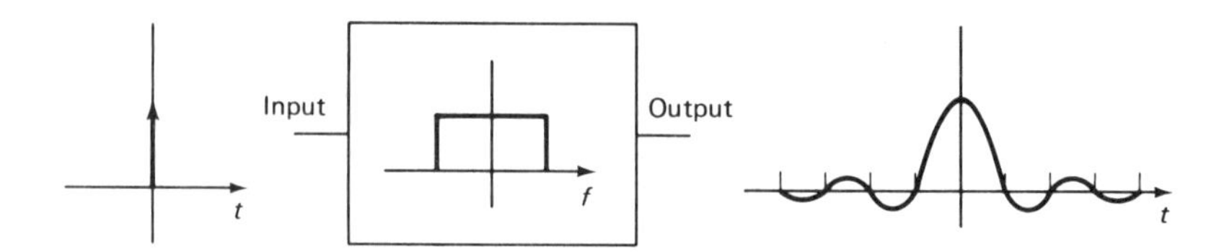

Figure 7.12 Pulses formed by ideal low-pass filter

The assumed pulse is therefore of the form

$$\frac{\sin \frac{\pi t}{T}}{\frac{\pi t}{T}}$$

This pulse has the property that it goes through zero at every sampling point except for the point at time zero. Therefore, in the intersymbol-interference analysis, $s_m = 0$ for all $m \neq 0$, and we have no intersymbol interference if samples are taken at equally spaced points in time. Figure 7.13 illustrates this for three pulses. This observation is actually a restatement of the sampling theorem. If the channel passes frequencies up to $f_m = 1/2T$, then the individual sample values are independent. One sample can be chosen independently of adjacent sample values.

The ideal bandlimited pulse is difficult to achieve because of the sharp corners in the frequency spectrum. This implies an impulse response that covers all time. Even if we could achieve this channel characteristics, the relatively significant sidelobes of the impulse response place demands upon the detection system. If sampling is not done at the precise multiples of T, intersymbol interference becomes a significant factor. Thus, synchronization must be precise, and time jitter must be kept to a minimum.

A desirable compromise is the *raised cosine* characteristic. The Fourier transform of this pulse is similar to the square transform presented above, but the transition from maximum to minimum follows a sinusoidal curve. The transform is shown in Fig. 7.14.

The value of K determines the width of the constant portion of the transform. If $K = 0$, the transform becomes that of the ideal bandlimited pulse. If $K = 1$, the flat portion is reduced to a point at the origin and the bandwidth is $2f_0$. Because of

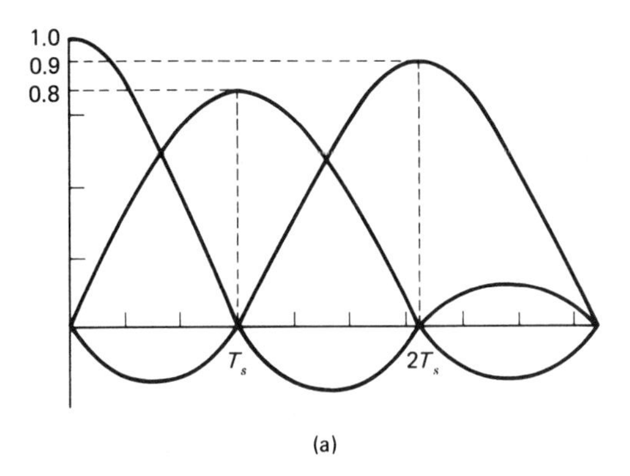

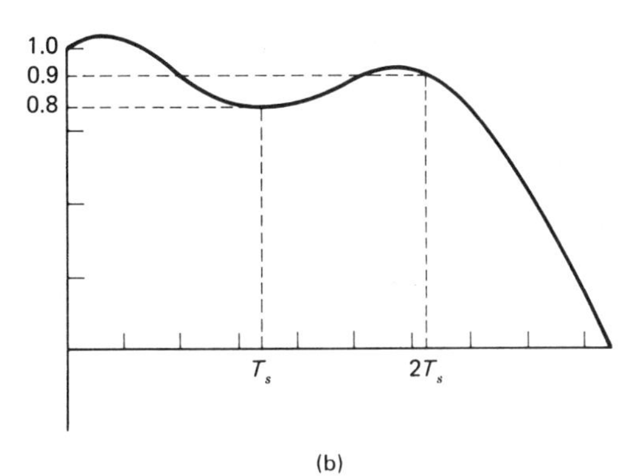

Figure 7.13 Sequence of pulses formed by ideal low-pass filter

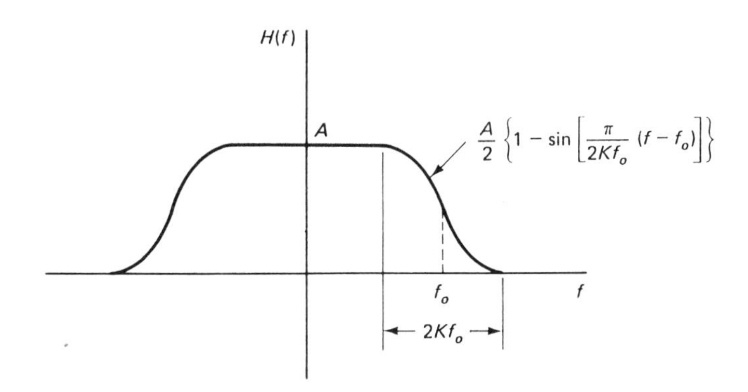

Figure 7.14 Fourier transform of raised cosine waveform

the rounded corners, we can approximate this characteristic to a far greater degree of accuracy than is possible for the square transform.

The time function corresponding to this Fourier transform is given by

$$h(t) = A\,\frac{\sin(2\pi f_0 t)}{2\pi f_0 t}\,\frac{\cos(KT2\pi f_0)}{1 - \dfrac{8K^2 f_0 t^2}{\pi}} \tag{7-11}$$

This impulse response is sketched for several representative values of K in Fig. 7.15. The second factor in Eq. (7-11) attenuates the sidelobes of the impulse response. This reduces the intersymbol interference and reduces the demands upon the synchronization system when compared to those of the ideal low-pass channel.

Note that at a value of $K = 1$, the response goes to zero not only at the zeros of $\sin(2\pi f_0 t)/2\pi f_0 t$ but also at points midway between these sample values. It is therefore possible to sample at the same rate as for the ideal channel, with no resulting intersymbol interference. We note that the price being paid is increased bandwidth. For $K = 1$, the bandwidth is twice what it was for the ideal channel. In reality, the raised cosine characteristic forms a goal that data communication systems aim toward.

We note that the received wave characteristic is a function of both the transmitted waveform and the channel characteristics. In the limiting case, the transmitted pulse can be a raised cosine, and the channel a perfect band-pass. Alternatively, the channel can be a raised cosine, and the transmitted pulse can have a flat spectrum in the passband of the channel (i.e., an impulse, or an impulse filtered by an ideal low-pass filter). In reality, the true situation is somewhere between these two, with the channel contributing part of the response characteristic. The next section discusses methods of compensating for channel characteristics that differ from this desired form.

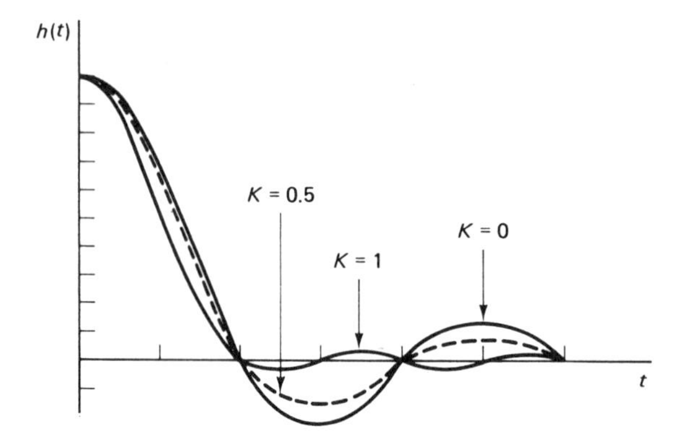

Figure 7.15 Impulse response corresponding to raised cosine channel

7.3 BASEBAND EQUALIZATION

The data communication receiver must do two separate filtering jobs. The first is filtering to decrease the effects of additive noise. We found that the matched filter is well suited to this task. The second filter decreases the effects of intersymbol interference. This second filter attempts to undue the channel distortion and is called an *equalizer*. The two filtering applications are usually addressed separately in a design problem. If strong additive noise is present, more attention is given to the design of the matched filter. If the additive-noise effect is small compared to intersymbol interference, the bulk of design attention goes to the equalizer. It is possible to combine the filters into one, but for practical reasons they are usually two different types (implementations) of filters. In any case, we shall find it conceptually simpler to separate the two functions.

Equalization in the most general case requires construction of a filter having the property that the product of the channel transfer function with the filter transfer function yields the desired composite effect. For example, if the channel has transfer function $H_c(f)$, and the transfer function of the equalizing filter is denoted $H_e(f)$, we choose the filter such that

$$H_e(f)H_c(f) = H_d(f)$$

where $H_d(f)$ is the desired overall characteristic. This might be a raised cosine, or it might represent the ideal band-pass filter characteristic. Synthesis of the equalizing filter is not simple in the general case. However, a simplification is possible, since the data systems we have been discussing work with time samples of the received waveform. That is, we were not concerned with the overall time function of the interference, only in its values at the midpoints of each sampling period. We can thus construct a discrete form of the filter using a *tapped delay line.* The tapped delay line filter is known as a *transversal equalizer* and is illustrated in Fig. 7.16. The impulse response of the transversal equalizer is given by

$$h(t) = \sum_{n=0}^{\infty} a_n\delta(t - nT) \tag{7-12}$$

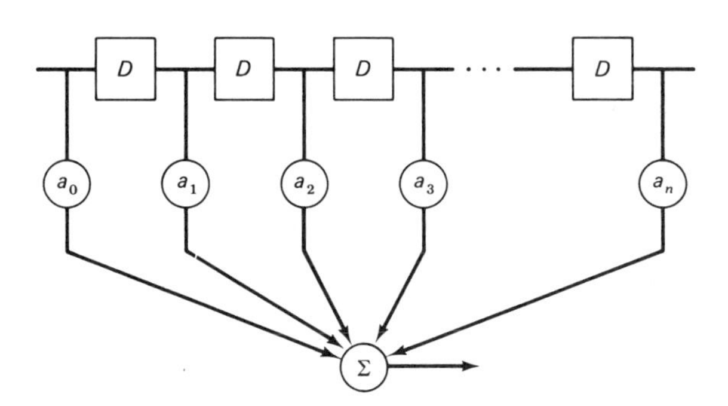

Figure 7.16 Transversal equalizer

This can be seen by inspection, letting $s(t) = \delta(t)$ in Fig. 7.16. The transfer function is the Fourier transform of this impulse response,

$$H_e(f) = \sum_{n=0}^{\infty} a_n e^{-jn2\pi fT} \tag{7-13}$$

We can think of the finite transversal filter as an approximation to the continuous equalizer where the first N Fourier series harmonics are substituted for the continuous transfer function.

Example 7-3

Approximate an even periodic function, $s_p(t)$, by the first three terms in a Fourier series expansion. Choose the coefficients for minimum mean-square error.

Solution. The approximation is given by

$$\hat{s}_p(t) = a_0 + a_1 \cos 2\pi f_0 t + a_2 \cos 4\pi f_0 t \tag{7-14}$$

The error is the difference between the function and the approximation.

$$e(t) = \hat{s}_p(t) - s_p(t)$$

We learned in Sec. 3.4 that a_0, a_1, and a_2 are chosen to make the error uncorrelated with the approximating functions. Therefore,

$$\begin{aligned} E\{e(t) \times 1\} &= 0 \\ E\{e(t) \cos 2\pi f_0 t\} &= 0 \\ E\{e(t) \cos 4\pi f_0 t\} &= 0 \end{aligned} \tag{7-15}$$

If we now interpret the expectations as time averages and denote these by a bar over the function, Eq. (7-15) becomes

$$\begin{aligned} \overline{s_p(t)} - a_0 - a_1 \overline{\cos 2\pi f_0 t} - a_2 \overline{\cos 4\pi f_0 t} &= 0 \\ \overline{s_p(t) \cos 2\pi f_0 t} - a_0 \overline{\cos 2\pi f_0 t} - a_1 \overline{\cos^2 2\pi f_0 t} - a_2 \overline{\cos 2\pi f_0 t \cos 4\pi f_0 t} &= 0 \\ \overline{s_p(t) \cos^2 4\pi f_0 t} - a_0 \overline{\cos 4\pi f_0 t} - a_1 \overline{\cos 2\pi f_0 t \cos 4\pi f_0 t} - a_2 \overline{\cos^2 4\pi f_0 t} &= 0 \end{aligned} \tag{7-16}$$

We now observe that

$$\overline{\cos 2\pi f_0 t} = \overline{\cos 4\pi f_0 t} = \overline{\cos 2\pi f_0 t \cos 4\pi f_0 t} = 0$$

$$\overline{\cos^2 2\pi f_0 t} = \overline{\cos^2 4\pi f_0 t} = \tfrac{1}{2}$$

Using these relationships, Eq. (7-16) becomes

$$a_0 = \overline{s_p(t)} = \frac{1}{T}\int_{t_0}^{t_0+T} s_p(t)\, dt$$

$$a_1 = \overline{2s_p(t) \cos 2\pi f_0 t} = \frac{2}{T}\int_{t_0}^{t_0+T} s_p(t) \cos 2\pi f_0 t\, dt$$

$$a_2 = \overline{2s_p(t) \cos 4\pi f_0 t} = \frac{2}{T}\int_{t_0}^{t_0+T} s_p(t) \cos 4\pi f_0 t\, dt$$

Note that the optimum choice of coefficients is exactly the same as that used in the Fourier series expansion. This is no accident. You should be able to prove that this is the case for any orthogonal expansion.

Example 7-3 illustrates that the optimum values of the a_n in Eq. (7-13) are the values that would obtain if we were expanding the continuous $H(f)$ in a Fourier series expansion. Thus, the a_n are given by

$$a_n = T \int_{-1/2T}^{1/2T} H_e(f) e^{-jn2\pi fT}\, df$$

where $H_e(f)$ is the ideal continuous equalizing filter transfer function.

This function can be found by observing the incoming signal due to a single pulse at the transmitter. We decide what the ideal received waveform would be, and compare this to the actual received waveform. As a bonus, if the transmitted pulse approximates an impulse, the received signal waveform is approximately equal to the impulse response of the channel, $h(t)$. In this case, we can find the transfer function of the channel by taking the Fourier transform of the received waveform. We use the discrete form of the transform to find

$$H_c(f) = T \sum_{n=0}^{\infty} x_n e^{-jn2\pi fT}$$

where x_n are the samples of the received signal waveform. We then form $H_e(f)$ as $H_d(f)/H_c(f)$, so that the composite is the desired characteristic. Hopefully, the a_n approach zero rapidly so that we can limit the length of the tapped delay line. This effects significant cost savings. Putting this all together, and assuming that $H(f) = 1$ for $|f| < f_m$, the delay-line taps are given by

$$a_k = T \int_{-1/2T}^{1/2T} \frac{1}{T \sum_n x_n e^{-jn2\pi fT}} e^{jk2\pi fT}\, df$$

$$= \int_{-1/2T}^{1/2T} \frac{1}{\sum_n x_n e^{-j2\pi fT(k-n)}}\, df$$

Since

$$\int_{-1/2T}^{1/2T} e^{-j2\pi fT(m-k)}\, df = 0 \qquad \text{if } m \neq k$$

a series expansion can be used to show that the tap weights are approximately given by

$$a_k = \int_{-1/2T}^{1/2T} \frac{1}{x_k}\, df = \frac{1}{Tx_k} \tag{7-17}$$

and the *tap weights are proportional to the reciprocals of the measured samples.* One way to look at this is that the tapped delay line is acting as a filter matched to the distorted impulse response of the channel.

An alternative method of equalization is to place a filter at the transmitting end. This predistorts the signal so that when acted upon by the channel, the composite result is the desired characteristic. The predistortion method could lead to simpler equalizers than those used at the receiver. This is so because the distorting function is performed on a simple pulse signal which has not encountered any time dispersion yet, and therefore the filter might be constructed using fewer delay elements. There are some disadvantages, however. If the transmitter is sending data to more than one receiver, different channel distortions may be applied in the different transmission paths. In addition, the channel characteristic must be known in advance, while an equalizer at the receiver can be designed based upon measurements of a received test signal.

Receiver equalizers can even be constructed to adapt to changes in the channel over time. To do so, the receiver must make some assumptions about what it is expecting to receive. For example, in binary bipolar communications, the output sample should be $\pm V$ volts at each sampling point. If it is not one of these values, the cause must be a combination of intersymbol interference and channel noise. By averaging over time, an *adaptive equalizer* can recognize that part of the error due to intersymbol interference. Using a feedback loop, the equalizer can alter tap weights to try to make this interference component equal to zero.

7.4 PERFORMANCE

7.4.1 Single-Sample Detector

Let us first look at an idealized baseband system in the absence of any distortion and using the simple $+V$- or $-V$-volt bipolar pulse to send 1's and 0's. We assume that the received signal is sampled at some intermediate point within each sampling interval. Since the noise-free pulse has constant amplitude over the interval, it will turn out that performance is independent of the point at which we sample. In most practical situations, the sample will be taken either at the midpoint or at the endpoint of the interval to achieve the best performance.

Each sample is composed of a signal ($\pm V$) plus a sample of the additive noise, $n(t_0)$, where t_0 is the sampling point. We assume that the noise is a sample of a Gaussian noise process and has zero mean. The result of adding a signal, $\pm V$, to a zero mean Gaussian random variable is a Gaussian random variable with mean $\pm V$ and variance equal to the variance of the noise sample. We therefore find that under each of the two signal hypotheses, the sample value is Gaussian distributed, and the two densities are as sketched in Fig. 7.17. $p_1(y)$ is the density of the sample under hypothesis 1 (a binary 1 is transmitted) and $p_0(y)$ is the density assuming that a 0 is sent. The notation introduced in Section 5.5 will be followed in this analysis. The probability of the two types of error is given by the integral of the appropriate tail of the Gaussian density. Thus, the probability of an error of the first kind or false alarm, P_{FA}, is the area indicated as P_{FA} in Fig. 7.17. The probability of an error

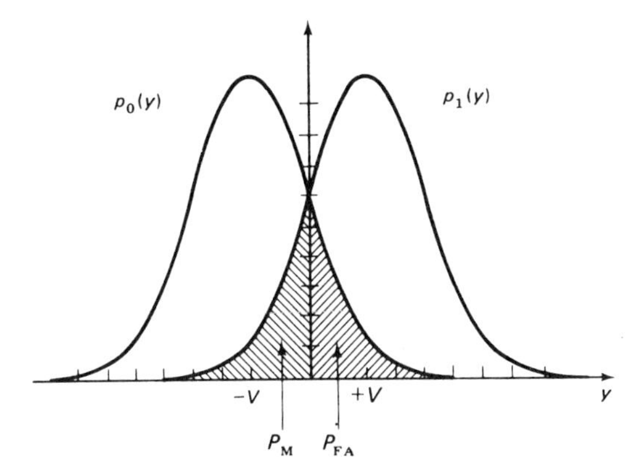

Figure 7.17 Conditional probabilities for single-sample detector

of the second kind, or miss, is also labeled on the diagram. These are conditional error probabilities, and the total probability of error is given by the conditional error probabilities weighted by the a priori probabilities of the two hypotheses.

$$\begin{aligned} P_e &= \Pr(D_1, H_0) + \Pr(D_0, H_1) \\ &= P_{FA}P(H_0) + P_M P(H_1) \end{aligned}$$

Because of the symmetry of the densities, the two conditional error probabilities are equal, and therefore the probability of error is the same as either of the conditional error probabilities. This is so because the probability of a 1 plus the probability of a 0 is unity. That is,

$$\begin{aligned} P_e &= P_{FA}P(H_0) + P_M P(H_1) \\ &= P_{FA}[P(H_0) + P(H_1)] \\ &= P_{FA} = P_M \end{aligned}$$

The case above, although extremely simple to analyze, is not very realistic. For the received signal to consist of perfectly square pulses, the channel would have to admit frequencies up to infinity. If the channel did so, the received noise, assuming additive white channel noise, would have infinite variance, and the probability of error would be 0.5! If we instead consider a channel with minimum Nyquist bandwidth, $f_m = 1/2T$, we can assume that the signal is of the form

$$\pm \frac{TV \sin \dfrac{\pi t}{T}}{\pi t}$$

This is the result of starting with an impulse and passing this through an ideal low-pass filter. We assume that sampling is done at the midpoint of each interval, yielding the peak signal value of $\pm V$. The noise power, or variance, would be Kf_m, where K is the noise power in the channel per hertz of bandwidth. Therefore, the probability of error is given by

$$P_e = \frac{1}{\sqrt{2\pi}\sqrt{Kf_m}} \int_0^\infty \exp\left[\frac{-(y+V)^2}{2(Kf_m)}\right] dy$$

We can make a change of variables to simplify this expression. Letting

$$x = \frac{y+V}{\sqrt{2}\sqrt{Kf_m}}$$

the integral becomes

$$P_e = \frac{1}{\sqrt{\pi}} \int_{V/(\sqrt{2}\sqrt{Kf_m}}^\infty e^{-x^2}\,dx = \tfrac{1}{2}\,\mathrm{erfc}\left(\frac{V}{\sqrt{2Kf_m}}\right)$$

We now identify $V^2/(Kf_m)$ as the signal-to-noise ratio, so finally

$$P_e = \tfrac{1}{2}\,\mathrm{erfc}\left(\sqrt{\frac{S}{2N}}\right) \tag{7-18}$$

This is plotted as a function of the signal-to-noise ratio in Fig. 7.18.

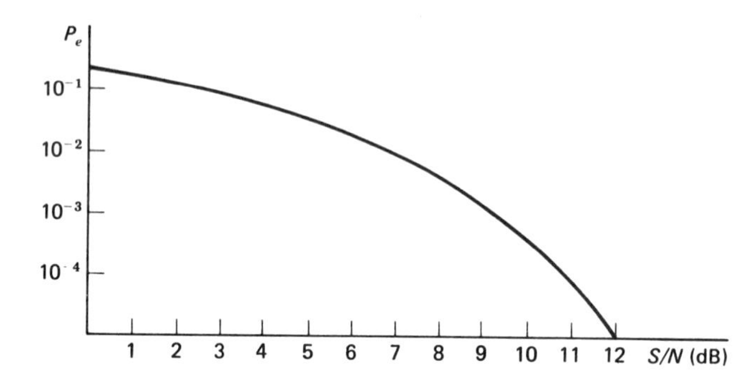

Figure 7.18 Error probability as a function of signal-to-noise ratio.

Example 7-4

A 1-V bipolar pulsed signal at 1 kbps goes through a channel with ideal low-pass filter characteristic, and a cutoff frequency of 4 kHz. Noise of 10^{-5} W/Hz is added in the channel. Find the probability of error using a receiver that samples at the midpoint of each interval.

Solution. The signal portion of the output due to a single pulse is as shown in Fig. 7.19. Since the ripple has a frequency of about 4 kHz, we can assume that it has essentially died out at the next sample point. That is, the intersymbol interference is approximately zero. The signal portion of each sample is therefore ± 1 V, with average power of 1 W. The noise power is 4000×10^{-5} or 0.04 W, yielding a signal-to-noise ratio of 25 or +14 dB. The corresponding probability of error can be found from Eq. (7-18) and is equal to

$$P_e = 3 \times 10^{-7}$$

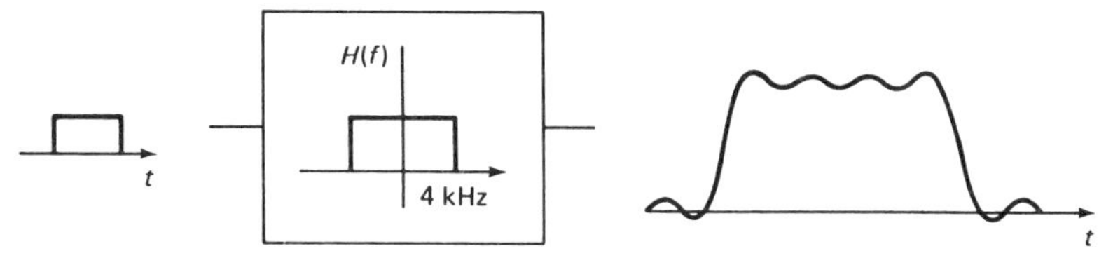

Figure 7.19 Signal portion of output for Example 7-4

7.4.2 Matched Filter Detector

We now include a matched filter in the receiver to improve the signal-to-noise ratio. Continuing with the idealized case where the received signal is a square pulse, we would build the matched filter detector of Fig. 7.20. The output of the integrator consists of a signal part and a noise part. The signal part is $\pm V^2T$. The noise part is given by

$$n = \pm \int_0^T Vn(t)\,dt$$

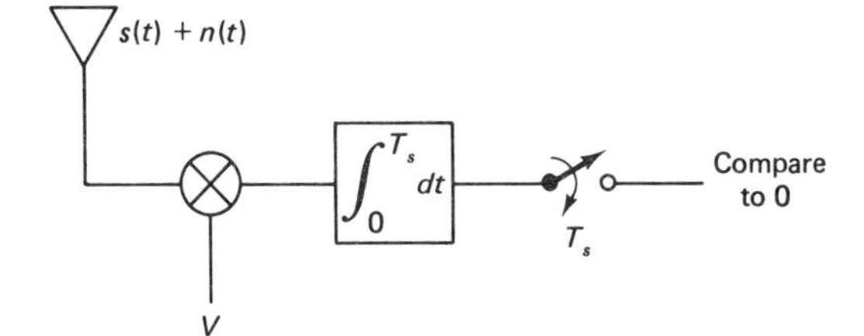

Figure 7.20 Matched filter detector for square pulse waveform

Since $n(t)$ is assumed to be white Gaussian noise with zero mean, n will be a Gaussian random variable. This is so because integration is a linear operation. The mean value of n is zero, and the variance is given by

$$\sigma^2 = E\left\{\pm \int_0^T Vn(t)\,dt\right\}^2$$

$$= E\left\{V^2 \int_0^T \int_0^T n(t)n(\tau)\,dt\,d\tau\right\}$$

We now move the expected-value operation inside the integral, since the expected value of a sum is the sum of the expected values. The expected value of $n(t)n(\tau)$ is the autocorrelation of the noise, which, for white noise, is an impulse. Thus,

$$\sigma^2 = V^2 \int_0^T \int R_n(t-\tau)\,dt\,d\tau$$

$$= V^2 \int_0^T \int \frac{N_0}{2}\,\delta(t-\tau)\,dt\,d\tau \tag{7-19}$$

$$= V^2 \frac{N_0}{2} \int_0^T 1\, dt = V^2 \frac{N_0 T}{2}$$

$N_0/2$ is the height of the power spectral density of the noise. Thus, the noise power in a band from 0 to f_m is $N_0 f_m$. The signal-to-noise ratio at the output is

$$\text{SNR} = \frac{V^4 T^2}{V^2 N_0 T/2} = \frac{2V^2 T}{N_0}$$

V in the detector of Fig. 7.20 multiplies both the signal and the noise components. Therefore if the receiver had been modified by eliminating the preintegration multiplier, the signal-to-noise ratio would not change.

We now examine the symbol error performance. As before, the input to the comparator is a Gaussian random variable. Under hypothesis 1, its mean is V^2T, and under hypothesis 0, its mean is $-V^2T$. In both cases the variance is $V^2N_0T/2$. The densities therefore look the same as those sketched in Fig. 7.17, with a change in the means and variances (spreading). This is sketched in Fig. 7.21. The probability of error is now given by

$$P_{\text{M}} = P_{\text{FA}} = \int_0^\infty \frac{1}{\sqrt{2\pi}\sigma} \exp\left[\frac{-(x + V^2T)^2}{2\sigma^2}\right] dx$$

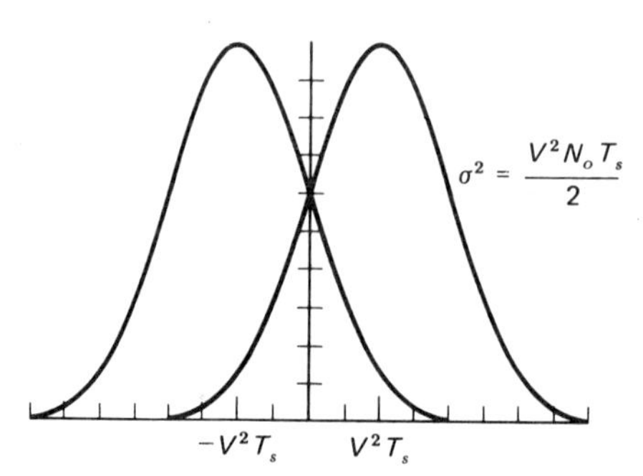

Figure 7.21 Conditional probabilities for matched filter detector

Let

$$y = \frac{x + V^2T}{\sqrt{2}\sigma}$$

Then

$$P_{\text{FA}} = \frac{1}{\sqrt{\pi}} \int_{V^2T/\sqrt{2}\sigma}^\infty e^{-y^2}\, dy$$

$$= \tfrac{1}{2} \operatorname{erfc}\left(\frac{V^2T}{\sqrt{2}\sigma}\right)$$

$$= \tfrac{1}{2} \operatorname{erfc}\left(\sqrt{\frac{V^2T}{N_0}}\right)$$

This is sketched in Fig. 7.22.

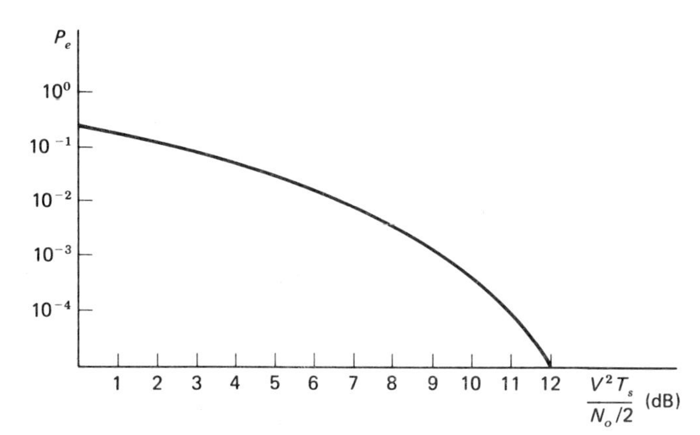

Figure 7.22 Error probability as function of signal-to-noise ratio

If the input noise to the matched filter is not white, the autocorrelation will not be an impulse and the result in Eq. (7-19) would have to be modified accordingly.

If the noise-free signals are not square pulses, the matched filter would have to be modified. Rather than handle this modification for every possible signal shape, the next section derives a general result for binary communication. This result will prove very useful in assessing performance of systems presented in the following chapters.

7.5 BINARY MATCHED FILTER DETECTOR

We now consider the matched filter receiver for binary signals in noise. We shall not restrict the two signal waveforms to any particular format. Therefore, the results of this section will be applicable to any of the modulation schemes to be discussed in subsequent chapters.

Figure 7.23 shows the binary matched filter detector; $s_0(t)$ and $s_1(t)$ are the two signals that are assumed to be known completely. We illustrate the receiver using correlators, but an exact equivalent receiver is that using matched filters to replace the multiplication and integration. The details of the comparison will be discussed first, and then the performance of this receiver will be derived.

The input to the comparator is a random variable. That is, the output of each integrator consists of a portion due to the signal plus a portion due to the noise. Since the input noise is assumed to be Gaussian, the integrator outputs will be Gaussian, and we need simply calculate the mean and variance. The mean is given by

$$E\{y\} = \int_0^T s_i(t)[s_1(t) - s_0(t)]\, dt \tag{7-20}$$

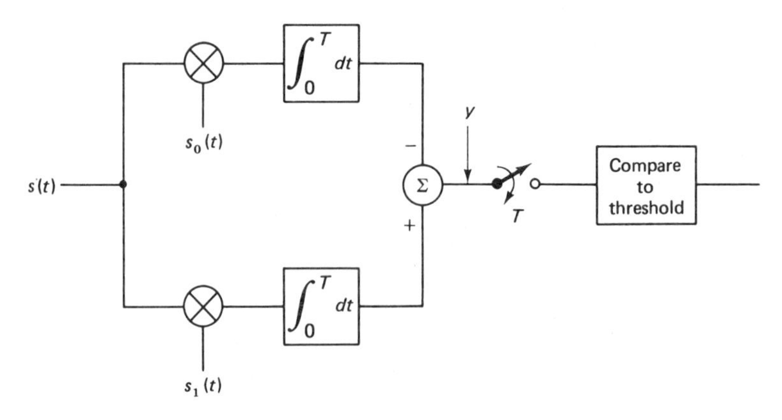

Figure 7.23 Binary matched filter detector

where $s_i(t)$ is the received signal; $i = 0$ if a zero is being sent and $i = 1$ if a one is being sent. In deriving Eq. (7-20), we have assumed that the noise is zero-mean. The variance of y is found as we did in Section 7.4 as follows:

$$\begin{aligned}
\operatorname{var}(y) &= E\left\{\int_0^T s(t)[s_1(t) - s_0(t)]\,dt - E\left\{\int_0^T s(t)[s_1(t) - s_0(t)]\,dt\right\}\right\}^2 \\
&= E\left\{\int_0^T\!\!\int n(t)n(\tau)[s_1(t) - s_0(t)][s_1(\tau) - s_0(\tau)]\,dt\,d\tau\right\} \\
&= \int_0^T\!\!\int \frac{N_0}{2}\delta(t-\tau)[s_1(t) - s_0(t)][s_1(\tau) - s_0(\tau)]\,dt\,d\tau \\
&= \frac{N_0}{2}\int_0^T [s_1(t) - s_0(t)]^2\,dt
\end{aligned} \tag{7-21}$$

We now use techniques from Section 5.5 to design the detector. The likelihood ratio is

$$\lambda(y) = \frac{p_1(y)}{p_0(y)} = \frac{\left(\dfrac{1}{\sqrt{2\pi}\sigma}\right)\exp\left[-\dfrac{(y-m_1)^2}{2\sigma^2}\right]}{\left(\dfrac{1}{\sqrt{2\pi}\sigma}\right)\exp\left[-\dfrac{(y-m_0)^2}{2\sigma^2}\right]}$$

where m_i is the mean of y under hypothesis i and σ^2 is the variance found in Eq. (7-21). This likelihood ratio is compared to the threshold. Recall that the value of the threshold depends upon the decision criterion being employed. For example, in Bayes detection, the threshold depends upon the a priori probabilities of the two hypotheses and upon the costs of the various errors. We shall denote the threshold by λ_0 and therefore not restrict this derivation to a particular criterion. We choose hypothesis 1 if

$$\lambda(y) = \exp\left[-\frac{(y - m_1)^2}{2\sigma^2} + \frac{(y - m_0)^2}{2\sigma^2}\right] > \lambda_0$$

or, performing the squaring operations, if

$$\exp\left[\frac{1}{\sigma^2}[y(m_1 - m_0)] + \frac{m_0^2 - m_1^2}{2\sigma^2}\right] > \lambda_0 \tag{7-22}$$

To use Eq. (7-22) in the real world, we measure y and plug this measurement into the equation. All other quantities are known, so the inequality can then be tested.

It would be preferable to have a decision rule that operates more directly upon the observation rather than upon a complicated function of this observation. We would therefore like to solve Eq. (7-22) explicitly for y.

Since the logarithm is a monotonic increasing function, taking logs of both sides of an inequality does not change the direction of the inequality. That is, if

$$A > B$$

then

$$\log A > \log B$$

Taking the log of both sides of Eq. (7-22) and doing appropriate manipulation yields the following decision rule: Choose H_1 if

$$y > \frac{m_1 + m_0}{2} + \frac{\sigma^2 \ln \lambda_0}{m_1 - m_0} \tag{7-23}$$

Equation (7-23) has some intuitive interpretation. Note that if $\lambda_0 = 1$, the decision rule compares y to the midpoint between the two means. As λ_0 increases, the threshold increases, thus making it less likely that H_1 will be chosen. Recall that λ_0 increases as hypothesis 0 becomes more likely and/or the cost of incorrectly choosing H_1 increases.

Now that we have set the threshold, it is possible to evaluate the *performance* of this receiver. The input to the comparitor is Gaussian. Under hypothesis 0, the mean of this Gaussian variable is

$$\int_0^T s_0(t)[s_1(t) - s_0(t)]\,dt = m_0$$

and under hypothesis 1, the mean is

$$\int_0^T s_1(t)[s_1(t) - s_0(t)]\,dt = m_1$$

Under either hypothesis, the variance is given by

$$\frac{N_0}{2}\int_0^T [s_1(t) - s_0(t)]^2\,dt = \sigma^2$$

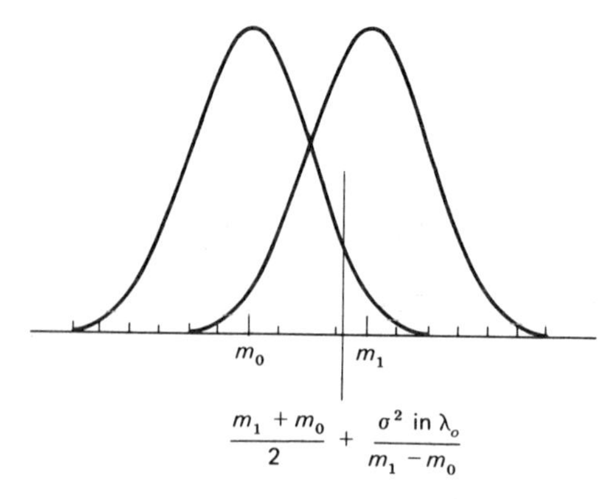

Figure 7.24 Conditional densities and threshold for binary matched filter detector

The two densities, together with the threshold, are drawn in Fig. 7.24.

In order to arrive at an intuitively meaningful result, we shall now assume that $\lambda_0 = 1$. This will obtain if (for Bayes) the a priori probabilities are equal and the costs of the various errors are equal. It can also hold if the costs and a priori probabilities are such that the ratio given in Eq. (5-17) reduces to unity. Under this condition, the threshold against which the observation is compared becomes

$$\frac{m_1 + m_0}{2}$$

If we now define two new variables, E and ρ, the expressions for the mean and variance can be simplified, and intuitive meaning can be given to the result. We define E as the *average energy* of the two signals and ρ as the *correlation coefficient:*

$$E = \frac{1}{2}\int_0^T [s_0^2(t) + s_1^2(t)]\, dt \tag{7-24}$$

$$\rho = \frac{\int_0^T s_0(t)s_1(t)\, dt}{E} \tag{7-25}$$

Example 7-5

Show that ρ is bounded by 1.

Solution. We begin with the inequality

$$\int_0^T [s_0(t) \pm s_1(t)]^2\, dt \geqslant 0$$

This is so because the integrand is nonnegative. Expanding this, we find that

$$\int_0^T [s_0^2(t) + s_1^2(t)]\, dt \pm \int_0^T s_0(t)s_1(t)\, dt \geqslant 0$$

and solving for ρ, we find that

$$|\rho| \leq 1$$

The decision rule is to choose H_1 if

$$y > \frac{m_0 + m_1}{2}$$

Equivalently, a new variable, z, can be defined as

$$z = y - \frac{m_0 + m_1}{2}$$

The decision rule is then to choose H_1 if

$$z > 0$$

The mean of this new random variable, z, becomes

$$\begin{aligned} E\{z\} &= \pm \frac{m_1 - m_0}{2} \\ &= \pm E(1 - \rho) \end{aligned}$$

with the plus sign under hypothesis 1 and minus sign under hypothesis 0. The variance of z is the same as the variance of y and is given by

$$\begin{aligned} \operatorname{var}(z) &= \frac{N_0}{2} \int_0^T [s_1(t) - s_0(t)]^2 \, dt \\ &= \frac{N_0}{2} \int_0^T [s_1^2(t) + s_0^2(t) - 2s_1(t)s_0(t)] \, dt \\ &= N_0 E(1 - \rho) = N_0(E - E\rho) \end{aligned}$$

The probability of error is given by

$$\begin{aligned} P_e &= P_{FA} = \Pr(z > 0/H_0) \\ &= \int_0^\infty \frac{1}{\sqrt{2\pi}\sigma} \exp\left\{-\frac{[z - E\{z\}]^2}{2\sigma^2}\right\} dz \\ &= \tfrac{1}{2} \operatorname{erfc}\left[\frac{E(1 - \rho)}{\sqrt{2}\sqrt{N_0 E(1 - \rho)}}\right] \\ &= \tfrac{1}{2} \operatorname{erfc}\left[\sqrt{\frac{E(1 - \rho)}{2N_0}}\right] \end{aligned} \qquad (7\text{-}26)$$

This is plotted as a function of signal-to-noise ratio for various values of ρ in Fig. 7.25. The results presented in Fig. 7.25 will be used later in this text. As expected, for constant S/N, performance improves as ρ decreases. This is intuitively satisfying,

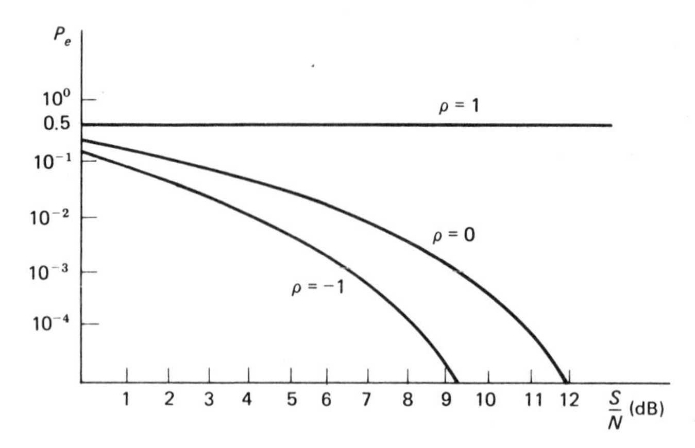

Figure 7.25 Error probability as a function of signal-to-noise ratio

since decreasing ρ indicates that the two signals are more distinct from each other.

Example 7-6

Given two signals, $s_0(t)$ and $s_1(t)$, as shown in Fig. 7.26, find the probability of error using a matched filter detector. The additive channel noise has two-sided spectral density, $N_0/2 = 10^{-1}$.

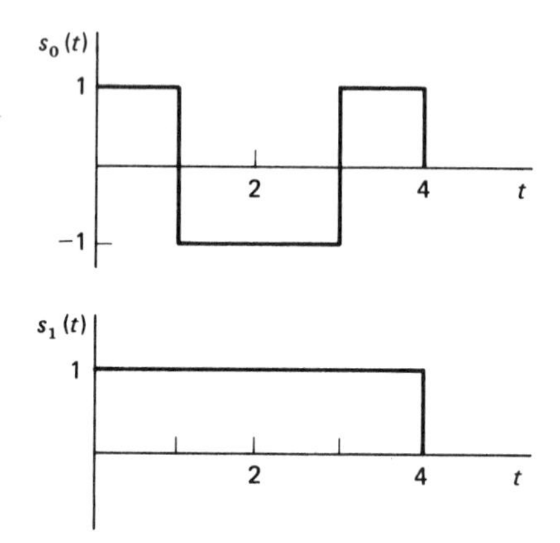

Figure 7.26 Signals for Example 7-6

Solution. Since no information was given to set the threshold, we will assume that $\lambda_0 = 1$. We need therefore only calculate the average energy, E, and the correlation coefficient, ρ. Using Eqs. (7-24) and (7-25), these are easily found to be

$$E = 4$$

$$\rho = 0$$

and the probability of error is

$$P_e = \frac{1}{2}\,\text{erfc}\left(\sqrt{\frac{2}{N_0}}\right)$$

$$= \frac{1}{2} \text{erfc}\,(3.16) = 4 \times 10^{-6}$$

7.5.1 Integrate-and-Dump Detector

The integrate-and-dump detector, which was shown in its simplest form in Fig. 7.9, is a special case of the binary matched filter detector. Two forms of this detector are illustrated in Figure 7.27. Note that the second form of the detector is used for biphase-L transmissions. The incoming signal is first multiplied by a pulse train with period equal to the sampling period.

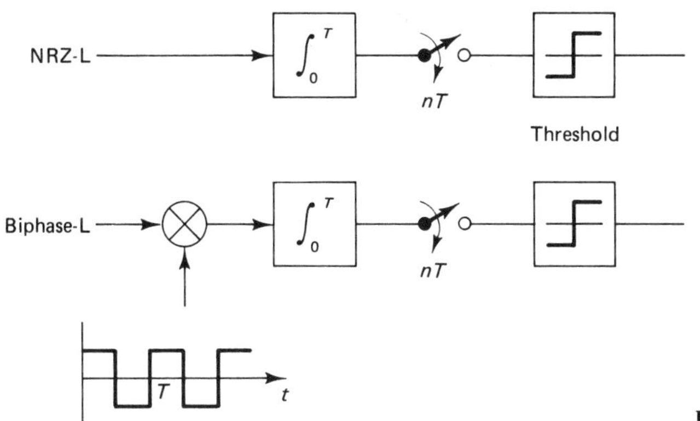

Figure 7.27 Integrate-and-dump detector

The integrate-and-dump is a *suboptimal* detector, usually exhibiting performance inferior to that of the binary matched filter detector. However, the integrate-and-dump is identical to the matched filter detector for rectangular signal shapes.

7.6 M-ARY BASEBAND

The communication techniques described so far in this chapter have centered upon the concept of sending information in binary form. Thus, we have coded our analog signals into a train of 1's and 0's.

Although we will see that binary communication has significant advantages in terms of noise immunity and equipment simplicity, it is sometimes desirable to code information into a format that has more than two possible elements. We do this in everyday life using 26 possible elements (the letters of the alphabet). We will discuss techniques for communicating with three and four possibilities. The more possibilities, the less unique (separated) the various transmitted words will be, and the more likely it becomes that an error will be made. Thus, as we increase the number of elements, performance is degraded and the complexity of equipment increases. There must be good news somewhere.

The good news is that we decrease transmission rate. For example, assume that a PCM system requires that 1000 bps be transmitted. If we now replace each pair of bits with a base-4 number, only 500 of these new numbers must be transmitted each second. This is equivalent to decreasing bandwidth.

As a simple example of a 4-ary system, consider the coding of Fig. 7.28. A pulse of $+V$ volts represents the 2-bit number 11; $+\frac{1}{3}V$ volts represents 10; $-\frac{1}{3}V$, 01; and $-V$, 00. Note that the levels are separated by $\frac{2}{3}V$. Although the present section is not concerned with performance, it should be mentioned that nonuniform error performance results from this four-level choice. For example, if a $+V$-volt pulse is transmitted, positive noise cannot cause an error, while negative noise could make this look like any of the three lower voltage values. It would take a noise of $\frac{1}{3}V$ in the negative direction to mistake a 11 for a 10. On the other hand, if a 10 is transmitted using $+\frac{1}{3}V$ volts, noise in the positive direction could cause this to be mistaken for a 11, while in the negative direction, the 10 might be mistaken for a 01. We will thus see that the probability of error when either 01 or 10 is transmitted is twice that when 11 or 00 is transmitted.

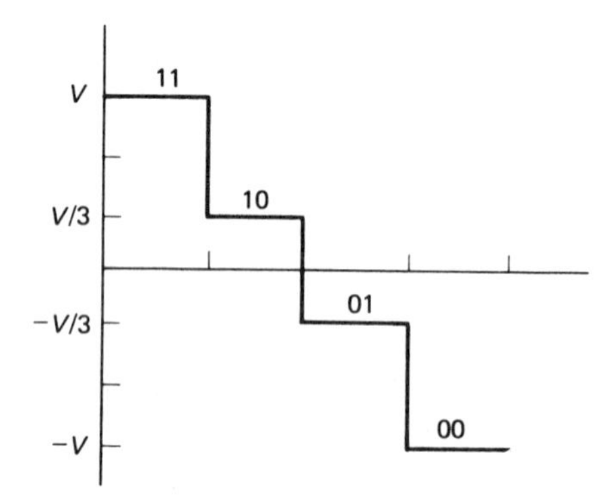

Figure 7.28 4-ary system

Example 7-7

Consider a 3-ary (ternary) communication system with the following three signals:

$$H_0\colon\ s_0(t) = 0$$

$$H_1\colon\ s_1(t) = 5\cos 100t, \qquad 0 < t < 2\pi$$

$$H_2\colon\ s_2(t) = -5\cos 100t, \qquad 0 < t < 2\pi$$

Assume that noise with power spectral density, $N_0/2 = 1.75$ is added. Further assume that the hypotheses are equally likely. Find the probability of error.

Solution. It is first necessary to design the receiver. Using the techniques of Chapter 6, the receiver is as shown in Fig. 7.29. If $s_0(t)$ is the input, the output will be 0. If $s_1(t)$ is the input, the output will be 5π. Finally, if $s_2(t)$ is the input, the output would be -5π.

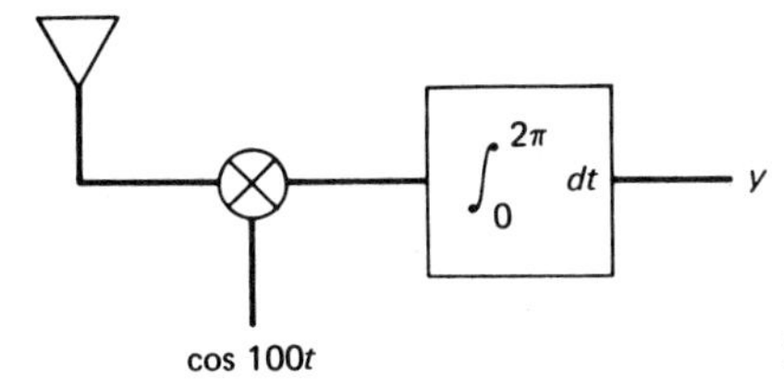

Figure 7.29 Receiver for Example 7-7

The detector decides

$$\begin{array}{ll} s_1 & \text{if } y > \tfrac{5}{2}\pi \\ s_2 & \text{if } y < -\tfrac{5}{2}\pi \\ s_0 & \text{otherwise} \end{array}$$

The likelihood functions are given by

$$p(y \mid H_i) = \frac{1}{\sqrt{2\pi}\sigma} \exp\left[\frac{-(y-m)^2}{2\sigma^2}\right]$$

where

$$\sigma^2 = \frac{N_0\pi}{2} = 5.5$$

$$m = 0 \quad \text{or} \quad \pm 5\pi \qquad \text{depending upon the signal}$$

The probability of error is then given by

$$\begin{aligned} P_e = {} & [P(D_1 \mid H_0) + P(D_2 \mid H_0)]P(H_0) \\ & + [P(D_0 \mid H_1) + P(D_2 \mid H_1)]P(H_1) \\ & + [P(D_0 \mid H_2) + P(D_1 \mid H_2)]P(H_2) \end{aligned}$$

It is a little simpler to first find the probability of a correct decision and then to subtract this result from 1 to get the probability of error. The probability of a correct decision is given by

$$P_c = P(D_0 \mid H_0)P(H_0) + P(D_1 \mid H_1)P(H_1) + P(D_2 \mid H_2)P(H_2)$$

where

$$\begin{aligned} P(D_0 \mid H_0) &= \int_{-(5/2)\pi}^{(5/2)\pi} p_0(y)\, dy \\ &= \frac{1}{\sqrt{2\pi}\sigma} \int_{-(5/2)\pi}^{(5/2)\pi} e^{-y^2/2}\, dy \\ &= \frac{2}{\sqrt{\pi}} \int_0^{5\pi/2\sqrt{2}\sigma} e^{-z^2}\, dz \\ &= \text{erf}\left(\frac{5\pi}{2\sqrt{2}\sigma}\right) \end{aligned}$$

and

$$
\begin{aligned}
P(D_1 | H_1) = P(D_2 | H_2) &= \int_{(5/2)\pi}^{\infty} p_1(y)\, dy \\
&= \frac{1}{\sqrt{2\pi}\sigma} \int_{(5/2)\pi}^{\infty} \exp \frac{-(y - 5\pi)^2}{2\sigma^2}\, dy \\
&= \frac{1}{\sqrt{\pi}} \int_{-5\pi/2\sqrt{2}\sigma}^{\infty} e^{-z^2}\, dz \\
&= 1 - \frac{1}{2} \operatorname{erfc}\left(\frac{5\pi}{2\sqrt{2}\sigma}\right) \\
&= \frac{1}{2} + \frac{1}{2} \operatorname{erf}\left(\frac{5\pi}{2\sqrt{2}\sigma}\right)
\end{aligned}
$$

so

$$
\begin{aligned}
P_c &= \frac{1}{3}\left[\operatorname{erf}\left(\frac{5\pi}{2\sqrt{2}\sigma}\right) + 1 + \operatorname{erf}\left(\frac{5\pi}{2\sqrt{2}\sigma}\right)\right] \\
&= \frac{1}{3}\left[1 + 2 \operatorname{erf}\left(\frac{5\pi}{2\sqrt{2}\sigma}\right)\right]
\end{aligned}
$$

and the probability of error is

$$
P_e = 1 - P_c = \frac{2}{3} - \frac{2}{3} \operatorname{erf}\left(\frac{5\pi}{2\sqrt{2}\sigma}\right) = 5 \times 10^{-4}
$$

7.7 PARTIAL RESPONSE SIGNALING

We have shown that in order to eliminate intersymbol interference at a sampling rate that is twice the highest frequency of the signal (the Nyquist rate), we must have a square ideal distortionless transfer function. Zero intersymbol interference can also be achieved with a raised cosine channel but at the expense of increased bandwidth over the equivalent square-channel response

The ideal square passband channel is very difficult to approach in the real world. For that reason we now examine whether there are any nonsquare response channels limited in frequency to $1/2T$ which permit intersymbol interference-free transmission at a rate of $1/T$. In fact, it is impossible to eliminate intersymbol interference completely in the non-square-channel case. Thus one pulse must affect the waveform in the following sampling interval(s). It is, however, possible in some cases to predict the nature of the interference sufficiently that it can be completely canceled. This will be true in cases where the intersymbol interference is limited to sample times not widely separated from the current sample. For example, if one sample value affects only the immediately following sample, once the present sample is decoded, the necessary amount can be subtracted from the following value.

As a practical, although hopefully not typical, example, consider a row of students in a class that had a history of cheating on exams. Each student copied only from the student to his or her left. Experience has shown that the copying student's grade is improved by one-tenth of the honest grade of the student whose paper was copied from. Although this represents interstudent interference extending to adjacent students, the instructor can completely cancel this effect during the grading operation. The first student's paper is graded. Then 10% of that grade is subtracted from the next student's grade to arrive at the honest score for the second student. That score is weighted by 10%, and subtracted from the third student's grade, and so on. In theory, we could handle this for higher orders of cheating (i.e., a student looks two papers away), but beyond a few sampling periods the processing becomes cumbersome. We also point out that in a practical situation, it is possible for one to lower one's grade by cheating, but this was meant only as a hypothetical example anyway.

We shall examine one class of partial response (or controlled intersymbol interference) techniques known as *duobinary*. At first glance, it will appear that we are transmitting at twice the rate allowed to avoid intersymbol interference. Let us start with some handwaving. If a system is being operated at twice the Nyquist frequency, we can intuitively reason that the system is being given only enough time to respond halfway to each input pulse before the next pulse arrives. Thus, if the system is sitting at the $-V$ level corresponding to a 0 binary input, a 1 will not take the system all the way to $+V$, but only halfway, to 0 volts. If this is followed by another binary 1, the output will rise the rest of the way to $+V$ volts. If, instead, the first 1 is followed by a 0, the output falls back down to $-V$ volts. It should therefore be clear that at any sampling point, the output has one of three values, $-V$, 0, and $+V$. If the output is $-V$ or $+V$, there is no ambiguity and we know the input bit had to be a 0 or 1, respectively. If the output is 0, it is necessary to examine the previous detected bit to determine whether the input is 1 (bringing the output from $-V$ to 0) or 0 (bringing the output from $+V$ to 0). We have therefore in essence coded each *pair* of input bits into three possible outputs, and the signaling rate can be increased by abandoning binary transmission. The disadvantage of this system is an increase in noise susceptibility, since we must now distinguish between levels separated by half as much. In addition, a bit error has the potential for causing another error in the following interval.

We now take a more detailed approach to analyzing the duobinary system. Suppose that the channel has a raised cosine characteristic with $K = 1$. The system function and associated impulse response are repeated as Fig. 7.30.

We see that sampling at $1/2T$ samples per second eliminates intersymbol interference. Suppose, instead, that sampling is done at $1/T$ samples per second, effectively communicating at twice that bit rate. The sampling points are indicated as dashed lines on Fig. 7.30. Looking at the intersymbol interference, we find that the previous sample interferes with the present sample, but samples prior to that contribute zero to the present value. Therefore, having detected the previous sample, we can subtract its effect from the current value to effectively cancel the effects of any intersymbol interference.

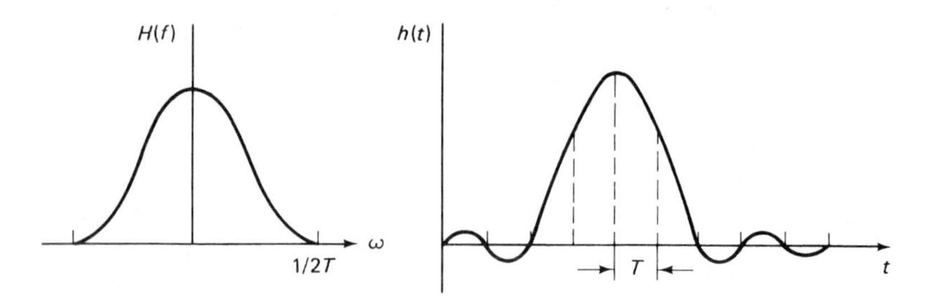

Figure 7.30 Raised cosine channel characteristic

Different filter characteristics would lead to additional interactions, but if the transfer function is such that after a limited number of sampling periods, the pulse response is zero for all future sampling points, the partial response approach can be effectively used.

Example 7-8

Compare the following filter characteristic to the raised cosine for use in partial response signaling.

$$H(f) = \cos^2 \pi f T, \qquad |f| < \frac{1}{2T}$$

Solution. We rewrite the transfer function and then evaluate the impulse response:

$$H(f) = \frac{1 + \cos 2\pi f T}{2}$$

so

$$h(t) = \frac{1}{4}\left[2\delta(t) + \delta(t - T) + \delta(t + T)\right] * \frac{\sin \frac{\pi t}{T}}{\pi t}$$

where the asterisk denotes convolution. We derived this by noting that the given transfer function is the product of a continuous unlimited function with a gating function. The corresponding time function is therefore a convolution of the corresponding time functions. The resulting impulse response is

$$h(t) = \frac{1}{4}\left[\frac{2 \sin \frac{\pi t}{T}}{\pi t} + \frac{\sin \frac{\pi(t - T)}{T}}{\pi(t - T)} + \frac{\sin \frac{\pi(t + T)}{T}}{\pi(t + T)}\right]$$

which is sketched in Fig. 7.31. Viewing Fig. 7.31, we see that each pulse in the input produces three pulses in the output, two of which are in the immediately adjacent sampling intervals. This filter can therefore be used in the same way as the raised cosine since intersymbol interference is limited to the immediately adjacent samples.

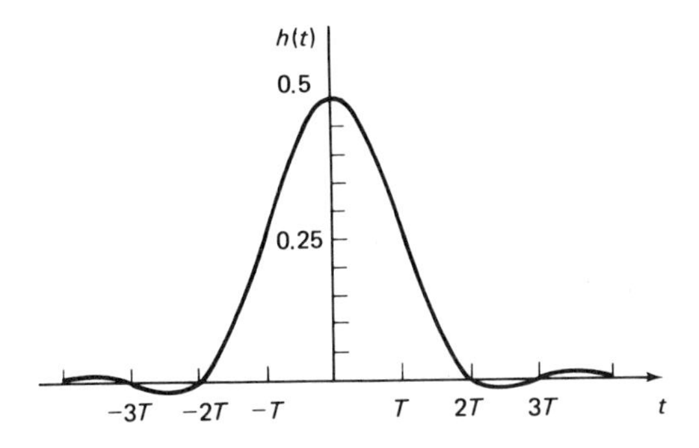

Figure 7.31 Impulse response of channel of Example 7-8

A duobinary type of system is used in *Telefax* systems to compress bandwidth. The basic Telefax machine scans a page of text and sends binary information to communicate whether each resolution point is black or white. With reasonable resolution, and using a phone line, transmission is relatively slow, with each page taking well over 1 minute to send. Adaptive bandwidth compression techniques take advantage of long strings of zeros or ones to reduce bandwidth. If a relatively long section of a particular trace is white (or black), the system transmits only boundary information about the area. The sophistication necessary to code and decode this information increases the system cost.

Duobinary transmission yields a simple technique approximately to double the speed of transmission for a given bandwidth channel. The baseband polarity of the "black" signal is altered. Figure 7.32 shows a typical signal and its coded version. The bandwidth of the coded signal is approximately one-half that of the original signal. You can convince yourself of that using a checkerboard as an example. Therefore, the transmission rate can be doubled, thereby sending a page of text in about half the time.

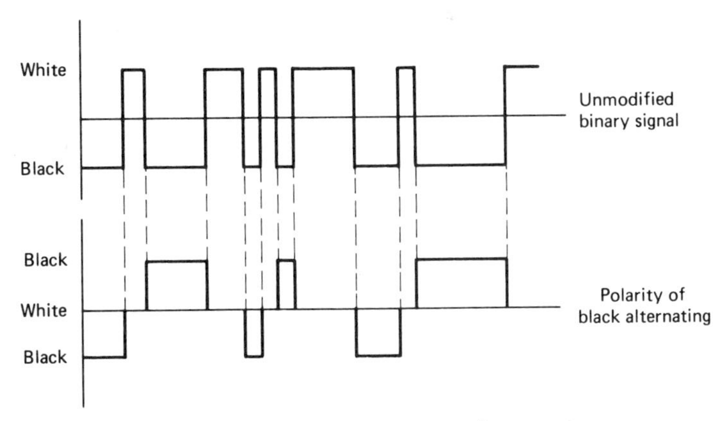

Figure 7.32 Duobinary coding

7.8 THRESHOLD IN PCM

We now have almost all the tools we need to start designing a baseband transmission system. Chapter 3 taught us how to start with an analog signal and to encode this into a digital signal. The current chapter has discussed the transmission of baseband signals and the performance of the matched filter detector. We are therefore in a position to design for a particular signal-to-quantization-noise ratio, and a particular bit error rate.

The current section relates these two parameters. That is, it would not make much sense to significantly increase the numbers of bits of quantization, thereby raising the signal-to-quantization-noise ratio, if indeed the major limitation on system performance were bit error rate. In fact, as the resolution is improved by increasing the number of quantization bits, the bit transmission rate will normally increase. Doing so would increase the bit error rate. This is so because faster transmission rates normally lead to less energy per bit, so the signal-to-noise ratio decreases. This increases the bit error rate. Alas, we have come upon another significant trade-off design consideration.

Let us first express both performance indicators in the same terms. The important output parameter is the signal-to-noise ratio. That is, we compare the final recovered signal with the transmitted signal, and take the power ratio of the signal to the error term. The error, or noise, will be composed of two parts. One of these is the quantization noise, the other is the time signal resulting from bit errors. We will denote the quantization noise as N_q, and the noise due to bit errors as N_{BER}. The signal-to-noise ratio is then

$$\mathrm{SNR}_{\mathrm{out}} = \frac{S}{N_q + N_{\mathrm{BER}}}$$

The bit-error-rate noise is approximated by assuming that errors can occur with equal probability in any bit position. This is an important assumption, since a bit error in the most significant bit position is much more damaging than a bit error in the least significant bit. The output signal-to-noise ratio can be shown to be given by

$$\mathrm{SNR}_{\mathrm{out}} = \frac{2^{2N}}{1 + 2^{2(N+1)}\mathrm{BER}} \tag{7-27}$$

Let us examine the limiting cases. If the bit error rate is zero, this expression reduces to 2^{2N}, which is identical to what we derived in Section 3.2.12. In that case, all of the output noise is due to quantization noise. In the other limiting case, suppose N approaches infinity. For that assumption, the resolution is infinite, and there is zero quantization noise. The signal-to-noise ratio reduces to $1/(4 \times \mathrm{BER})$. For a bit error rate of $\frac{1}{2}$ (the worst-case situation, where you may as well eliminate the communication system and replace it with a coin which you flip), this expression indicates a signal-to-noise ratio of $\frac{1}{2}$. Obviously the expression of Eq. (7-27) is an approximation. In fact, the approximation improves as the bit error rate ap-

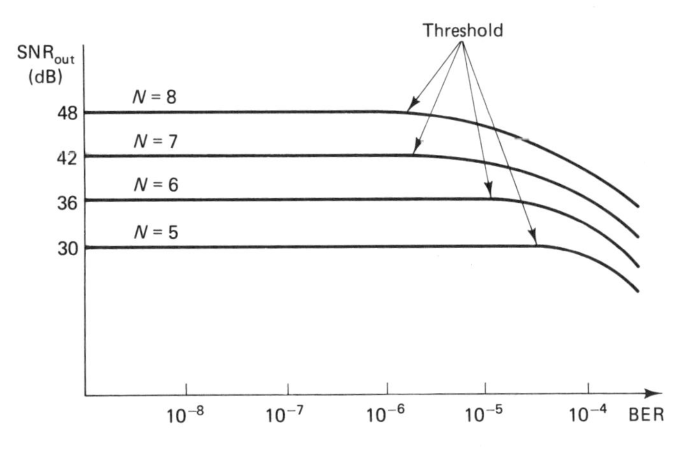

Figure 7.33 Threshold effect in PCM

proaches zero. Since the systems we will design have bit error rates of the order of 10^{-5} or less, this equation will prove to be sufficiently accurate.

Figure 7.33 is a plot of Eq. (7-27) for various values of N. Note that for bit error rates below a certain value, the output signal-to-noise ratio does not change appreciably. That is, for a given resolution (value of N), there is no substantial benefit in lowering the bit error rate below a certain threshold value.

Figure 7.33 is an extremely useful design tool. For example, suppose you design a system with a bit error rate of 5×10^{-7} using 6 bits of quantization. You measure an output signal-to-noise ratio of 36 dB. You are told that this SNR does not meet the specifications, and you must raise it to at least 40 dB. A glance at Fig. 7.33 shows that you are limited not by the bit error rate, but by the resolution. The clear choice would be to raise the number of bits of quantization to 7.

On the other hand, suppose you design a system using 6 bits of quantization and measure a signal-to-noise ratio of 30 dB. A glance at the curve shows that you are to the right of the threshold and can therefore realize an improvement by concentrating upon bit error rate. The bit error rate can be reduced by decreasing the noise power, increasing the signal power, or decreasing the correlation between the two signals used to send 1's and 0's. You can decrease noise by shielding the system (if wire transmission is used) or by using other frequency bands, directional antennas, or different transmission paths if transmission is through the air. If the system constraints permit, it is a simple matter to increase signal power. If the correlation is already -1, no further improvement is possible in this parameter.

We shall be using these observations in system design in the next section.

7.9 DESIGN EXAMPLES

It is still quite early in our study of digital communication for us to perform a total system design, including the various trade-off decisions. In later chapters we will study alternatives to baseband transmission. However, we can examine the basic considerations in a system design, and we do so in the following three discussion examples.

Example 7-9

You wish to communicate binary information at a rate of 500 bps. The channel consists of a wire that passes frequencies between 300 Hz and 3 kHz. White noise of power 10^{-7} watts/Hz adds to the signal during transmission. The channel can support voltages between -3 V and $+3$ V. Design the transmitter and receiver.

Discussion. The first step in the design process is to choose the two signal waveforms used to send a zero and a one. Since this channel will not support frequencies down to DC, the unipolar or bipolar pulses studied earlier would not represent a good choice. Significant signal energy would not reach the receiver, and performance would suffer. Since the bit transmission rate is quite small, we do not have to go through a great deal of work in this design. We will probably choose sinusoidal bursts to send the binary information. Suppose we choose

$$3 \cos(2\pi f_0 t)$$

and

$$3 \cos(2\pi f_1 t)$$

to send a binary 0 and 1, respectively. Since the sinusoidal bursts are 2 msec long, the Fourier transform of each burst has a $(\sin f)/f$ shape, with the first zero 500 Hz away from the center frequency. This is sketched in Fig. 7.34.

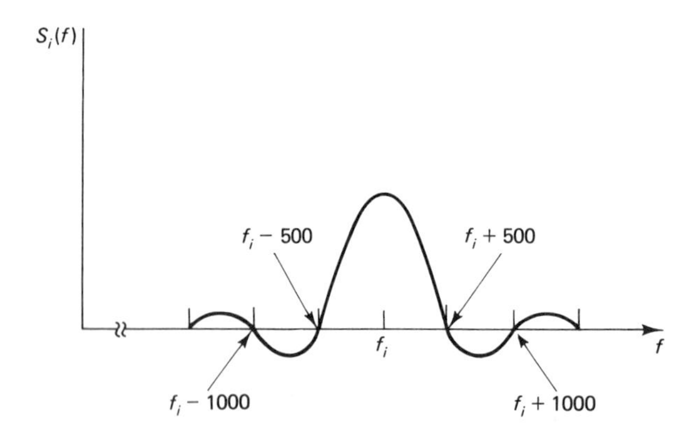

Figure 7.34 Fourier transform of sinusoidal burst

No matter where we place these frequencies within the given transmission bandwidth, distortion will occur. The reason is that the Fourier transform of each signal extends over infinite frequency while the channel passes the band only between 300 and 3000 Hz. This distortion will cause intersymbol interference and increase the bit error rate beyond that which applies without distortion. To minimize this effect, we would like to place the two frequencies as close to the center of the band as possible. We wish to do this while maintaining the minimum correlation between signals.

Therefore, to place the frequencies symmetrically around the center of the band, we shall choose

$$f_0 = 1650 - \Delta f \text{ Hz}$$
$$f_1 = 1650 + \Delta f \text{ Hz}$$

We shall attempt to choose Δf for minimum correlation. (It will be necessary to double-check distortion after this minimization. For example, if we have the best correlation occurring at $\Delta f = 1.5$ kHz, obviously this would be a poor choice).

The correlation is given by

$$\rho = 2\int_0^{2 \text{ msec}} [\cos(2\pi f_0 t) \times \cos(2\pi f_1 t)]\, dt$$
$$= 2\int_0^{2 \text{ msec}} [\cos(2\pi \times 3300t) + \cos(2\pi \times 2\,\Delta f t)]\, dt$$
$$= 5.66 \times 10^{-5} + \frac{\sin(2\pi \times 2\,\Delta f \times 0.002)}{\pi \times 2\,\Delta f}$$

We shall discuss the best possible choice of Δf in Chapter 9. For now, we note that if Δf is 250 Hz, the second term is zero and the correlation coefficient is 5.66×10^{-5}. For all practical purposes, we can call this a zero correlation. The two frequencies are then at 1400 and 1900 Hz. In viewing the distortion and the Fourier transform of Fig. 7.29 we see that the center lobe and the first sidelobe of the transform are completely passed by the channel. This represents more than 98% of the total power of the signal, so we will assume that the distortion and intersymbol interference are essentially zero.

Now that we have determined the two waveforms used to send the binary information, the remainder of the design is trivial. We can generate these two waveforms in a number of ways. Two sinusoidal generators can be used, with enable circuitry used to switch between the two. Alternatively, a VCO can be fed by a two-level voltage signal.

The receiver is a matched filter detector. Bit synchronization can be obtained using the feedback control-loop techniques described in Chapter 6. If the data are often static, we may have to modify the system to use RZ or biphase transmission. This will increase the bandwidth of the signal and thereby increase the distortion. If doing so leads to significant intersymbol interference, we may have to resort to signal shaping and partial response signaling.

We now evaluate the performance of this system. The bit error rate is given by

$$\text{BER} = \frac{1}{2}\,\text{erfc}\sqrt{\frac{E(1-\rho)}{2N_0}}$$
$$= \frac{1}{2}\,\text{erfc}\sqrt{\frac{0.009(1-0)}{2 \times 10^{-3}}} = 0.5\,\text{erfc}\sqrt{4.5} = 6 \times 10^{-12}$$

Example 7-10

You wish to communicate a voice signal using PCM. The channel passes frequencies between 300 Hz and 200 kHz. White noise of power 3×10^{-6} watts/Hz adds to the signal during transmission. The channel can support voltages between -3 V and $+3$ V. The required signal-to-noise ratio at the receiver is 40 dB. Design the transmitter and receiver.

Discussion. The problem statement specifies PCM, so we must assume that some hardware constraints exist that do not permit more sophisticated modulation techniques. We must determine the number of bits of quantization and the two waveforms used to transmit 1's and 0's. Since the required signal-to-noise ratio is 40 dB, it is clear (see Fig. 7.28) that we need to use at least 7 bits of quantization. This is so because, with 6 bits, the maximum signal-to-noise ratio is 36 dB even if no bit errors are made during transmission (i.e., due to quantization noise alone). In order to achieve an SNR of 40 dB using 7 bits of quantization, the bit error rate must be less than about 10^{-5}.

Since we start with a voice signal, we must sample at a frequency of at least 7 kHz (assuming maximum voice frequencies of 3.5 kHz). Suppose we choose an 8-kHz sampling rate, since this is used in the T1 system, and off-the-shelf hardware is easier to obtain. Each sample must be coded into 7 bits, so assuming no frame sync or overhead information is needed, we must transmit 56 kbps down the channel. Since the channel bandwidth is approximately 200 kHz, we should be able to accomplish this without sophisticated techniques for dealing with intersymbol interference.

As in the previous problem, suppose we use sinusoidal bursts to send the information, and we separate the two frequencies by the bit rate, or 56 kHz. If we center these within the band, we could choose 72 kHz and 128 kHz for the two frequencies. The channel would then pass the center lobe and part of the first sideband of each signal, for a total power of about 95%. The correlation coefficient is very close to zero, and the distortion (and intersymbol interference) would be negligible. The remainder of the design follows that of the previous example. The bit error rate is then given by

$$\text{BER} = \frac{1}{2}\,\text{erfc}\sqrt{\frac{E(1-\rho)}{2N_0}} = \frac{1}{2}\,\text{erfc}\,(3.65) = 3 \times 10^{-7}$$

This meets the specifications for the problem.

Example 7-11

Redesign the system of Example 7-10, except for a channel with an upper frequency cutoff of 50 kHz.

Discussion. The previous design cannot be used, since the channel bandwidth has been cut by a factor of 4. Since the first zero of the Fourier transform of a sinusoidal burst of duration $\frac{1}{56}$msec falls 56 kHz away from the frequency of the sinusoid, use of that technique would result in significant distortion and intersymbol interference.

If we could somehow lower the bit transmission rate by a factor of 4, we could borrow the solution to the previous design. Perhaps we can approach this savings by going to more sophisticated modulation techniques. That is, in the previous example we sampled at 8 kHz and coded each sample into 8 bits. We thus were transmitting at 56 kbps. It is not unreasonable to expect acceptable performance at 28 kbps using a combination of adaptive techniques and memory systems, such as differential PCM.

The actual performance levels are difficult to calculate (indeed, they cannot be found in general) and depend upon the specific form of the voice signal being transmitted. You might, however, find results of some studies in the technical literature. For example, if you go to the library, as you should at the start of any design project, you would find that the (former) Bell System has published a massive amount of literature on delta modulation as used in speech [see *Bell Laboratories Record* and *Bell System* (now *AT&T*) *Technical Journal*]. In fact, in the early 1970s, the Subscriber Loop Multiplex (SLM) system was developed and installed. This is an adaptive delta modulation system used to multiplex 24 voice channels, just as in the T1 PCM system. However, the sampling rate for each channel is 57,200 samples per second vs. 56,000 bits per second required for the T1 samples of a single channel (8000 samples per second and 8 bits per sample). The SLM therefore system does not provide bandwidth efficiency, and in fact it was designed for rural telephone service, taking into account constraints related to hardware interfacing. However, in reading about this system (see the references), you can gain valuable insight into the potential for bandwidth conservation using ADM. You can also often get clues to locating other articles, once you find one source of information.

If we achieve a factor-of-2 bandwidth reduction through the modulation technique, we can achieve an additional factor of 2 by using controlled ISI (partial response signaling). With this overall bandwidth reduction by a factor of 4, the remainder of the design follows that of the previous example.

7.10 SUMMARY

Baseband transmission is the simplest form of digital communication. In the binary case, we simply replace the bits, 0 and 1, with two relatively low frequency waveforms. The selection of waveform shapes and the effects of these shapes upon performance are the major focus of this chapter.

The first section presents the concepts of unipolar and bipolar transmission and discusses techniques for detecting the waveforms at the receiver. This analysis concentrates upon the integrate-and-dump detector. We also examine the frequency spectra of various baseband signals.

Section 7.2 introduces the concept of intersymbol interference and develops the eye pattern. We derive equations for the peak and mean-square values of interference and discuss the effects of pulse shape upon intersymbol interference. The raised cosine pulse is presented and analyzed.

Section 7.3 explores baseband equalization as a technique for offsetting the effects of channel distortion.

Section 7.4 analyzes the performance of baseband detectors, starting with the sampling detector and progressing to matched filter detectors. The following section generalizes upon the performance results for the binary matched filter detector. This is done under the assumption of equal a priori probabilities and equal costs for the various errors. The resulting equations will be used throughout the analysis of the next three chapters.

Section 7.6 extends the results of binary baseband communication to M-ary signaling sets. We derive equations for the probability of error.

Section 7.7 presents a technique known as partial response signaling, which allows controlled intersymbol interference. In this manner, we can communicate at a speed which appears to be higher than the Nyquist rate. We present the duobinary system and discuss applications.

When designing a PCM system, the engineer must first perform an analog-to-digital conversion upon the continuous signal and then configure a system to transmit the resulting digital information. Errors occur in both parts of this process. That is, quantization noise is generated by the A/D conversion, and bit errors occur in the digital transmission system. Section 7.8 explores the relationship of these two types of errors and gives guidelines for the best possible design of a system.

The final section presents several design examples in order to reinforce the concepts presented in this chapter.

PROBLEMS

7.1. A binary sequence

$$1 \quad 1 \quad 0 \quad 0 \quad 1 \quad 1 \quad 0 \quad 0$$

is transmitted using a baseband bipolar signal. The bit rate is 1 Mbps. Find the frequency distribution of the transmitted waveform. Repeat the analysis assuming that unipolar transmission is used, and compare the results.

7.2. The eight possible 3-bit code words are block-encoded using a single parity-bit code. Assume that even parity is used. We wish to send five code words per second. Choose any sequence of five words and sketch the resulting bipolar waveform. Label the time axis.

7.3. The signal

$$s(t) = \cos(4\pi t + 5°) + \sin(2\pi t + 10°)$$

is sampled at 4 samples per second and encoded into 7-bit PCM. The resulting signal is transmitted via bipolar baseband. Find the bandwidth of the transmitted signal. Sketch the form of a receiver that could be used to decode this signal.

7.4. Derive the expression of Eq. (7-11) starting with the Fourier transform of Fig. 7.14. [*Hint:* Attempt to write the Fourier transform as the convolution of two transforms.]

7.5. The response of a channel to a pulse is as shown below, where T is the period of transmission. Sketch the eye pattern.

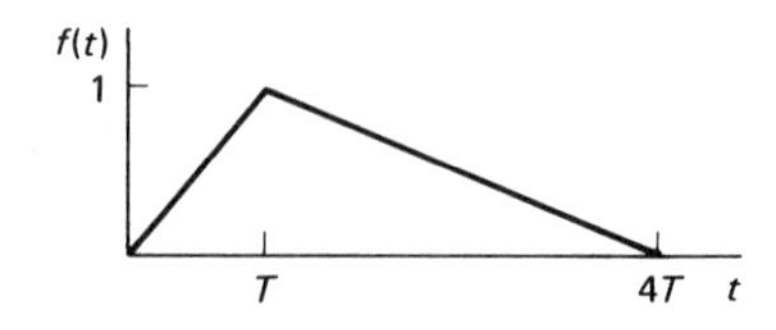

7.6. A channel has a square passband characteristic. That is, it can modeled as an ideal low-pass filter. The baseband signal is bipolar triangular with $T = 1/2f_m$, where f_m is the channel cutoff frequency.

(a) Find the peak intersymbol interference.

(b) Find the mean-square intersymbol interference.

7.7. For Problem 7.6, assume that 1's and 0's are equally likely. Find the density of the intersymbol interference (use only four terms in the expansion).

7.8. A bipolar baseband signal using $\pm A$ is transmitted. In the absence of additive noise, the sample value at the receiver takes on one of three values. It is equal to the transmitted value with probability $\frac{1}{2}$, and it is greater or less than this value by Δ with probabilities $\frac{1}{4}$ and $\frac{1}{4}$, respectively. The system is now corrupted by additive Gaussian noise of zero mean and variance σ^2.

(a) Find the probability of error in terms of A, Δ, and σ^2.

(b) For $A = 5$, $\sigma^2 = 1$, and $\Delta = 1$, find the probability of error.

(c) Compare your answer to part (b) with the probability of error that would result if there were no intersymbol interference.

7.9. An ideal low-pass channel with $f_m = 1$ MHz receives a square bipolar signal at 1 Mbps. The output is sampled at the *end* of each interval (not in the middle).

(a) Find the peak intersymbol interference.

(b) Find the mean-square intersymbol interference.

(c) Find the probability of error using the techniques of Example 7-2.

7.10. Repeat Problem 7.9 using a raised cosine pulse with $K = 0.5$ and $1.5f_0 = 1$ MHz.

7.11. The received pulse due to a 2-s duration unit height square pulse is given by

$$\frac{\sin \pi t}{\pi t}$$

The received waveform is sampled at points displaced to the right of the mid-point by 0.2 s. That is, a timing error has occurred. Find the peak and mean-square intersymbol interference. Also find the intersymbol interference if a long sequence of alternating 1's and 0's is transmitted.

7.12. Approximate $s(T + \Delta)$ as a linear combination of $s(T)$ and the derivative, $s'(T)$. Find the mean-square error in the approximation.

7.13. Approximate $s(NT + \Delta)$ as a linear combination of $s(nT)$ for all n. Find the mean-square error in the approximation. Show that if $T < 1/2f_m$, the error goes to zero (i.e., prove the sampling theorem).

7.14. A single pulse input to a system results in received samples of

$$0.2, \quad 0.5, \quad 1, \quad 0.5, \quad 0.2$$

where the 1 is in the sampling interval corresponding to the transmitted pulse and the other entries are in adjacent intervals. Design a transversal equalizer (choose tap weights) that will eliminate intersymbol interference.

7.15. We desire a received pulse to be of the shape $(\sin 1000t)/t$. The actual pulse shape is $(\sin 500t)/t$. Can you design an equalizer to eliminate intersymbol interference? If so, do it. If not, explain why.

7.16. You are given the two baseband waveshapes shown below. Binary 1's are sent three times as often as binary 0's, and a matched filter detector is used, as shown.

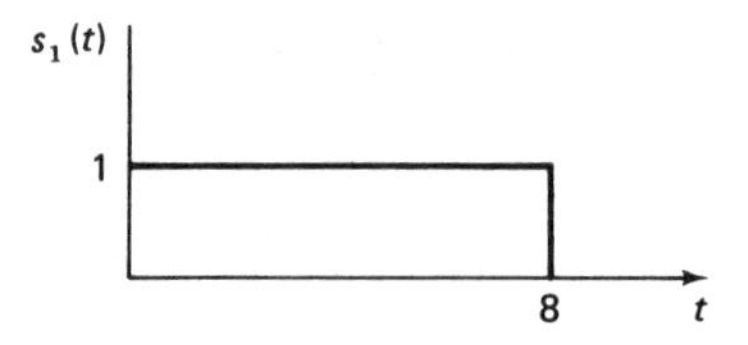

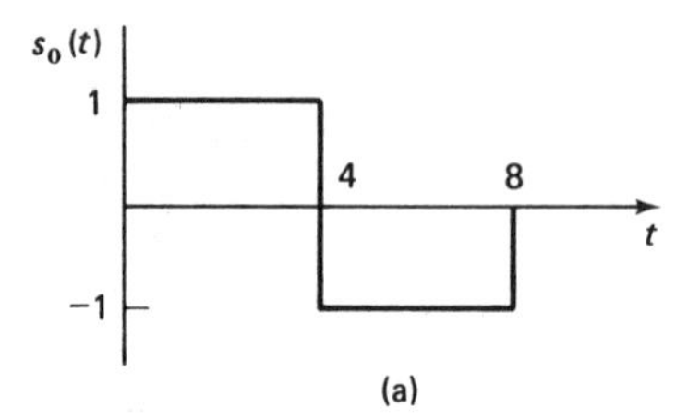

(a)

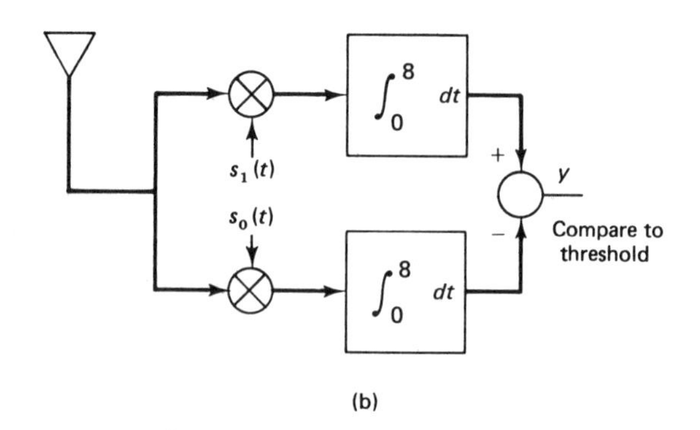

(b)

(a) Find the probability density of the output variable, y, under each of the two hypotheses.

(b) Using the minimum-error criterion, find the threshold.

(c) Find the probability of error.

7.17. Design a matched filter detector to choose between the two baseband signals shown below. (Note that this is a Manchester code.) Assume that white Gaussian noise of power spectral density $N_0/2$ is added, and that the two possible messages are equally probable. Find the probability of error for $E/N_0 = 3$.

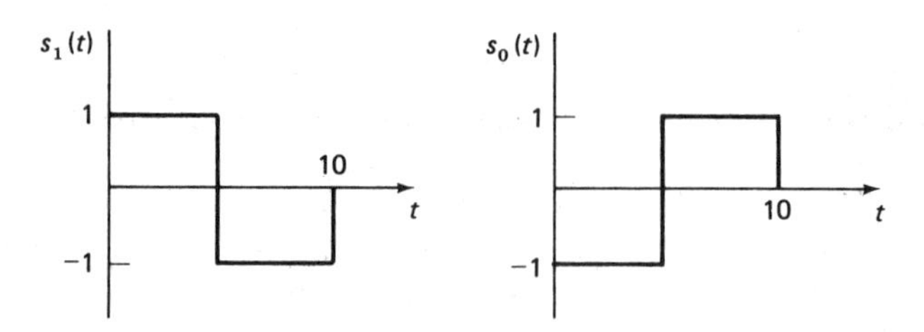

7.18. **(a)** Design a detector for the two signals shown below. Assume that white Gaussian noise is added. Find the probability of error.

(b) Redesign this detector for $s_0(t) = 0$.

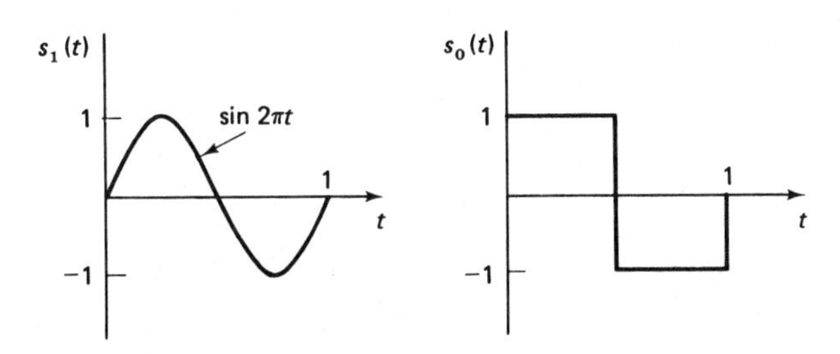

7.19. You are given the two baseband waveshapes as shown below. A matched filter detector is used to decide between the two possible transmitted signals. The additive white Gaussian noise has power spectral density $N_0/2 = 0.1$.

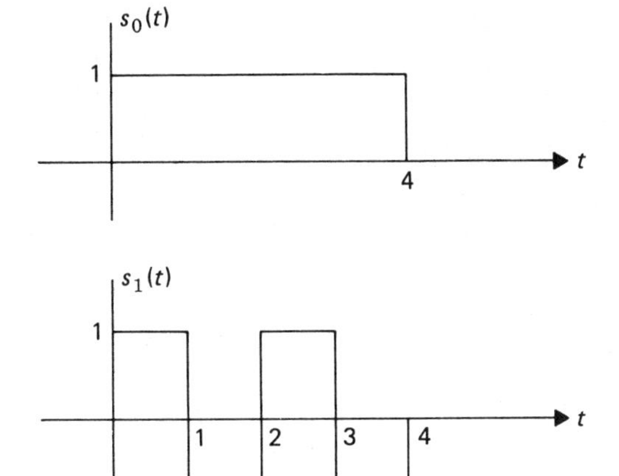

(a) Find the probability density of the detector output, assuming that a 1 is being sent.
(b) Repeat part (a) for the 0 transmission.
(c) Find the output signal-to-noise ratio under each transmission assumption.
(d) Find the bit error rate.

7.20. Given the two signals shown below, design a matched filter detector and evaluate its performance as a function of Δ.

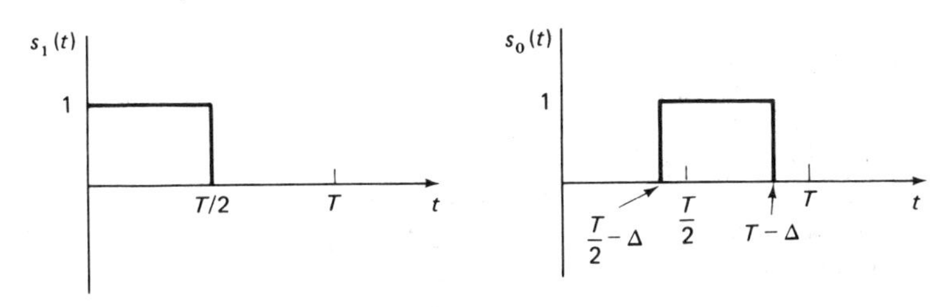

7.21. In a ternary communication system, the three signals are as shown below. The signals are equally probable. Design a minimum-error detector and find the probability of error.

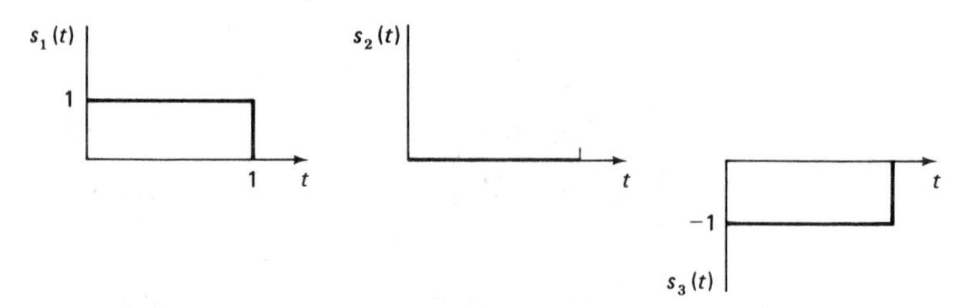

7.22. Design a minimum-error detector for the four signals shown below. Find the probability of error for this detector assuming equal a priori probabilities.

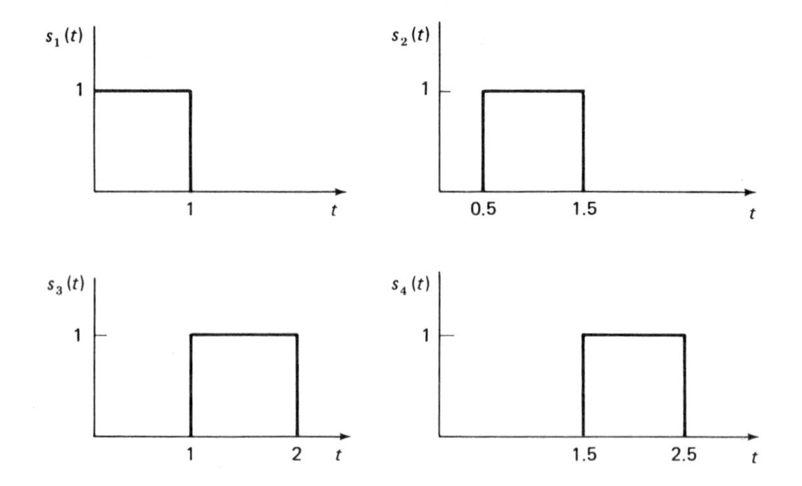

7.23. Design a minimum-error detector for the ternary system with three signals as shown below. Assuming equal a priori probabilities, find the probability of error for this detector.

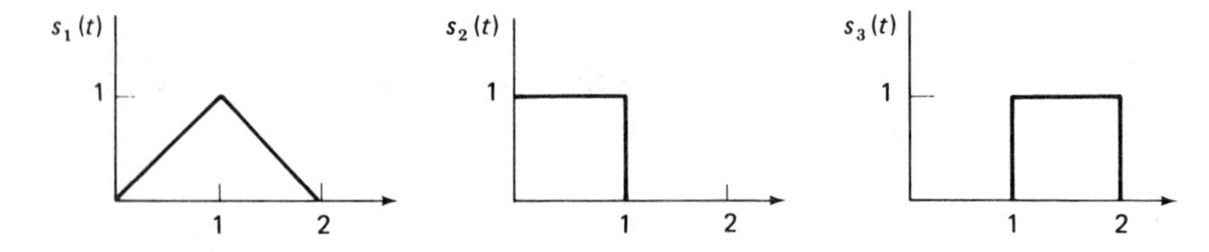

7.24. Binary information is to be transmitted at a rate of 1 kbps on a telephone channel. Assume that the channel passes frequencies between 300 Hz and 3 kHz. White noise of power 10^{-6} watts/Hz adds to the signal during transmission. The channel can support voltages between -10 V and $+10$ V. Design the transmitter and receiver and evaluate error probabilities.

7.25. A PCM system is to be used to transmit a monochrome video signal. Assume that each frame contains 100,000 picture elements and that 32 levels of grayness are desired. At least 50 frames per second are required to prevent "flicker." Additive white Gaussian noise of power 4×10^{-6} watts/Hz is present in the channel. The channel can support voltages between -3 V and $+3$ V. The required signal-to-noise ratio at the receiver is 30 dB. Design the system and specify the required bandwidth for the channel.

7.26. You are given the two signals shown below, which are transmitted in the presence of additive Gaussian white noise.

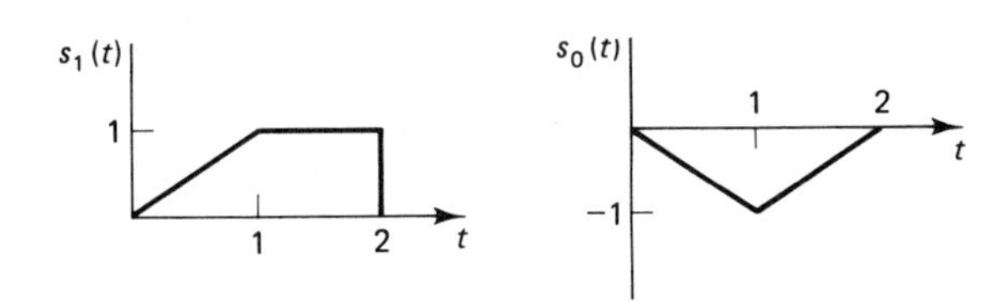

(a) Design a binary matched filter detector to decide which of these two signals is being sent. Your design must include the value of threshold against which the output is compared.

(b) Find the maximum noise power per Hertz such that the bit error rate does not exceed 10^{-5}.

7.27. A bi-phase baseband communication system uses the two signals shown below in order to transmit 1's and 0's.

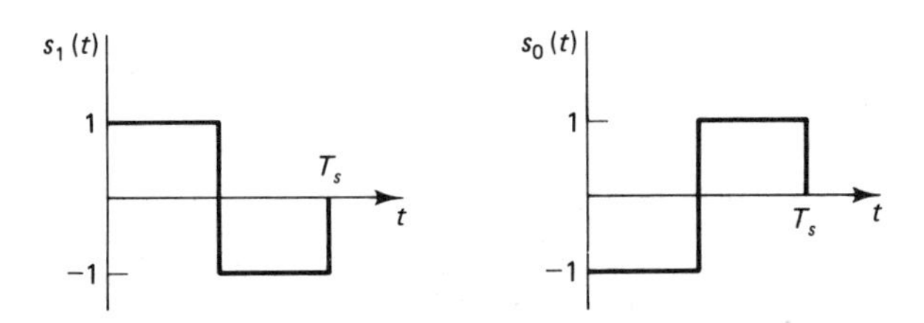

The additive white noise has a power of 10^{-4} watts/Hz. Assume that the two signals are equally likely.

(a) Design a detector to decide between 1's and 0's.

(b) Find the maximum bit rate (bits/sec) that can be used in order to achieve a bit error rate below 10^{-4}.

(c) Now assume that $P(H_0) = \frac{3}{4}$. Find the bit error rate for this condition.

8

Amplitude Modulation

Chapter 7 developed the concept of baseband signaling. Since the baseband signal consists of low frequencies, it cannot be efficiently transmitted through a channel with band-pass characteristics.

There are three general methods of reformatting the baseband signal so that it can be efficiently transmitted through band-pass channels. Besides affording increased efficiency of transmission, these techniques will also open up the second dimension of multiplexing known as *frequency-division multiplexing* (*FDM*). Using this type of multiplexing, we will be able to utilize wideband channels to send many more signals than would otherwise be possible.

The three distinct techniques are *amplitude modulation* (*AM*), *frequency modulation* (*FM*), and *phase modulation* (*PM*). We discuss the overall system, modulators and demodulators, and performance of the AM system in the following sections. Chapter 9 deals with FM, and Chapter 10 discusses PM.

8.1 INTRODUCTION

Chapter 7 introduced the concept of using multiple-amplitude square pulses in order to send digital information. This technique is called *pulse-amplitude modulation* (PAM), since it can be visualized as taking a periodic pulse train and modulating the individual pulse amplitudes. The Fourier transform of a PAM waveform is concentrated in the band around DC, or zero frequency. Since most real communication channels do not pass very low frequency signals, frequency-translation techniques are commonly used to shift frequencies into a desirable range. Amplitude modulation, as used extensively in analog communication, finds application in digital communication as one simple way of translating frequencies. The reasons for using this technique for digital communication are essentially the same as those for analog communication.

In this section we discuss generation and reception of AM signals in general terms. Specifics of modulator and demodulator design are discussed in Section 8.3.

Generation of the AM waveform can result from two distinct approaches. One technique starts with the baseband signal as discussed in Chapter 7 and uses this signal to amplitude-modulate a sinusoidal carrier. Since the baseband signal consists of distinct waveform segments, the AM wave will also consist of distinct waveform segments. Another approach to modulation therefore exists, since it is possible to generate the AM wave directly without first forming the baseband signal. Thus, in the binary case, the generator would have to be capable of formulating only one of two distinct AM wave segments. This second technique leads to the name *amplitude shift keying* (*ASK*) as a term that describes digital AM.

Let us examine the baseband approach. The baseband signal is multiplied by a sinusoid at the carrier frequency f_c. This system is illustrated in Fig. 8.1. In this system we have started with impulse samples so that the filter, $H_1(f)$, can be used to develop any baseband signal shape. Section 8.3 discusses methods of performing the actual modulation. Note that we are illustrating a bipolar baseband signal. As shown in the figure, the modulated signal will have a constant amplitude. We shall see that such a transmission scheme requires a *coherent* detector, since the digital information is contained in the *phase* of the modulated signal. The *envelope* of this signal does not contain the information, so an incoherent (e.g., envelope) detector could not be used. On the other hand, if the baseband signal is unipolar, incoherent detection can be used.

The filter, $H_2(f)$, following the modulator is used to shape the signal prior to transmission through the channel. For example, it can be used to shape the square pulses shown in the figure in order to limit the bandwidth. The filter can also be used to eliminate portions of the signal, thus producing *single-sideband* or *vestigial-sideband* waveforms. Part of the filter function can be moved in front of the mod-

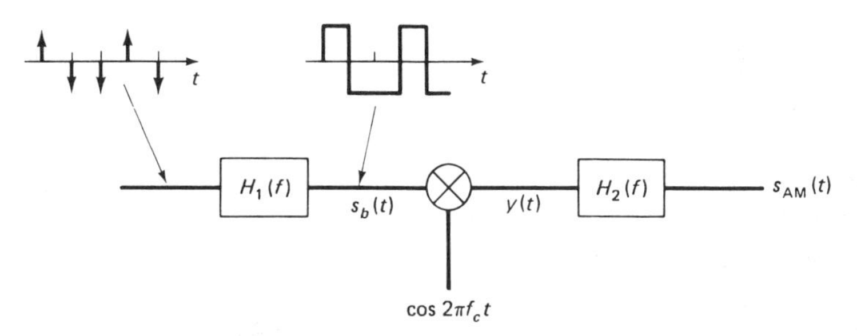

Figure 8.1 Generation of ASK from baseband

ulator. That is, the baseband signal can be shaped prior to modulation. The required filter would be a frequency-shifted version of $H_2(f)$ and could be combined with $H_1(f)$. For that reason, we assume in the following analysis that $H_2(f) = 1$.

We can analyze the class of situations where the baseband signal is piecewise constant by using the following equation for the transmitted signal segments:

$$s_i(t) = \frac{A}{2}[1 + md_i(t)] \cos(2\pi f_c t) \tag{8-1}$$

The two signal segments result where $i = 0$ and $i = 1$ to send a binary 0 and 1, respectively. $d_i(t)$ is either $+1$ or -1, so it can be considered as normalized bipolar data. The index of modulation is m. Thus, for example, if $m = 0$, we send a pure carrier sinusoid. If $m = \frac{1}{2}$, we send a sinusoidal burst of amplitude $A/4$ for a 0 and $3A/4$ for a 1.

The case of $m = 1$ is commonly used. With this choice of modulation index, we transmit a zero-amplitude signal to send a binary 0 and a sinusoid of amplitude A to send a binary 1. This is known as *on-off keying* (OOK) and is illustrated in Fig. 8.2. It is identical to modulating the carrier with a unipolar baseband signal. If T is the bit transmission period (the reciprocal of the bit rate), and 1's and 0's are equally likely, the average energy per bit is

$$E_b = \frac{A^2T}{4}$$

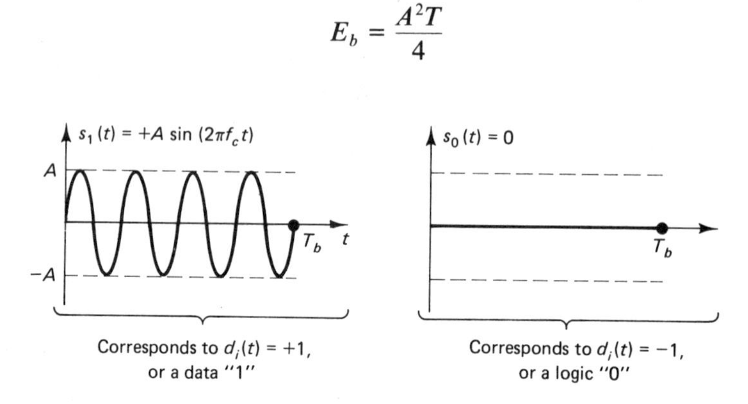

Figure 8.2 Waveforms for on-off keying

Just as there are two broad classes of modulators, there are also two classes of demodulators. One approach to demodulating the AM waveform is to recover the baseband signal. This process can be performed using any of the analog demodulation techniques, some of which are described in Section 8.3. Once the baseband signal has been recovered, the techniques of Chapter 7 can be used to decode the resulting signal into a data signal.

The second class of demodulators combines the demodulation and decoding into a single operation. Since the communication is digital, the received AM waveform consists of discrete signal segments. The receiver need simply recognize which of the possible signal segments is being received during each sampling period. In Section 8.3, we show that the matched filter detector is well suited to this approach.

8.2 BASK SPECTRUM

The Fourier transform of the output of the amplitude modulator is

$$S_{AM}(f) = \frac{H(f)}{2}[S_b(f - f_c) + S_b(f + f_c)] \tag{8-2}$$

where $S_b(f)$ is the Fourier transform of the baseband signal. This transform is sketched in Fig. 8.3.

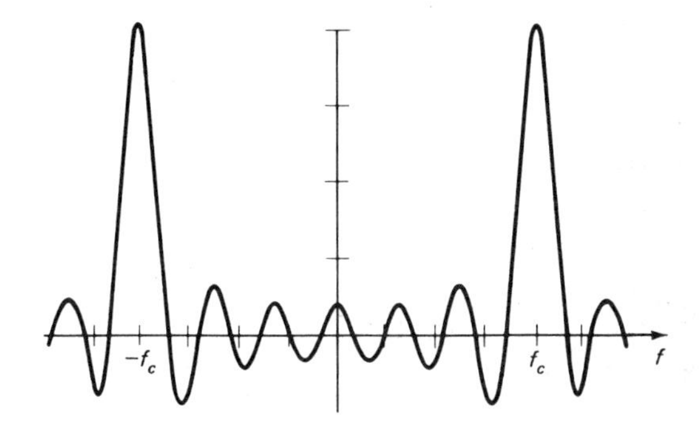

Figure 8.3 Fourier transform of AM waveform

We can find the power spectral density of the BASK (binary amplitude-shift keying) signal in the same manner as in Chapter 7. We assume random data and on-off keying. The average amplitude of the carrier is then $A/2$, so the carrier power, P_c, is $A^2/8$. Since the total average power of the ASK signal is $A^2/4$, the total power in the sidebands (information), P_i, is $A^2/8$. This must equal the area under the power spectral density curve. The power spectral density is sketched in Fig. 8.4.

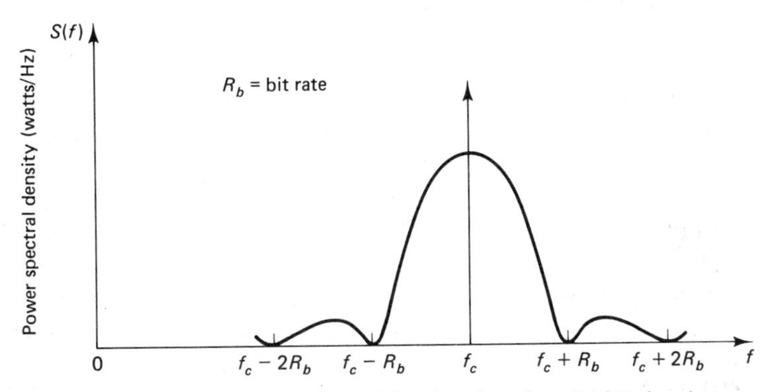

Figure 8.4 Power spectral density of random BASK signal

Example 8-1

Find the bandwidth of a unipolar ASK signal where the bit rate is 1000 bits/sec and the carrier frequency is 1 MHz. Assume that an alternating series of 1's and 0's is sent.

Solution. The ASK waveform consists of 1-ms bursts of the carrier alternating with 1-ms segments of zero signal as shown in Fig. 8.5. The Fourier transform of this waveform is most easily found by first evaluating the transform of the baseband signal—that is, a periodic pulse train. By first finding the Fourier series, we evaluate the transform as shown in Fig. 8.6. The Fourier transform of the ASK waveform is found by translating the transform of Fig. 8.6 to a center frequency of 1 MHz. Since the transform never goes identically to zero, the exact bandwidth depends upon the definition used.

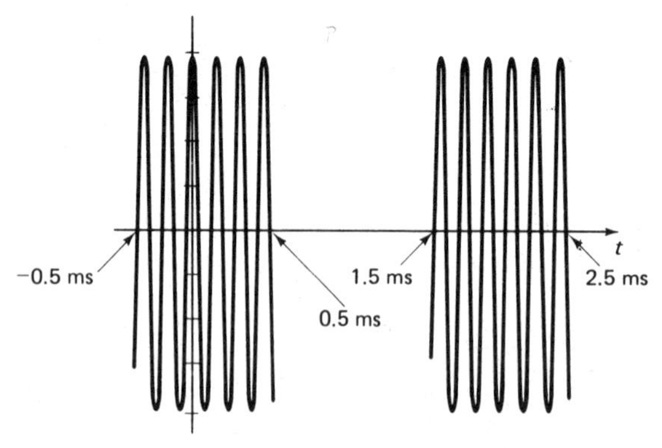

Figure 8.5 ASK waveform for Example 8-1

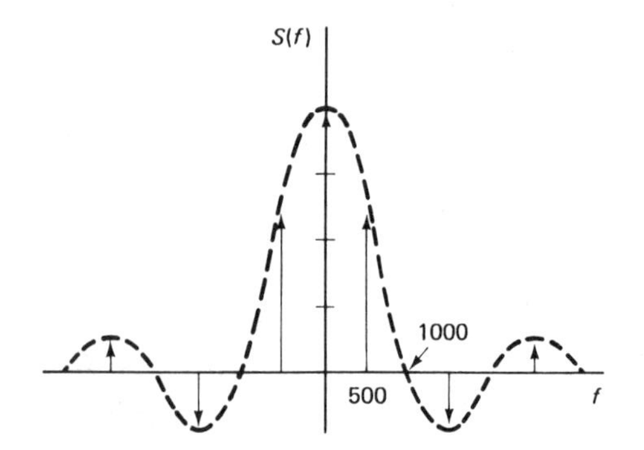

Figure 8.6 Transform of ASK waveform for Example 8-1

8.3 MODULATORS AND DEMODULATORS

8.3.1 Modulators

Generation of the amplitude-shift-keyed waveform can be accomplished in two distinct ways. One technique is to first generate the baseband signal and then use any of the amplitude-modulation techniques that are common to analog forms of AM communication. We briefly review these techniques below. The second way is to actually do a keying operation. For example, in binary communication the transmitted waveform is one of two possible signal waveforms. In the baseband case, these two possible signals are low-frequency, whereas in the ASK case, they are modulated sinusoidal bursts. A particularly simple realization occurs if the baseband signal before modulation is *unipolar*. That is, the baseband signal (before shaping) is a positive constant to transmit a binary 1 and zero to transmit a binary 0. The ASK waveform then becomes a sinusoidal burst to transmit a 1 and zero to transmit a 0. We therefore have an oscillator that is turned on for a specified length of time in order to send a 1 and is left off to send a 0. This technique is known as *on-off keying,* and is illustrated in Fig. 8.7.

If a level other than zero is used to transmit a binary 0, we can resort to an oscillator feeding a step-variable attenuator. A conceptualized version of this using resistive voltage dividers is shown in Fig. 8.8.

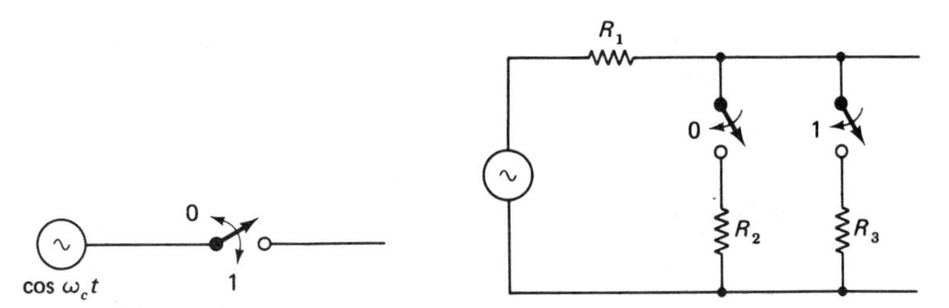

Figure 8.7 On-off keying

Figure 8.8 ASK with nonzero voltage levels

In a sense, the typical amplitude modulator from AM applications represents an overkill for ASK, since these modulators are capable of handling any shape of information signal, while in the present data application, only discrete cases occur. However, if the baseband pulse has already been shaped for minimum intersymbol interference, and if the AM modulators are readily available, this may prove to be the better route to take. Let us therefore briefly examine the various methods of analog amplitude modulation. The student who has had a course in analog communication should already be familiar with this material.

We start with a description of the *gated modulator.* Note that if the baseband signal is multiplied by *any* periodic signal, the result is a series of AM waves with carrier frequencies that are harmonic multiples of the fundamental frequency of the periodic signal. Figure 8.9 illustrates the special case where the multiplying signal is a periodic pulse train. The signal at the output of the multiplier can be expanded in a Fourier series to yield

$$s(t)p(t) = s(t)\left(a_0 + \sum_{n=1}^{\infty} a_n \cos n\ 2\pi f_c t \right) \tag{8-3}$$

Since we assume $s(t)$ to be bandlimited (e.g., the raised cosine spectrum), the Fourier transform of the product in Eq. (8-3) consists of translations of the baseband spectrum by multiples of f_c. The bandpass filter can then be used to extract any of the harmonics, thus generating an AM signal.

The multiplication operation portrayed in Fig. 8.9 is equivalent to a switching operation. The circuit shown in Fig. 8.10 can accomplish this gating if the switch, S, is opened and closed at the desired rate. Obviously, with desired carrier frequencies in the vicinity of 1 MHz, or higher, mechanical switches are ruled out. Electronic switches, including the diode bridge, are the obvious alternatives.

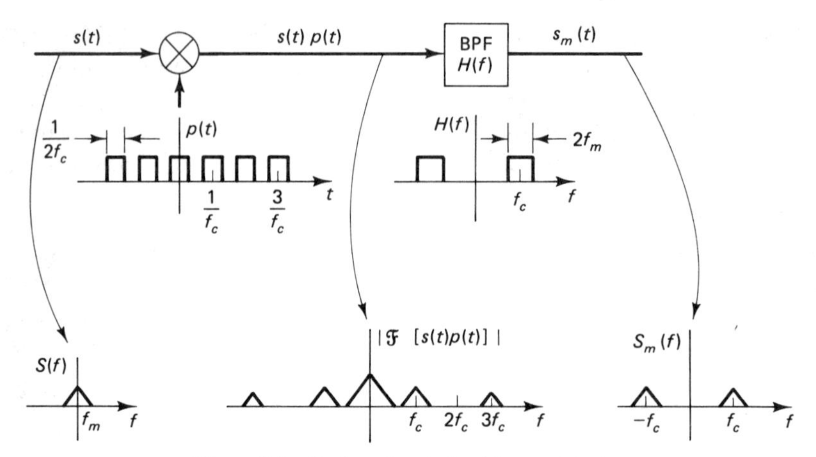

Figure 8.9 Gating circuit for AM modulation

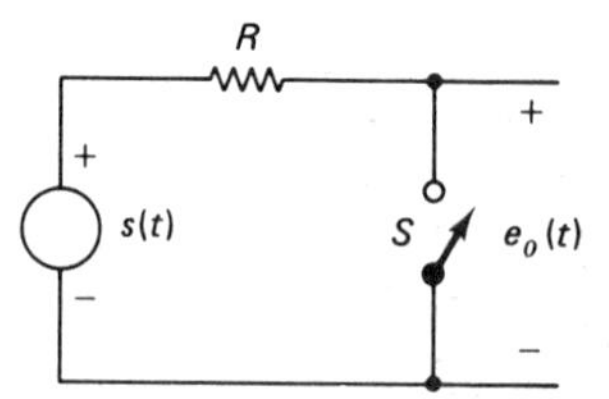

Figure 8.10 Switch used instead of multiplier in gated modulator

A second generic form of modulator utilizes the *square-law device.* If we add the baseband signal to the carrier oscillations and square this sum, the cross product is the desired modulation term. That is,

$$[s_b(t) + \cos 2\pi f_c t]^2 = s_b^2(t) + \cos^2 2\pi f_c t + 2s_b(t) \cos 2\pi f_c t \tag{8-4}$$

The last term in Eq. (8-4) is the desired one. $s_b^2(t)$ occupies frequencies up to twice the highest frequency in the baseband, while $\cos^2 2\pi f_c t$ consists of a DC term and a term at $2f_c$ frequency. Thus, a band-pass filter can separate out the desired term, which resides at frequencies surrounding f_c. This yields the modulator of Fig. 8.11. Nonlinear devices with predominant square terms are readily available. Indeed, a diode is a reasonably good squaring device. We note (see Problem 8.3) that any nonlinear device will produce modulation terms. The modulated term around the carrier frequency, f_c, would be a combination of powers of the baseband signal. This would not be acceptable for analog systems, where the modulation term must

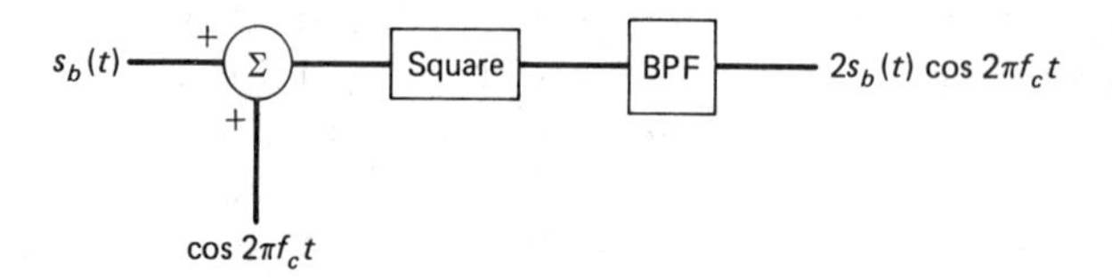

Figure 8.11 Square-law modulator

be linear in $s(t)$ to avoid distortion. However, for digital communication, it is possible that we could accept terms of the type $s_b^n(t) \cos 2\pi f_c t$. The major requirement is that the two signals being transmitted (for the binary case) be as different as possible.

8.3.2 Single Sideband

We have seen that the bandwidth occupied by the ASK waveform is twice the maximum frequency component of the baseband signal. Viewing Eq. (8-2), we see that redundancy is occurring, since the frequency content of the ASK waveform is symmetrical about the carrier frequency. It is therefore possible to eliminate one of the terms in Eq. (8-2) without losing any signal information. This is done by filtering out either the upper or the lower sideband, as shown in Fig. 8.12.

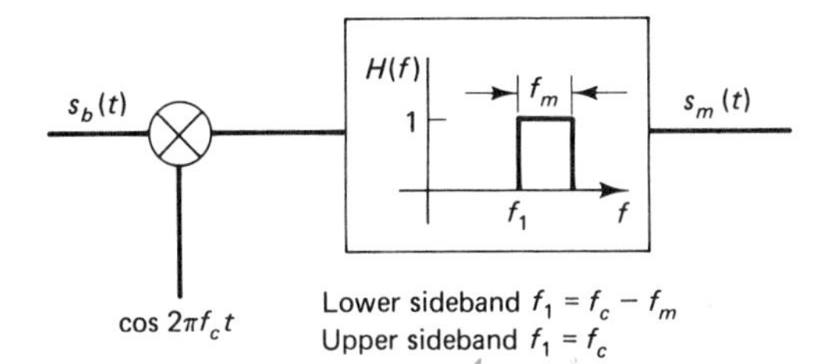

Figure 8.12 Single-sideband generation

It can be shown that the resulting time function at the output of the system in Fig. 8.12 can be expressed as

$$s_m(t) = \tfrac{1}{2}s_b(t) \cos 2\pi f_c t + \tfrac{1}{2}\tilde{s}_b(t) \sin 2\pi f_c t$$

where

$$\tilde{s}_b(t) = \int_{-\infty}^{\infty} \frac{s_b(\tau)}{\pi(t-\tau)} d\tau \tag{8-5}$$

8.3.3 Demodulators

Demodulation can be either *coherent* or *incoherent*. Coherent demodulators maintain precise timing (phase) of the incoming carrier. Incoherent demodulators do not maintain this phase, and essentially perform a nonlinear operation on the modulating signal to retrieve the baseband amplitude.

The *synchronous demodulator* is an example of coherent detection. It simply retranslates the frequencies of the incoming waveform back down to baseband. This is done by multiplying, or *heterodyning*, the incoming AM waveform with a local oscillator matched to the carrier. This is illustrated in Fig. 8.13. We see in the next section what the consequences are of not perfectly matching the carriers.

Although the multiplication by a sinusoid could be viewed as a shifting of frequencies, we need only resort to simple trigonometry to verify the operation of the

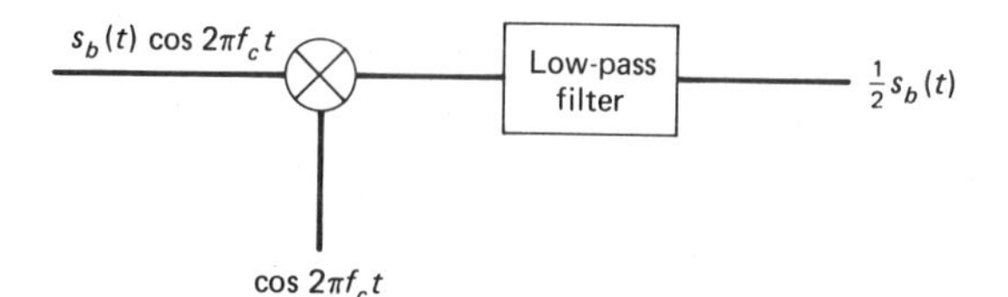

Figure 8.13 Synchronous demodulator

synchronous demodulator. The output of the multiplier in Fig. 8.13 is given by

$$s_b(t)\cos^2 2\pi f_c t = \tfrac{1}{2}s_b(t) + \tfrac{1}{2}s_b(t)\cos 4\pi f_c t \tag{8-6}$$

If we now low-pass-filter this, the baseband signal is the only term that reaches the filter output, since the second term in Eq. (8-6) has a frequency content around $2f_c$.

The synchronous demodulator can also be used for *single-sideband* demodulation. Modifying Eq. (8-6) accordingly, we find that

$$s_b(t)\cos^2 2\pi f_c t + \tilde{s}_b(t)\cos 2\pi f_c t \sin 2\pi f_c t = \tfrac{1}{2}s_b(t) + \tfrac{1}{2}s_b(t)\cos 4\pi f_c t + \tfrac{1}{2}\tilde{s}_b(t)\sin 4\pi f_c t$$

Again, the only term that gets through the low-pass filter is the baseband signal.

The exact carrier frequency can be *derived* from the incoming signal in the case of double sideband. This is so because the upper sideband is a mirror image of the lower sideband. Thus, the carrier can be identified as the "foldover" point for symmetry. A system for performing this carrier extraction is illustrated in Fig. 8.14. We now trace the signal at various points in this system. Starting with $s_b(t)\cos 2\pi f_c t$, we find that

$$a(t) = s_b^2(t)\cos^2 2\pi f_c t = \tfrac{1}{2}s_b^2(t) + \tfrac{1}{2}s_b^2(t)\cos 4\pi f_c t$$

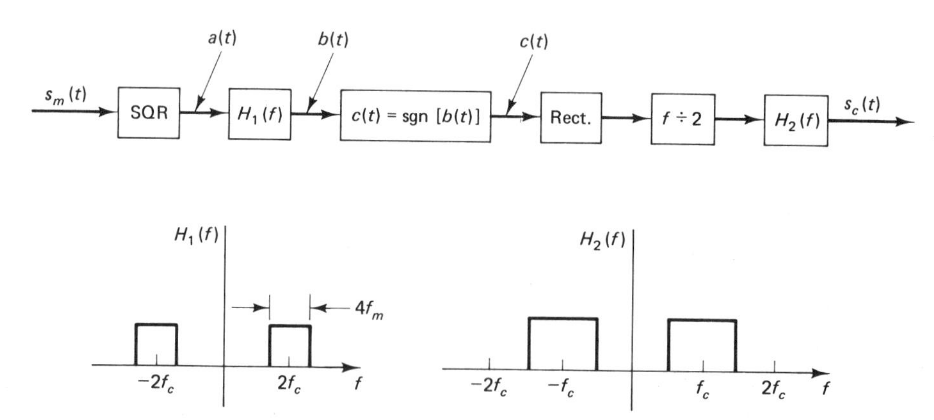

Figure 8.14 Carrier extraction for synchronous demodulator

$$b(t) = \tfrac{1}{2}s_b^2(t) \cos 4\pi f_c t$$

$$c(t) = \text{sgn}(\cos 4\pi f_c t)$$

Note that $c(t)$ is a square wave with value +1 when $\cos 4\pi f_c t$ is positive, and −1 otherwise.

It is now necessary to halve the fundamental frequency of this square wave. We do this by erasing every other pulse. This is a gating operation and is done in a way similar to the operation of a flip-flop. Each incoming pulse triggers a binary device to change state. It takes two pulses for the device to go through one entire cycle. The output of the device therefore has a fundamental frequency which is one-half that of the input. Another way to view frequency division is by looking at the operation of a binary counter. If the counter is controlled by an oscillator at a certain frequency, viewing each bit position from least to most significant bit affords division by powers of 2. For example, the 3-bit counter cycles through the eight counts

000

001

010

011

100

101

110

111

The least significant bit cycles at one-half the clock rate. The next bit to the left cycles at one-fourth of the clock rate, and the most significant bit oscillates at one-eighth of the clock rate. The counter can therefore be used to divide an input frequency by any power of 2. A 3-bit counter is shown in Fig. 8.15 together with the waveforms at each bit output. Note that the frequencies are related by powers of 2.

Returning to the system of Fig. 8.14, it is now only necessary to extract the fundamental frequency of the device output. This is done with a filter that is designed to pass frequencies near f_c and reject all higher harmonics of that frequency. The output of the filter is a carrier whose frequency exactly matches that of the transmitter carrier oscillator.

If the carrier cannot be extracted, as is the case in SSB, synchronous demodulation is possible only if a pilot carrier is transmitted together with the AM wave. This pilot can be either at the exact carrier frequency or at a related frequency, such as a subharmonic. If the pilot is at the exact carrier, extraction is sometimes complicated by the fact that components of the signal are immediately

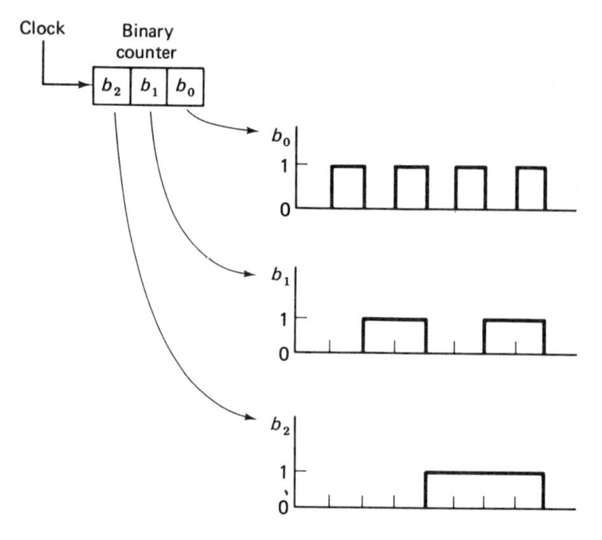

Figure 8.15 Three-bit counter

adjacent to the carrier frequency. A very narrowband filter is required to minimize distortion. If, instead, a harmonic or subharmonic is used, the pilot can often be placed at a portion of the spectrum in which it is not competing with a strong signal. The local carrier can then be reconstructed from the pilot by either multiplying up in frequency or dividing down. This is illustrated in Fig. 8.16.

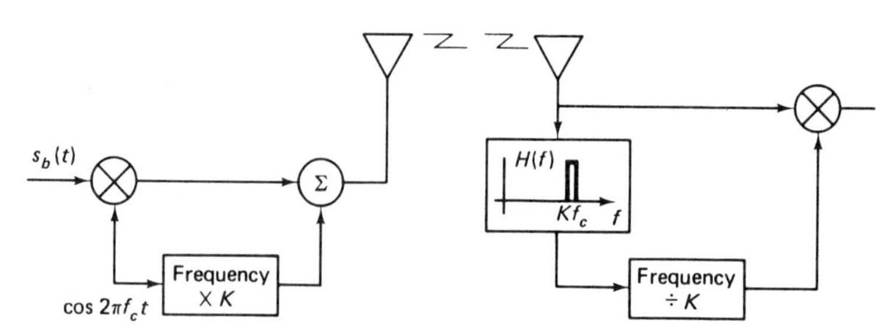

Figure 8.16 Pilot carrier for carrier reconstruction

If the pilot is large in amplitude, such that it is of the same or greater amplitude than the baseband signal, incoherent techniques of detection become available. We write this transmitted carrier AM signal in the form

$$[A + s_b(t)] \cos 2\pi f_c t$$

If A is sufficiently large that $A + s_b(t)$ never goes negative, either square-law or envelope demodulation becomes available.

Envelope demodulation takes advantage of the frequency relationship between the carrier and the multiplying information baseband signal. If these two frequencies are sufficiently widely separated, a circuit can be constructed that will respond

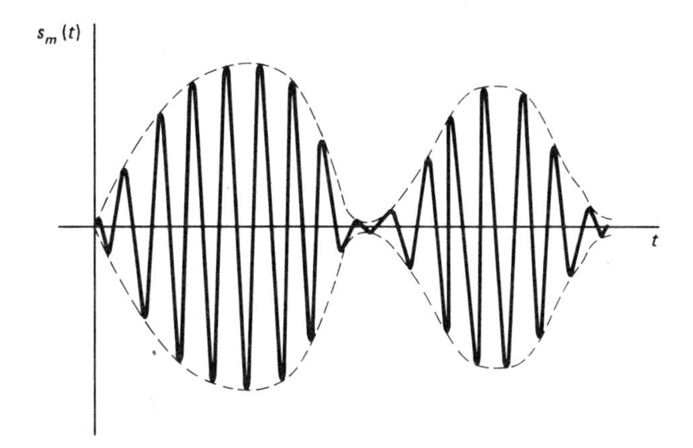

Figure 8.17 AM waveform

to the relatively slow changes of the baseband signal while not responding to the fast changes of the carrier. The AM wave is illustrated in Fig. 8.17. An envelope detector consists, in its simplest form, of a diode, resistor, and capacitor, as shown in Fig. 8.18. The capacitor charges during increases in the AM wave, since the diode is then forward-biased. During decreases, the diode opens and the capacitor discharges through the resistor. The discharge time constant (RC) is such that the downward adjustment can follow the envelope but not the rapid carrier oscillations. The envelope detector is analogous to the shock-absorber system of a car, which follows the envelope of the road but not the high-frequency bumps.

We note that the addition of the carrier is critical, since, if we took a bipolar baseband signal modulating a carrier and fed this AM wave into an envelope detector, the 1's and 0's would be indistinguishable. That is, the output of the envelope detector is $|s_b(t)|$, so the sign information is lost.

A second type of incoherent demodulator is the *square-law demodulator.* This is sketched in Fig. 8.19. The low-pass filter is sometimes replaced by a finite time integrator.

The output of the square device is

$$[A + s_b(t)]^2 \cos^2 2\pi f_c t = [A + s_b(t)]^2(\tfrac{1}{2} + \tfrac{1}{2} \cos 4\pi f_c t)$$

The output of the low-pass filter will then be

$$\tfrac{1}{2}[A + s_b(t)]^2$$

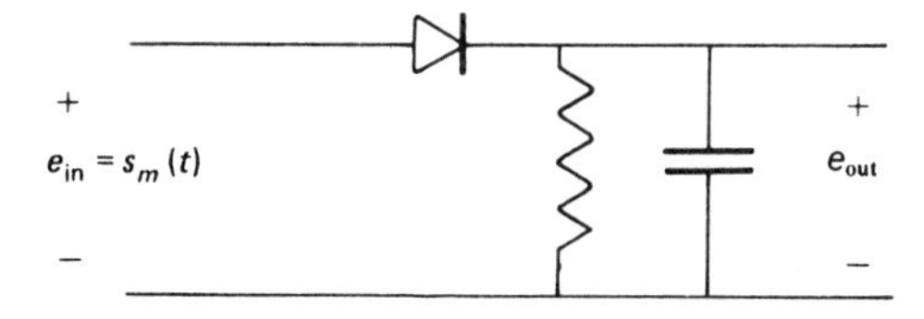

Figure 8.18 Envelope detector

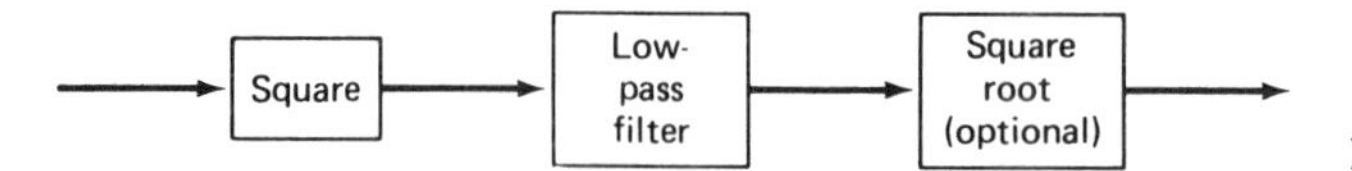

Figure 8.19 Square-law detector

We note that this has frequency content up to twice the maximum frequency of the baseband signal, and the low-pass filter is adjusted accordingly. We then use a nonlinear device to take the square root of this signal to get $A + s_b(t)$. Note that $A + s_b(t)$ had to be nonnegative to permit this unambiguous square root to be taken. Also note that since this is a digital system, and it is only necessary to have two distinguishable signals upon which to base a decision; the square-root operation might not be necessary.

8.3.4 Quadrature Detector

A variation of the incoherent detector is the *quadrature detector,* which forms an incoherent sum of two basically coherent signals. The receiver is illustrated in Fig. 8.20. It is used when the frequency of the carrier is known but the phase is not known.

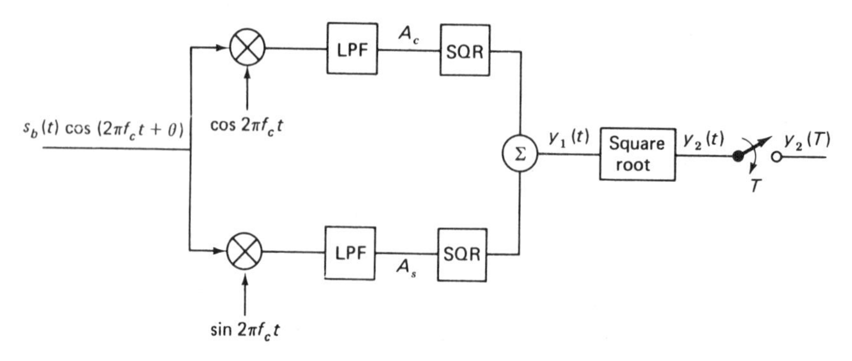

Figure 8.20 Quadrature detector

The signals at points A_c and A_s are found in the same manner as for the synchronous demodulator.

$$A_c(t) = \tfrac{1}{2}s_b(t)\cos\theta$$

$$A_s(t) = -\tfrac{1}{2}s_b(t)\sin\theta$$

$A_c(t)$ is often called the *in-phase* (or I) part of the baseband signal and $A_s(t)$ is called the *quadrature* or (Q) part. The word "quadrature" is used since cosine and sine are in phase quadrature (90° difference). The receiver of Fig. 8.20 is known as a *quadrature receiver.*

The signal at $y_1(t)$ is

$$\begin{aligned} y_1(t) &= \tfrac{1}{4}s_b^2(t)(\cos^2\theta + \sin^2\theta) \\ &= \tfrac{1}{4}s_b^2(t) \end{aligned}$$

As in the case of the square-law detector, the final square-root operation may prove unnecessary in the case of data transmission. It will create an ambiguity if $s_b(t)$ goes negative.

8.3.5 Carrier Recovery Using Control Loops

We showed earlier in this section that carrier recovery is sometimes possible using the nonlinear system of Fig. 8.14. We now explore the replacing of many of the blocks in that system with feedback control systems.

The received signal is of the form

$$\frac{A}{2}[1 + md_i(t)] \cos (2\pi f_c t + \theta)$$

or simply

$$A(t) \cos (2\pi f_c t + \theta)$$

We can use the squaring loop of Fig. 8.21 to recover the carrier signal. The squarer creates a low-frequency term plus a modulated term at twice the carrier frequency. The band-pass filter eliminates the low-frequency term. The modulated double-frequency carrier then forms the input to the PLL (within the dashed box in the figure). The output of the PLL is a sinusoid at twice the required frequency. We divide this frequency by 2 to recover the original carrier, which can then be used in a synchronous demodulator. The PLL was described in Chapter 6, where we used it for timing recovery. It will be examined in greater detail in Section 9.4.

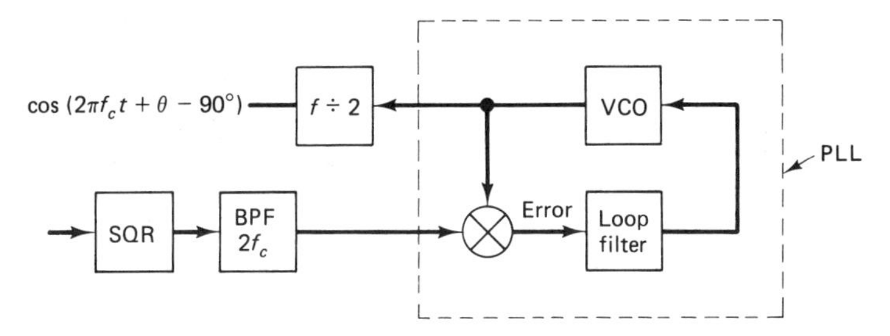

Figure 8.21 Squaring loop for carrier recovery

Another method of carrier recovery uses a variation of the PLL known as a *Costas loop*. This is illustrated in Fig. 8.22. Note that the VCO operates at the carrier frequency, f_c, rather than at $2f_c$ as in the system of Fig. 8.21. The loop is basically equivalent to the squarer followed by a PLL.

The output of the upper low-pass filter is given by

$$0.5A(t) \sin (\theta - \phi)$$

This output is therefore proportional to the sine of the phase difference. Note that if

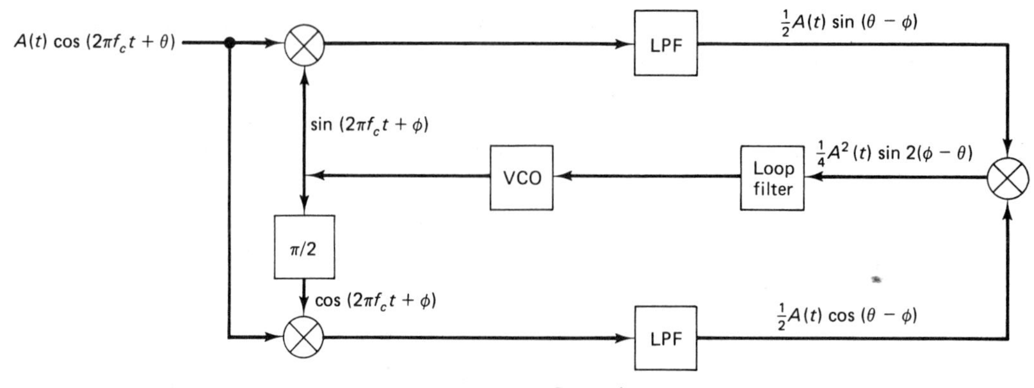

Figure 8.22 Costas loop

the two frequencies are not matched, the phase difference includes a linearly increasing term.

The output of the lower lowpass filter is given by

$$0.5A(t)\cos(\theta - \phi)$$

This output is therefore proportional to the cosine of the phase difference. When these two terms are multiplied together, the result is the error term,

$$\begin{aligned} e(t) &= 0.25A^2(t)\sin(\theta - \phi)\cos(\theta - \phi) \\ &= 0.25A^2(t)\sin[2(\theta - \phi)] \end{aligned}$$

In many systems, $A^2(t)$ will be a constant. If this is not the case, the loop filter can be used to extract the average value of this term. The error term is therefore proportional to the sine of twice the phase difference, and the loop drives this term toward zero.

The Costas loop can be thought of as comprising two PLLs fed by a single VCO. The inputs to the multiplier in this loop are the sine and cosine of the phase difference between the input carrier and the VCO output.

8.4 PERFORMANCE

8.4.1 Coherent Detection

In the case of coherent reception of ASK, the receiver's first task is often to recover the baseband signal. If noise has been added in the channel, the received AM signal will contain a noise term. Upon demodulation, the demodulated signal will consist of the baseband signal plus the translated noise term. The additive noise of concern in AM is that channel noise which is in a band of frequencies around the carrier frequency. By way of contrast, recall that in baseband transmission, the additive noise of concern was in the low-frequency band around dc. The statistics of the noise may dictate that one of these bands is more noisy than the other. This depends upon the actual sources of the noise and the nature of the channel. Most common sources produce noise that is approximately white up to frequencies beyond those of the carrier. In such cases, we can assume that the same noise magnitudes exist in both the baseband and the carrier band.

We will show that for synchronous forms of demodulation, there is no difference in performance between the baseband system and the amplitude-modulated system. On the other hand, we will see that incoherent forms of demodulation lead to poorer performance than that of coherent detectors. This is intuitively satisfying, since it would appear that the more that is known about the arriving signal, the better the detection capability. In the coherent case, it is assumed that the arriving phase is exactly known, whereas incoherent detectors do not use this phase information.

There is a second source of error in synchronous demodulation in addition to additive channel noise. If the local carrier oscillator is not perfectly synchronized to the transmitter oscillator, demodulation errors occur.

Let us consider multiplication of the AM wave by a cosine waveform that deviates from the desired values of frequency and phase by Δf and $\Delta\theta$, respectively. Performing the necessary trigonometric operations, we have

$$\begin{aligned} s_{AM}(t) \cos [2\pi(f_c + \Delta f)t + \Delta\theta] \\ &= s_b(t) \cos 2\pi f_c t \cos [2\pi(f_c + \Delta f)t + \Delta\theta] \qquad (8\text{-}7) \\ &= \tfrac{1}{2}s_b(t)\{\cos (2\pi \Delta f t + \Delta\theta) + \cos [2\pi(2f_c + \Delta f)t + \Delta\theta]\} \end{aligned}$$

Since the expression in Eq. (8-7) forms the input to the low-pass filter of the synchronous demodulator, the output of this filter will be that given in Eq. (8-8). This is true since the second term in Eq. (8-7) has a Fourier transform that is centered about a frequency of $2f_c$, and it is therefore rejected by the low-pass filter:

$$\text{filter output} = \tfrac{1}{2}s_b(t) \cos (2\pi \Delta f t + \Delta\theta) \qquad (8\text{-}8)$$

If Δf and $\Delta\theta$ are both equal to zero, Eq. (8-8) simply becomes $\frac{1}{2}s_b(t)$, and the demodulator gives the desired output. Assuming that these deviations cannot be made equal to zero, we are stuck with the output given in Eq. (8-8).

To see the manifestations of this error, let us first assume that $\Delta\theta = 0$. The output then becomes

$$\tfrac{1}{2}s_b(t) \cos (2\pi \Delta f t)$$

The amount of frequency deviation will probably change or drift with time. It is unlikely that it would be a harmonic of the sampling rate. Therefore, different message bits would be affected by different portions of the low-frequency cosine, thus making compensation all but impossible.

If the frequency of the local oscillator is perfectly synchronized but the phase is mismatched by $\Delta\theta$, the output can be found from Eq. (8-8) to be

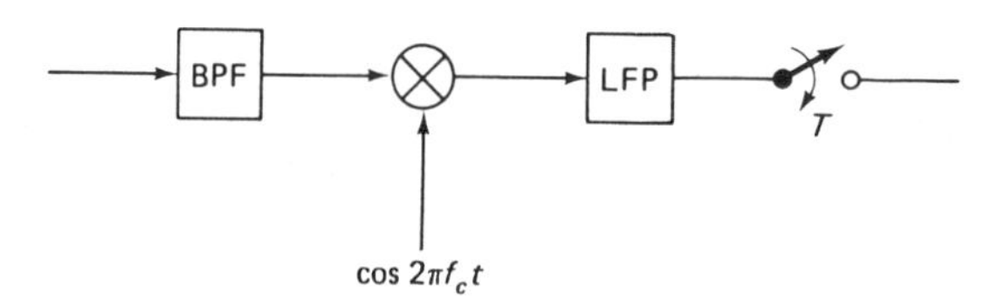

Figure 8.23 Synchronous demodulator

$$\tfrac{1}{2}s_b(t)\cos(\Delta\theta)$$

This mismatch does not distort the signal but will have the effect of attenuating it and thereby raising the probability of error.

We now turn our attention to the effects of *additive channel noise.* We look first at the synchronous demodulator of Fig. 8.23. We will calculate performance for double-sideband modulation, but the result will be identical for single sideband. The input to the multiplier will be

$$[x(t) + s_b(t)]\cos 2\pi f_c t - y(t)\sin 2\pi f_c t$$

In this equation, the narrowband noise has been expanded in the manner discussed in Section 2.5. The output of the multiplier is of the form

$$[x(t) + s_b(t)]\cos^2 2\pi f_c t - y(t)\sin 2\pi f_c t \cos 2\pi f_c t =$$
$$[x(t) + s_b(t)][\tfrac{1}{2} + \tfrac{1}{2}\cos 4\pi f_c t] - \tfrac{1}{2}y(t)\sin 4\pi f_c t$$

The output of the low-pass filter is then

$$\tfrac{1}{2}[x(t) + s_b(t)]$$

The power spectral density of $x(t)$ is found by adding the positive and negative portions of the spectrum of $n(t)$ together after shifting each term to DC. That is,

$$G_x(f) = G_n(f - f_c) + G_n(f + f_c) \tag{8-9}$$

Thus, the power per hertz of this baseband output noise $[x(t)]$ is twice the power per hertz of the noise signal incident on the receiver.

It might therefore appear that the AM system would have poorer performance than the baseband system, since we are effectively doubling the contribution of the additive noise. However, before evaluating system performance, we must examine the signal power. In the AM system, it was assumed that the transmitted signal was of the form $s_b(t)\cos 2\pi f_c t$. The average power of this signal is one-half of the average power of $s_b(t)$. The signal power must therefore be doubled in order to compare these two systems of transmission fairly. When this doubling is done, the baseband and carrier system become *identical* in signal-to-noise ratio. Therefore, the error performance will also be identical.

Example 8-2

Calculate the signal-to-noise ratio for an ASK communication system where the baseband is a $(\sin 2000\pi t)/2000\pi t$ pulse with duration 1 ms, and noise of power 10^{-4} W/Hz is added. The carrier is at 1 MHz. Assume that a synchronous detector is used. Also find the probability of bit error.

Solution. The detector is as shown in Fig. 8.23. The band-pass filter has a bandwidth of 2 kHz centered at 1 MHz. The output of the low-pass filter is of the form

$$\tfrac{1}{2}[x(t) + s_b(t)]$$

The signal-to-noise ratio is the power of the baseband sample divided by the power of $x(T)$, the in-phase narrowband component of the noise.

The power of $x(T)$ is

$$2 \times 10^{-4} \times 1000 = 0.2 \text{ W}$$

The signal sample value is 1 V, yielding a signal power of 1 W. We are assuming that the signal is sampled at the midpoint of the interval, thus yielding the peak of the $(\sin 2000\pi t)/2000\pi t$ curve.

Finally, the signal-to-noise ratio is given by

$$\frac{S}{N} = 5$$

The probability of error is given by Eq. (7-18) as

$$P_e = \frac{1}{2}\operatorname{erfc}\sqrt{\frac{S}{2N}} = \frac{1}{2}\operatorname{erfc}(1.58) = 0.013$$

Example 8-3

Replace the synchronous detector of Example 8-2 with the detector shown in Fig. 8.24. Note that the final low-pass filter has been replaced by an integrator. Find the signal-to-noise ratio and the probability of error.

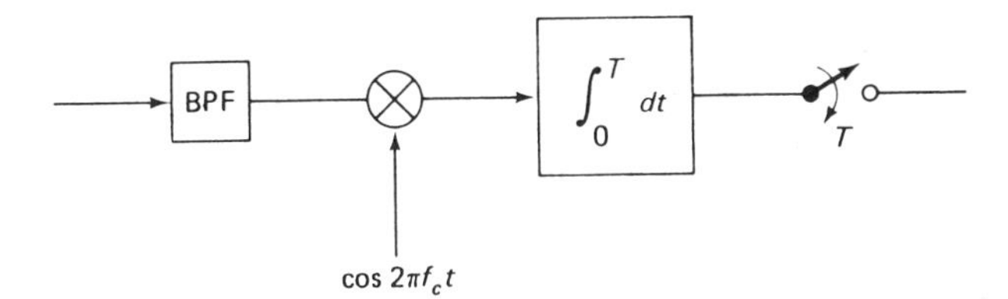

Figure 8.24 Detector for Example 8-3

Solution. The signal at the input to the integrator is

$$\tfrac{1}{2}[x(t) + s_b(t)][1 + \cos 4\pi f_c t] - \tfrac{1}{2}y(t)\sin 4\pi f_c t$$

The output of the integrator is therefore of the form

$$\tfrac{1}{2}\int_0^T x(t)\,dt + \tfrac{1}{2}\int_0^T s_b(t)\,dt$$

We have assumed that the terms of $2f_c$ frequency integrate approximately to zero. This assumption requires that T be much larger than $1/2f_c$. The output is sampled at $t = T$, so the signal portion of the output is

$$\frac{1}{2}\int_0^T s_b(t)\,dt$$

and the noise portion is

$$\frac{1}{2}\int_0^T x(t)\,dt$$

The signal power is found by integrating the baseband signal to yield, for $T = 1$ ms,

$$P_s = \frac{1}{2}\left[\int_{-0.5\times10^{-3}}^{0.5\times10^{-3}} \frac{\sin 2000t}{2000t}\,dt\right]^2 = (2.94 \times 10^{-4})^2$$
$$= 8.64 \times 10^{-8}\text{ W}$$

The noise power is most easily found by recognizing that the finite time integrator is a linear system with an impulse response that is a pulse of height 1 and duration T. Thus, the power spectral density at the output with $x(t)$ as input is given by

$$G_y(f) = G_x(f)\,|H(f)|^2$$

$$= 10^{-4}\left[\frac{4\sin^2(2\pi f \times 10^{-3})}{4\pi^2 f^2}\right]$$

and the average output noise power is

$$\int_0^\infty G_y(f)\,df = \frac{2\times10^{-7}}{\pi}\int_0^\infty \frac{\sin^2 x}{x^2}\,dx$$

$$= 10^{-7}\text{W}$$

Finally, the signal-to-noise ratio is

$$\frac{S}{N} = 0.864$$

and the probability of error is

$$P_e = \tfrac{1}{2}\,\text{erfc}\sqrt{\frac{S}{2N}} = \tfrac{1}{2}\,\text{erfc}\,(0.657) = 0.179$$

8.4.2 Incoherent Detection

Analysis of the incoherent forms of demodulation is complicated by the non-linearities involved. The results of the analysis will show that performance of the incoherent detector is worse than that of the coherent detector. We then show that

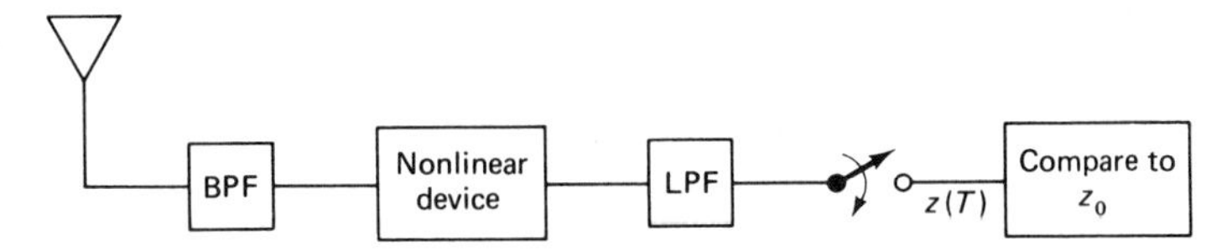

Figure 8.25 Incoherent detector

the incoherent square-law detector can be used for data communication without the extra step of taking square roots.

The performance of the incoherent detector of Fig. 8.25 will be examined first. The output of this detector is independent of the exact phase of the input. The nonlinear element can be an envelope detector, rectifier, or square-law device. Looking at the input to the nonlinear device, we have (as in the coherent demodulator case),

$$[s_b(t) + x(t)] \cos 2\pi f_c t - y(t) \sin 2\pi f_c t$$

This can be rewritten as a single sinusoid:

$$\sqrt{[s_b(t) + x(t)]^2 + y^2(t)} \cos\left[2\pi f_c t - \tan^{-1}\frac{y(t)}{x(t) + s_b(t)}\right] \tag{8-10}$$

Note that we are assuming that no intermediate frequency (IF) is being used. If there is a frequency translation, as for example if the nonlinear device cannot operate at the radio frequency (RF), we know from the analysis of synchronous coherent detectors that the frequency translation has no effect upon performance.

Let us first assume that the nonlinear device is an envelope detector. Assuming that the carrier frequency is much higher than the baseband frequencies, the output of the envelope detector is given by

$$z(t) = \sqrt{[s_b(t) + x(t)]^2 + y^2(t)} \tag{8-11}$$

This output is now sampled and the sample is compared to a threshold to decide whether a binary 1 or 0 is being transmitted. We assume that $x(T)$ and $y(T)$ are zero-mean Gaussian variables, where T is the sampling time; $s_b(T)$ is a deterministic binary signal. Therefore, $s_b(T) + x(T)$ is Gaussian with nonzero mean. Thus, the detector is taking the square root of the sum of the squares of two Gaussian densities. If both densities are zero mean, the result is a *Rayleigh density.* If one or both are not zero mean, the result is what is known as a *Ricean density* and is given by

$$p(z) = \frac{z}{\sigma^2} \exp\left[-\frac{1}{2\sigma^2}(z^2 + s_b^2(T))\right] I_0\left[\frac{s_b(T)z}{\sigma^2}\right] \tag{8-12}$$

σ^2 is the variance of either x or y (we assume them to be equal) and I_0 is the modified Bessel function of zero order. If we set $s_b(T) = 0$, Eq. (8-12) reduces to the Rayleigh density.

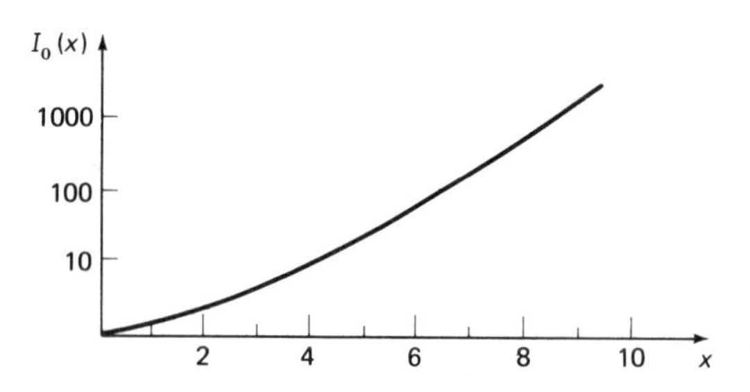

Figure 8.26 Bessel function

The Bessel function is sketched in Fig. 8.26.

Example 8-4

A binary communication system uses the two signals

$$s_0(t) = 0$$

$$s_1(t) = \sin \frac{\pi t}{T_s}$$

where $1/T_s$ is the data rate in bits per second. These are transmitted using amplitude modulation, and an envelope detector is used at the receiving end. Find the probability density of the sampled output of the detector under the two hypotheses.

Solution. The output of the envelope detector is of the form

$$\sqrt{[s_b(t) + x(t)]^2 + y^2(t)}$$

The variance of $x(t)$ or $y(t)$ is

$$\Delta f N_0$$

where the power of the additive noise is assumed to be N_0 watts/Hz, and the bandwidth of the band-pass filter is Δf. The density of a single sample is then

$$p_0(z) = \frac{z}{\sigma^2} \exp\left(\frac{-z^2}{2\sigma^2}\right)$$

under hypothesis 0, where $\sigma^2 = N_0 \, \Delta f$. Under hypothesis 1, the density is

$$p_1(z) = \frac{z}{\sigma^2} \exp\left[\frac{-1}{2\sigma^2}(z^2 + 1)\right] I_0\left(\frac{z}{\sigma^2}\right)$$

where we have assumed that the sample is taken at the peak of $s_1(t)$ (the middle of the interval).

Once the conditional densities are known, we can use any of the detection criteria of Section 5.5 to find the threshold. We illustrate this process with another example.

Example 8-5

For the signals of Example 8-4, design an envelope detector using

(a) the minimum-error criterion assuming equal a priori probabilities.
(b) the Neyman–Pearson criterion with a false-alarm probability of 10^{-5}.

Solution
(a) Using the minimum-error criterion, we choose H_1 if

$$\frac{p_1(z)}{p_0(z)} > 1$$

The likelihood ratio is found by taking the ratio of the two conditional probabilities

$$\frac{p_1(z)}{p_0(z)} = e^{-1/2\sigma^2} I_0\left(\frac{z}{\sigma^2}\right)$$

This exceeds the threshold if

$$I_0\left(\frac{z}{\sigma^2}\right) > e^{1/2\sigma^2}$$

or

$$z > \sigma^2 I_0^{-1}(e^{1/2\sigma^2}) = z_0 \tag{8-13}$$

Given σ^2 we can find z_0. For example, if $N_0 \Delta f = 0.1$, we have a variance of $\sigma^2 = 0.1$, and the argument of the inverse Bessel function is exp (5) = 148. We can read the inverse Bessel function directly from Fig. 8.26, and Eq. (8-13) becomes: Choose H_1 if

$$z > 0.1 I_0^{-1}(148) = 0.685$$

(b) From part (a), we see that since the Bessel function is monotonic increasing, the form of the decision rule is to choose H_1 if

$$z > z_0$$

The false-alarm probability is

$$P_{FA} = \int_{z_0}^{\infty} p_0(z)\, dz = \int_{z_0}^{\infty} \frac{z}{0.1} e^{-5z^2}\, dz$$
$$= e^{-5z_0^2}$$

where we have again assumed that $N_0\,\Delta f = 0.1$. Setting this false-alarm probability equal to 10^{-5}, we find that

$$z_0^2 = -0.2 \ln 10^{-5} = 2.3$$

and

$$z_0 = 1.52$$

The receiver therefore chooses H_1 if the detector output sample exceeds this threshold.

The probability of an error of the second kind, or miss, is given by the integral of the Ricean density over the region of received values for which H_0 is chosen. Thus, denoting the decision threshold as z_0, we have

$$P_M = \int_{-\infty}^{z_0} \frac{z}{\sigma^2} \exp\left[-\frac{1}{2\sigma^2}[z^2 + s_b^2(T)]\right] I_0\left[\frac{s_b(T)z}{\sigma^2}\right] dz$$

This integral cannot be evaluated in closed form. It is tabulated under the name *Marcum-Q function* as

$$Q(\alpha, \beta) = \int_\beta^\infty v \exp\left(\frac{v^2 + \alpha^2}{-2}\right) I_0(\alpha v)\, dv$$

Example 8-6

Given a binary communication system with

$$s_0(t) = 0$$

$$s_1(t) = \cos\frac{\pi t}{T_s}$$

where the data rate is $1/T_s$ bits per second. Amplitude modulation is used with a carrier frequency of 1 MHz; $T_s = \pi$ seconds. Noise of power spectral density, $G_n(f) = N_0/2 = 0.3$ is added during transmission. If a coherent matched filter detector is used, find the probability of a miss. Repeat the analysis for an incoherent detector.

Solution. We have shown that the performance of the coherent detector is the same with or without modulation. The matched filter detector will therefore be examined at baseband for simplicity. The output of the matched filter has a signal portion that is zero under hypothesis 0 and $\pi/2$ under hypothesis 1. This value is found by integrating the square of $s_1(t)$ over the observation interval. The noise at the output has variance of $N_0\pi/4$ (see the examples in Chapter 7 if you have trouble calculating this). Since no information was given to set a value for the threshold, the midvalue between the two output means should be used. The threshold is thus set at $\pi/4$. The probability of a miss is then given by

$$P_M = \int_{-\infty}^{\pi/4} \frac{1}{\sqrt{N_0\pi^2/2}} \exp\left[-\frac{\left(r - \frac{\pi}{2}\right)^2}{\frac{N_0\pi}{2}}\right] dr$$

$$= \frac{1}{2}\operatorname{erfc}\left(\frac{1}{4}\sqrt{\frac{\pi}{N_0/2}}\right) = 0.129$$

Now using an incoherent detector, the likelihood function under hypothesis 1 is given by

$$p_1(z) = \frac{z}{\sigma^2} \exp\left[\frac{-1}{2\sigma^2}(z^2 + 1)\right] I_0\left(\frac{z}{\sigma^2}\right)$$

with

$$\sigma^2 = N_0 f_m = 0.095$$

Note that f_m is the maximum baseband frequency, which is $1/2\pi$. The probability of a miss is then given by the integral

$$P_M = \int_0^{z_0} p_1(z)\, dz$$

where the threshold is found from

$$z_0 = \sigma^2 I_0^{\,1} e^{1/2\sigma^2} = 0.675$$

By making the following change of variables, the integral reduces to a Marcum-Q function, which can be found in a table.

$$v = \frac{z}{\sigma}, \qquad \alpha = \frac{1}{\sigma}$$

Making the substitution yields

$$P_M = \int_0^{z_0/\sigma} v \exp\left(\frac{v^2 + \alpha^2}{-2}\right) I_0(\alpha v)\, dv$$

$$= 1 - Q\left(\alpha, \frac{z_0}{\sigma}\right)$$

and, plugging in all values, we have

$$P_M = 1 - Q(3.24, 2.19) = 1 - 0.86 = 0.14$$

This is higher than the probability of miss for the coherent detector.

We now turn attention to the *square-law detector.* This is illustrated in Fig. 8.27. At $a(t)$, the signal is

$$a(t) = [s_b(t) + x(t)]^2 \cos^2 2\pi f_c t + y^2(t) \sin^2 2\pi f_c t$$
$$- 2[s_b(t) + x(t)]y(t) \sin 2\pi f_c t \cos 2\pi f_c t$$

We recognize the trigonometric identities:

$$\cos^2 2\pi f_c t = \tfrac{1}{2} + \tfrac{1}{2} \cos 4\pi f_c t$$
$$\sin^2 2\pi f_c t = \tfrac{1}{2} - \tfrac{1}{2} \cos 4\pi f_c t$$
$$\sin 2\pi f_c t \cos 2\pi f_c t = \sin 4\pi f_c t$$

Therefore, the output of the low-pass filter is given by

$$\tfrac{1}{2}[(s_b(t) + x(t))^2 + y^2(t)]$$

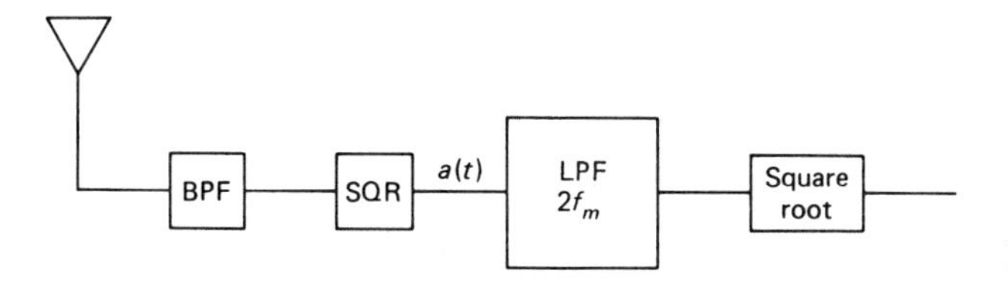

Figure 8.27 Square-law detector

and the square root of this is the envelope. Thus, the previous analysis obtains, and *the error probabilities at the output of the square-root device are the same as those derived earlier for the envelope detector.*

Let us now take a few moments to examine the modified square-law detector of Fig. 8.28. Note that this differs from the square-law detector presented earlier, since there is no square-root operation following the low-pass filter. We wish to evaluate the performance of this detector and compare it to the case where a final square-root operation is performed. In actuality, this is an unnecessary effort since it can be easily reasoned that the elimination of the square-root operation has no effect upon performance. We compare the output of this detector to a threshold. If that threshold is the square of the threshold found earlier for the square-law detector (including the square-root operation), the performance of the two detectors must be identical. That is, since the square-root operation is monotonic, comparing the square root of the output to z_0 is exactly the same as comparing the raw output to z_0^2. In any case, we shall calculate the likelihood functions of the output of the detector of Fig. 8.28 as an exercise in probability. Indeed, I hope it can be agreed that the purpose of studying this material is not simply to get results. If only results are wanted, performance curves can be found in any handbook.

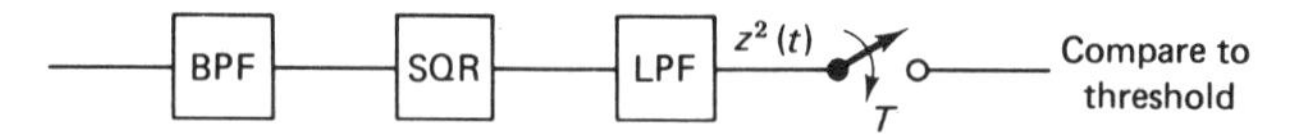

Figure 8.28 Modified square-law detector

Returning to Fig. 8.28, the output of the squaring device is of the form

$$z^2(T) = [s_b(T) + x(T)]^2 + y^2(T) \tag{8-14}$$

$s_b(T)$ is a constant and $x(T)$ and $y(T)$ are zero-mean Gaussian variables. The density of $z^2(T)$ is a *noncentral chi-squared* density with 2 degrees of freedom. The expressions for the density become much simpler if the component Gaussian densities all have the same mean. We can accomplish this by going back to the beginning and assuming that the incoming AM signal is of the form

$$s_b(t)\cos\left(2\pi f_c t + \frac{\pi}{4}\right)$$

This phase shift will not affect the performance of an incoherent detector. Equation (8-14) then becomes

$$z^2(T) = \left[\frac{s_b(T)}{\sqrt{2}} + x(T)\right]^2 + \left[\frac{s_b(T)}{\sqrt{2}} + y(T)\right]^2$$

The density of $z^2(T)$ is given by

$$p_1(q) = \frac{1}{2}\exp\left(-\frac{\lambda}{2} - \frac{q}{2\sigma^2}\right) I_0(\sqrt{q\lambda}) \qquad \text{for } q \geqslant 0$$

where

$$q = z^2(T), \qquad \lambda = \frac{2A^2}{\sigma^2}, \qquad A = \frac{s_b(T)}{\sqrt{2}}$$

Under hypothesis 0, we simply set $A = 0$ to get

$$p_0(q) = \tfrac{1}{2}e^{-q/2\sigma^2}, \qquad q \geqslant 0$$

This is a *chi-squared* density with 2 degrees of freedom, or an *exponential density.*

If we were to continue and design the detector, we would calculate the likelihood ratio and compare it to a threshold:

$$\text{If } \frac{p_1(q)}{p_0(q)} > \lambda_0, \qquad \text{choose } H_1$$

Substituting the likelihood functions, we find that

$$e^{-\lambda/2} I_0(\sqrt{q\lambda}) > \lambda_0$$

and solving for q yields

$$q > \frac{[I_0^{-1}(\lambda_0 e^{\lambda/2})]^2}{\lambda}$$

The probability of false alarm is then given by

$$P_{\text{FA}} = \int_{q_0}^{\infty} \tfrac{1}{2}e^{-q/2\sigma^2}\, dq = \sigma^2 e^{-q_0/2}$$

When all the numbers are plugged in, this will give the same result as we obtained earlier for the square-law detector, including the final square-root operator.

Since the performance is not affected by eliminating the square-root operation, it is clearly advantageous (i.e., cheaper) to use the simpler detector.

8.4.3 Quadrature Detection

The *quadrature detector* is an alternative implementation of incoherent detection. As with other forms of incoherent receivers, it essentially operates upon the envelope of the incoming wave. It should therefore not be surprising to find its performance to be the same as that of the other incoherent detectors. We shall now go through considerable effort to illustrate this last point.

The quadrature detector was illustrated in Fig. 8.20. The signals at points A_c and A_s on that figure are given by

$$A_c(t) = \tfrac{1}{2}[s_b(t) + x(t)] \cos\theta$$

$$A_s(t) = \tfrac{1}{2}[s_b(t) + y(t)] \sin\theta$$

and the output is given by

$$y_1(t) = \tfrac{1}{4}[s_b(t) + x(t)]^2 \cos^2\theta + \tfrac{1}{4}[s_b(t) + y(t)]^2 \sin^2\theta$$

$A_c(T)$ and $A_s(T)$ are Gaussian, and the final output, $y_2(T)$ is the square root of the sum of these two variables. Therefore, $y_2(T)$ will be Ricean, and we need only find the means and variances of $A_c(T)$ and $A_s(T)$ in order to characterize the output. Under hypothesis 0, both means are 0, and under hypothesis 1, they are given by

$$E\{A_c\} = \tfrac{1}{2}s_b(T)\cos\theta$$

$$E\{A_s\} = \tfrac{1}{2}s_b(T)\sin\theta$$

Given any particular value of θ, the variance of either variable is given by

$$\operatorname{var}[A_c] = \operatorname{var}[A_s] = \tfrac{1}{2}E\{x^2(T)\cos^2\theta\}$$

$$= \frac{N_0 f_m}{2}\cos^2\theta$$

where the power spectral density of the noise is $G_n(f) = N_0/2$ and f_m is the maximum frequency passed by the low-pass filters. Averaging this result over θ (i.e., assuming that θ is uniformly distributed), we find that

$$\text{variance} = \frac{N_0 f_m}{4}$$

Now since the means of both $x(T)$ and $y(T)$ are zero, we find that

$$E\{A_c(T)A_s(T)\} = E\{A_c\}E\{A_s\}$$

Therefore, A_c and A_s are *uncorrelated,* and since they are Guassian, this makes them *independent.* The joint density is then the product of the individual densities:

$$p(A_c, A_s) = \frac{1}{2\pi\sigma_1\sigma_2}\exp\left[-\frac{(A_c - m_1)^2}{2\sigma_1^2} - \frac{(A_s - m_2)^2}{2\sigma_2^2}\right] \quad (8\text{-}15)$$

where

$$m_1 = \tfrac{1}{2}s_b(T)\cos\theta, \qquad m_2 = \tfrac{1}{2}s_b(T)\sin\theta$$

$$\sigma_1^2 = \sigma_2^2 = \frac{N_0 f_m}{4}$$

The square root of the sum of the squares is a distance function and can be arrived at by converting Eq. (8-15) to polar coordinates. That is, we let

$$r = \sqrt{A_c^2 + A_s^2}$$

$$\theta_0 = \tan^{-1}\left(\frac{A_s}{A_c}\right)$$

This yields

$$p(r, \theta_0) = \frac{r}{\frac{N_0 f_m}{2}} \exp\left\{-\frac{1}{\frac{N_0 f_m}{4}}[r^2 + s_b^2(T) - 2s_b(T)r\cos(\theta - \theta_0)]\right\}$$

We are only interested in r at the output, so we integrate over θ_0. Also since θ is the phase of the incoming signal, we can again integrate over this variable, assuming that it is uniformly distributed, to get

$$p_1(r) = \frac{r}{\frac{N_0 f_m}{4}} \exp\left[\frac{r^2 + s_b^2(T)}{-\frac{N_0 f_m}{2}}\right] I_0\left[\frac{r s_b(T)}{\frac{N_0 f_m}{4}}\right]$$

$p_0(r)$ is found by setting $s_b(T) = 0$ to get

$$p_0(r) = \frac{r}{\frac{N_0 f_m}{4}} \exp\left(\frac{-r^2}{\frac{N_0 f_m}{2}}\right)$$

But this is identical to what we found for the square-law detector (see Example 8-4), so the performance of the quadrature detector is identical to that of the square-law detector.

Example 8-7

A unipolar baseband signal transmits a half-cycle of sin πt for a binary 1, and 0 for a binary 0. The sampling period is $T_s = 1$. ASK is used with a carrier frequency of 1 MHz. Noise with power of N_0 watts/Hz is added during transmission, and a quadrature detector is used. Find the probability of bit error.

Solution. We will assume that the sampling is performed at the midpoint of each interval, so $s_b(T) = 1$ or 0 depending upon whether the transmitted bit is 1 or 0, respectively. Further assume $f_m = 0.5$, the frequency of the continuous sinusoid.

The probability density of y_2 is given by

$$p_1(y_2) = \frac{y_2}{N_0\pi/16} \exp\left(\frac{-(y_2^2 + 1)}{N_0\pi/8}\right) I_0\left(\frac{y_2}{N_0\pi/16}\right)$$

$$p_0(y_2) = \frac{y_2}{N_0\pi/16} \exp\left(\frac{-y_2^2}{N_0\pi/8}\right)$$

Unfortunately, these densities are not symmetric as in our previous analyses. This has two consequences. The decision threshold is no longer in the middle of the interval, and both conditional error probabilities are needed to find P_e.

Results drawn from decision theory are needed in order to set the threshold. For simplicity, we shall assume that the decision threshold is set in the middle. Its value is then 0.5. The conditional error probabilities are then given by

$$P_{\mathrm{FA}} = \int_{1/2}^{\infty} p_0(y_2)\, dy_2$$

$$= \int_{1/2}^{\infty} \frac{y_2}{\frac{N_0\pi}{16}} \exp\left(-\frac{y_2^2}{\frac{N_0\pi}{8}}\right) dy_2$$

$$= -\exp\left(\frac{-y_2^2}{\frac{N_0\pi}{8}}\right)\Bigg|_{1/2}^{\infty} = \exp\left(-\frac{2}{N_0\pi}\right)$$

$$P_{\mathrm{M}} = \int_{0}^{1/2} p_1(y_2)\, dy_2$$

Since $p_1(y_2)$ contains the Bessel function, this last integral cannot be evaluated in closed form. It is tabulated under the name *Marcum-Q function.* Once this is found, the probability of bit error is given by

$$P_e = P_{\mathrm{FA}} \Pr(0) + P_{\mathrm{M}} \Pr(1)$$

If the probabilities of transmitting a 1 and 0 are equal, the bit error rate is the average of the two conditional error probabilities:

$$P_e = \frac{1}{2}(P_{\mathrm{FA}} + P_{\mathrm{M}})$$

8.5 MASK

So far, this chapter has dealt with binary amplitude-shift keying (BASK), where there were only two possible amplitudes. Therefore, each modulated pulse is capable of transmitting one bit of information. In order to increase the bit rate, we must decrease the pulse width, thereby requiring a greater bandwidth.

A second option for increasing the bit transmission rate is to allow each modulated pulse to transmit more than one bit of information. This leads to *M-ary ASK,* or *MASK.* This is analogous to M-ary baseband as discussed in the previous chapter.

Since we have already studied encoders which change an analog signal into a binary digital signal, we shall approach MASK starting with a binary signal. Figure 8.29 illustrates a MASK modulator.

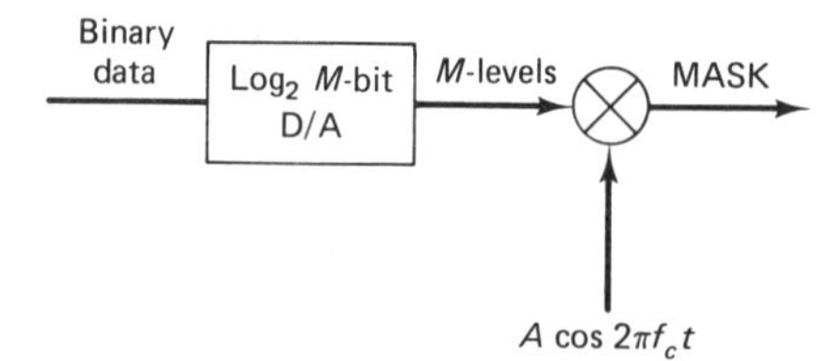

Figure 8.29 MASK modulator

We start by combining bits in groups of $\log_2 M$. Thus, if $M = 4$, we combine bits in pairs; if $M = 8$, we combine bits in groups of 3. We then code each combination into an analog voltage level, using a $\log_2 M$-bit D/A converter. This yields a signal with M possible voltage levels. This discrete signal then modulates a carrier to produce MASK.

The demodulator is the same as that used for BASK. Thus, we can use a coherent detector to synchronously demodulate the MASK signal, yielding an M-level baseband signal. Let us denote the various voltage levels as s_i, where i varies from 1 to M. We can then use a matched filter detector to recover the original M-ary numbers. Since we are assuming that the baseband signal is piecewise constant, the matched filter detector is simply an integrate-and-dump detector. The demodulator/detector is shown in Fig. 8.30.

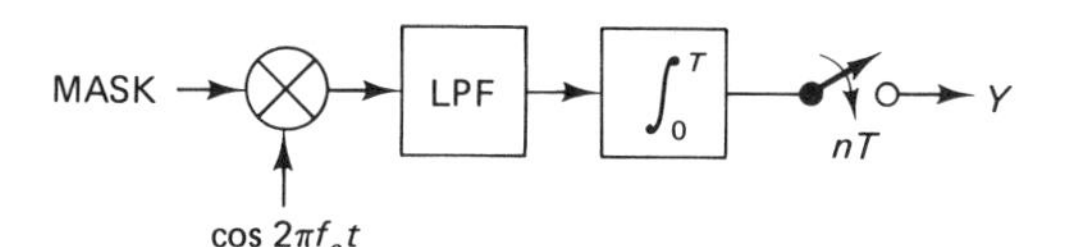

Figure 8.30 MASK demodulator

We now evaluate the performance of this system. The noise at the output of the detector is Gaussian, with variance of $N_0T/2$. (You should take the time now to prove this.) The various outputs of the detector are therefore Gaussian with mean value of s_iT and variance $N_0T/2$. The densities are therefore as shown in Fig. 8.31.

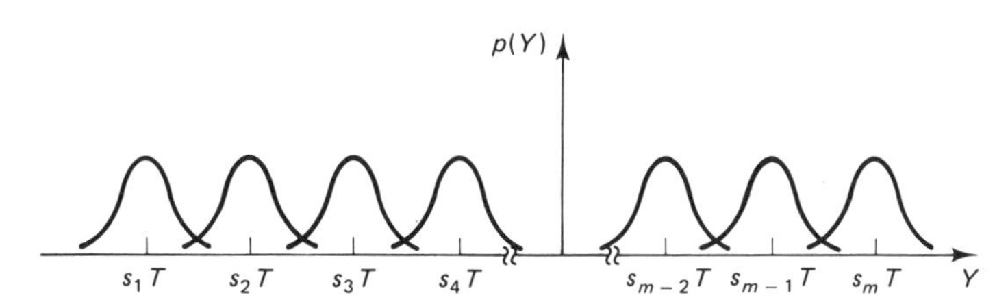

Figure 8.31 Probability densities at output of MASK detector

The probability of symbol error is the probability that the noise is large enough to make one symbol look like another. If we assume that the s_i are separated by ΔS, there will be an error provided that the noise exceeds $\Delta S/2$ in magnitude. This is true except at the two end points. That is, if s_1 is transmitted, noise of negative amplitude does not cause an error. Similarly, if s_M is transmitted, noise of positive amplitude does not cause an error. If the various signal levels are equiprobable (i.e., each has probability $1/M$), the probability of error is then given by

$$P(e) = \frac{1}{M}\left[2(M-2)\int_{\Delta sT/2}^{\infty} N\left(0, \frac{N_0T}{2}\right) + 2\int_{\Delta sT/2}^{\infty} N\left(0, \frac{N_0T}{2}\right)\right]$$

where $N(m, \sigma^2)$ is a Gaussian density with mean m and variance σ^2. This expression reduces to

$$P(e) = \frac{2(M-1)}{M}\left[\frac{1}{2}\operatorname{erfc}\sqrt{\frac{(\Delta S)^2T}{4N_0}}\right] \tag{8-16}$$

We would like to relate the probability of error to the signal-to-noise ratio. The noise already appears in Eq. (8-16), but the signal energy does not. Therefore, we now evaluate the average signal energy. Again, if the signals are equally probable, this average energy is

$$E_{\text{avg}} = \frac{(\Delta S)^2T(M^2-1)}{12}$$

We will want to compare the final result with that of BASK, so we shall divide this by $\log_2 M$ to get the average energy per bit, E_b.

$$E_b = \frac{(\Delta S)^2T(M^2-1)}{12\log_2 M} \tag{8-17}$$

Finally, substituting this into the argument in Eq. (8-16), we obtain

$$P(e) = \frac{M-1}{M}\operatorname{erfc}\sqrt{\frac{3(\log_2 M)E_b}{(M^2-1)N_0}} \tag{8-18}$$

This expression for probability of symbol error is plotted for various values of M in Fig. 8.32.

Note that for a fixed signal-to-noise ratio, as M increases, the probability of symbol error also increases. This is so because the various signal values must get closer together, thus increasing the correlation. We shall refer back to this result later.

8.6 DESIGN EXAMPLES

Example 8-8

An OOK ASK system is to be designed where the peak amplitude of the sinusoidal bursts is 0.5 volt. The additive white Gaussian noise has a power spectral density of 10^{-5} watts/Hz. A control loop is used to recover the carrier, so a coherent detector can be used.

What is the maximum bit transmission rate if the specifications call for a maximum bit error rate of 10^{-4}?

Solution. We assume that the control loop recovers the carrier exactly. If this were not the case, there would be some degradation in the detected signal energy, which would lead to an increase in the bit error rate.

The probability of bit error for a coherent OOK system is given by

$$P_e = \frac{1}{2}\operatorname{erfc}\sqrt{\frac{E(1-\rho)}{2N_0}} = \frac{1}{2}\operatorname{erfc}\sqrt{\frac{E}{2N_0}}$$

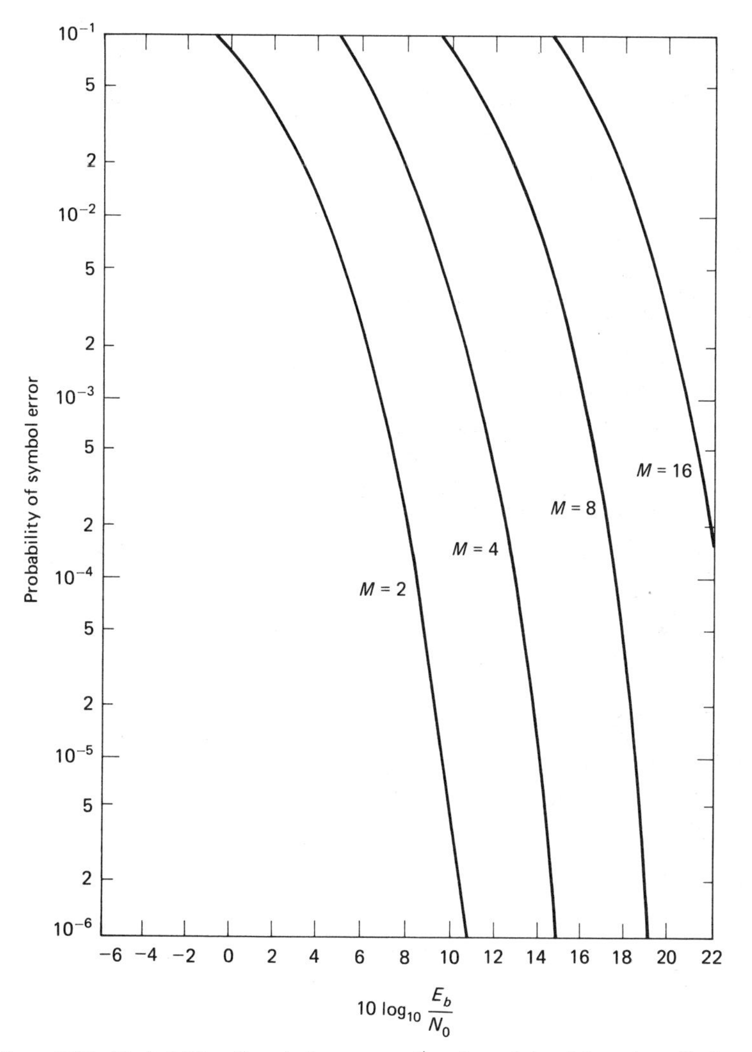

Figure 8.32 Probability of symbol error as a function of signal-to-noise ratio for MASK detector

This is so because the correlation coefficient is zero. The average energy is

$$\frac{(0.5)^2}{4f}$$

where f is the number of bits/sec. We are assuming that the two possible transmitted signals, corresponding to a 0 and a 1, are equally likely. In order to achieve the specified bit error rate, we have

$$\frac{1}{2}\,\text{erfc}\sqrt{\frac{(0.5)^2}{4f \times 2N_0}} = 10^{-5}$$

We use a table of error functions to find

$$\frac{1}{32f(10^{-5})} = 9$$

$$f = 347$$

Thus, the fastest transmission rate is only 347 bits/sec.

Example 8-9

(a) Design a binary matched filter detector for bipolar ASK where the baseband consists of 1-V flat pulses, $T = 1$ ms, and the carrier frequency is 1 MHz. Find the probability of error if the a priori probabilities of the two signals are equal. The additive noise has a power of 2×10^{-3} watts/Hz.

(b) Now assume that we find the error probability of part (a) to be too large to be acceptable in our application. In order to decrease this probability, a "dead space" is defined for which no decision is made. That is, if the output is greater than some threshold Δ, we decide that a 1 was transmitted. If the output is less than $-\Delta$, a decision is made in favor of a 0. However, if the output is between $-\Delta$ and Δ, no decision is made (in some systems, a request for retransmission is sent). Sketch the probability of error and the probability of no decision as a function of Δ.

Solution. (a) The matched filter detector is as shown in Fig. 8.33(a). Since the two signals are the negative of each other, the detector can be simplified to that shown in Fig. 8.33(b). The error probability is given by

$$P_e = \frac{1}{2}\,\text{erfc}\sqrt{\frac{E(1-\rho)}{2N_0}} = \frac{1}{2}\,\text{erfc}\sqrt{\frac{10^{-3}}{N_0}} = 0.24$$

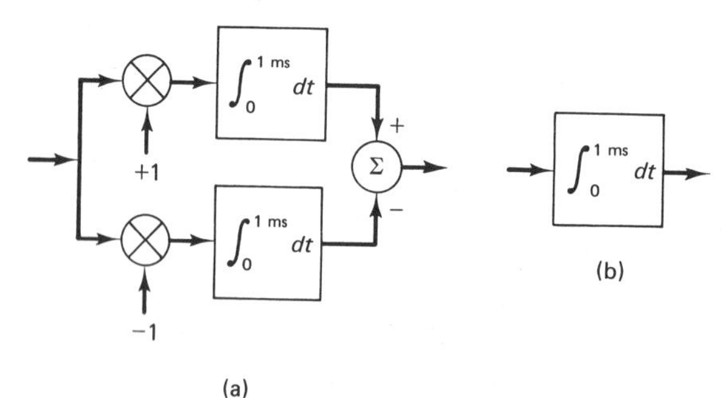

Figure 8.33 Matched filter detector for Example 8-9

(b) We now examine the "soft decision" system. The output of the single correlator detector of Fig. 8.33(b) has a mean value given by $\pm 10^{-3}$ and a variance of $0.5N_0 \times 10^{-3}$. The probability of no decision is given by the probability that a Gaussian density with mean value of 10^{-3} and variance of $0.5N_0 \times 10^{-3}$ is between $-\Delta$ and $+\Delta$. We are taking advantage of the fact that the two conditional densities are symmetrical about the zero axis. This probability is given by

$$P_{ND} = \int_{-\Delta}^{\Delta} N(10^{-3}, 0.5N_0 \times 10^{-3})$$

$$= \frac{1}{2}\,\text{erfc}\sqrt{\frac{(10^{-3} - \Delta)^2}{10^{-3}N_0}} = \frac{1}{2}\,\text{erfc}\sqrt{\frac{(10^{-3} + \Delta)^2}{10^{-3}N_0}}$$

The probability of bit error is given by

$$P_e = \int_{-\infty}^{-\Delta} N(10^{-3}, 0.5N_0 \times 10^{-3})$$

$$= \frac{1}{2}\,\text{erfc}\sqrt{\frac{(10^{-3} + \Delta)^2}{10^{-3}N_0}} = \frac{1}{2}\,\text{erfc}\sqrt{\frac{10^{-3}}{N_0} + \frac{2\Delta}{N_0} + \frac{\Delta^2}{10^{-3}N_0}}$$

Note that the last two terms in the P_e expression reduce the probability of error when compared to the "hard decision" situation analyzed in part (a) of this example.

The probability of error and of no decision are as sketched in Fig. 8.34. Since the specification called for an error probability less than 0.1, the value of Δ is 1.1×10^{-4}. At this choice, the probability of no decision is given by about 0.2. We thus reject approximately 1 out of every 5 receptions in order to achieve the lower bit error rate.

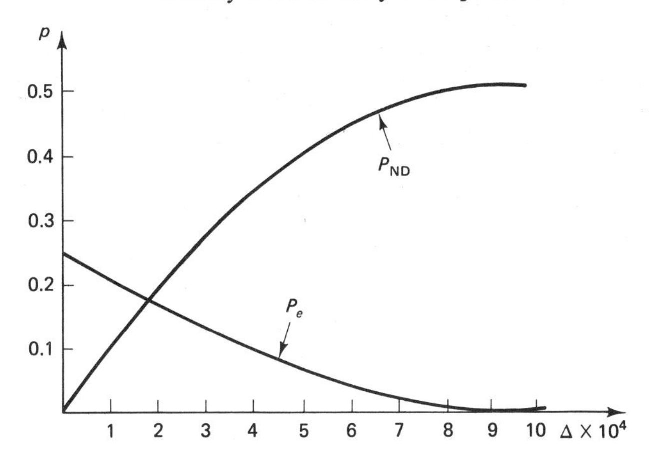

Figure 8.34 Error probabilities for Example 8-9

Example 8-10

When we wish to detect the OOK signal, $A\cos(2\pi f_c t + \theta)$, where θ is known, the matched filter is optimum. Now suppose that θ is unknown but that we can guess at a value of $\theta = \theta_0$.

We wish to explore the advisability of this approach by examining the probability of correct detection as a function of the accuracy of our guess. This is done to decide at which point incoherent detection becomes preferable.

Write the necessary equations in order to decide the best decision technique.

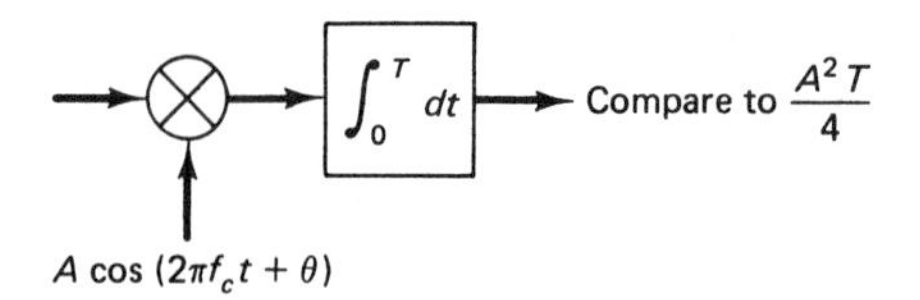

Figure 8.35 Detector for Example 8-10

Solution. The detector is as shown in Fig. 8.35. We shall assume that the phase is mismatched by Δ. That is,

$$\Delta = \theta - \theta_0$$

When a binary 1 is being sent, the output of the detector is a Gaussian random variable with variance $N_0A^2T/4$ and mean value

$$0.5A^2T\cos(\theta - \theta_0)$$

Therefore, the probability of correctly detecting a 1 is given by

$$P(D_1 | H_1) = \int_{A^2T/4}^{\infty} N\left(\frac{A^2T\cos\Delta}{2}, \frac{N_0A^2T}{4}\right)$$

$$= 0.5 \operatorname{erfc}(A\sqrt{T/N_0}[1 - 2\cos\Delta])$$

Note that as Δ increases, this quantity decreases. A typical curve of performance as a function of phase mismatch is shown in Fig. 8.36. Exact values cannot be placed upon this figure until the noise power and bit rate are given.

When an incoherent detector is used, we have

$$P(D_1 | H_1) = Q\left(\frac{1}{\sqrt{N_0f_m}}, \frac{z_0}{\sqrt{N_0f_m}}\right)$$

where

$$z_0 = N_0f_mI_0^{-1}\left[\exp\left(\frac{1}{2N_0f_m}\right)\right]$$

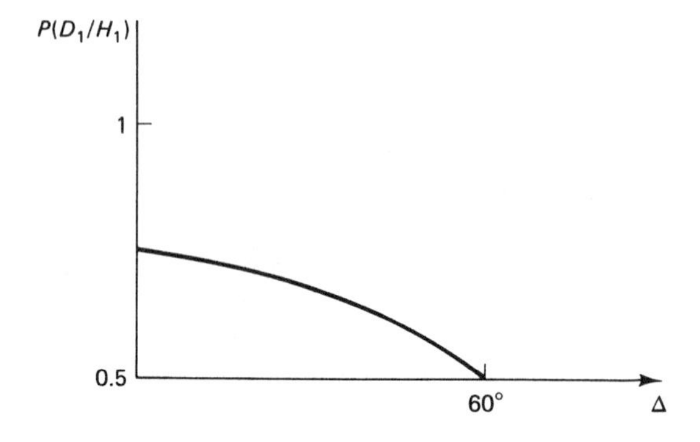

Figure 8.36 Performance in the presence of phase mismatch

We can approximate f_m as the first zero of the baseband frequency characteristic, or $1/T$. Therefore, given values for T and N_0, we can find the probability of correct detection for the incoherent detector. We can then compare this to the curve of Fig. 8.36 in order to find the point at which the incoherent detector represents a better choice. Examples are presented in the problems at the back of this chapter.

8.7 SUMMARY

The presentation of this chapter parallels that of the previous chapter. In this case, we introduce a carrier frequency in order to aid transmission.

We begin with a discussion of the types of amplitude-shift keying, including suppressed-carrier, transmitted-carrier, and on-off keying.

Section 8.2 derives the frequency spectrum of the ASK waveform. We show that this is a shifted version of the spectrum of the baseband signal.

Section 8.3 explores the generic forms of modulators and demodulators for both double sideband and single sideband. Coherent and incoherent forms of demodulation are presented. The section also includes a discussion of carrier recovery using feedback control loops. Squaring loops and Costas loops are presented.

Performance of ASK systems is the topic of Section 8.4, where equations are derived for the probability of bit error. This is done for both the coherent (matched filter) and the incoherent (square-law, quadrature, and envelope) detector.

M-ary ASK is presented in Section 8.5. This presentation includes a derivation of performance curves.

The final section presents several design examples. In the first example we explore the trade-off between bit error rate and transmission rate. The second example presents a method for decreasing the bit error rate by utilizing a soft decision rule. In this situation, we create a region wherein no decision is made. In the final example, we explore the advisability of using coherent detection when not all of the signal parameters are known precisely. In particular, we attempt to guess at the phase of the carrier and compare the coherent detector performance using this guess with the performance of the incoherent detector.

PROBLEMS

8.1. Find the bandwidth of a bipolar ASK signal where the sampling period is 2 ms and a carrier of 10 MHz is used. Do this for several representative data sequences.

8.2. Can an ASK signal with carrier of 1 MHz be formed by multiplying the baseband by a periodic triangular waveform of frequency 500 kHz? 200 kHz?

8.3. Consider a general nonlinear device where

$$y = \sum_{n=1}^{\infty} a_n x^n$$

Can this be used to form ASK as shown below? If not, give restrictions that would make this approximate ASK.

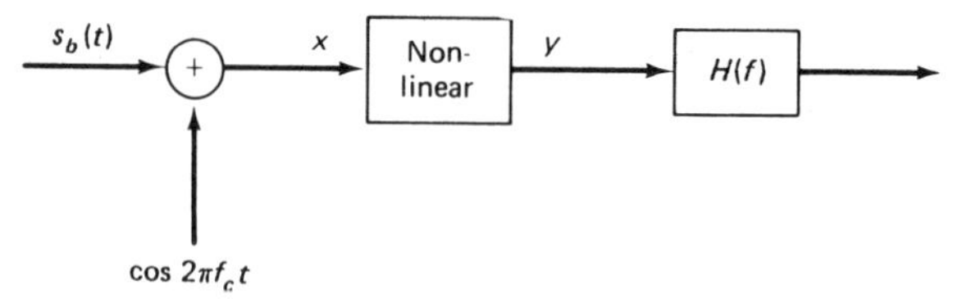

8.4. Prove that the signal given by

$$s_m(t) = \tfrac{1}{2}s_b(t) \cos 2\pi f_c t + \tfrac{1}{2}\tilde{s}_b(t) \sin 2\pi f_c t$$

where

$$\tilde{s}_b(t) = \int_{-\infty}^{\infty} \frac{s_b(\tau)}{\pi(t - \tau)}\, d\tau$$

is a single-sideband signal. What happens if the + is replaced by a −?

8.5. Discuss whether a gated modulator can be used to demodulate ASK. That is, will the system shown below, where $s(t)$ is a square wave of frequency f_c, demodulate for any choice of $H(f)$?

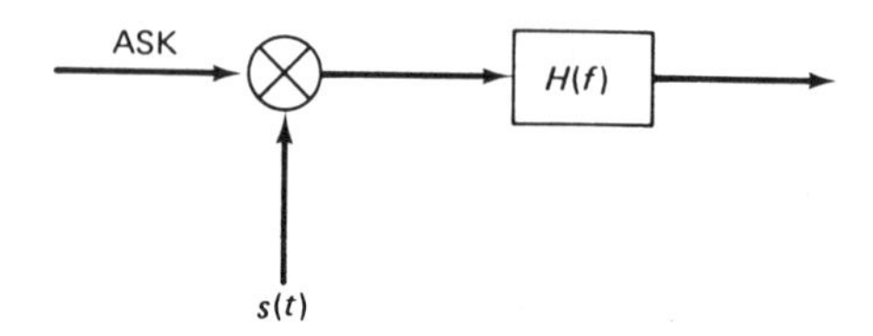

8.6. Suppose that an ASK waveform,

$$[A + s_b(t)] \cos 2\pi f_c t$$

forms the input to the square-law demodulator shown below. Analyze the systems and find the output.

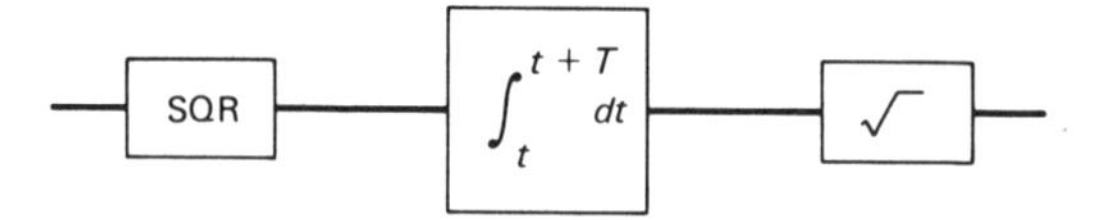

8.7. In the detector of Fig. 8.20, suppose that each of the locally generated signals is off in phase. That is, replace

$$\cos 2\pi f_c t \quad \text{by} \quad \cos(2\pi f_c t + \theta_1)$$

and

$$\sin 2\pi f_c t \quad \text{by} \quad \sin(2\pi f_c t + \theta_2)$$

What effect does this have upon the output? If $\theta_1 = \theta_2$, what effect does this have upon the output?

8.8. Calculate the signal-to-noise ratio for ASK, where $s_b(t)$ is a $(\sin 1000\pi t)/1000\pi t$ pulse with period 2 ms, the noise has power spectral density of 10^{-5}, and the carrier is at 2 MHz. Assume that synchronous detection is used. Find the probability of bit error.

8.9. Repeat the analysis in Problem 8.8 for single-sideband transmission.

8.10. Find the probability of bit error for an ASK system where the carrier is at 1 MHz and the baseband signal is a bipolar series of unit-height triangles with the peak at the midpoint of each sampling interval. Let the sampling period be 0.05 ms and the additive noise be white Gaussian with a power spectral density of 10^{-6} W/Hz. A synchronous detector is used, and $P(H_1) = 0.8$.

8.11. A baseband signal is as shown below. This baseband signal modulates a carrier of 1 MHz. Noise of power spectral density 10^{-3} is added, and coherent detection is used. Assume that the a priori probabilities of the two hypotheses are equal.

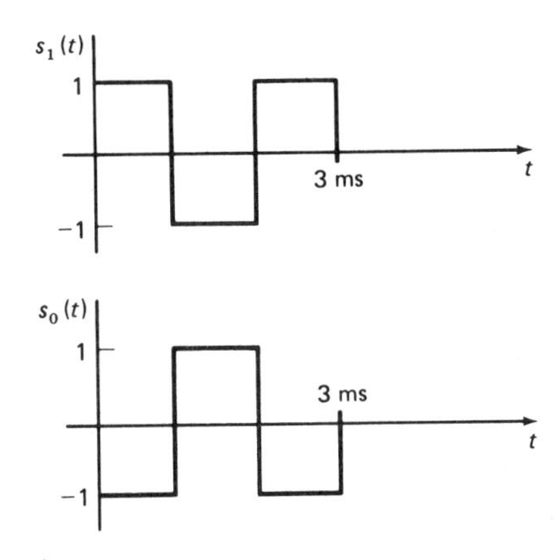

(a) Find the probability of error.

(b) Now assume that the detector treats each received binary signal as a composite of 3 bits and processes (detects) these bit by bit. Majority logic is then used to decide which hypothesis is true. That is, if at least two received bits match a particular signal, the corresponding hypothesis is chosen. Find the probability of error and compare this to your answer in part (a).

8.12. Design a minimum-error incoherent etector using an envelope detector which will choose between

$$s_0(t) = 0$$

$$s_1(t) = \sin(1000t + \theta)$$

$$P(H_0) = \tfrac{1}{4}$$

Find the probability of error if the additive noise has power spectral density of 10^{-3}. The bit rate is 100 bps.

8.13. Design a Neyman–Pearson incoherent detector for the signals of Problem 8.12, and find the probability of correctly deciding H_1. Assume a false-alarm rate of 3×10^{-6}.

8.14. Given a binary communication system with

$$s_0(t) = 0$$

$$s_1(t) = \frac{4}{\pi} \cos(100t + \theta)$$

$$T = \tfrac{1}{2}\pi$$

and where θ is random and uniformly distributed. The additive noise has power spectral density of $N_0/2 = 0.04$. Find the probability of each type of error using a quadrature detector, and find the overall probability of bit error assuming equal a priori probabilities.

8.15. Given a binary system with

$$s_1(t) = \cos 2\pi f_c t$$

$$s_0(t) = 0$$

$$T = \frac{5}{f_c}$$

Assume equal a priori probabilities, and additive noise with a power spectral density in the range between 10^{-2} and 10^{-1}.

(a) Plot the probability of error for coherent detection as a function of $N_0/2$.
(b) Repeat for incoherent detection and compare to your answer for part (a).

8.16. The signals of Problem 8.15 are detected using a square-law detector. The power spectral density of the noise is 0.01. If Neyman–Pearson detection criterion is used with a probability of false alarm of 0.01, find the threshold setting. Repeat the analysis for the quadrature detector.

8.17. Find the maximum bit transmission rate for an OOK system where the peak amplitude of the sinusoidal bursts is 1 volt, the additive white Gaussian noise has a power of 10^{-7} watts/Hz, and the maximum bit error rate is 10^{-5}.

8.18. A binary matched filter detector is used in OOK where the peak amplitude of the sinusoidal burst is 0.5 volt and the additive white Gaussian noise has a power of 10^{-6} watts/Hz. Information is transmitted at 10 kbps.

(a) Find the bit error rate.
(b) A "soft decision" rule, such as that discussed in Example 8-9, is to be used to reduce the bit error rate of part (a) by a factor of 2. Design the detector.

8.19. A coherent matched filter detector is used to detect an OOK signal. The carrier frequency is 1 MHz, the additive noise has a power of 10^{-5} watts/Hz, and the signal amplitude is 1 volt. The bit rate is 10 kbps. In designing the matched filter detector, there is a phase mismatch of $K°$. Find the maximum value of K such that the coherent detector performs better than an incoherent detector.

8.20. Repeat Problem 8.19 for a bit transmission rate of 1 kbps.

9

Frequency Modulation

Chapter 7 developed the concept of baseband signaling, where digital communication was accomplished using low-frequency signals. Chapter 8 extended this concept to band-pass signals, which were developed using amplitude modulation.

The current chapter parallels the development in Chapter 8, but for frequency instead of amplitude modulation. *Frequency modulation* is a method of frequency translation where the frequency shifting is done in a manner that changes the relative frequency content of the signal in a nonlinear manner. The bandwidth of the modulated signal also changes from that of the baseband signal.

In analog communication, FM is often used in place of AM because of resulting performance improvement in the presence of additive noise. We will find other reasons to use FM for data communication. Among these will be some extremely simple implementations for the encoders and decoders.

9.1 INTRODUCTION

We start with a system which generates the baseband signal. Let us assume that the baseband signal is composed of piecewise-constant segments. That is, a binary 1 is transmitted with a square pulse of voltage V_1 and a binary 0 is transmitted with a pulse of V_0 volts.

In *frequency modulation* it is the frequency of the carrier that varies in accordance with the information baseband signal. This contrasts with amplitude modulation, wherein the amplitude of the carrier is controlled by the baseband signal. If we denote the modulated carrier as

$$s_m(t) = A \cos \theta(t) \tag{9-1}$$

then the frequency of this is given by the derivative of the phase,

$$f(t) = \frac{d\theta/dt}{2\pi}$$

This definition of time-varying frequency coincides with the usual definition of frequency in cases where the frequency is a constant. The modulation is now performed by varying $f(t)$ in accordance with the baseband signal as follows:

$$f(t) = f_c + Ks_b(t)$$

f_c is the carrier frequency and K is a constant which, together with the amplitude of the baseband signal, determines the size of the swings in frequency.

Since we are currently assuming that the baseband signal takes on only one of two values, the frequency of the modulated waveform also will take on one of two values, and the modulation process can be thought of as a keying operation. Figure 9.1 illustrates a simple modulator consisting of two oscillators and a switch (key). This form of FM for data information transmission is referred to as *frequency-shift keying* (*FSK*).

An alternate method of producing FSK is shown in Fig. 9.2. Here, an FM modulator is used with a bipolar baseband signal as input. Let us assume that the input signal is either +1 or −1 depending upon whether a 1 or 0 is being sent. Further assume that the modulator constant is set so the maximum frequency deviation is Δf. The output signal is then given by

$$s_m(t) = A \cos \left[2\pi f_c t + 2\pi \, \Delta f \int s_b(t) \, dt \right]$$

$$= A \cos \{2\pi t [f_c + \Delta f s_b(t)]\}$$

The instantaneous frequency is then

$$f(t) = f_c \pm \Delta f$$

So far, we have been dealing with piecewise-constant baseband signals, which lead to FSK signals where the frequency is constant over the bit period. The abrupt transitions in frequency lead to wide bandwidths. In an attempt to decrease the bandwidth, we can perform two types of shaping upon the FSK signal. First, the amplitudes of the sinusoidal bursts can be shaped. An illustration of this shaping is presented in Example 9-2. It is also possible to shape the frequency of each burst.

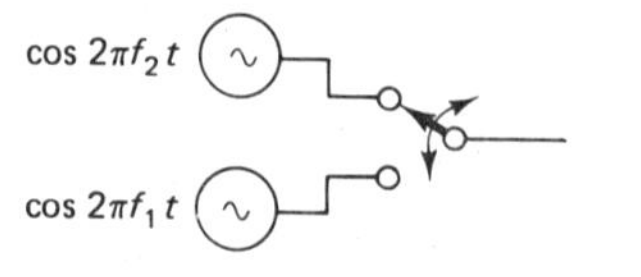

Figure 9.1 FSK modulator

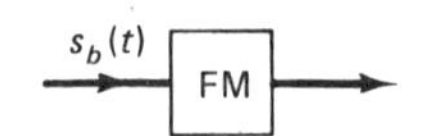

Figure 9.2 FM modulator used to generate FSK waveform from baseband

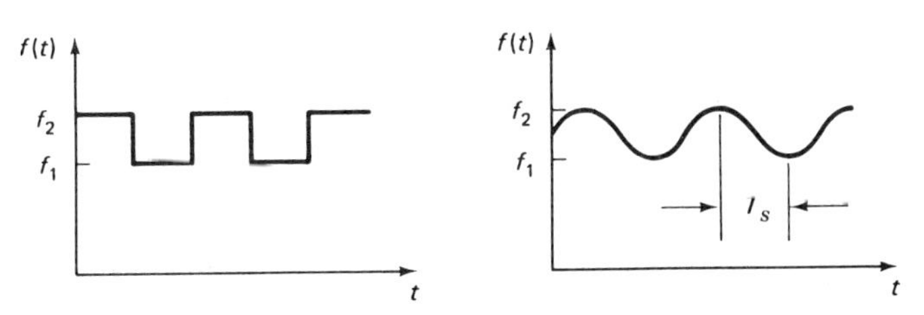

Figure 9.3 FSK with continuous frequency

This is comparable to the analog FM case. Thus, we might replace the abrupt frequency transition by sinusoidal variations, as shown in Fig. 9.3.

If the baseband signal is shaped prior to frequency modulation in order to decrease intersymbol interference, the keying concept is no longer directly applicable. It can still be used if the sources are changed from sinusoidal to shaped sinusoids, where the shaping function is the pulse shape. Since we have already analyzed the shaping in the baseband case, we shall build upon that result.

The instantaneous frequency is given by

$$f(t) = f_c + \Delta f \cos 2\pi f_0 t$$

where f_c is the carrier frequency, and we are assuming sinusoidal frequency shaping as in Fig. 9.3.

The frequency deviation is given by

$$\Delta f = \frac{f_2 - f_1}{2}$$

and $f_0 = 1/2T$.

The resulting FM waveform is

$$\begin{aligned} s_m(t) &= A \cos\left(2\pi f_c t + 2\pi \int \Delta f \cos 2\pi f_0 t \, dt\right) \\ &= A \cos\left(2\pi f_c t + 2\pi \frac{\Delta f}{f_0} \sin 2\pi f_0 t\right) \end{aligned} \tag{9-2}$$

We shall use the expression in Eq. (9-2) to evaluate the bandwidth.

9.2 BFSK SPECTRUM

The output of the modulator presented in Fig. 9.1 can be considered as the superposition of two amplitude-shift-keyed signals. One of these is the ASK signal resulting from modulating a carrier of frequency f_1 using on-off keying from the baseband signal. The other results from amplitude-shift-keying a carrier of frequency f_2 with the complement of the baseband signal. Figure 9.4 shows the system that would accomplish this.

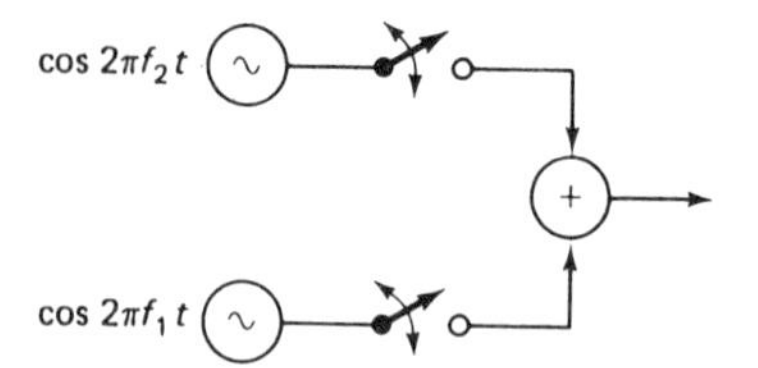

Figure 9.4 FSK as the superposition of two ASK waveforms

We can therefore borrow the AM results of Section 8.2 to find the spectrum of the FSK signal. Each of the two components has a frequency spectrum which is of the general shape of $(\sin^2 f)/f^2$ shifted to the carrier frequency. Thus, the power spectrum of the FSK signal is as shown in Fig. 9.5.

We are assuming the 0's and 1's are equally probable. The power of each carrier is $A^2/8$. The power in the sidebands around each carrier is also $A^2/8$. The total transmitted power is $A^2/2$.

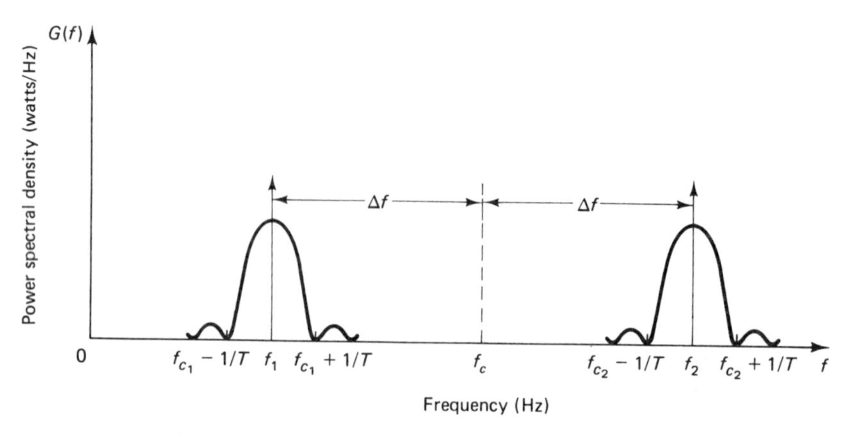

Figure 9.5 Power spectral density of random FSK waveform

Figure 9.6 shows a modified version of Fig. 9.5, where we assume that the spacing between the two carrier frequencies is equal to the bit transmission rate. This is known as *orthogonal tone spacing.* Its importance will become clear when we view the performance of FSK systems.

Example 9-1

An FSK signal consists of bursts of frequency 999 kHz and 1 MHz, the higher frequency being used to send a binary 1. The bit rate is 1000 bps. Find the bandwidth of the FSK signal.

Solution. Looking first at the bursts representing the 1's in the binary signal, the frequency spectrum is centered at 1 MHz and has a $(\sin f)/f$ shape. Since the bursts are of 1-ms duration, the first zero of the spectrum is at 1 MHz $\pm$ 1 kHz. The spectrum of the

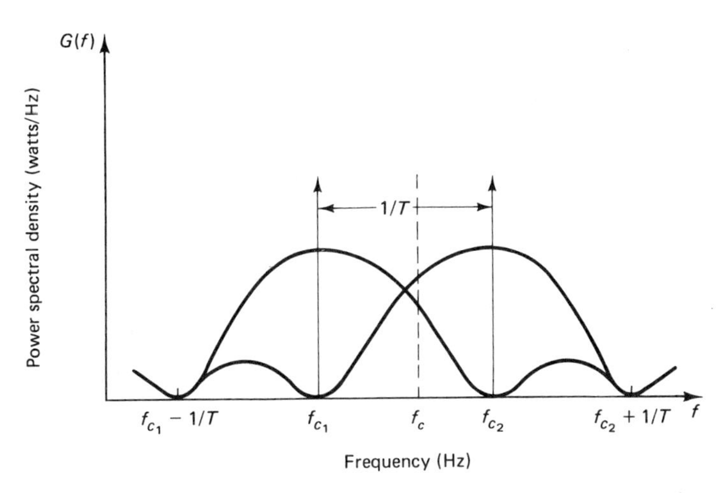

Figure 9.6 Frequency spectrum of FSK signal with orthogonal tone spacing

binary zeros is centered at 999 kHz, with the first zeros of the spectrum at 999 kHz $\pm$ 1 kHz. The bandwidth out to the first zero is therefore 1001 kHz $-$ 998 kHz, or 3 kHz.

Example 9-2

Find the bandwidth of the FM signal where the pulses are shaped to be $\frac{1}{2}$ cycle of a sine wave. Assume that an alternating train of 0's and 1's is transmitted and that bipolar transmission is used.

Solution. We shall assume that the shaping is done on the amplitude and not on the frequency of the modulated waveform. Thus, the signal is a sinusoidal burst during each period, where the envelope is sinusoidally shaped. The signal can therefore be thought of as a superposition of two waveforms differing only in the carrier frequency, as shown in Fig. 9.7. The frequency spectrum of this waveform is found by first

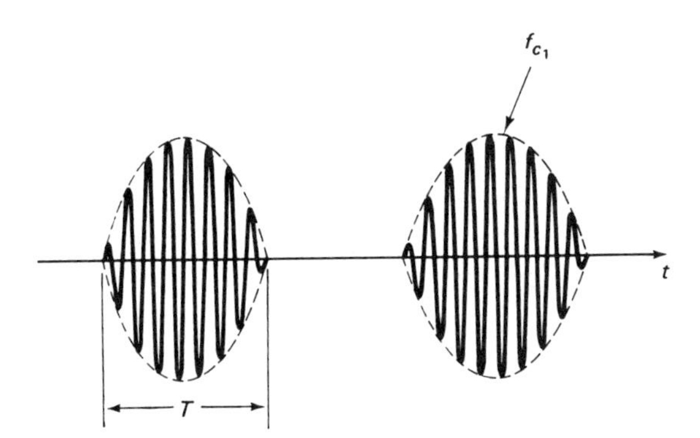

Figure 9.7 Waveform for Example 9-2

evaluating the Fourier transform of the unmodulated waveform. This wave is periodic with period $2T$. The Fourier series can be evaluated as

$$s_{b1}(t) = a_0 + \sum_{n=1}^{\infty} a_n \cos n \frac{\pi t}{T}$$

where

$$a_n = \frac{2}{T} \int_0^{T/2} \cos \frac{\pi t}{T} \cos \frac{n\pi t}{T} \, dt$$

$$a_0 = \frac{1}{T} \int_0^{T/2} \cos \frac{\pi t}{T} \, dt = \left. \frac{\sin \pi t/T}{T(\pi/T)} \right|_0^{T/2} = \frac{1}{\pi}$$

Evaluating the integral, we find that

$$a_n = \frac{\sin (n-1)\pi/2}{(n-1)\pi} + \frac{\sin (n+1)\pi/2}{(n+1)\pi}$$

and

$$a_0 = \frac{1}{\pi}$$

$$a_n = \frac{(-1)^{n/2}}{\pi} \left(\frac{1}{n+1} - \frac{1}{n-1} \right)$$

The spectrum of the wave shown in Fig. 9.7 consists of lines at multiples of $2\pi/T$ away from f_c, where the height of each line is the corresponding value of a_n. Looking at the magnitudes of the a_n for n even, we have

$$\{|a_n|\} = \left\{ \frac{1}{\pi} [1, 0.667, 0.133, 0.06, 0.03, 0.014, 0.010, 0.008] \right\}$$

If we define bandwidth by the point at which the amplitude has decreased to 1% (a very rigid definition), we see that it is necessary to go out six harmonics of the double frequency. The bandwidth of the FSK waveform is therefore

$$\text{BW} = f_2 - f_1 + 12 \times \frac{1}{T}$$

We now examine the bandwidth in the shaped-frequency case. We start with Eq. (9-2). We wish to separate the f_c and f_0 portions of this equation as a step in finding the frequency spectrum. We can do this either by resorting to trigonometric expansions, which will result in two terms, or by using exponential (preenvelope) format for a single term. Using the latter approach, we have

$$s_m(t) = A \operatorname{Re} \left[e^{j2\pi f_c t} \exp \left(j 2\pi \frac{\Delta f}{f_0} \sin 2\pi f_0 t \right) \right] \tag{9-3}$$

We now concentrate upon the second exponential term in Eq. (9-3). This term is periodic with period $1/f_0$. It can therefore be expanded in a Fourier series to yield

$$\exp\left(j2\pi \frac{\Delta f}{f_0} \sin 2\pi f_0 t\right) = \sum_{n=-\infty}^{\infty} c_n \exp(jn2\pi f_0 t)$$

The c_n are proportional to Bessel functions of the first kind of order n. Thus,

$$c_n = J_n\left(\frac{\Delta f}{f_0}\right)$$

Finally, the FM waveform is given by

$$s_m(t) = \sum_{n=-\infty}^{\infty} c_n A \cos 2\pi(f_c + nf_0)$$

If the argument of the Bessel function is very small, the c_n are very small for $n > 1$, so the bandwidth is approximately $2f_0$. For large arguments it is necessary to examine the c_n more closely. Asymptotic expansions exist which clearly show that as n approaches infinity, the c_n approach zero. But we need something that applies in the intermediate region. It turns out that for $n > \Delta f/f_0$, the c_n rapidly approach zero. The bandwidth is therefore often approximated as

$$\text{BW} \approx 2\left(\frac{\Delta f}{f_0}\right) f_0 = 2\,\Delta f$$

The bandwidth of FSK with continuous $f(t)$ therefore ranges from twice the maximum frequency of $f(t)$ to twice the maximum amplitude swing of $Kf(t)$.

9.3 MODULATORS AND DEMODULATORS

9.3.1 Modulation

Frequency modulation can be performed using the keying system of Fig. 9.1, provided that the baseband is composed of square pulses. For continuous baseband signals, we can borrow the results from analog FM. We wish to form the wave

$$s_m(t) = A \cos\left[2\pi f_c t + K \int s_b(t)\,dt\right]$$

Trigonometric identities can be used to rewrite this as

$$s_m(t) = A \cos 2\pi f_c t \cos\left[K \int s_b(t)\,dt\right]$$
$$- A \sin 2\pi f_c t \sin\left[K \int s_b(t)\,dt\right]$$

Now if K is small, the cosine can be approximated by unity and the sine by its argument, yielding

$$s_m(t) = A \cos 2\pi f_c t - A \int s_b(t)\, dt \sin 2\pi f_c t \tag{9-4}$$

Equation (9-4) gives rise to the modulator of Fig. 9.8(a). If K is not small enough to make the approximation of Eq. (9-4), the modulated waveform can still be formed using the small K modulator followed by a frequency multiplier. The system shown in Fig. 9.8(b) can thus be used as the modulator.

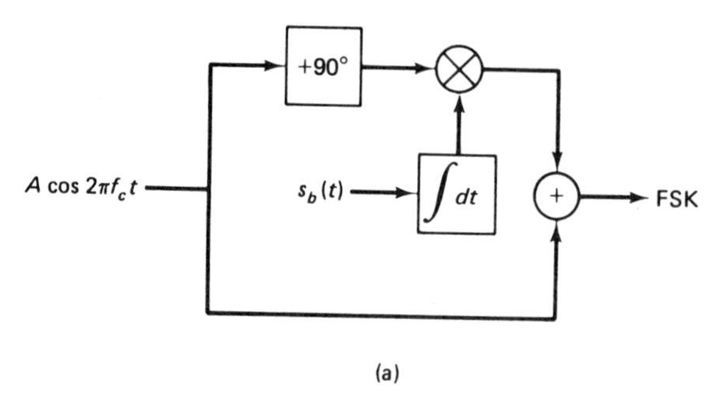

(a)

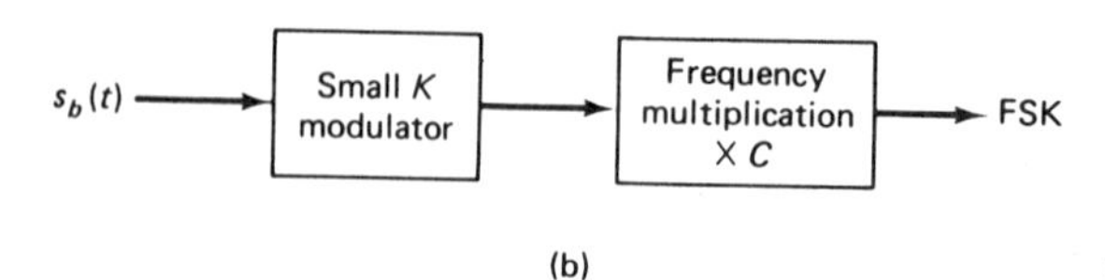

(b)

Figure 9.8 FSK modulator

The frequency modulation can be performed in a direct manner using a *voltage-controlled oscillator* (VCO). The VCO is a system which produces a periodic waveform whose frequency is a function of the input voltage. A typical device is the 566 VCO, which is illustrated in block-diagram form in Fig. 9.9.

C_T is a timing capacitor which is charged or discharged with the constant current at the output of the *current source/sink*. The timing resistor, R_T, controls this current. The voltage across resistor R_T depends upon the voltage at pin #6, which is constrained to be the same as that at pin #5. Thus, the input voltage on pin #5 controls the timing of the system. The capacitor voltage is applied to the *Schmitt trigger* through a buffer amplifier. The Schmitt trigger produces a square-wave output.

The output frequency of the VCO is given by

$$f = \frac{2(V_{CC} - V_{\text{in}})}{C_T R_T V_{CC}} \tag{9-5}$$

In order to use this system to generate FSK, we need to condition the baseband signal so that the voltage swings are within the range required by the VCO. In particular, for proper operation, the input to pin #5 of the VCO should have voltages

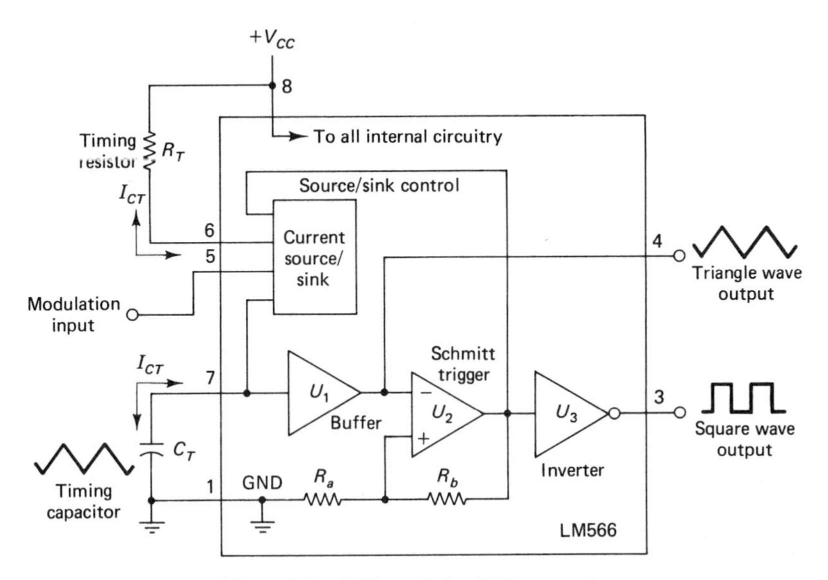

Figure 9.9 VCO used for FSK generation

centered about $0.875V_{CC}$. Thus if we start with either unipolar or bipolar baseband, the signal must be shifted and scaled prior to input to the VCO.

9.3.2 Demodulation

There are two distinct ways of approaching demodulation. The first is to handle the demodulation and decoding separately, the first operation yielding the baseband signal. The second approach combines the two operations by simply applying tests to see which of the two FM bursts is most likely present in each sampling interval.

The general analog FM demodulator can be used to retrieve the baseband signal from the FM waveform. This modulator first changes the FM to AM using time differentiation. In this application, the differentiator is known as a *discriminator.* The time derivative of the FM waveform is given by

$$\frac{d}{dt}A\cos\left[2\pi f_c t + K\int s_b(t)\,dt\right] = A2\pi[f_c + Ks_b(t)]\sin\left[2\pi f_c t + K\int s_b(t)\,dt\right] \tag{9-6}$$

This is a waveform that is both amplitude- and frequency-modulated. However, the frequency range of the sinusoid is usually much higher than the frequency content of the multiplying function (the baseband signal). Thus, the envelope of the function is well defined. The AM signal can then be demodulated with an envelope detector, yielding the receiver of Fig. 9.10. The output of the demodulator would have to be sampled and compared to the threshold in the manner applied to baseband signals in Chapter 7.

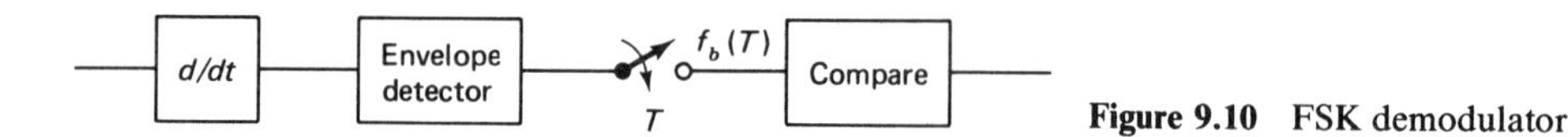

Figure 9.10 FSK demodulator

9.3.3 Coherent Detection

The analog demodulator approach is a bit of an overkill, since the information signals are drawn from a discrete set. In the binary case, there are only two possible baseband signal segments, so it should not be necessary to first demodulate and then make a decision. In addition, if the frequencies shift rapidly at transitions, the analog demodulator is difficult to design. Indeed, a *matched filter detector* can be applied directly to the FSK signal. The matched filter, or correlator detector, requires exact knowledge of the phase of the incoming signal, and is thus referred to as a *coherent detector.* This detector is illustrated in Fig. 9.11. Figure 9.11 illustrates the correlator implementation of the detector. Indeed, each multiplier-integrator combination can be replaced by the corresponding matched filter.

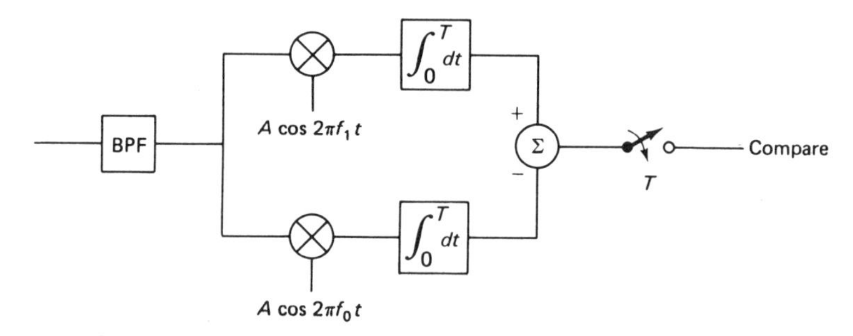

Figure 9.11 Coherent FSK detector

9.3.4 Incoherent Detection

If the exact phase of the incoming wave is not known, we must resort to *incoherent* forms of detection. An incoherent demodulator is illustrated in Fig. 9.12. The system is shown employing envelope detectors, although any of the nonlinear devices discussed earlier could be used.

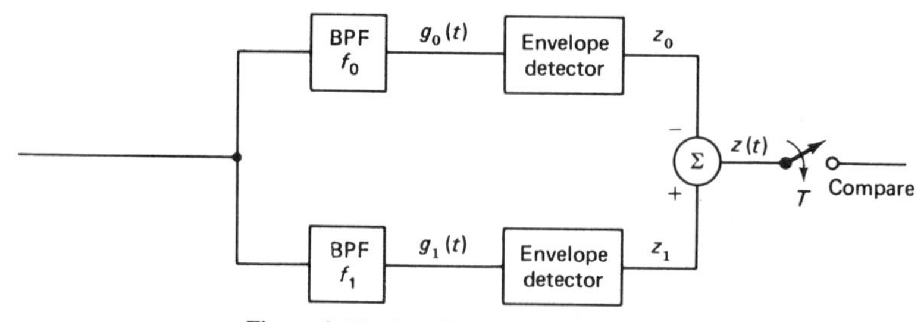

Figure 9.12 Incoherent FSK detector

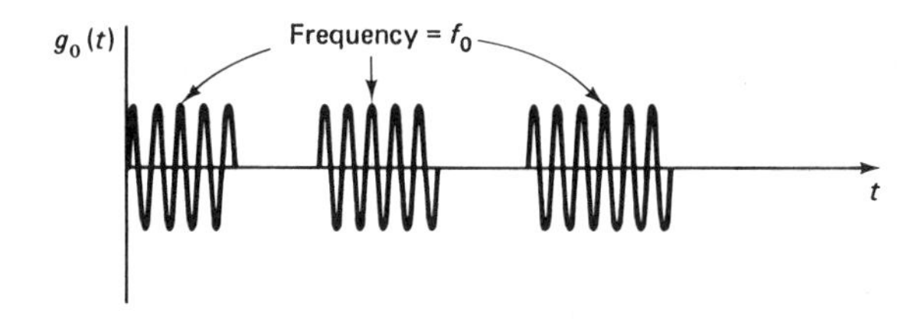

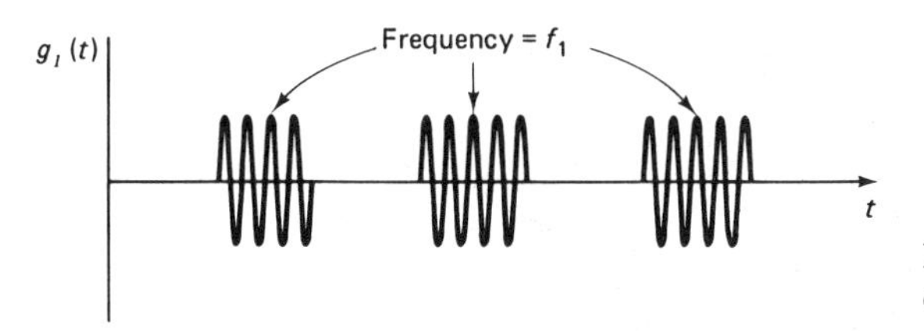

Figure 9.13 Signal components in incoherent FSK detector

Suppose now that the input to the demodulator of Fig. 9.12 is an alternating FSK signal where the amplitude is constant and the frequency jumps instantaneously between f_0 and f_1 at each sampling point. We wish to find the steady-state output.

We look first at $g_0(t)$, which results from putting the FSK waveform through a filter centered at f_0. Recall that the FSK wave can be considered as a superposition of two waveforms. One of these has a transform centered at f_0 with sidebands following a $(\sin f)/f$ envelope. The second component is centered at f_1 with a similar envelope. The specific form of $g_0(t)$ and $g_1(t)$ depends upon the relationship between the sampling period (T_s) and the spacing between f_0 and f_1. If $(f_1 - f_0)T \gg 2$, then $g_0(t)$ and $g_1(t)$ are approximately equal to the two signal components, as illustrated in Fig. 9.13. Now $g_0(t)$ and $g_1(t)$ go through envelope detectors and the difference is taken. The resulting waveform at the output is as shown in Fig. 9.14, where the exponential rounding is due to the time constant of the envelope detectors. As $T_s(f_1 - f_0)$ gets smaller, the band-pass filters no longer reproduce the precise

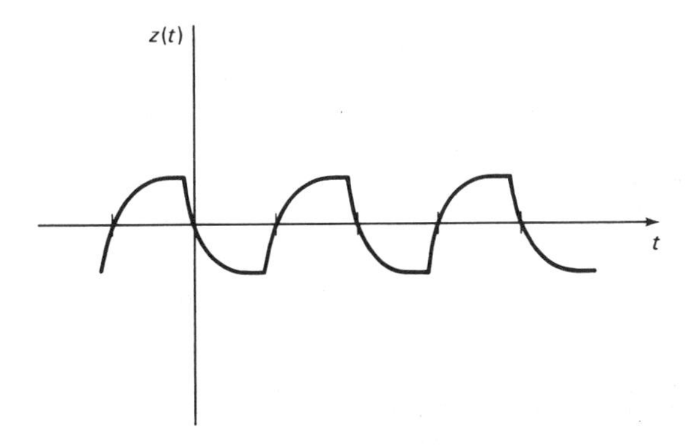

Figure 9.14 Output waveform for incoherent FSK detector

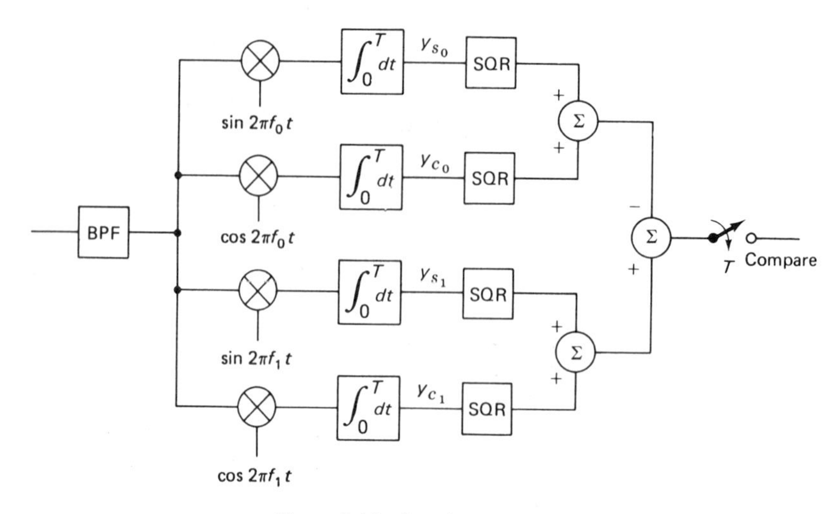

Figure 9.15 Quadrature detector

gated waveforms shown in Fig. 9.13. This is so because the bandwidth of these signals increases beyond the bandwidth of the filters. Additionally, crosstalk occurs as the wider bandwidth causes part of $g_1(t)$ to pass through the filter at f_0, and vice versa.

The incoherent system shown in Fig. 9.15 is a *quadrature detector.*

The output of the quadrature detector will be of the same form as that of the envelope detector of Fig. 9.12. We showed in Chapter 8 that the result of multiplying by sine and cosine, integrating, and taking the square root of the sum of the squares is the same as taking the envelope of the original function. The output of the quadrature detector is therefore comparable to

$$\sqrt{|z_0^2 - z_1^2|}$$

in the envelope detector. The only difference is that the rounded corners (as in Fig. 9.14) will now be caused by the finite integration time rather than by the time constant of the envelope detector.

9.3.5 Modems

FSK is the most common form of digital communication in use with telephone transmission systems. When a voice channel is used for transmitting digital information, the form of modulation must be compatible with the characteristics of the voice channel.

Modulator-demodulators, or *MODEMS,* are used to transform the baseband signal into a modulated signal which is capable of transmission through the channel. The data input to the modem is converted into audio-frequency signals which are coupled into the phone lines. Modems are classified as either *asynchronous* or *synchronous*. Asynchronous modems do not have a clock, and data rates need not be constant. This is the least complex type of modem. In fact, most modems operating below 1800 bps are asynchronous. Alternatively, synchronous modems send data at a fixed periodic rate. They can operate at higher data rates than asynchronous modems.

As may be expected, standards have been developed in an attempt to make systems compatible. "Type 103" modems (originated as part of the Bell System 103 and 113 Series) include asynchronous modems which operate up to 300 baud (bits per second for binary communication). These are *full-duplex modems.* the two directions of transmission are known as *originate* and *answer*. The frequencies are assigned as follows:

	Space	Mark
Originate	1070 Hz	1270 Hz
Answer	2025 Hz	2225 Hz

A different but related standard is the *CCITT recommendation V.21.* (We will discuss CCITT Recommendations in Chapter 13. For now, it is sufficient to view CCITT as an international standards recommending committee.) This standard provides for the following frequency assignments:

	Space	Mark
Originate	1180 Hz	980 Hz
Answer	1850 Hz	1650 Hz

In addition to the different frequencies, note that the lower frequency in each pair is used for the mark, whereas in the 103 series the reverse is true.

Figure 9.16 shows a block diagram of a low-speed asynchronous modem. The FSK modulator can be a VCO with signal conditioning at the input. The band-pass filter restricts out-of-band signals prior to coupling the signal to the phone line.

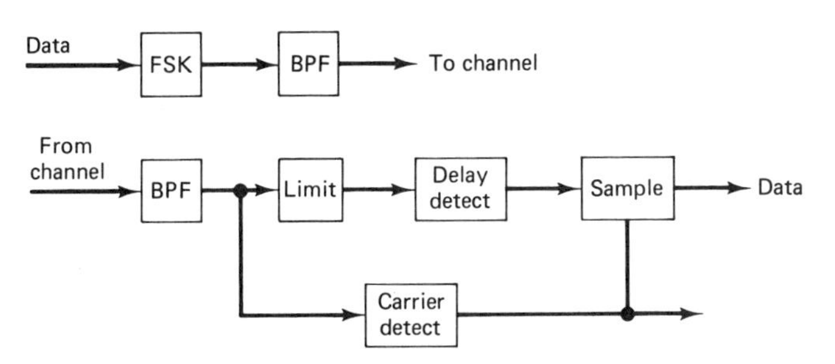

Figure 9.16 FSK modem

The receive portion starts with a *band-pass filter* to decrease noise outside of the band of signal frequencies. An *amplitude limiter* is used to decrease noise effects, since the signal information is contained in the frequency of the incoming waveform. The *delay detector* compares the signal to a delayed version of itself. In this manner, it approximates the derivative of the signal, which is proportional to the frequency. It therefore acts as a discriminator, yielding an output voltage which is proportional to frequency. The *carrier detect* portion of the system is required, since the modem is asynchronous. We would not want the system deciding between 1's and 0's when nothing is being sent.

In order to achieve higher speeds, modems can be made synchronous instead of asynchronous. Additional speed improvements are possible by going from duplex to half-duplex, and by changing from binary to M-ary transmission. Some improvements are also possible by using forms of modulation other than FM. These are discussed in the next chapter.

The *Bell 202 series modem* is a half-duplex modem capable of higher speeds than the 103 series. This modem can send data at 1200 bps on dial-up lines, and at 1800 bps on leased lines with C2 conditioning. A space is sent using 2200 Hz, and a mark is transmitted using 1200 Hz.

9.4 PERFORMANCE

9.4.1 Coherent Detection

We begin by analyzing performance of the coherent matched filter, or correlation detector. The performance of the correlation detector for any binary communication system is given by (see Section 7.5):

$$P_e = \frac{1}{2}\,\text{erfc}\,\sqrt{\frac{(1-\rho)E}{2N_o}} \tag{9-7}$$

where E is the average energy of the two signals and ρ is the correlation coefficient. The derivation of this result assumes that a priori probabilities of the two signals are equal and that the costs associated with the various errors are also equal. Now we apply this to the FSK case, where

$$s_0(t) = A\cos 2\pi f_0 t, \qquad 0 < t < T$$

$$s_1(t) = A\cos 2\pi f_1 t, \qquad 0 < t < T$$

Therefore, E and ρ are given by

$$E = \int_0^T A^2 \cos^2 2\pi f_0 t\, dt$$

$$= \int_0^T \left(\frac{A^2}{2} + \frac{A^2}{2}\cos 4\pi f_0 t\right) dt$$

$$= \frac{A^2T}{2} \qquad \text{if } T \gg \frac{1}{f_0} \tag{9-8}$$

$$\rho = \frac{A^2 \int_0^T \cos 2\pi f_0 t \cos 2\pi f_1 t \, dt}{A^2T/2}$$

$$= \frac{\sin 2\pi(f_0 + f_1)T}{2\pi(f_0 + f_1)T} + \frac{\sin 2\pi(f_1 - f_0)T}{2\pi(f_1 - f_0)T} \tag{9-9}$$

It is desirable to make ρ as close to -1 as possible in order to realize the best possible performance. The first term in Eq. (9-9) is usually much smaller than the second, since $f_0 + f_1 \gg f_1 - f_0$. We can therefore approximately minimize ρ by minimizing the second term in Eq. (9-9). Letting $f_1 - f_0 = \Delta f$ and setting the derivative equal to zero, we have

$$\frac{\partial \rho}{\partial \Delta f} = 2\pi \Delta f T^2 \cos(2\pi \Delta f)T - T \sin(2\pi \Delta f)T = 0$$

$$2\pi \Delta f T = \tan(2\pi \Delta f T)$$

and

$$2\pi \Delta f T \approx 0.715 \times 2\pi \text{ radians}$$

This yields a correlation coefficient,

$$\rho \approx -0.22$$

Note that ρ can be made equal to zero to setting

$$f_1 - f_0 = \frac{n}{T} \tag{9-10}$$

It would therefore appear that choosing $2\pi\Delta f T = 0.715 \times 2\pi$ yields better performance than $\Delta f T = n$. Caution is, however, advised in a practical situation. Suppose we are more realistic and modify s_0 and s_1 as follows:

$$s_0(t) = A \cos(2\pi f_0 t + \theta_0)$$

$$s_1(t) = A \cos(2\pi f_1 t + \theta_1) \tag{9-11}$$

That is, we admit that matching phases is virtually impossible. Equation (9-9) then becomes

$$\rho \approx \frac{1}{T} \int_0^T \cos(2\pi \Delta f t + \Delta\theta) \, dt$$

$$= \frac{1}{2\pi \Delta f T} [\sin(2\pi \Delta f T + \Delta\theta) - \sin(\Delta\theta)] \tag{9-12}$$

If ΔfT is set equal to 0.715, there is no longer any assurance that ρ will be a minimum. It may even be positive.

However, if $\Delta fT = n$, the expression in Eq. (9-12) still yields $\rho = 0$. Therefore, with this choice of frequency separation, performance is independent of the exact phase.

If we use $\rho = 0$, the probability of error is given by

$$P_e = \frac{1}{2}\,\text{erfc}\sqrt{\frac{E}{2N_0}} = \frac{1}{2}\,\text{erfc}\left(\frac{A}{2}\sqrt{\frac{T}{N_0}}\right) \tag{9-13}$$

Again, $N_0/2$ is the height of the power spectral density of the noise. The error is plotted as a function of signal-to-noise ratio (E/N_0) in Fig. 9.17.

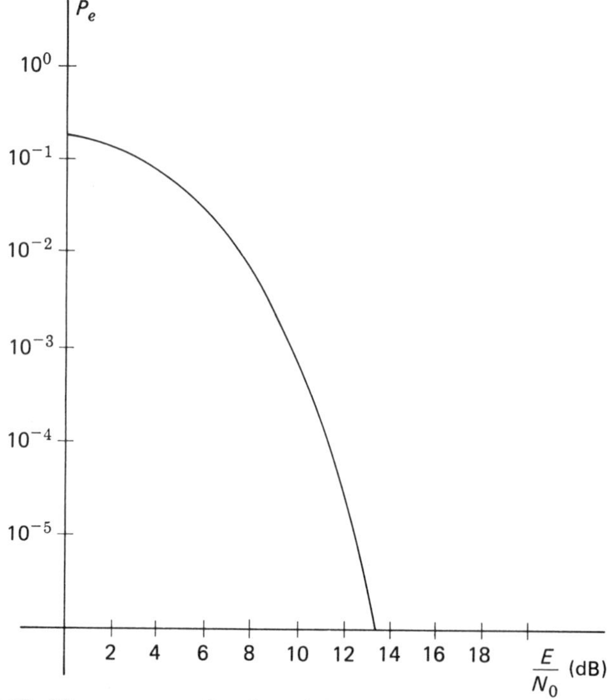

Figure 9.17 Bit error rate as function of signal-to-noise ratio for coherent FSK detector

Example 9-3

Find the probability of error for the FSK system where

$$s_0(t) = 1.414 \cos 1000t$$

$$s_1(t) = 1.414 \cos 1010t$$

Noise of power spectral density

$$\frac{N_0}{2} = 0.01$$

is added, and a matched filter detector is used. Assume that the phase can be exactly reproduced and that the sampling period is equal to 1 second.

Solution. The expression for error is

$$P_e = \frac{1}{2}\,\text{erfc}\sqrt{\frac{E(1-\rho)}{2N_0}}$$

where

$$E = \tfrac{1}{2}(1.414^2) \times 1 \text{ second} = 1$$

Since $T \gg 1/f_0$, ρ is given by

$$\rho = \frac{\sin 2010}{2010} + \frac{\sin 10}{10}$$
$$= -0.054$$

The probability of error is then

$$P_e = \frac{1}{2}\,\text{erfc}\sqrt{\frac{1.054}{2(0.02)}}$$
$$= \tfrac{1}{2}\,\text{erfc}\,(5.1)$$

Most tables of the error function do not have arguments going above 5, since the erfc(5) is extremely small. For larger arguments, asymptotic expansions can be used. One such expansion is

$$\sqrt{\pi z}\, e^{z^2}\,\text{erfc}\,(z) \approx 1 + \sum_{m=1}^{\infty} (-1)^m \frac{1 \times 3 \times \cdots \times (2m-1)}{(2z^2)^m}$$

For $z = 5.1$, this yields

$$\text{erfc}\,(5.1) = 5.5 \times 10^{-13}$$

and the probability of error becomes

$$P_e = 2.75 \times 10^{-13}$$

9.4.2 Incoherent Detection

We now turn our attention to the performance of the incoherent FSK detector. The nonlinearities in the incoherent detector make the analysis more complex than for the coherent detector. This is so because the various random quantities are no longer Gaussian distributed.

We repeat the block diagram of the incoherent FSK detector as Fig. 9.18. Performance of this detector is the same as that of the quadrature FSK detector. Owing to symmetry, the probability of erroneously detecting a transmitted 1 as a received 0 (miss) is the same as the probability of detecting a transmitted 0 as a received 1 (false alarm). Thus, the probability of bit error is equal to either one of these probabilities. We therefore need only calculate one of the error probabilities. If a binary 1 is transmitted, the received signal is

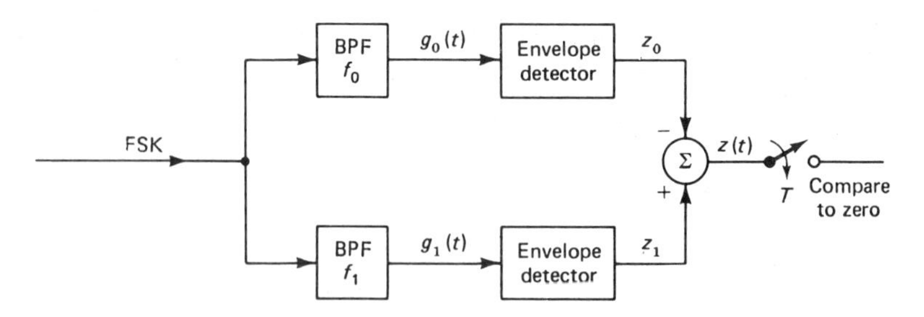

Figure 9.18 Block diagram of incoherent FSK detector

$$A(t) \cos 2\pi f_1 t$$

where $A(t)$ is the pulse shape. In most cases we will assume $A(t)$ to be a constant over the bit interval, T. The output of the band-pass filters is given by

$$g_0(t) = x_0(t) \cos 2\pi f_0 t - y_0(t) \sin 2\pi f_0 t$$
$$g_1(t) = [x_1(t) + A(t)] \cos 2\pi f_1 t - y_1(t) \sin 2\pi f_1 t$$

We have performed the orthogonal expansion in the direction of the signal and have therefore omitted a phase angle in the cosine terms. Thus, the output of the filter tuned to f_0 is simply bandlimited noise, while the output of the other filter is the signal plus bandlimited noise.

The outputs of the envelope detectors are given by

$$z_0(t) = \sqrt{x_0^2(t) + y_0^2(t)}$$
$$z_1(t) = \sqrt{[x_1(t) + A(t)]^2 + y_1^2(t)}$$

As we showed in the previous chapter, when we sample $z_0(t)$ and $z_1(t)$, the resulting random-variable distributions are Rayleigh and Ricean, respectively. That is,

$$p(z_0) = \frac{z_0}{\sigma^2} \exp\left(\frac{-z_0^2}{2\sigma^2}\right) \tag{9-14}$$

$$p(z_1) = \frac{z_1}{\sigma^2} \exp\left(-\frac{z_1}{2\sigma^2}[z_1^2 + A^2(T)]\, I_0\left(\frac{A(T)z_1}{\sigma^2}\right)\right) \tag{9-15}$$

σ^2 is the variance of either $x(t)$ or $y(t)$ and is given by

$$\sigma^2 = N_0 \, \text{BW} \tag{9-16}$$

where BW is the bandwidth of the filters, and N_0 is the power per herz of the noise. The probability of saying a 0 was sent when in fact a 1 was transmitted, or equivalently the probability of error under equal a priori probabilities, is given by

$$P_e = \Pr\{z_0 > z_1 \mid 1 \text{ sent}\}$$
$$= \int_0^\infty \int_{z_1}^\infty p(z_0, z_1)\, dz_0\, dz_1$$

Since z_0 is independent of z_1, the error probability can be rewritten as

$$= \int_0^\infty dz_1 \int_{z_1}^\infty \frac{z_0 z_1}{\sigma^4} \exp\left[\frac{z_0^2 + z_1^2 + A^2(T)}{-2\sigma^2}\right] I_0\left[\frac{z_1 A(T)}{\sigma^2}\right] dz_0$$

If the portion containing z_0 is isolated, the inner integral becomes

$$\int_{z_1}^\infty z_0 e^{-z_0^2/2\sigma^2}\, dz_0 = \sigma^2 e^{-z_1^2/2\sigma^2}$$

so

$$P_e = \int_0^\infty \frac{z_1}{\sigma^2} \exp\left[\frac{2z_1^2 + A^2(T)}{-2\sigma^2}\right] I_0\left[\frac{z_1 A(T)}{\sigma^2}\right] dz_1$$

Now with a change of variables,

$$v = \frac{\sqrt{2} z_1}{\sigma}, \qquad \alpha = \frac{A(T)}{\sqrt{2}\sigma}$$

we find that

$$P_e = \frac{1}{2} \exp\left[\frac{-A^2(T)}{2\sigma^2}\right] \int_0^\infty v \exp\left(\frac{v^2 + \alpha^2}{-2}\right) I_0(\alpha v)\, dv \tag{9-17}$$

The integral in Eq. (9-17) is a Marcum-Q function, and it is equal to unity. Therefore,

$$P_e = \frac{1}{2} \exp\left[\frac{-A^2(T)}{2\sigma^2}\right] \tag{9-18}$$

Example 9-4

Find the probability of error of the incoherent detector of Fig. 9.18, where $A = 1$, $f_1 = 1000, f_2 = 1010$, and the sampling period is 2 s. Noise of $N_0/2 = 10^{-2}$ is added during transmission.

Solution. The bandwidth of the filters was not specified, so we must first assume a value for this. Since the sampling period is 2 s, the frequency spectrum of a single sinusoidal burst has a $(\sin f)/f$ shape, with the bandwidth out to the first zero being 1. This would make the lower-frequency (f_0) filter range from $1000 - \frac{1}{2}$ to $1000 + \frac{1}{2}$ and the higher-frequency filter from $1010 - \frac{1}{2}$ to $1010 + \frac{1}{2}$. Since these passbands do not overlap, we shall assume that a filter bandwidth of 1 is used. Note that some signal distortion will occur with this choice.

Plugging this into Eq. (9-16), we have

$$\sigma^2 = N_0\,\text{BW} = N_0 = 0.02$$

$$P_e = \frac{1}{2}\exp\left[-\frac{1}{2(2 \times 10^{-2})}\right]$$
$$= 6.9 \times 10^{-12}$$

We can therefore expect about 7 bit errors out of every 10^{12} bits on the average.

We wish to compare the performance of the incoherent detector to that of the coherent detector. That is, we look at Eqs. (9-13) and (9-16). Figure 9.17 was presented as a plot of the coherent detector performance as a function of the signal-to-noise ratio, E/N_0. We therefore wish to introduce the signal-to-noise ratio, E/N_0, into Eq. (9-18). We do this by assuming that the band-pass filter has a bandwidth which goes to the first zero of the spectrum of the pulsed sinusoidal signal. Thus, we assume that the bandwidth is $2/T$. Equation (9-18) then reduces to

$$P_e = \frac{1}{2}\exp\left[-\frac{\frac{TA^2(T)}{2}}{2N_0}\right]$$

but $TA^2(T)/2$ is the transmitted signal energy per bit, so we have

$$P_e = \frac{1}{2}\exp\left(-\frac{E}{2N_0}\right) \tag{9-19}$$

Figure 9.19 compares the performance of the incoherent detector with that of the coherent detector. The curve for the coherent detector is repeated from Fig. 9.17.

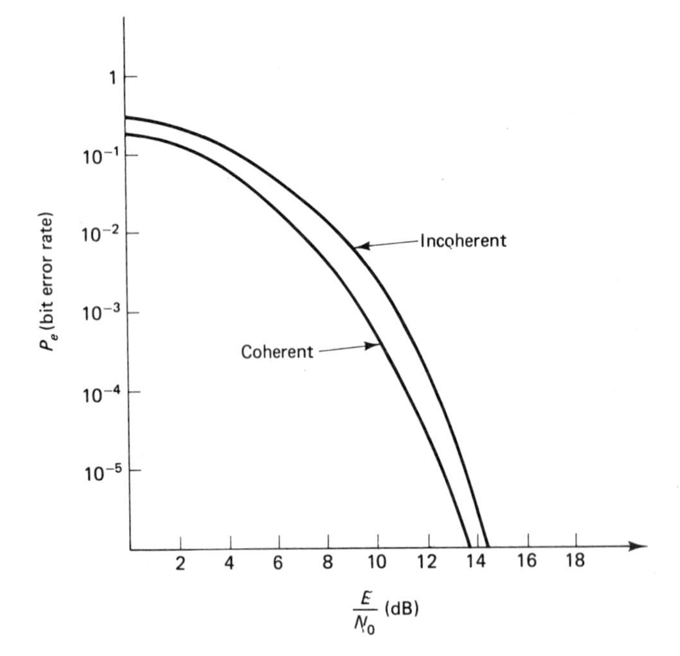

Figure 9.19 Performance comparison of incoherent and coherent FSK detector

Example 9-5

Find the probability of error of the incoherent detector when used for FSK where the bit period is 2 s, the two signals have amplitude of 0.4 volt, and the frequencies are 1 kHz and 2 kHz. The additive noise has power of 10^{-2} watts/Hz. Compare the performance with that of the coherent detector.

Solution. We must first determine that the results of Fig. 9.19 apply to this problem. Within the derivation of those results, the only assumptions (in the incoherent case) were that the signal frequencies are sufficiently separated that the band-pass filters essentially totally reject the alternate frequency. The bandwidth of the filters will be 1 Hz, and the frequencies are separated by 1 kHz, so this assumption is clearly valid.

The signal-to-noise ratio, E/N_0, is given by $A^2T/2N_0$, or 16 (12 dB). We then read from Fig. 9.19 that the probability of error is 1.7×10^{-4}.

We now view the curve for the coherent case. The assumption to derive that curve was orthogonal tone spacing. The zeros of the spectrum of each signal pass through zero at spacings of $\frac{1}{2}$ Hz. Since the spacing between the two frequencies is 1 kHz, this is a multiple of $\frac{1}{2}$, so the orthogonal assumption is valid. Actually, the spacing is so large that, even if it were not a multiple of $1/T$, the correlation between signals is essentially zero. Thus, the probability of error is read from Fig. 9.19 [or Eq. (9-13)] as 3.8×10^{-5}. This is 4.5 times better than that of the incoherent detector.

We now briefly examine the FM discriminator detector as shown in Fig. 9.20. The signal at the output of the band-pass filter is

$$v(t) = A \cos\left[2\pi f_c t + K \int s_b(t)\, dt\right]$$
$$+ x(t) \cos 2\pi f_c t - y(t) \sin 2\pi f_c t$$

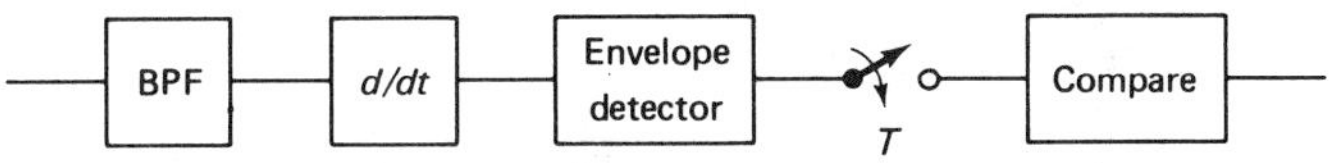

BPF — d/dt — Envelope detector — T — Compare

Figure 9.20 Discriminator detector

Letting

$$g(t) = \int s_b(t)\, dt$$

$v(t)$ can be rewritten as

$$v(t) = A(t)[\cos(2\pi f_c t + \theta(t)]$$

where

$$A(t) = \sqrt{[A \cos Kg(t) + x(t)]^2 + [A \sin Kg(t) - y(t)]^2}$$

$$\theta(t) = \tan^{-1}\left[\frac{A \sin Kg(t) - y(t)}{A \cos Kg(t) + x(t)}\right]$$

The derivative of this is

$$v'(t) = A'(t) \cos [2\pi f_c t + \theta(t)] + A(t)[2\pi f_c t + \theta'(t)] \sin [2\pi f_c t + \theta(t)]$$

and the output of the envelope detector would be

$$z(t) = \sqrt{\left(\frac{dA}{dt}\right)^2 + A^2(t)[2\pi f_c t + \theta'(t)]^2}$$

In the analog case, $\theta'(t)$ is the desired output of the envelope detector, since this is linearly related to $s_b(t)$. For the digital application, as long as $A(t)$ is distinctly different under the two hypotheses, a decision can be made at the receiver. It is not necessary to recover the exact baseband signal.

To carry the analysis further, some simplifying assumptions are required. First, we assume that the detector contains a limiter preceding the differentiator. This limiter decreases the effects of additive noise. The input to the differentiator then becomes

$$v(t) = \cos (2\pi f_c t + \theta)$$

and the envelope detector output would be

$$\begin{aligned} z(t) &= |2\pi f_c + \theta'(t)| \\ &= \left|2\pi f_c + \frac{d}{dt} \tan^{-1} \frac{\sin Kg(t) + y(t)}{\cos Kg(t) + x(t)}\right| \end{aligned} \qquad (9\text{-}20)$$

If the carrier frequency is large compared to the rate of change of instantaneous frequency, the absolute-value sign in Eq. (9-20) can be omitted. If we further assume that $y(t)$ and K are much less than unity, only the first two terms in a series expansion need be retained to yield

$$\begin{aligned} z(t) &= 2\pi f_c + \frac{d}{dt} [Kg(t) + y(t)] \\ &= |2\pi f_c + Kf_b(t) + y'(t)| \end{aligned}$$

Since $y(t)$ is Gaussian, $y'(t)$ will also be Gaussian. The power spectral density of $y'(t)$ is f^2 times the power spectral density of $y(t)$. Thus, assuming white noise in the channel, the variance of the noise becomes

$$\begin{aligned} E\{y'^2(t)\} &= N_0 \int_0^{\text{BW}/2} f^2\, df \\ &= \frac{N_0(\text{BW})^3}{24\pi} \end{aligned}$$

Once this variance is known, the various probabilities of error can be found.

If the noise is not sufficiently small to allow the various approximations to be made, additional terms in the series expansion of the inverse tangent must be retained. This would result in a complex analysis problem.

9.5 M-ARY FSK

M-ary FSK (MFSK) is a way to trade bandwidth for signaling speed. Instead of sending data using binary signals with one of two frequencies, the signaling alphabet is expanded to include M possible frequencies. This process will normally increase the spread between the lowest and highest frequency, and therefore the bandwidth can be expected to increase. However, since increased information is sent with each signal element, the *baud rate* can be decreased to partially counteract the increase in bandwidth. For example, if it were necessary to send 1000 bps of data, this could be done by sending a binary FSK pulse every millisecond. Alternatively, a 4-ary FSK burst could be sent every 2 ms, representing a decrease in baud rate by a factor of 2.

We consider the signal set to be composed of M signals,

$$s_i(t) = \begin{cases} A \cos 2\pi f_i t, & 0 < t < T \\ 0 & \text{otherwise} \end{cases} \qquad (i = 1, M)$$

We know from our studies of detector performance that, all other quantities remaining equal, performance improves as the correlation between signals decreases. That is, the various signals should be as different as possible. Extending this concept and assuming that the correlation is nonnegative (recall the dangers described in Section 9.5, where we discussed the best choice of frequency spacing), the best possible performance results if the various signals used form an orthogonal set. We then call this *orthogonal tone spacing* or *orthogonal signaling.* We therefore assume that the frequencies and the period are chosen such that

$$\int_0^T s_i(t)s_j(t)\,dt = \frac{A^2T}{2}\delta_{ij}$$

where δ_{ij} is the *Kronecker delta,* which is equal to 1 if $i = j$ and zero otherwise. It is now necessary to reexamine the relevant results from decision theory, since, so far, these have been developed only for the binary hypothesis case.

9.5.1 M-ary Decision Theory

Assume that there are M hypotheses and that *Bayes detection criterion* is to be used. Recall that all the other criteria are derivable as special cases of the Bayes result. The average cost will be

$$\bar{C} = \sum_j \sum_i C_{ij} P(D_i, H_j) = \sum_j \sum_i C_{ij} P(D_i | H_j) P(H_j) \tag{9-21}$$

Note that, as in Section 5.5, $P(D_i | H_j)$ is the probability of deciding that hypothesis i is true given that hypothesis j is actually true. If i does not equal j, this represents an error. We now define a conditional cost associated with choosing hypothesis j. This is given by the average of the costs of choosing this hypothesis, where each cost is weighted by the probability of the appropriate hypothesis being true conditioned on the observation. The conditional cost is therefore given by

$$C_j = \sum_{i=1}^{M} C_{ji} P(H_i | y)$$

where y is the observation. Since

$$P(H_i | y) = \frac{P(H_i, y)}{p(y)}$$

this conditional cost can be rewritten as

$$C_j = \sum_{i=1}^{M} C_{ji} p_i(y) \frac{P(H_i)}{p(y)} \tag{9-22}$$

where $p_i(y)$ is the conditional density of the observation, y, based upon the assumption that hypothesis i is true. The detection rule can now be stated as follows: Given the observation y, find the value of j that minimizes the conditional cost C_j. We thus find the decision that leads to the minimum cost. Since only the numerator of Eq. (9-22) depends upon the decision rule, the criterion is simplified to: *Choose j to minimize*

$$\sum_{i=1}^{M} C_{ji} p_i(y) P(H_i)$$

We now consider the special case of the minimum-error detector. As before, if we let the cost of an error be 1 and the cost of a correct decision be 0, Eq. (9-21) becomes

$$\bar{C} = \sum_j \sum_{\substack{i \\ i \neq j}} P(D_i, H_j)$$

but this is simply the probability of error. The rule therefore becomes that of choosing j to minimize

$$\sum_{\substack{i=1 \\ i \neq j}}^{M} P(H_i) p_i(y) = \sum_{i=1}^{M} P(H_i) p_i(y) - P(H_j) p_j(y) \tag{9-23}$$

Since the summation on the right side of Eq. (9-23) is not a function of j, minimizing the entire expression is equivalent to maximizing $P(H_j)p_j(y)$. The decision rule can thus be restated as: *Choose j to maximize $P(H_j)p_j(y)$.*

If we now assume that all of the hypotheses are equally probable, the $P(H_j)$ are each equal to $1/M$, and the decision rule takes on a very simple form. Given an observation, y, we choose the hypothesis for which the conditional probability, $p_j(y)$, is a maximum.

Example 9-6

Three hypotheses have probability density functions as shown in Fig. 9.21. The a priori probabilities are given as

$$P(H_1) = P(H_2) = \tfrac{1}{4}, \qquad P(H_3) = \tfrac{1}{2}$$

and the cost matrix is

$$[C] = \begin{bmatrix} 0 & 1 & 1 \\ 1 & 0 & 1 \\ 1 & 1 & 0 \end{bmatrix}$$

Design a Bayes detector to minimize the average cost.

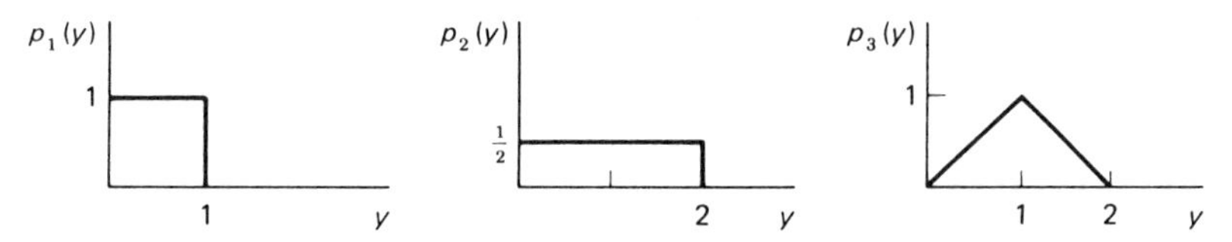

Figure 9.21 Probability density functions for Example 9-6

Solution. Since the costs of the various errors are equal, the Bayes detector is equivalent to the minimum-error detector. Therefore, we choose j to maximize

$$P(H_j)p_j(y)$$

Figure 9.22 is a plot of these three functions for the three values of j. Viewing Fig. 9.22, the detector can be designed by inspection:

$$\text{For } y < 0.5, \text{ choose } H_1.$$

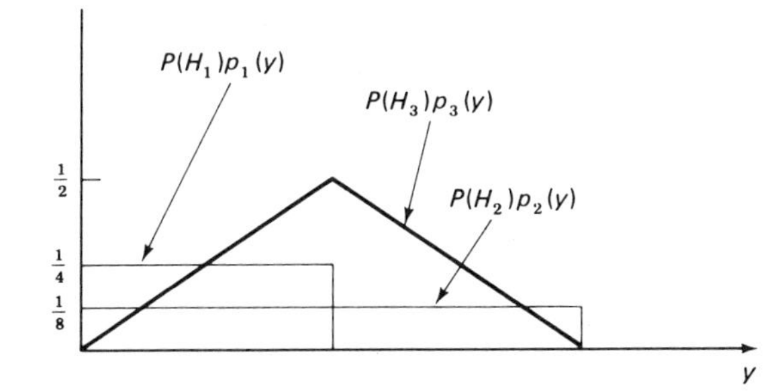

Figure 9.22 Functions to be maximized in 3-ary detector

For $0.5 < y < 1.75$, choose H_3.

For $y > 1.75$, choose H_2.

Example 9-7

Consider a single-sample detector where the output of the sampler is

$$y = s_i(T) + n(T)$$

$n(T)$ is Gaussian noise and $s_i(T)$ is a sample of the ith signal. Suppose that there are M equally probable hypotheses. Design a detector.

Solution. The likelihood functions are given by

$$p_i(y) = \frac{1}{\sqrt{2\pi\sigma^2}} \exp\left\{-\frac{[y - s_i(T)]^2}{2\sigma^2}\right\}$$

Since the $P(H_i)$ are all equal, we choose i to maximize $p_i(y)$. Since all the variances are equal, this reduces to choosing i to maximize

$$\exp\left\{-\frac{[y - s_i(T)]^2}{2\sigma^2}\right\}$$

or taking logs, to maximize

$$-\frac{y^2}{2\sigma^2} + \frac{s_i^2(T)}{2\sigma^2} + \frac{2ys_i(T)}{2\sigma^2} \tag{9-24}$$

The detector measures y, plugs the measured value into the expression above, and finds the value of i that maximizes the expression in Eq. (9-24).

If the detector of Example 9-7 now bases its decision upon a continuous observation of $y(t)$, the single sample would be replaced by an integral. (We are handwaving through a reasonable amount of theory and derivation here. The interested student is referred to any standard text on detection theory. For our applications, the detailed progression from single sample to multiple sample to continuous observation would represent a more serious interruption than does this extended parenthetical comment.) Replacing single samples with integrals, we would choose i to maximize

$$2\int_0^T y(t)s_i(t)\,dt - \int_0^T [s_i^2(t) + y^2(t)]\,dt \tag{9-25}$$

Now if the signals are of equal energy, the second integral in Eq. (9-25) is not a function of i, so we need only choose i to maximize

$$\int_0^T y(t)s_i(t)\,dt$$

Implementing this leads to the familiar *matched filter detector.*

9.5.2 Performance of Matched Filter Detector

As the previous discussion showed, the optimum minimum-error detector is the matched filter configuration, as illustrated in Fig. 9.23. We assume the $s_i(t)$ are chosen to be orthogonal; that is, each pair of signals is uncorrelated.

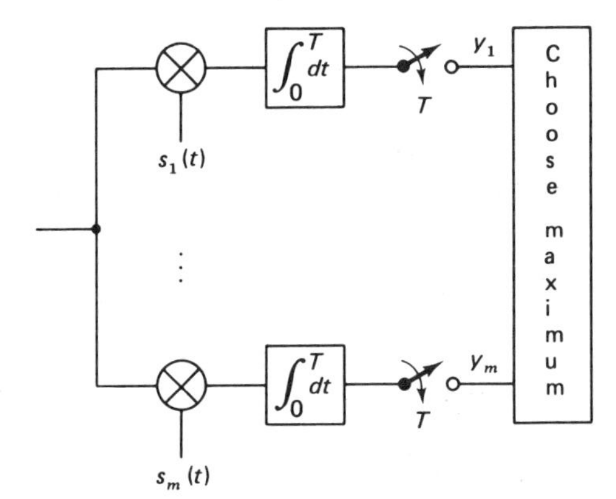

Figure 9.23 Matched filter detector

To evaluate performance, let us assume that a particular hypothesis, H_k, is true. Under this assumption and our usual assumption of white Gaussian noise added in the channel, each input to the comparitor is a Gaussian random variable. Because of the orthogonality of the signals, the mean of each of these is 0 *except* for y_k, which has mean value of $A^2T/2$. The variance of each input is found as before.

$$\begin{aligned} E\{y_i^2\} &= E\left\{\int_0^T s_i(t)n(t)\,dt\right\}^2 \qquad \text{for } i \neq j \\ &= \int_0^T \int \frac{N_0}{2}\delta(t-\tau)\,dt\,d\tau \\ &= \frac{N_0 T}{2} \end{aligned}$$

The M variables are therefore jointly normal. If we can show that the expected value of the product of any two of them is zero, they will then be independent. Taking this expected value, we obtain

$$E\{y_i y_j\} = E\left\{\int_0^T \int_0^T s_i(t)s_j(\tau)[n(t) + s_k(t)][n(\tau) + s_k(\tau)]\,dt\,d\tau\right\}$$

Separating the various terms and assuming white zero-mean noise, this becomes

$$E(y_i y_j) = \frac{N_0}{2}\int_0^T s_i(t)s_j(t)\,dt$$

$$+ \int_0^T s_k(t)s_i(t)\,dt \int_0^T s_k(t)s_j(t)\,dt$$

If $i \neq j$, the first integral is zero, owing to the orthogonality of the signals. The last term in this expression is the product of two integrals, at least one of which must be zero, since both i and j cannot simultaneously equal k. The entire expression is therefore equal to zero, and the y_i are orthogonal and independent. This greatly simplifies the analysis, since the joint density is then the product of the individual densities.

The probability of a correct decision is equivalent to the probability that y_k is the largest value. This probability is given by the M-dimensional integral of Eq. (9-26):

$$P_c = \int_{-\infty}^{\infty} \cdots \int_{-\infty}^{y_k} p(y_1, y_2, \ldots, y_M)\,dy_1 \cdots dy_{k-1}\,dy_{k+1} \cdots dy_M\,dy_k \tag{9-26}$$

This integral can be simplified, since the variables are independent:

$$P_c = \int_{-\infty}^{\infty} p_k(y_k)\left[\int_{-\infty}^{y_k} p_i(x)\,dx\right]^{M-1} dy_k \tag{9-27}$$

This expression can be calculated numerically. If the probability of error is the desired parameter, one need simply subtract the probability of correct decision as given by Eq. (9-27) from 1. That is,

$$P_e = 1 - P_c$$

It is interesting to note that the performance of the M-ary detector does not depend upon the actual choice of frequencies, only upon the assumption of orthogonality.

Equation (9-27) reduces to the familiar

$$P_c = 1 - \frac{1}{2}\operatorname{erfc}\sqrt{\frac{E}{2N_0}}$$

for the case of $M = 2$. It is not a simple computational job to perform this reduction. In fact, we used a different approach earlier toward analyzing the binary case. We were able to form a single Gaussian variable as the difference between the two Gaussian distributed outputs. That is the reason that the two results appear so different.

Figure 9.24 shows a plot of Eq. (9-27) for various values of M. These results are obtained numerically. Note that the ordinate is the *symbol error* probability, not the bit error probability. This is an important distinction, since a single symbol error can cause more than one bit error. We expand upon this in a moment. You should also note that constant energy (E) does not imply constant signal power. As M increases, the symbol period increases, so proportionately less signal power is required to achieve the same signal-to-noise ratio. Also shown on the figure is a

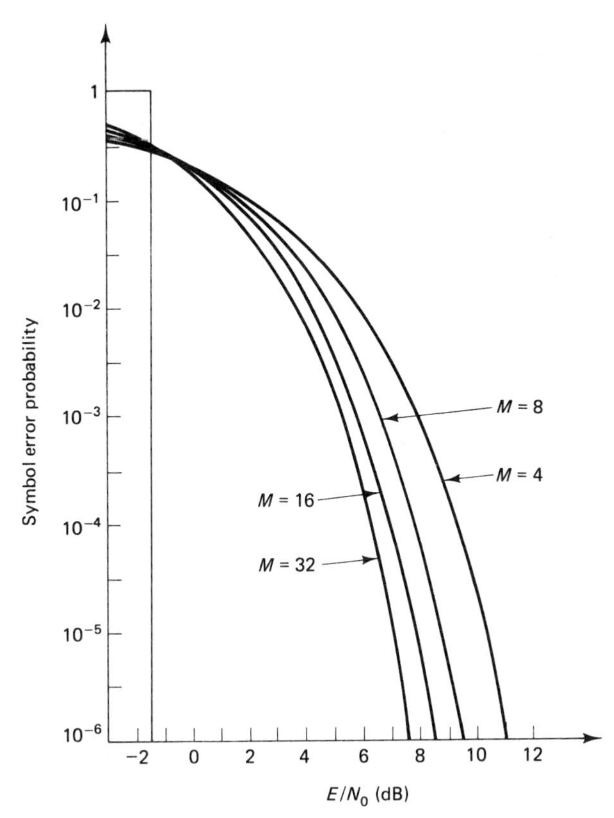

Figure 9.24 Symbol error performance of matched filter detector for MFSK

theoretical bound for $M \to \infty$, which is obtained from the Shannon channel capacity theorem. We begin with Eq. (9-28) which is a repeat of Eq. (4-3).

$$C = BW \log_2\left(1 + \frac{S}{N}\right) \tag{9-28}$$

The signal power, S, is the energy per bit multiplied by the number of bits per second. The noise power is N_0 multiplied by the system bandwidth. If we take the limit as the bandwidth approaches infinity, we can assume operation at capacity, so the equation becomes

$$C = BW \log_2\left(1 + \frac{EC}{BWN_0}\right)$$

Taking the limit as the bandwidth approaches infinity yields

$$E/N_0 = -1.6 \text{ dB}$$

This is shown in Fig. 9.24 labeled as $M \to \infty$, since the infinite bandwidth assumption coincides with the infinite value of M.

In attempting to make design tradeoff decisions, we must be careful to examine the entire picture. As an example, one might look at Fig. 9.24 and say that with a signal-to-noise ratio of 5 dB, you can achieve an error rate of 0.08 with an $M = 4$ system, and 0.04 with $M = 8$. Thus you might conclude the $M = 8$ is better. This is not necessarily true. Other considerations include the following. The $M = 8$ system combines bits in groups of 3, and a symbol error can result in up to 3 bit errors. The bandwidth of the $M = 8$ system is nominally $\frac{3}{2}$ that of the $M = 4$ system. The transmission rate of the $M = 8$ system is nominally $\frac{2}{3}$ that of the $M = 4$ system. Thus, the symbol energy is spread over less time, so the power must increase.

Because of all of these tradeoffs and conflicting quantities, we will find it preferable to reference performance to the bit error rate and to the bit-energy-to-noise ratio. In order to convert symbol error probability to bit error probability, we must first examine the propagation of symbol errors into bit errors. For example, consider the case of $M = 8$. If any one of the 8 possibilities is transmitted, an error can be made in one of seven possible ways. Of these, a particular bit will change in four of them. If we assume that the probability of each type of symbol error is the same (this is an approximation except in the orthogonal case, where the assumption is exactly correct), the probability of any particular bit being in error is $\frac{4}{7}$ of the probability of symbol error. In general, the probability of bit error is related to the probability of symbol error by

$$P_B = \frac{2^{n-1}}{2^n - 1} P_S \tag{9-29}$$

where n is the number of bits of information in each symbol, or

$$n = \log_2 M$$

This can be seen by starting with an n-bit word. There are $2^n - 1$ possible ways in which this word can be received in error. If we now examine any particular bit of the n-bit word, that bit is reversed in 2^{n-1} of these $2^n - 1$ error words. If we now assume that each error word is equally likely, the result of Eq. (9-29) applies.

As an example, suppose we transmit the 3-bit word, 110. There are seven possible words that could be received if a detection error is made:

000, 001, 010, 011, 100, 101, 111

If we now examine any one of the three bits in these error words, we see that an error is made in four of these seven words. For example, the middle bit changes from 1 to 0 in four words (000, 001, 100 and 101). Thus the bit error rate is $\frac{4}{7}$ of the symbol error rate.

The energy per bit, E_b, is simply the symbol energy divided by n.

When we take these conversions into account, Fig. 9.24 is modified to yield the results in Fig. 9.25. The changes from one figure to the other are hardly noticeable.

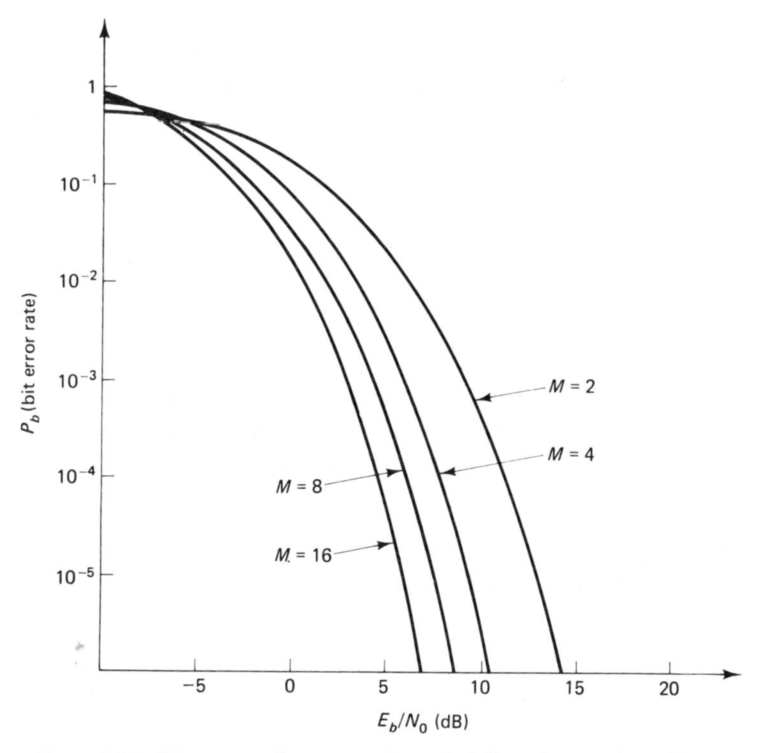

Figure 9.25 Bit error performance of matched filter detector for MFSK

Figure 9.25 clearly illustrates the design tradeoffs in using MFSK. These are summarized as follows:

1. For a given bit-energy-to-noise ratio, the probability of error is inversely related to the bandwidth of the system (i.e., the number of frequencies used).
2. For a fixed bandwidth, probability of bit error is inversely related to the signal-to-noise ratio.
3. For a fixed probability of bit error, the signal-to-noise-ratio is inversely related to the bandwidth.

We shall use these tradeoff considerations in making decisions regarding system design.

9.6 DESIGN EXAMPLES

Example 9-8

A binary FSK system transmits the following two signal bursts:

$$s_1(t) = A \cos (2\pi \times 1100t + 30°)$$

$$s_0(t) = A \cos(2\pi \times 1000t + 30°)$$

The two-sided noise power spectral density is $N_0/2 = 10^{-4}$. The two signals can be considered to be equally likely. The maximum acceptable bit error rate is 3×10^{-7}.
(a) Design a coherent detector and find the maximum achievable bit rate.
(b) Design an incoherent detector using the same bit rate as that found in part (a). Can you still achieve the specified bit error rate?

Solution. (a) The coherent matched filter detector is as shown in Fig. 9.11. The performance is given by

$$P_e = \frac{1}{2} \operatorname{erfc} \sqrt{\frac{E(1-\rho)}{2N_0}} \tag{9-29}$$

If we assume that the channel bandwidth is sufficient to pass the entire transmitted signal burst, and that the attenuation along the transmission path is characterized by a multiplicative factor, K, then the energy at the receiver is

$$E = 0.5K^2A^2T$$

where T is the bit period.

As long at the bit period is a multiple of 10 msec, the two signals are orthogonal, and the correlation coefficient is zero. If, after solving the problem, we find this assumption is not correct, we will return to this step and modify the approach. Recall that the advantage of choosing orthogonal tone spacing is that perfect phase coherence is no longer required.

Using these values and the specified performance, Eq. (9-29) becomes

$$P_e = \frac{1}{2} \operatorname{erfc} \sqrt{\frac{0.5K^2A^2T}{4 \times 10^{-4}}} = 3 \times 10^{-7}$$

We consult a table of error functions to find

$$\frac{0.5K^2A^2T}{4 \times 10^{-4}} = 3.55$$

$$A^2K^2T = 0.01$$

$$T = \frac{0.01}{K^2A^2}$$

Thus, for example, if there were no attenuation in the transmission path and the signals were unit-amplitude, $AK = 1$ and the bit period is 10 msec. We could therefore transmit at a rate of 100 bps.

(b) The performance of the incoherent detector depends upon the bandwidth of the signal entering the receiver. If the bit rate is 100 bps, then the bandwidth of each signal burst, out to the first zeros of the spectrum, is 200 Hz.

We use Eq. (9-18) where the variance is given by Eq. (9-16). Therefore,

$$\sigma^2 = N_0\,\text{BW} = 4 \times 10^{-2}$$

The bit error rate is then given by

$$P_e = \frac{1}{2}\exp\left[\frac{-A^2(T)}{2\sigma^2}\right] = 0.5e^{-12.5} = 1.86 \times 10^{-6}$$

This error rate is about five times that specified in the problem. Thus, the incoherent detector cannot meet the specifications given in this design.

Example 9-9

N independent samples of a received signal are taken. The received signal under the three possible hypotheses is given by

$$H_1\text{:}\quad r(t) = n(t)$$

$$H_2\text{:}\quad r(t) = 10 + n(t)$$

$$H_3\text{:}\quad r(t) = -10 + n(t)$$

where $n(t)$ is Gaussian with mean 0 and variance of 100. The a priori probabilities of the three hypotheses are equal.

Design a minimum-error detector and find the resulting probability of error.

Solution. The best detector performs the following operation:

$$2\sum_k y_k s_{ik} - \sum_k (s_{ik}^2 + y_k^2)$$

That is, we choose the maximum of

$$0 \qquad \text{(If maximum, choose } H_1\text{)}$$

$$20\sum_k y_k - 100N \qquad \text{(If maximum, choose } H_2\text{)}$$

$$-20\sum_k y_k - 100N \qquad \text{(If maximum, choose } H_3\text{)}$$

The variance of the sum, $\Sigma_k\, y_k$, is $100N$ under each hypothesis, and the mean is 0, $10N$, and $-10N$ under H_1, H_2, and H_3, respectively. Thus,

$$P(D_2 \mid H_1) = \Pr\left\{20\sum_k y_k - 100N > 0\right\} = 0.5 \text{ erfc}\,(0.35\sqrt{N})$$

$$P(D_3 \mid H_1) = P(D_2 \mid H_1) = 0.5 \text{ erfc}\,(0.35\sqrt{N})$$

$$P(D_1 \mid H_2) = P(D_1 \mid H_3)$$

$$= \Pr\left\{20\sum_k y_k - 100N < 0 \text{ AND } -20\sum_k y_k - 100N < 0\right\}$$

$$= \int_{-5N}^{5N} N(100N,\, 100N) = 0.5 \text{ erfc}\,(0.35\sqrt{N}) - 0.5 \text{ erfc}\,(1.05\sqrt{N})$$

$$P(D_2 | H_3) = P(D_3 | H_2)$$

$$= \Pr\left\{ -20 \sum_k y_k - 100N > 20 \sum_k y_k - 100N \text{ AND } -20 \sum_k y_k - 100N > 0 \right\}$$

$$= \int_{-\infty}^{-5N} N(10N, 100N) = 0.5 \operatorname{erfc}(1.05\sqrt{N})$$

Putting this all together, we find

$$P_e = 0.33[P(D_2 | H_1) + P(D_3 | H_1) + P(D_1 | H_2) + P(D_3 | H_2) + P(D_1 | H_3) + P(D_2 | H_3)]$$

$$= 0.66 \operatorname{erfc}(0.35\sqrt{N})$$

9.7 SUMMARY

This chapter presents the important transmission technique known as FSK. Like ASK, FSK involves a carrier form of modulation, so it is well suited to transmission through band-pass channels. Incoherent detection is possible without sacrificing transmission energy. This is contrasted with ASK, where we use on-off keying to permit incoherent detection.

We begin the chapter with an analysis of frequency modulation. The results are applied to both the discrete and continuous modulation forms of FSK. In the discrete case, the modulated signal possesses one of a number of fixed frequencies (two in the binary case), while in the continuous form, the frequency of the modulated signal can vary within a particular range.

Section 9.2 presents a derivation of the frequency spectrum for both the discrete and continuous case. The resulting spectrum is wider than that of the ASK form of modulation.

Modulators and demodulators are presented in Section 9.3. We explore both coherent and incoherent detectors. The combined modulator-demodulator, or MODEM, is discussed, and practical examples are given. These include the CCITT recommendation V.21.

Section 9.4 presents performance, where we compare the coherent detector to the incoherent detector. The discriminator detector is also analyzed. We discuss the optimum spacing between frequencies and the benefits of choosing orthogonal tone spacing, even though this choice is suboptimal.

M-ary FSK is discussed in Section 9.5. We use this opportunity to extend the decision theory results originally discussed in Section 5.5 beyond the binary applications. Performance curves for MFSK are developed. These curves can be used to optimize system design.

The final section of the chapter presents two design examples. The first compares incoherent and coherent detection. The second explores detectors for 3-ary communication.

PROBLEMS

9.1. A unit-amplitude bipolar baseband signal is shaped by a low-pass filter that passes frequency up to $1/T_s$. The sampling period, T_s, is 1 ms. The shaped signal then frequency-modulates a carrier of frequency 1 MHz. Find the bandwidth of the resulting signal.

9.2. Using a table of Bessel functions, verify the statements made about the magnitude of c_n following Eq. (9-3).

9.3. Design a Bayes matched filter detector for binary FSK with frequencies

$$f_1 = 1 \text{ kHz}, \qquad f_2 = 1.1 \text{ kHz}$$

$$T_s = 10 \text{ ms}, \qquad N_0/2 = 10^{-2}, \qquad P(H_1) = \tfrac{1}{4}$$

$$C_{10} = 1, \qquad C_{01} = 5, \qquad C_{11} = C_{00} = 0$$

Find the average cost, $\bar{C}$.

9.4. An FSK signal consists of bursts of frequency 800 kHz and 990 kHz, the higher frequency being used to transmit a binary 1. The bit rate is 2 kbps. Find the bandwidth of the FSK signal.

9.5. Find the bandwidth of an FM signal resulting from binary modulation where the pulses are shaped according to a raised cosine. Assume that an alternating train of 0's and 1's is transmitted and that bipolar transmission is used.

9.6. Find the steady-state output of the quadrature detector of Fig. 9.15 when the input is an alternating binary FSK data signal.

9.7. What happens to the output of the system of Fig. 9.8(a) if the 90-degree phase-shift block is omitted from the system?

9.8. Find the probability of bit error for an FSK system with

$$s_1(t) = \cos(1100t + 30°)$$

$$s_0(t) = \cos(1000t + 30°)$$

$$N_0/2 = 0.1, \qquad T_s = 10 \text{ s}, \qquad P(H_0) = 0.5$$

(a) using a coherent detector.
(b) using an incoherent detector.

9.9. Redo Problem 9.8 for $P(H_0) = 0.25$.

9.10. Find the probability of false alarm, P_{FA}, for the incoherent detector shown on page 398, where

$$s_1(t) = \cos(1100t + 30°)$$

$$s_0(t) = \cos(1000t + 30°)$$

$$N_0/2 = 0.1, \qquad T_s = 10 \text{ s}$$

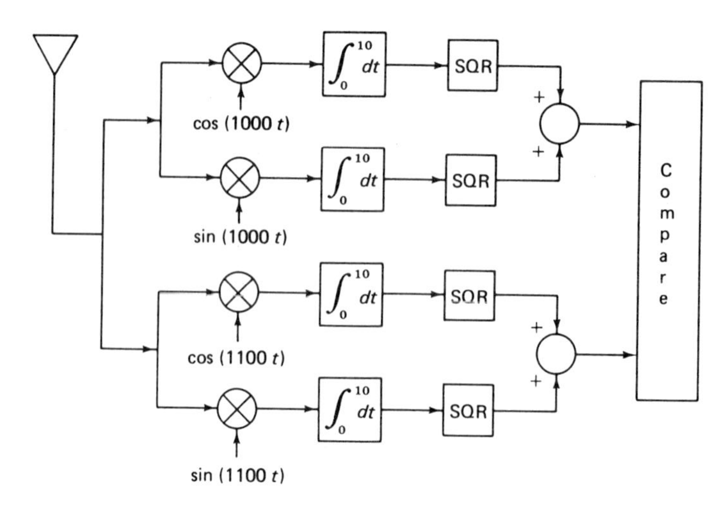

9.11. Derive an expression for the maximum allowable phase mismatch between two frequency bursts such that a tone spacing of $0.715/T$ is preferable to orthogonal tone spacing. Assume that a coherent detector is used.

9.12. A communication channel passes frequencies between 300 Hz and 3 kHz. Voltage amplitudes of up to 5 V can be permitted on the channel. Assume that simplex (one-way) transmission is used, and that the noise has spectral density of 10^{-5} watts/Hz.

(a) Design a binary coherent communication system and evaluate its performance.

(b) If the channel is used to transmit a voice signal of maximum frequency $f_m = 3$ kHz, how many bits of PCM quantization are permitted?

(c) Repeat part (a) for incoherent detection.

9.13. Find the probability of error for the incoherent detector of Fig. 9.18, where $A = 5$, $f_1 = 1200$ Hz, $f_2 = 2200$ Hz, and the sampling period is 1 msec. Noise of $N_0/2 = 10^{-3}$ is added during transmission.

9.14. Find the probability of error of the incoherent detector when used for FSK where the bit period is 1 msec. The two signals have amplitude of 2 V, and the frequencies are 1.2 kHz and 2.2 kHz. The additive noise has a power of 10^{-4} watts/Hz. Compare the performance with that of the coherent detector.

9.15. In a ternary communication system, the signal is given by

$$H_1: \quad s_1(t) = 0$$

$$H_2: \quad s_2(t) = 10 \sin 1000t$$

$$H_3: \quad s_3(t) = -10 \sin 1000t$$

$$N_0/2 = 1, \qquad T_s = 25 \text{ ms}$$

Design a minimum-error detector for equal a priori probabilities. Find the probability of error and the probability of a correct transmission of a single message.

9.16. Find the probability of error for a quaternary transmission system where the signals are given by

$$s_i(t) = \cos(1000 + 10i)t \qquad \text{for } i = 1 \text{ to } 4$$

Assume that the a priori probabilities are equal, that $T = \pi$, and the power spectral density of the additive noise is 0.1. Use a matched filter detector.

10

Phase Modulation

10.1 INTRODUCTION

The present chapter extends the discussion of Chapters 8 and 9 by introducing phase modulation. In the case of *analog communication,* a great deal of similarity exists between frequency modulation and phase modulation. This is so because the frequency of a waveform is defined as the time derivative of the instantaneous phase.

In the case of *digital communication,* the distinction between frequency modulation and phase modulation is much more significant. The reason is that digital information signals are drawn from a discrete set of waveforms.

In phase modulation, just as in amplitude and frequency modulation, we start with a sinusoidal carrier of the form

$$A \cos [\theta(t)]$$

In frequency modulation, the derivative of $\theta(t)$, which is proportional to the instantaneous frequency, follows the baseband or information signal. In phase modulation, the phase itself follows this baseband signal. Thus,

$$\theta(t) = 2\pi[f_c t + K s_b(t)] \tag{10-1}$$

Note that there is no difference between phase-modulating a carrier with $s_b(t)$ and frequency-modulating the same carrier with the time derivative ds_b/dt. This is the reason why the form of *analog* phase modulation does not look significantly different from that of *analog* frequency modulation.

Digital communication starts with a baseband signal that is piecewise constant, and the resulting simplification makes phase modulation look considerably different from frequency modulation.

In *binary frequency-shift keying* (*BFSK*), we switch back and forth between sinusoids of two different frequencies, depending upon whether a 1 or a 0 is being transmitted. In *binary phase-shift keying* (*BPSK*), the frequency of the carrier stays constant while the phase shift takes on one of two constant values.

The two signals used to transmit a 0 and 1 are expressed as:

$$s_0(t) = A \cos(2\pi f_0 t + \theta_0) \qquad 0 < t \leqslant T$$
$$s_1(t) = A \cos(2\pi f_0 t + \theta_1)$$

where θ_0 and θ_1 are constant phase shifts and T is the reciprocal of the bit rate. Another way to express this signal is as follows:

$$s_i(t) = A \cos[2\pi f_0 t + \beta d_i(t)] \qquad 0 < t \leqslant T \qquad (10\text{-}2)$$

where $d_i(t)$ is a signal taking on one of two possible values, +1 or −1, depending upon whether a zero or one is being transmitted, and β is the phase deviation, also known as the *modulation index*. We can use simple trigonometric expansions to express Eq. (10-2) as follows:

$$s_i(t) = A \cos(2\pi f_0 t) \cos[\beta d_i(t)] - A \sin(2\pi f_0 T) \sin[\beta d_i(t)]$$

We use the even and odd properties of the cosine and sine to reduce this further to

$$s_i(t) = A \cos(\beta) \cos(2\pi f_0 t) - A d_i(t) \sin(\beta) \sin(2\pi f_0 t) \qquad (10\text{-}3)$$

The first term in Eq. (10-3) is the *residual carrier*, and it has a power of

$$P_c = \frac{A^2 \cos^2 \beta}{2}$$

The second term represents the *modulated information signal*, or sidebands, and the power of this term is

$$P_d = \frac{A^2 \sin^2 \beta}{2}$$

The total transmitted power is the sum of these two terms, or $A^2/2$. This result is as expected, since we transmit a sinusoid of amplitude A.

If the modulation index, β, is set equal to $\pi/2$, then Eq. (10-3) reduces to

$$s_i(t) = A d_i(t) \cos(2\pi f_0 t) \qquad (10\text{-}4)$$

This represents the *suppressed-carrier* case, and the two signals are the negative of each other:

$$s_1(t) = -s_0(t)$$

We shall see (when performance is analyzed later in this chapter) that this is often

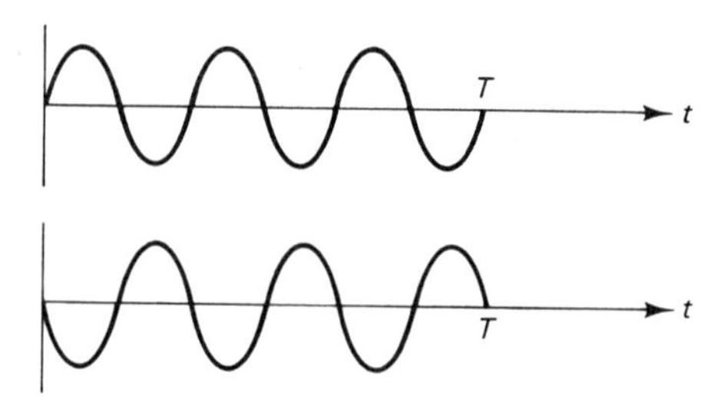

Figure 10.1 Time signals for suppressed-carrier BPSK

the best choice of phase angle, since it achieves a minimum bit error rate. In this case, the two signals are 180 degrees out of phase with each other. They can be represented as shown in Fig. 10.1. Suppressed-carrier phase-shift keying is seen to be identical to bipolar amplitude-shift keying.

As we study complex phase-shift keying systems, and in particular as we generalize from binary to M-ary, an alternate signal representation will often prove useful. This representation is known as *signal space.* A signal-space diagram is a vector representation which illustrates the complex projection of the transmitted signal in the direction of two orthogonal normal signals (i.e., generalized unit vectors). We define *projection* as the correlation between the signal and the normal signal taken over one bit period. Thus, for example, the projection of

$$A \cos (2\pi f_0 t + 45°)$$

on the $\cos 2\pi f_0 t$ axis is given by

$$\int_0^T A \cos (2\pi f_0 t + 45°) \cos 2\pi f_0 t \, dt = 0.707A$$

We have assumed that $T = n/f_0$ where n is an integer. That is, the bit period equals an integer multiple of the period of the sinusoid. The horizontal axis is used to display the $\cos (2\pi f_0 t)$ component, and the vertical axis represents the $\sin (2\pi f_0 t)$ component. This is so because the sine and cosine are 90 degrees out of phase—they are in *phase quadrature.* We illustrate the signal-space representation for the suppressed-carrier case in Fig. 10.2.

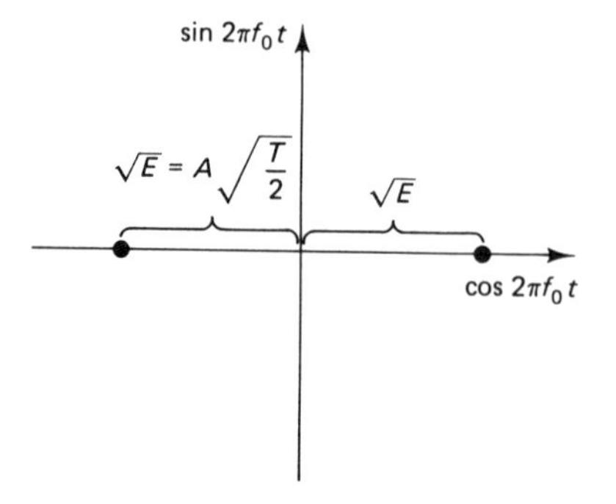

Figure 10.2 Signal-space representation of suppressed-carrier BPSK

The distance of each point from the origin is the square root of the signal energy per bit. Since the power of each signal is $A^2/2$, the energy per bit is $A^2T/2$. The distance between the two points is $2\sqrt{E}$, or $A\sqrt{2T}$. This will prove to be an important parameter in measuring system performance. As may be expected, the greater the distance between points in the signal-space representation, the smaller the probability of bit error. The distance represents the degree of dissimilarity between the two signals.

10.2 BPSK SPECTRUM

The BPSK signal can be considered as a superposition of two ASK waveforms, and the bandwidth of the resulting waveform can be found by examining the component parts. If the suppressed carrier case is considered, PSK is equivalent to *bipolar amplitude-shift keying*. The frequency spectrum is therefore found in the same manner as illustrated in Chapter 8. The power spectrum of the BPSK waveform is then as shown in Fig. 10.3. We have assumed that the data are random. That is, each bit has an equal probability of being either 0 or 1 independent of the value of previous bits. This represents a shifted version of the NRZ-L frequency spectrum. Note that the first zero of the spectrum occurs at a distance of R_b, the bit rate, away from the carrier frequency.

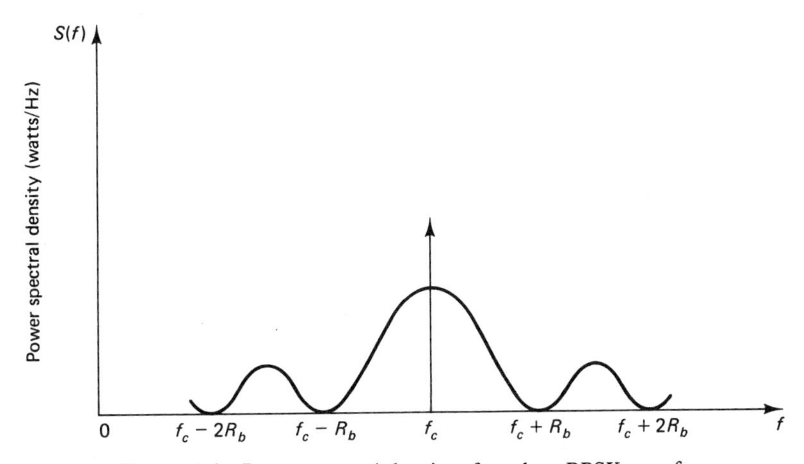

Figure 10.3 Power spectral density of random BPSK waveform

If a continuous baseband signal is used instead of the pulse waveform to reduce distortion (intersymbol interference), the bandwidth is given by

$$\text{BW} = 2\left(f_m + K \max \frac{ds_b}{dt}\right) \tag{10-5}$$

Equation (10-5) results from analog PM theory and Carson's approximation. f_m is the maximum frequency component of the baseband signal.

Example 10-1

Find the bandwidth of a PSK waveform used to transmit an alternating train of 0's and 1's. The carrier frequency is 1 MHz, and the phase varies sinusoidally from 0 to 180 degrees. The bit rate is 2000 bps. Compare the resulting bandwidth with the bandwidth that would obtain if the transitions between the two phase values were instantaneous rather than continuous. That is, assume that the baseband signal changes from sinuosidal to a square-wave function.

Solution. We need simply evaluate the expression in Eq. (10-5). f_m is the maximum frequency of the baseband. In this case, the baseband is a sinusoid of frequency 1000 Hz (one-half of the bit rate). The second term in Eq. (10-5) requires that we find an expression for the baseband signal. If we carefully match the given information to Eq. (10-5), we have

$$\theta(t) = 2\pi f_c t + \frac{\pi}{2} \sin 2\pi \times 10^3 t$$

Thus, $Ks_b(t)$ is given by

$$Ks_b(t) = \frac{\pi}{2} \sin 2\pi \times 10^3 t$$

and

$$\max K \frac{ds_b}{dt} = \pi^2 \times 10^3$$

Finally, the bandwidth is given by Eq. (10-5):

$$\begin{aligned} \text{BW} &= 2(2\pi \times 500 + \pi^2 \times 10^3) \quad \text{rad/s} \\ &= 2(3141 + 9859) \quad \text{rad/s} \\ &= 26 \times 10^3 = 4.14 \times 10^3 \times 2\pi \quad \text{rad/s} \\ &= 4.14 \text{ kHz} \end{aligned}$$

If the phase changes are now instantaneous rather than continuous, the frequency spectrum is as shown in Fig. 10.3, and the bandwidth out to the first zero would be 4 kHz. This seems to be an improvement over the continuous phase case, until we recall that about 9% of the power (in the abrupt phase-change situation) falls outside this band. The resulting distortion and intersymbol interference introduced by limiting to this frequency would increase the bit error rate.

10.3 MODULATORS AND DETECTORS

10.3.1 Modulation

The modulators for PSK exactly parallel those used in FSK. We start with the keyed modulator, which is applicable if the baseband signal is piecewise constant. This is illustrated in Fig. 10.4 and consists of two oscillators and a switch that cycles back

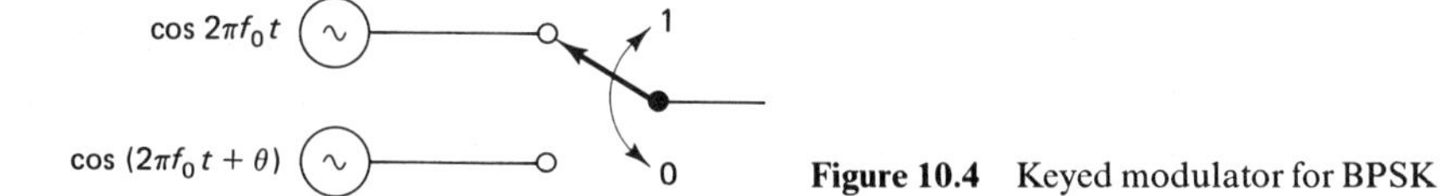

Figure 10.4 Keyed modulator for BPSK

and forth, depending upon which bit is to be transmitted. Alternatively, a single oscillator can be utilized with one delayed path and one direct path.

If the baseband signal is continuous, we use an analog phase modulator, which is illustrated in Fig. 10.5.

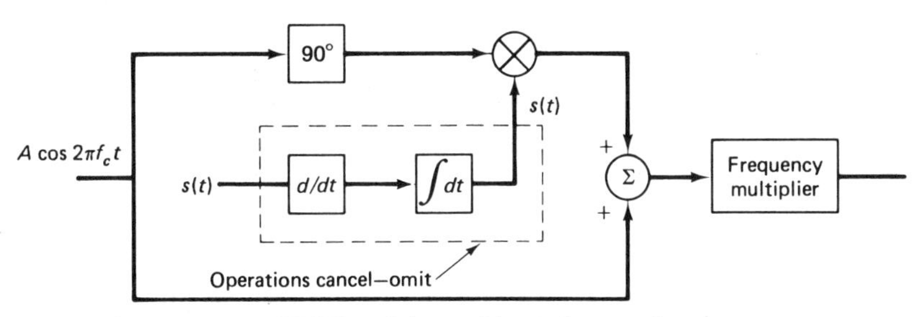

Figure 10.5 BPSK modulator with continuous phase

10.3.2 Detection

Once again, there are two generic ways of approaching detection. We can build a receiver that first demodulates the waveform to yield the reconstructed baseband signal. The baseband signal is then detected using methods discussed in Chapter 7. Alternatively, the receiver can combine the demodulating and detecting process into a single decision-making operation.

The combined process can be performed using a matched filter if the arrival phase is known (i.e., the carrier can be reconstructed). The matched filter detector is shown in Fig. 10.6.

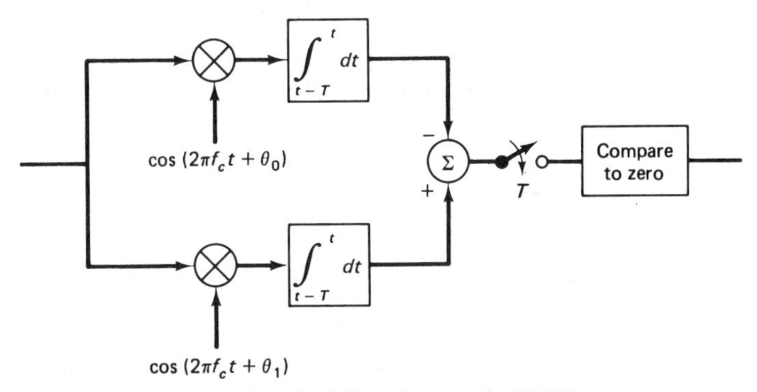

Figure 10.6 Matched filter detector for BPSK

If the PSK modulation is performed as transmitted carrier (i.e., $\beta \neq \pi/2$), the carrier can be recovered using a very narrowband band-pass filter. Alternatively, a phase-lock loop can be used to recover the carrier. This is illustrated in Fig. 10.7.

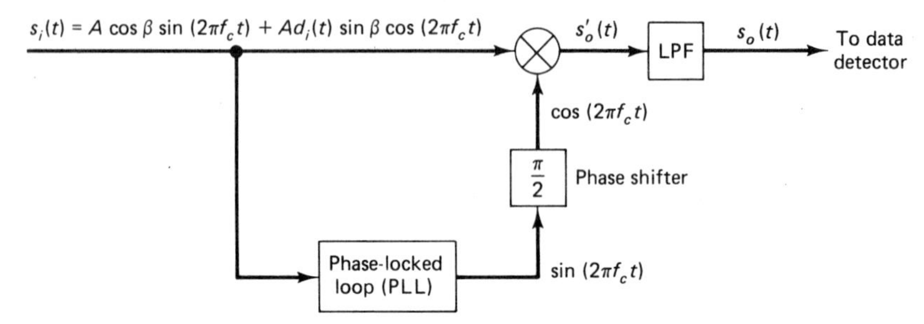

Figure 10.7 BPSK detector using phase-locked loop for carrier recovery

The output signal from the multiplier is given by

$$s_o'(t) = \frac{A}{2}\cos(4\pi f_0 t) + \frac{A}{2}d_i(t)\cos\beta + \frac{A}{2}d_i(t)\sin\beta\cos(4\pi f_0 t)$$

The output of the low-pass filter approximates

$$s_o(t) = \frac{A}{2}d_i(t)\sin\beta \tag{10-6}$$

and the output power is

$$P_o = \frac{A^2\sin^2\beta}{4} \tag{10-7}$$

This is an approximation since the low-pass filter rounds the corners of $d_i(t)$. If the modulation is performed as suppressed-carrier ($\beta = \pi/2$), the phase-lock loop has nothing to lock on to, and we must resort to using a squaring loop (or Costas loop) to reconstruct the carrier. This is shown in Fig. 10.8.

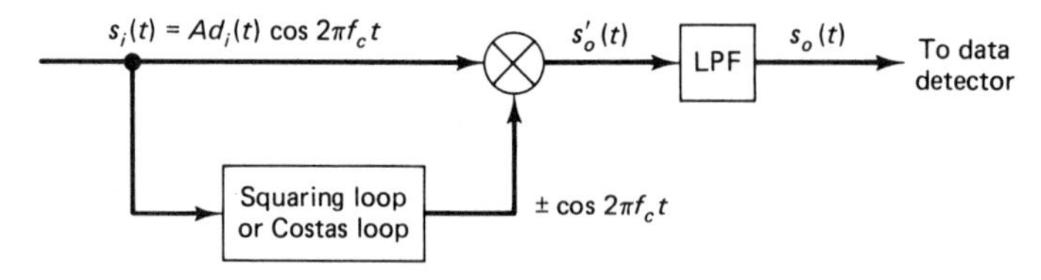

Figure 10.8 BPSK detector using squaring loop for carrier recovery

The Costas loop was discussed in Section 8.3. The loop block diagram is repeated as Fig. 10.9.

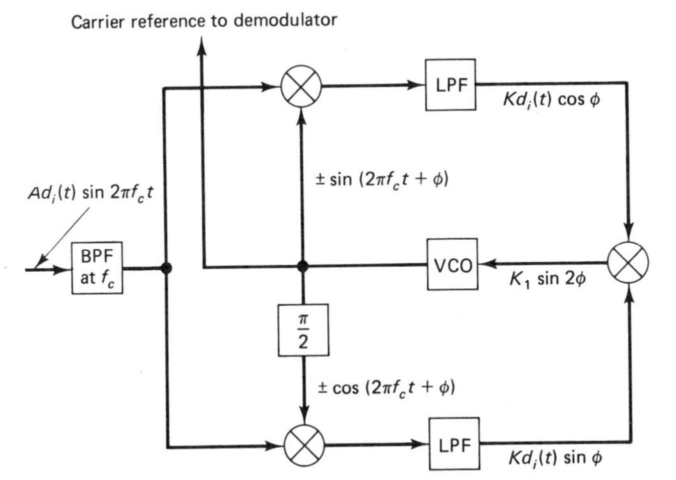

Figure 10.9 Costas loop

The Costas loop is in lock if the error angle, ϕ, approaches zero. This loop is subject to practical problems relating to *acquisition time* and *false lock*. The interested reader is referred to the references for details of the loop operation.

It should be obvious that *incoherent* detectors could not be used for PSK. Incoherent detectors lose all phase information. The output of an incoherent demodulator with PSK as input would be a constant for all time. That is, the output would be identical whether a zero or a one was being received.

Example 10-2

Verify that the output of a square-law detector would not contain sufficient information to permit PSK detection.

Solution. The square-law detector is shown in Fig. 10.10. If the input is $A\cos(2\pi f_0 t + \theta_0)$, the output of the squarer is

$$A^2\cos^2(2\pi f_0 t + \theta_0) = A^2[\tfrac{1}{2} + \tfrac{1}{2}\cos(4\pi f_0 t + 2\theta_0)]$$

The output of the low-pass filter is then $A^2/2$. If the input were instead equal to $A\cos(2\pi f_0 t + \theta_1)$, the output would still be $A^2/2$. Thus, viewing the square-law detector output yields no useful information in detecting a PSK signal.

Figure 10.10 Square-law detector for Example 10-2

In the case of continuous baseband signals, the demodulator is identical to that used in FM, except that the output signal must be integrated to yield the original baseband signal. This is illustrated in Fig. 10.11. Note that whereas the modulator of Fig. 10.5 had a *differentiator* at the very input, the detector of Fig. 10.11 has an *integrator* to cancel this operation. In the analog case, this difference helps to indicate the most desirable type of angle modulation to use for a particular situation.

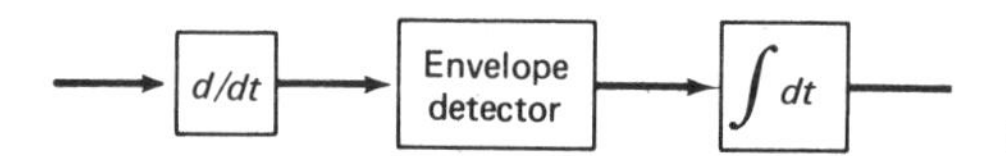

Figure 10.11 Phase demodulator

Differentiation is a process that enhances the higher frequencies. That is, when a signal is differentiated, frequency components get multiplied by f. Thus, if the additive noise were more annoying at the higher frequencies, differentiation at the transmitter would make the transmitted signal less susceptible to this additive noise. This is so because the associated integration at the receiver performs the opposite operation by dividing by the frequency. This division would attenuate the higher-frequency noise while restoring the signal to its original form. The effects of noise are decreased in a manner similar to that of the compression-expansion techniques we examined in the context of quantization.

In the case of pulsed baseband signals, we see additional differences between PSK and FSK. The bandwidth of the PSK waveform is less than that of the FSK waveform. We will find that if a coherent matched filter detector can be built, PSK is capable of yielding better performance than is FSK.

10.4 DIFFERENTIALLY COHERENT DEMODULATION OF BPSK

Unless transmitted carrier transmission is used, demodulation of PSK is rather difficult. Since we may have difficulty in reconstructing a local carrier at the receiver, we can instead compare the received waveform with a delayed version of itself. In this manner, we are using the coherence of the carrier in the modulator to observe changes from one bit interval to the next. We therefore are checking for *changes* in the phase of the received signal. This technique does not allow us to observe the absolute phase of the arriving signal. To use this technique, the information must be superimposed upon the carrier in a *differential* manner.

If the original signal is coded using one of the differential forms (e.g., NRZ-M or NRZ-S), then the digital information is contained in changes between adjacent bit intervals. Thus, if we can detect changes in phase between adjacent bit intervals, we can decode such differential information.

Figure 10.12 shows a differential PSK demodulator. If the signals being operated upon by the multiplier are identical, the input to the integrator is the square of the signal segment. Thus, provided that no change occurs between the two adjacent bit intervals, the output of the integrator is given by

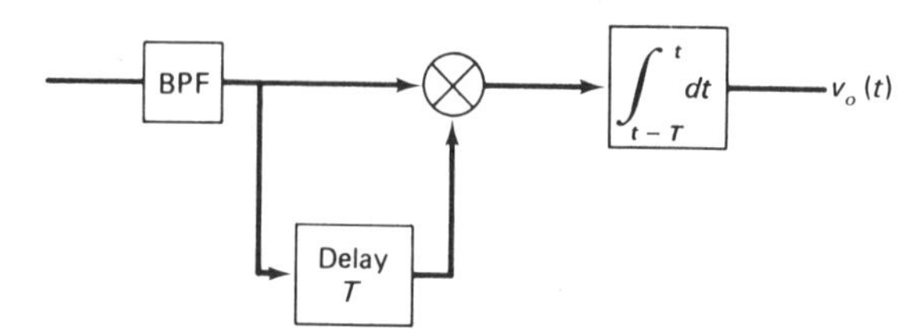

Figure 10.12. Differential PSK demodulator

$$\int_0^T A^2 \cos^2 (2\pi f_0 t + \theta_i)\, dt = \frac{A^2 T}{2} \tag{10-8}$$

If a change does occur between the two adjacent bit intervals, the output of the integrator is given by

$$A^2 \int_0^T \cos (2\pi f_0 t + \theta_0) \cos (2\pi f_0 t + \theta_1)\, dt \approx \frac{A^2 T}{2} \cos (\theta_0 - \theta_1) \tag{10-9}$$

If the phase difference between the two signals is 180 degrees (suppressed-carrier), the integrator output (under the two assumptions) becomes $\pm A^2T/2$. Note that the integrator could be replaced by a low-pass filter with no change in the results.

Thus, the output of the integrator is a bipolar signal, with the positive value obtaining if no change occurs between one interval and the next and the negative value obtaining if a change does occur. If the original data were coded using NRZ-S prior to phase modulation, the output of the demodulator would be the original uncoded baseband signal. If the NRZ-M format were used, the output would be the inverse of the original baseband signal. The resulting baseband signal then forms the input to a baseband detector, and decisions are made to recover the original data signal.

10.5 PERFORMANCE

10.5.1 Matched Filter Detector

The performance of the matched filter detector is completely specified by the three parameters ρ, N_0, and E. ρ is the correlation coefficient and E is the average energy of the signals used to transmit a 1 and a 0. The bit error rate of the matched filter detector is given by Eq. (10-10), which is derived in Section 7.5:

$$P_e = \frac{1}{2} \operatorname{erfc} \sqrt{\frac{(1 - \rho)E}{2N_0}} \tag{10-10}$$

The two signals are given by

$$s_0(t) = A \cos (2\pi f_0 t + \theta_0), \qquad 0 < t < T$$

and

$$s_1(t) = A \cos (2\pi f_0 t + \theta_1), \qquad 0 < t < T$$

The average energy is then $A^2T/2$, and the correlation coefficient is given by

$$\begin{aligned} \rho &= \frac{1}{\frac{A^2 T}{2}} \int_0^T A \cos (2\pi f_0 t + \theta_0) A \cos (2\pi f_0 t + \theta_1)\, dt \\ &\approx \cos (\theta_0 - \theta_1) \end{aligned} \tag{10-11}$$

The double frequency term has been omitted in Eq. (10-11), since we assume that T is large compared to the period of the sinusoidal signals. (Note that if $T = n/f_0$, the double frequency is exactly equal to zero.)

Let us pause for a moment to examine the PSK signals in the signal-space representation. The two signals are illustrated in Fig. 10.13. We have aligned the first signal with the horizontal axis.

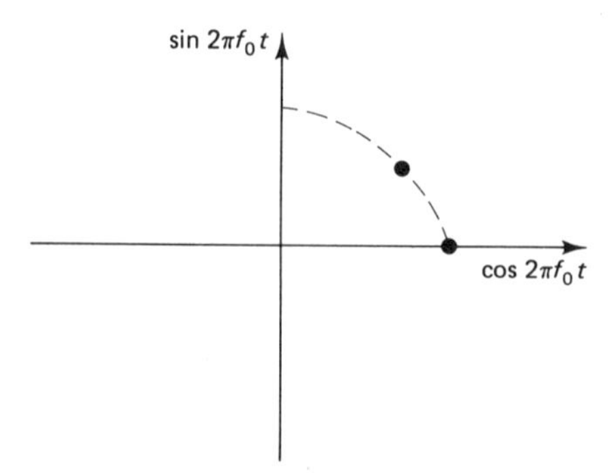

Figure 10.13 Signal-space representation of BPSK

The dot product of the two vectors (from the origin to the signal space points) is the product of the two vector magnitudes multiplied by the cosine of the angle between the two vectors. This is given by

$$S_0 \cdot S_1 = E \cos(\theta_1 - \theta_0) = E\rho \tag{10-12}$$

The quantity $E\rho$ appears with a negative sign in the numerator of the erfc argument of Eq. (10-10). We therefore wish to minimize this quantity in order to minimize the bit error rate. This occurs when the angle between the two vectors is 180 degrees.

The complementary error function decreases as its argument increases. Therefore, the probability of error is monotonic increasing in ρ, and we can minimize the error probability by making $\rho = -1$. This value of ρ will obtain if the phase difference between the two signals is 180 degrees, which occurs for the suppressed-carrier case. This result is not surprising, since the signals are then as different as possible—one is the negative of the other. The resulting error probability is given by

$$P_e = \frac{1}{2}\,\text{erfc}\sqrt{\frac{E}{N_0}} \tag{10-13}$$

The argument of the complementary error function in Eq. (10-13) is dimensionless. We can show this with the following substitution:

$$\frac{E}{N_0} = \frac{P_s T}{P_n/BW} = T(\text{BW})\,\frac{P_s}{P_n}$$

In this equation, T is the period of the signal, P_s is the power of the signal, P_n is the total noise power and BW is the bandwidth of the system. The units of T are seconds and the units of BW are Hertz, or 1/sec. Therefore, E/N_0 is dimensionless. That is, the units of N_0 are those of energy. This error probability is plotted in Fig. 10.14.

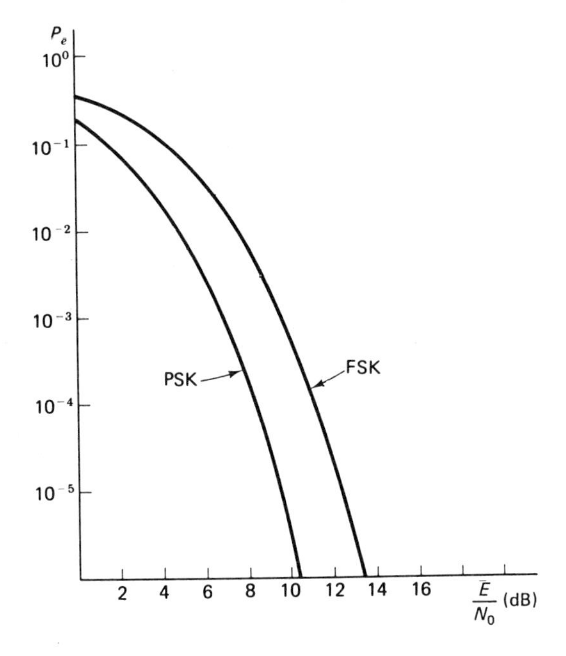

Figure 10.14 Error probability as a function of signal-to-noise ratio for BPSK and BFSK

We shall now make a slight change and derive a result that is more general. Suppose that a matched filter detector is used for PSK demodulation but that the local oscillators are mismatched in phase by $\Delta\theta$. The detector is then as shown in Fig. 10.15.

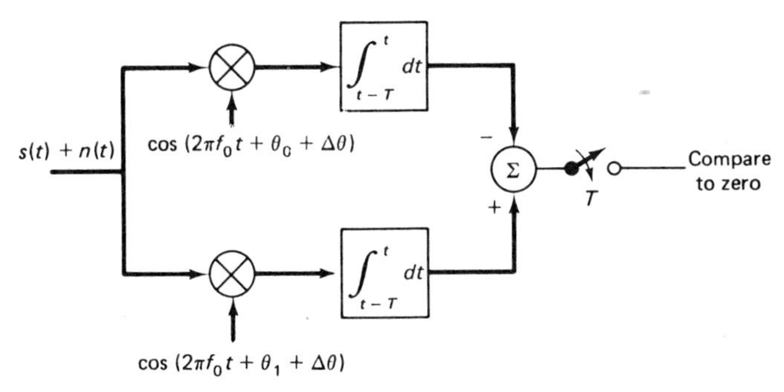

Figure 10.15 Matched filter detector with phase mismatch

The local oscillator phase shift will not affect the noise power at the output, so the variance of the output random variable can be shown to be

$$\begin{aligned}\text{var}\{n_o(T)\} &= E\left\{\int_0^T n(t)[\cos(2\pi f_0 t + \theta_1 + \Delta\theta) - \cos(2\pi f_0 t + \theta_0 + \Delta\theta)]\,dt\right\}^2 \\ &= \tfrac{1}{2}N_0 \int_0^T [\cos(2\pi f_0 t + \theta_1 + \Delta\theta) - \cos(2\pi f_0 t + \theta_0 + \Delta\theta)]^2\,dt \\ &\approx \tfrac{1}{2}N_0 T[\tfrac{1}{2} + \tfrac{1}{2} - \cos(\theta_1 - \theta_0)]\end{aligned}$$

The deterministic part of the output is of the form

$$s_o(T) = \int_0^T s(t)[\cos(2\pi f_0 t + \theta_1 + \Delta\theta) - \cos(2\pi f_0 t + \theta_0 + \Delta\theta)]\,dt$$

If the phases of the PSK signals now differ by 180 degrees, this becomes

$$s_o(T) = 2\int_0^T s(t)\cos(2\pi f_0 t + \theta_1 + \Delta\theta)\,dt$$

and the variance becomes

$$\text{var}\{n_o(T)\} = \tfrac{1}{2}N_0 T(\tfrac{1}{2} + \tfrac{1}{2} + 1) = N_0 T$$

If we now assume that

$$s(t) = \cos(2\pi f_0 t + \theta_1)$$

the output becomes

$$\begin{aligned}s_o(T) &= 2\int_0^T \cos(2\pi f_0 t + \theta_1)\cos(2\pi f_0 t + \theta_1 + \Delta\theta)\,dt \\ &= T\cos(\Delta\theta)\end{aligned}$$

Therefore, under hypothesis 1, the output mean is $T\cos(\Delta\theta)$, and under hypothesis 0, it is the negative of this quantity. For instructional purposes, we again repeat the derivation from this point. The probability of false alarm, probability of miss, and probability of error are all equal, because of symmetry. The probability of false alarm is the probability that the output exceeds zero under hypothesis zero. This is given by

$$P_{\text{FA}} = \int_0^\infty \frac{1}{\sqrt{2\pi N_0 T}} \exp\left[\frac{-(x + T\cos\Delta\theta)^2}{2N_0 T}\right] dx$$

Making the following change of variables,

$$y = \frac{x + T\cos\Delta\theta}{\sqrt{2N_0 T}}$$

the integral becomes

$$P_{\text{FA}} = \frac{1}{\sqrt{\pi}} \int_{T\cos\Delta\theta/\sqrt{2N_0 T}}^\infty e^{-y^2}\,dy \tag{10-14}$$

$$= \frac{1}{2}\,\mathrm{erfc}\left(\sqrt{\frac{T\cos^2\Delta\theta}{2N_0}}\right) \qquad \text{for}\,|\Delta\theta| < \frac{\pi}{2}$$

This result is plotted as a function of T/N_0 for representative values of $\Delta\theta$ in Fig. 10.16. Note that the curve for $\Delta\theta = 0$ is the perfectly matched case discussed at the beginning of this section.

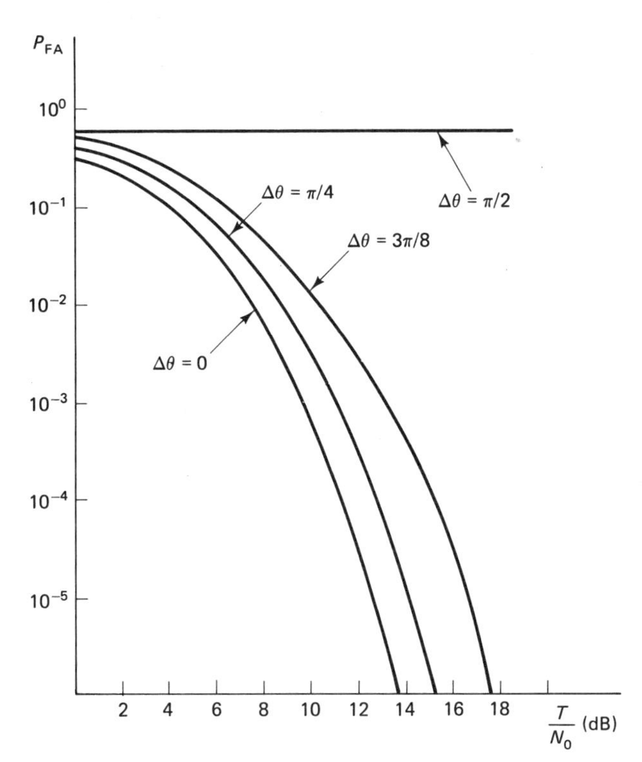

Figure 10.16 Performance of matched filter detector with phase mismatch

10.5.2 Discriminator Detector

The discriminator detector of Fig. 10.11 could be used for PSK reception. Starting with $s(t) + n(t)$ as input, the output of the differentiator is

$$\frac{ds}{dt} + \frac{dn}{dt}$$

If $s(t)$ is now given as $A\cos(2\pi f_0 t + \theta_i)$, and $n(t)$ is expanded in the usual narrowband format, the output of the envelope detector would be proportional to

$$z(t) = \sqrt{[x(t) - A\cos\theta_i]^2 + [y(t) - A\sin\theta_i]^2}$$

Performance will depend only upon the difference between the phases and not upon their actual values. Let us therefore suppose that $\theta_0 = 180°$ and $\theta_1 = 0°$. The envelope detector output then is simplified to

$$z(t) = \sqrt{[x(t) \pm A]^2 + y^2(t)}$$

The plus sign obtains if $\theta_i = \theta_0$; that is, a binary 0 is being sent. This output is a random process. Each sample is Ricean distributed, but the power spectral density is difficult to evaluate owing to the nonlinear operation of the envelope detector. For high input signal-to-noise ratio, we can approximate the envelope detector output as

$$z(t) = x(t) \pm A$$

This detector would then perform exactly like a coherent matched filter detector.

If the approximation cannot be made, calculation of the variance of the output (i.e., noise power) is rather complex.

A simplification is possible if we skip across the whole system and recognize that the result of the entire operation is to take an input of the form

$$A(t) \cos [2\pi f_0 t + \theta(t)]$$

and yield an output of the form, $\theta(t)$.

We can force the signal plus noise to be in this format.

$$\begin{aligned} s(t) + n(t) &= [x(t) - A \cos \theta_i] \cos 2\pi f_0 t - [y(t) - A \sin \theta_i] \sin 2\pi f_0 t \\ &= M(t) \cos [2\pi f_0 t + P(t)] \end{aligned}$$

where

$$M(t) = \sqrt{[x(t) - A \cos \theta_i]^2 + [y(t) - A \sin \theta_i]^2}$$

and

$$P(t) = \tan^{-1} \left[\frac{y(t) - A \sin \theta_i}{x(t) - A \cos \theta_i}\right]$$

The system output is then simply the phase, $P(t)$.

This cannot be conveniently carried further without some approximations. If the signal-to-noise ratio is large, this can be approximated by

$$P(t) \approx \theta_i + \frac{y(t)}{A \cos \theta_i}$$

With $\theta_0 = 0$ and $\theta_1 = \pi$, this becomes

$$P(t) \approx \theta_i \pm \frac{y(t)}{A}$$

with the plus sign under hypothesis 0. The variance of this output is

$$E\left\{\frac{y^2(t)}{A^2 \cos^2 \theta_i}\right\} = \frac{N_0 \text{ BW}}{A^2}$$

where the power of the input noise is N_0 watts/Hz and BW is the system bandwidth. If the phases are now shifted to be equally distributed around zero, the mean of the output becomes $\pm\pi/2$ with the plus sign under hypothesis 1. The variances do not change. Finally, the probability of error is the same as the false-alarm probability, and is given by

$$P_e = \int_0^\infty \frac{1}{\sqrt{\dfrac{2\pi N_0\,\mathrm{BW}}{A^2}}} \exp\left[\frac{-\left(x + \dfrac{\pi}{2}\right)^2}{\dfrac{2N_0\,\mathrm{BW}}{A^2}}\right] dx$$

$$= \tfrac{1}{2}\,\mathrm{erfc}\left(\frac{\pi}{2}\frac{A}{\sqrt{2N_0\,\mathrm{BW}}}\right)$$

We would probably like a basis for comparing this performance to that of the matched filter detector. As an attempt at this, we might set the bandwidth equal to $2/T$. This is the bandwidth of the signal out to the first zeros in the frequency spectrum. Using this value, the error probability is

$$P_e = \tfrac{1}{2}\,\mathrm{erfc}\left(\frac{\pi A}{2}\sqrt{\frac{T}{4N_0}}\right)$$

$$= \tfrac{1}{2}\,\mathrm{erfc}\left(0.79\,A\sqrt{\frac{T}{N_0}}\right)$$

This compares to an error probability of

$$P_e = \tfrac{1}{2}\,\mathrm{erfc}\left(A\sqrt{\frac{T}{N_0}}\right)$$

for the matched filter case. This result is intuitively satisfying, since it indicates a poorer performance for the discriminator detector than for the matched filter detector. The matched filter detector requires precise matching of phases, and a basic law of science says that the more information you have, the better job you can do of detecting a signal. The matched filter should therefore perform better (or at least as well) as the discriminator detector. There are two cautions that must be issued before interpreting this result. First, the choice of filter bandwidth as $2/T$ would distort the signal, and therefore would not lead to the exact performance derived. Second, the assumption of large signal-to-noise ratio may not be realistic.

10.5.3 Differential PSK

The output of the detector of Fig. 10.12 is

$$y_{n+1}(T) = \int_0^T [s_n(t) + n_n(t)]\,[s_{n+1}(t) + n_{n+1}(t)]\,dt$$

where $s_n(t)$ is the signal in the nth time interval and $n_n(t)$ is the noise in that interval. Expanding the brackets yields

$$y_{n+1}(T) = \int_0^T [s_n(t)s_{n+1}(t) + n_n(t)s_{n+1}(t) + n_{n+1}(t)s_n(t) + n_n(t)n_{n+1}(t)]\,dt$$

$n_n(t)$ and $n_{n+1}(t)$ are noise samples taken in two different sampling intervals. We can therefore assume that they are independent of each other. The mean of the output is then

$$E\{y_{n+1}(T)\} = \int_0^T s_n(t)s_{n+1}(t)\,dt \tag{10-15}$$

Now if $\theta_1 - \theta_0 = 180°$, Eq. (10-15) reduces to

$$E\{y_{n+1}(T)\} = \pm\int_0^T s_n^2(t)\,dt = \pm E \tag{10-16}$$

where E is the signal energy. The plus sign obtains when there is no phase change and the minus sign when there is a phase change from one interval to the next. Note that even though $n(t)$ is assumed Gaussian, $y_{n+1}(T)$ will not be Gaussian owing to the nonlinear operation wherein $n_n(t)$ is multiplied by $n_{n+1}(t)$.

We now find the variance of $y_{n+1}(t)$.

$$\begin{aligned}\operatorname{var}\{y_{n+1}(T)\} &= E\{y_{n+1}(T) - E\{y_{n+1}(T)\}\}^2 \\ &= E\left\{\int_0^T [n_n(t)s_{n+1}(t) + n_{n+1}(t)s_n(t) + n_n(t)n_{n+1}(t)]\,dt\right\}^2\end{aligned}$$

Owing to the assumption that $n_n(t)$ and $n_{n+1}(t)$ are independent and zero-mean, this variance simplifies to

$$\begin{aligned}\operatorname{var}\{y_{n+1}(T)\} = {} & 2E\left\{\int_0^T\int_0^T n_n(t)n_n(\tau)s(t)s(\tau)\,dt\,d\tau\right\} \\ & + E\left\{\int_0^T\int_0^T n_n(t)n_n(\tau)n_{n+1}(t)n_{n+1}(\tau)\,dt\,d\tau\right\}\end{aligned}$$

With white noise as input, this becomes

$$\begin{aligned}\operatorname{var}\{y_{n+1}(T)\} &= 2\int_0^T\int_0^T \frac{N_0}{2}\delta(t-\tau)s(t)s(\tau)\,dt\,d\tau + \left[\int_0^T \frac{N_0}{2}\,\delta(t)\,dt\right]^2 \\ &= N_0E + \frac{N_0^2}{4}\end{aligned}$$

The output of the detector is therefore a random variable with a mean value of $\pm E$ and a variance of $N_0E + N_0^2/4$. Unfortunately, unless N_0E is much larger than $N_0^2/4$, this random variable is not even approximately Gaussian, and further analysis is quite complex. It can be shown that the probability of error for DPSK is twice that for coherent PSK.

For the sake of an analysis exercise, we present a second approach to this difficult problem of evaluating performance in DPSK.

Let the output in the first interval be given by

$$[s_1(t) + x_1(t)] \cos 2\pi f_0 t - y_1(t) \sin 2\pi f_0 t$$

where $s_1(t) = \pm 1$ (we are assuming that $\theta_0 = 0$). This can be rewritten as a single sinusoid as follows:

$$R_1(t) \cos [2\pi f_0 t + \theta_1(t)]$$

where

$$R_1(t) = \sqrt{[s_1(t) + x_1(t)]^2 + y_1^2(t)}$$

and

$$\theta_1(t) = \tan^{-1}\left[\frac{y_1(t)}{x_1(t) + s_1(t)}\right]$$

Similarly, in the second interval the output can be written as

$$R_2(t) \cos [2\pi f_0 t + \theta_2(t)]$$

The integrator output is then approximately equal to

$$\begin{aligned} v_o(T) &= \frac{1}{2}\int_0^T R_1(t)R_2(t) \cos [\theta_1(t) - \theta_2(t)]\, dt \\ &= \frac{1}{2}\int_0^T R_1(t)R_2(t)[\cos \theta_1(t) \cos \theta_2(t) + \sin \theta_1(t) \sin \theta_2(t)]\, dt \\ &= \frac{1}{2}\int_0^T R_1(t)R_2(t)\left[\frac{[s_1(t) + x_1(t)][s_2(t) + x_2(t)] + y_1(t)y_2(t)}{R_1 R_2}\right] dt \end{aligned} \tag{10-17}$$

Now if this output is greater than zero, the detector assumes that no phase change has occurred from one interval to the next in the DPSK signal (i.e., a binary 1 is detected). Since $R_1(t)$ and $R_2(t)$ are nonnegative, the sign of the entire integral depends only upon the sign of the numerator in the brackets of Eq. (10-17). Thus, the probability of detecting a 1 (no phase change) is given by

$$\Pr\{z > 0\} = \Pr\left\{\int_0^T ([s_1(t) + x_1(t)][s_2(t) + x_2(t)] + y_1(t)y_2(t))\, dt\right\} > 0 \tag{10-18}$$

This problem would be finished if we could conveniently find the probability distribution of the integrand of Eq. (10-18). Unfortunately, this is not simple. By converting the cosine and sine portions (the in-phase and quadrature components) to complex numbers, some simplification is possible. Note that

$$\text{Re}\,\{R_1 e^{j\theta_1} R_2 e^{j\theta_2}\} = (s_1 + x_1)(s_2 + x_2) + y_1 y_2$$

where the time dependence has been omitted for simplicity. But this can be rewritten as

$$\text{Re}\,\{R_1 e^{j\theta_1} R_2 e^{j\theta_2}\} = |w_1|^2 - |w_2|^2$$

where

$$w_1 = \frac{R_1 e^{j\theta_1} + R_2 e^{j\theta_2}}{2}$$

and

$$w_2 = \frac{R_1 e^{j\theta_1} - R_2 e^{j\theta_2}}{2}$$

By plugging in values, we note that $|w_1|$ represents the magnitude of the signal plus noise while $|w_2|$ represents noise magnitude alone. With this observation, $|w_1|$ would be Ricean distributed while $|w_2|$ is Rayleigh. Finally, the probability of error is given by the probability that $z < 0$ in Eq. (10-18). This is the same as

$$P_e = \Pr\{|w_2| > |w_1|\}$$

The performance of this detector can therefore be evaluated by finding the critical parameters of these two distributions and integrating over the appropriate regions. This is rather complex, since w_1 is not independent of w_2.

The integral of the probability distribution is not simple to perform. The interested reader is referred for details to the references. However, since the differential detector multiplies the signal by a delayed version of itself, it is not surprising to find that we are dealing with nonlinear operations upon the noise. The results follow those of the other nonlinear detectors studied in previous chapters. In fact, the bit error rate reduces to the simple form

$$P_e = \frac{1}{2} \exp\left(-\frac{E}{N_0}\right)$$

This is plotted in Fig. 10.17. We have repeated the result for coherent suppressed-carrier BPSK on the same set of axes. Note that the probability of error is higher for the DPSK case than for the coherent BPSK case. This is expected, since additional information about the received signal is required for the coherent case. In general, the more you know about a signal, the better the job you can do detecting it.

10.6 QUADRATURE PHASE-SHIFT KEYING

We have shown that binary PSK with a phase difference of 180 degrees provides a high level of communication performance. For this case the signals are of the form $\pm A \cos 2\pi f_0 t$. We know that $\cos 2\pi f_0 t$ and $\sin 2\pi f_0 t$ are orthogonal signals. That is,

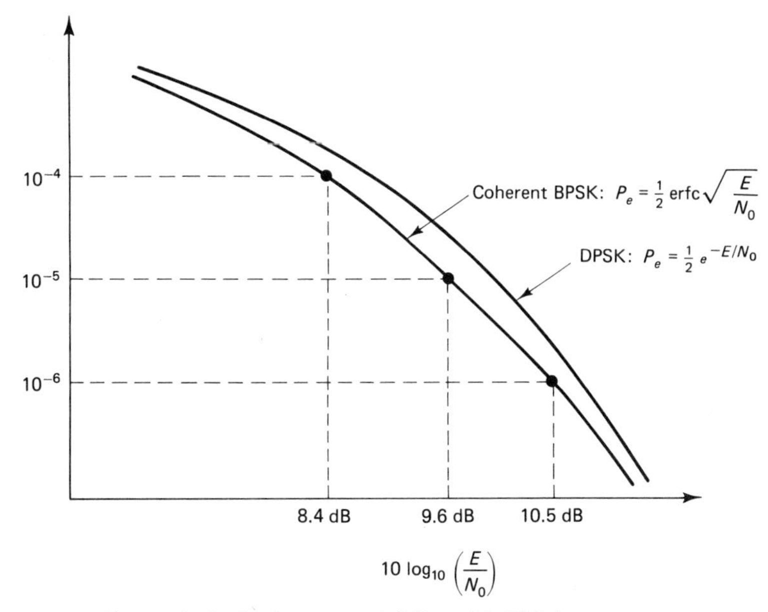

Figure 10.17 Performance of differential PSK detector

the integral of their product is zero. Thus, if binary information were transmitted using $\sin 2\pi f_0 t$ and $\cos 2\pi f_0 t$ as the two signals, we would expect approximately the same performance as in FSK systems. This is so because the energy is the same in both systems and the correlation coefficient is 0. These two quantities completely determine performance of the binary matched filter detector. Therefore, if coherent detection is possible, there is no difference in error performance between FSK and *orthogonal* PSK. In the case of PSK, the bandwidth of the transmitted signal is less than that of FSK, since only one frequency is used rather than two separated frequencies.

But we can go a step further. Since the cosine and sine are orthogonal, it is possible to completely separate them at the receiver using *coherent* detection techniques. It is therefore possible to send two completely separate PSK signals simultaneously with no increase in bandwidth. Figure 10.18 shows such a system. We assume that the baseband signals are bipolar pulses. Each leg of the modulator in Fig. 10.18 uses the fact that PSK with 180-degree phase shifts is equivalent to bipolar ASK.

The transmitted signal has one of four possible forms:

$$+\cos 2\pi f_0 t + \sin 2\pi f_0 t$$
$$+\cos 2\pi f_0 t - \sin 2\pi f_0 t$$
$$-\cos 2\pi f_0 t + \sin 2\pi f_0 t$$
$$-\cos 2\pi f_0 t - \sin 2\pi f_0 t$$

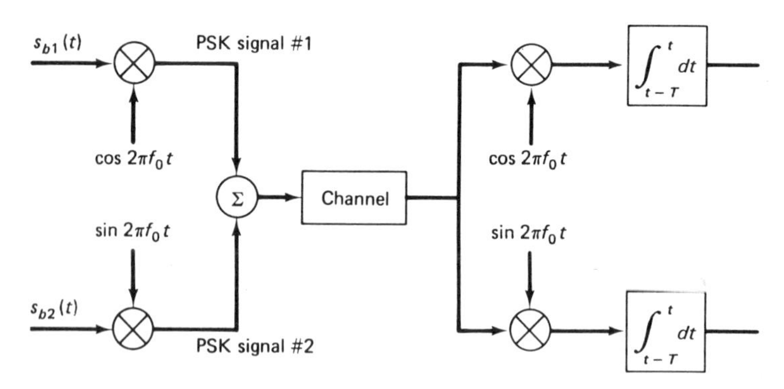

Figure 10.18 Simultaneous transmission of two BPSK signals

or equivalently,

$$\pm(\cos 2\pi f_0 t \pm \sin 2\pi f_0 t) \tag{10-19}$$

Using trigonometric identities, the terms in parentheses in Eq. (10-19) can be rewritten as

$$\cos 2\pi f_0 t \pm \sin 2\pi f_0 t = \sqrt{2}\cos\left(2\pi f_0 t \mp \frac{\pi}{4}\right)$$

so that the four possible signals can be denoted as

$$\sqrt{2}\cos\left(2\pi f_0 t + \frac{\pi}{4}\right)$$

$$\sqrt{2}\cos\left(2\pi f_0 t + \frac{3\pi}{4}\right)$$

$$\sqrt{2}\cos\left(2\pi f_0 t + \frac{5\pi}{4}\right)$$

$$\sqrt{2}\cos\left(2\pi f_0 t + \frac{7\pi}{4}\right)$$

The corresponding points are equally spaced around a unit circle as illustrated in Fig. 10.19. Thus, simultaneous transmission of orthogonal PSK signals is the same

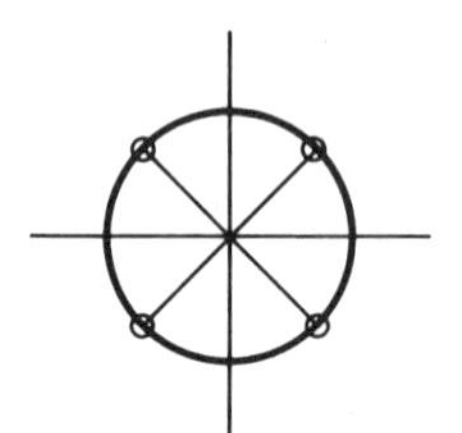

Figure 10.19 Phase distribution of QPSK signals

as PSK with four possible angles. This is known as *quadrature phase-shift keying* (QPSK).

We have derived this using two separate baseband signals. Alternatively, QPSK can be used to send data that are already in *quaternary* form. For example, if we start with a sampled analog signal and quantize to four levels, each level can correspond to one of the transmitted phases.

In the more general binary data case, bits can be combined in pairs to form a four-level signal. For example, the binary train

$$0 \quad 1 \quad 1 \quad 0 \quad 1 \quad 0 \quad 1 \quad 1 \quad 0 \quad 0 \quad 1 \quad 0$$

can be rewritten in base 4 as

$$1 \quad 2 \quad 2 \quad 3 \quad 0 \quad 2$$

where the first digit corresponds to 01, the second to 10, third to 10, and so on. We can then assign phases as follows:

$$00 \rightarrow \frac{\pi}{4}, \qquad 10 \rightarrow \frac{5\pi}{4} = \frac{-3\pi}{4}$$

$$01 \rightarrow \frac{3\pi}{4}, \qquad 11 \rightarrow \frac{7\pi}{4} = \frac{-\pi}{4}$$

Continuing the example using this assignment, the transmitted phases would then be

$$\frac{3\pi}{4} \quad \frac{-3\pi}{4} \quad \frac{-3\pi}{4} \quad \frac{-\pi}{4} \quad \frac{\pi}{4} \quad \frac{-3\pi}{4}$$

Note that the transmission baud rate is one-half of that needed for binary PSK of the same information content. Thus, the bandwidth would be one-half as much. However, without an increase in power, the error rate would increase, since the decision thresholds are closer together.

Example 10-3

Given a 4-ary PSK system with

$$\left.\begin{aligned} s_1(t) &= A\cos 2\pi f_0 t \\ s_2(t) &= A\cos\left(2\pi f_0 t + \frac{\pi}{2}\right) \\ s_3(t) &= A\cos(2\pi f_0 t + \pi) \\ s_4(t) &= A\cos\left(2\pi f_0 t + \frac{3\pi}{2}\right) \end{aligned}\right\} \quad 0 < t \leqslant T$$

The a priori probability of each hypothesis is $\frac{1}{4}$ and the error costs are equal. Design a matched filter Bayes detector and evaluate its performance.

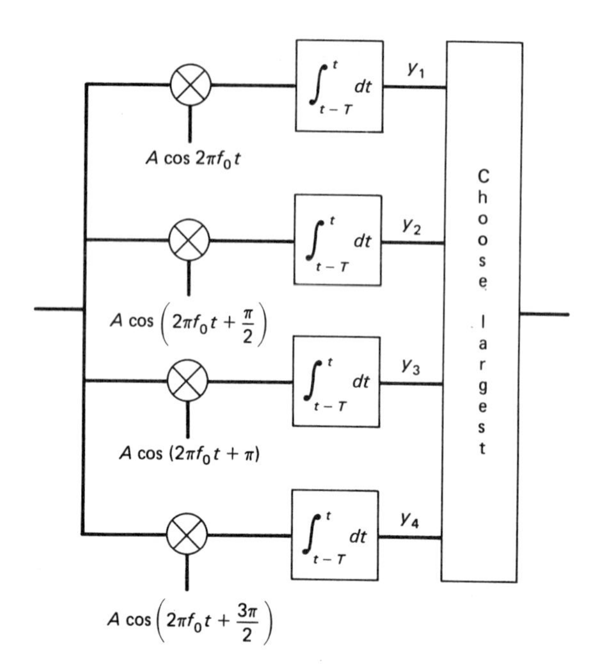

Figure 10.20 Matched filter detector for Example 10-3

Solution. The matched filter detector is as shown in Fig. 10.20. Note that since

$$A \cos\left(2\pi f_0 t + \frac{\pi}{2}\right) = -A \sin 2\pi f_0 t$$

and

$$A \cos\left(2\pi f_0 t + \frac{3\pi}{2}\right) = -A \cos\left(2\pi f_0 t + \frac{\pi}{2}\right) = A \sin 2\pi f_0 t)$$

the detector could have been simplified to contain only two correlators. In that case, a two-step comparison is required—first to find the largest output magnitude and second to test the sign of that output.

One might be tempted to use Eq. (9-27) to evaluate the performance of this detector, but that would yield an incorrect answer, since the four signals are not orthogonal.

Suppose we assume that $s_1(t)$ is being transmitted. The probability of a correct decision is the probability that

$$y_1 > y_2 \text{ and } y_1 > y_3 \text{ and } y_1 > y_4$$

But since $y_1 = -y_3$ and $y_2 = -y_4$, if

$$y_1 > y_2 \text{ and } y_1 > y_4$$

or equivalently, if

$$y_1 > y_2 \text{ and } y_1 > -y_2$$

then y_1 must be positive and therefore greater than y_3. This observation still does not make life simple, since y_4 and y_2 are not independent. In fact, they are completely correlated. Therefore, the probability of a correct decision can be rewritten as

$$\Pr\{y_1 > y_2 \text{ and } y_2 > 0\} + \Pr\{y_1 > -y_2 \text{ and } y_2 < 0\}$$

The two probabilities can be added together, since the events in parentheses are mutually exclusive. Finally, plugging in the probability density functions, we find that

$$\begin{aligned} P_c &= \int_0^\infty \frac{1}{\sqrt{2\pi}\sigma} \exp\left(\frac{-y_2^2}{2\sigma^2}\right) \int_{y_2}^\infty \frac{1}{\sqrt{2\pi}\sigma} \exp\left[\frac{-(y_1 - E)^2}{2\sigma^2}\right] dy_1\, dy_2 \\ &\quad + \int_{-\infty}^0 \frac{1}{\sqrt{2\pi}\sigma} \exp\left(\frac{-y_2^2}{2\sigma^2}\right) \int_{-y_2}^\infty \frac{1}{\sqrt{2\pi}\sigma} \exp\left[\frac{-(y_1 - E)^2}{2\sigma^2}\right] dy_1\, dy_2 \\ &= \int_0^\infty \frac{1}{\sqrt{2\pi}\sigma} \exp\left(\frac{-y_2^2}{2\sigma^2}\right)\left[\tfrac{1}{2}\operatorname{erfc}\left(\frac{y_2 - E}{\sqrt{2}\sigma}\right)\right] dy_2 \\ &\quad + \int_{-\infty}^0 \frac{1}{\sqrt{2\pi}\sigma} \exp\left(\frac{-y_2^2}{2\sigma^2}\right)\left[\tfrac{1}{2}\operatorname{erfc}\left(\frac{-y_2 - E}{\sqrt{2}\sigma}\right)\right] dy_2 \end{aligned} \tag{10-20}$$

With a simple change of variables it becomes obvious that the second integral in Eq. (10-20) is exactly the same as the first, so finally

$$P_c = \int_0^\infty \frac{1}{\sqrt{2\pi}\sigma} \exp\left(\frac{-y_2^2}{2\sigma^2}\right) \operatorname{erfc}\left(\frac{y_2 - E}{\sqrt{2}\sigma}\right) dy_2 \tag{10-21}$$

σ^2 is the variance of any one of the y_i and has previously been shown to equal

$$\sigma^2 = \frac{N_0 T}{4}$$

10.6.1 QPSK Transmitters

One way to generate QPSK is to use a keyed arrangement and either four sources or one source with precise delays. This is illustrated in Fig. 10.21.

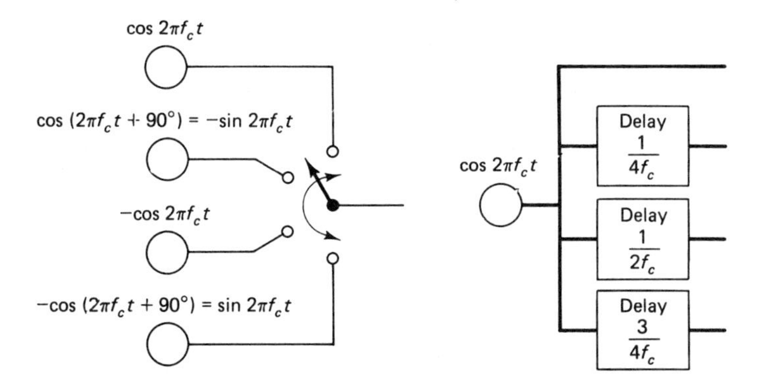

Figure 10.21 QPSK modulators

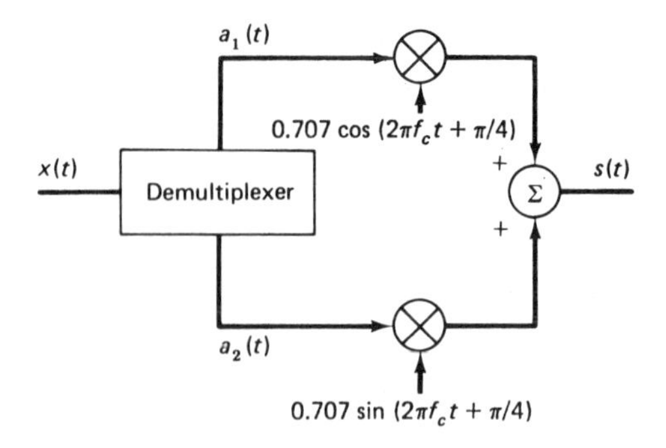

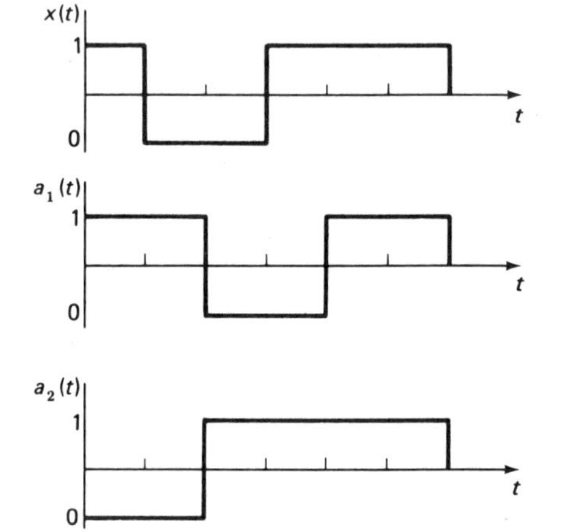

Figure 10.22 Alternate form of QPSK modulator

Another way of generating the QPSK signal is shown in Fig. 10.22. The incoming signal is sampled and held to generate an odd and an even version.

The samples are split by a demultiplexer. The top output is formed by sampling at every second sampling point and holding this for two sampling periods in a sample-and-hold circuit. The lower leg is formed with the other set (odd) of samples. Note that the two signals are shown with coinciding transitions. A delay would be necessary to accomplish this. The demultiplexer can be modeled as shown in Fig. 10.23. The following table gives the signal output for the four possible combinations of a_1 and a_2. This is derived as a simple exercise in trigonometry which you should perform.

a_1	a_2	$s(t)$
1	1	$+\cos 2\pi f_c t$
1	-1	$-\sin 2\pi f_c t$
-1	-1	$-\cos 2\pi f_c t$
-1	1	$+\sin 2\pi f_c t$

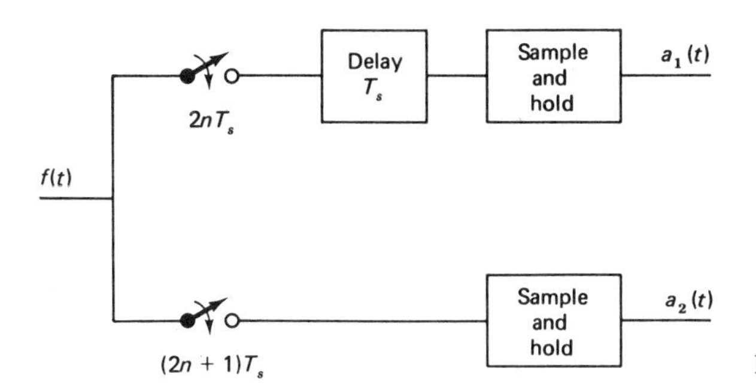

Figure 10.23 Demultiplexer for QPSK

The system therefore develops four possible phases for the four possible 2-bit input combinations.

There is a variation of this system where the odd samples are not delayed. This is shown in Fig. 10.24. This modification would seem to result in returning to the original bandwidth, since transitions can occur at each sampling point. However, by superposition, we can show that the bandwidth is still one-half of the original binary PSK waveform. That is, since the signal is composed of the sum of two signals, each of which has one-half the bandwidth of the original PSK, the summed signal must have the one-half bandwidth. We call this *orthogonal QPSK* (OQPSK).

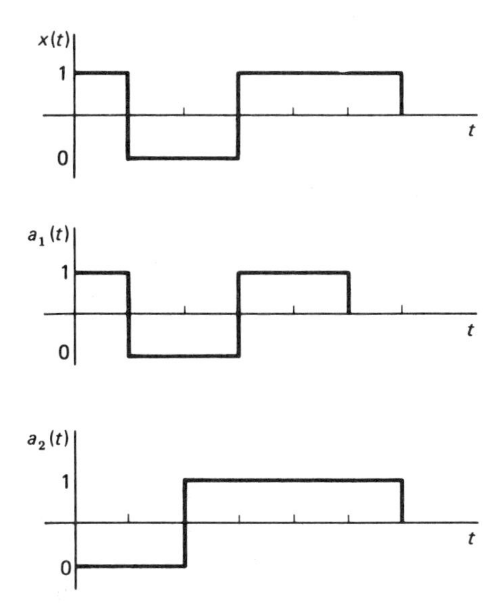

Figure 10.24 Waveforms for orthogonal QPSK generator

An alternate way of viewing the OQPSK modulator is shown in Fig. 10.25. Typical waveforms are illustrated in Fig. 10.26 for the data stream $d_i(t) = 1, 1, -1, -1, -1, 1, 1, 1$.

The advantage of OQPSK over QPSK is that performance improves under some circumstances. This is so because, at each transition, only one of the compo-

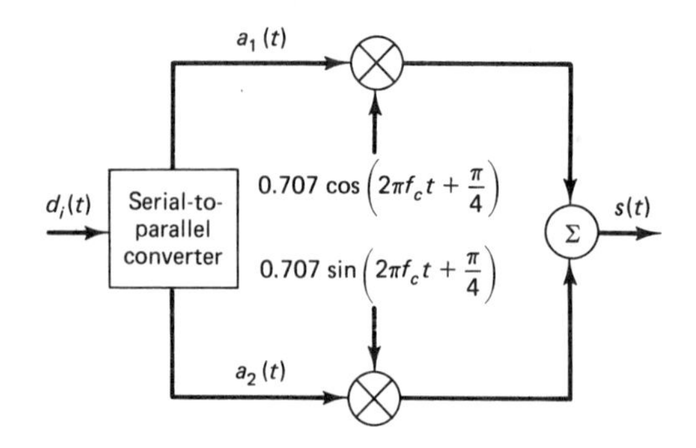

Figure 10.25 Alternate form of OQPSK generator

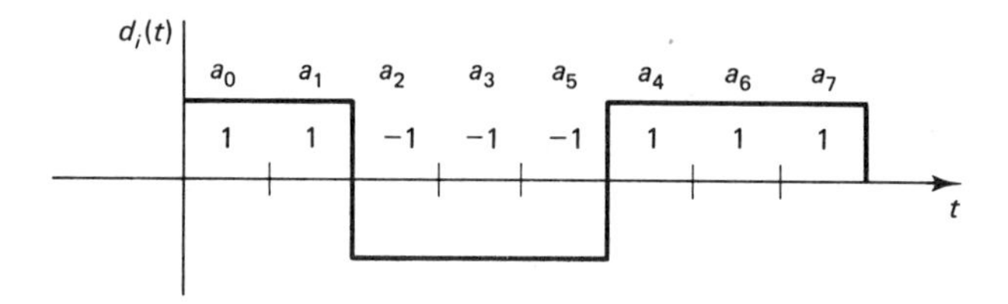

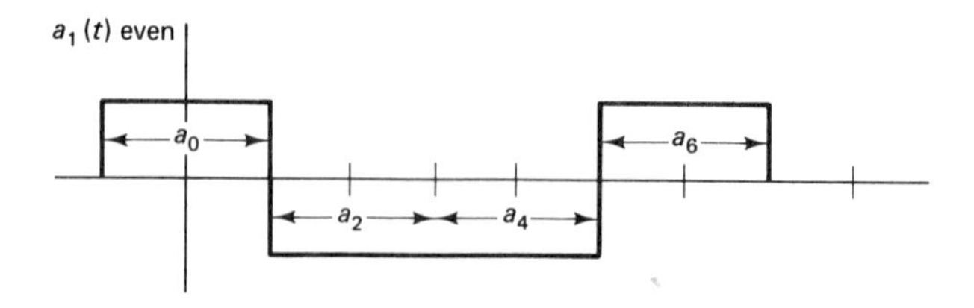

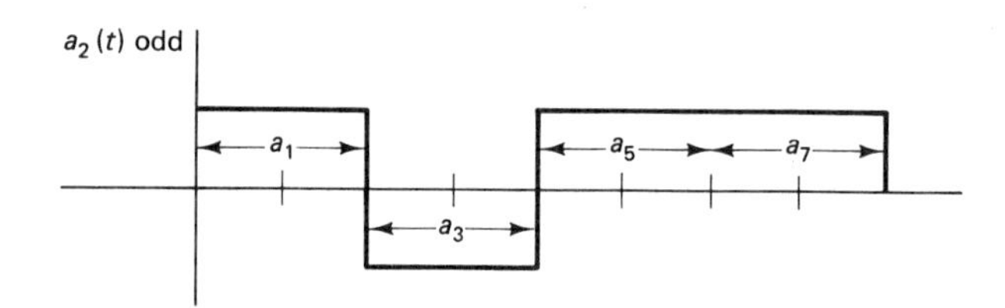

Figure 10.26 Typical waveforms for OQPSK generator

nent signals can change. Thus, the phase can change by only 90 degrees at a transition and not by 180 degrees. This suppression of large phase jumps leads to better performance under certain practical considerations. In particular, the modification can significantly improve performance if the channel is bandlimited or nonlinear (e.g., when a hard limiter is present). This is so because the envelope of the waveform does not go to zero in the OQPSK case, as it does in the QPSK case. The shape of the spectrum of OQPSK after limiting remains close to that of the original spectrum.

10.6.2 Minimum Shift Keying

In either QPSK or OQPSK, abrupt phase changes occur at bit interval transitions. In the case of QPSK, these changes are of 180-degree magnitude, and they occur at one-half of the bit rate. In OQPSK, the magnitude of the changes is 90 degrees, and they occur at a rate equal to the bit rate. These abrupt changes result in sidelobes of the power spectrum. Such signals are distorted by passing them through a band-limited channel (thereby causing intersymbol interference).

OQPSK exhibits improvement over QPSK because the phase transitions are reduced in magnitude. It would seem intuitively reasonable that further reduction in these abrupt phase changes would improve performance even more. *Minimum shift keying* (MSK) is a method of eliminating the abrupt phase changes occurring in QPSK or OQPSK. The instantaneous phase of the MSK signal is continuous. The method involves a trade-off between the width of main lobe of the power spectrum and the sidelobe power content. In fact, the power spectrum main lobe in MSK is 1.5 times as wide as the main lobe in QPSK, but the MSK sidelobes are much smaller than the QPSK sidelobes.

MSK applies sinusoidal weighting to the OQPSK baseband signals prior to modulation. The even and odd waveforms of Fig. 10.26 are therefore modified to become sinusoidal pulses. This is illustrated in Fig. 10.27 for the binary data signal of Fig. 10.26 (i.e., 1, 1, −1, −1, −1, 1, 1, 1).

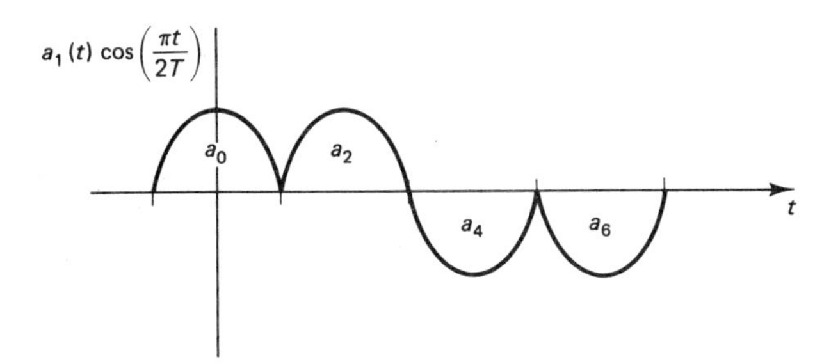

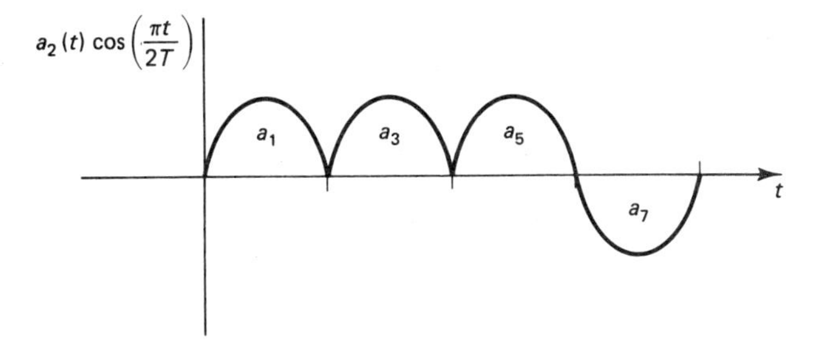

Figure 10.27 MSK generation

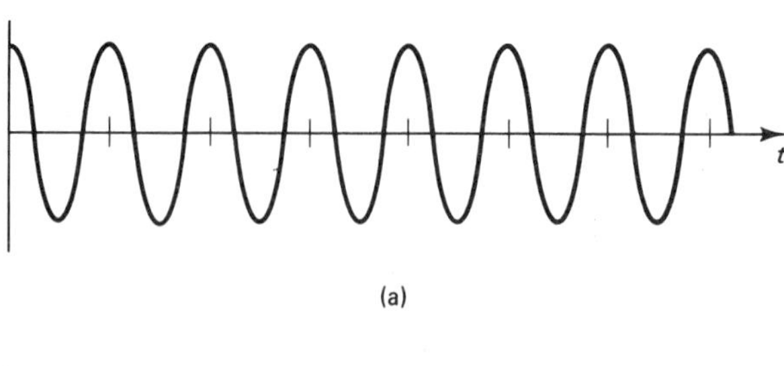

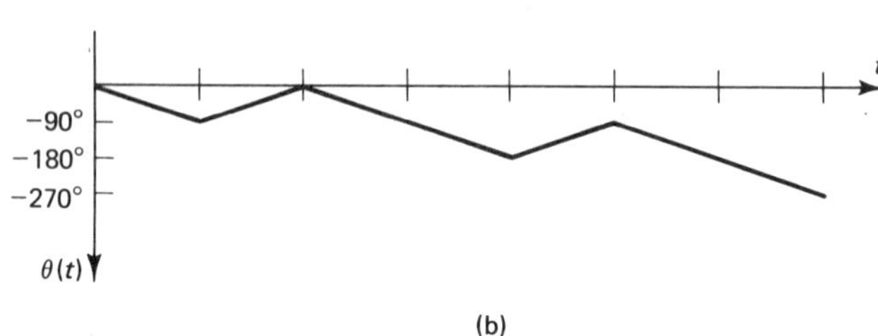

Figure 10.28 MSK waveform and instantaneous phase

We then modulate the two waveforms of Fig. 10.27 using the modulator of Fig. 10.25 to obtain the MSK waveform:

$$s(t) = d_e(t) \cos\left(\frac{\pi t}{2T}\right) \cos 2\pi f_0 t + d_0(t) \sin\left(\frac{\pi t}{2T}\right) \sin 2\pi f_0 t \qquad (10\text{-}22)$$

Figure 10.28 shows a sketch of the waveform represented by Eq. (10-22). Also shown in the figure is a sketch of the instantaneous phase of this time function. The instantaneous phase is found by combining the two terms of Eq. (10-22) into a single sinusoid:

$$s(t) = d_e(t) \cos\left[2\pi f_0 t - d_e(t)d_0(t)\frac{\pi t}{2T}\right] \qquad (10\text{-}23)$$

The instantaneous phase is therefore linear (a ramp), as shown in Fig. 10.28(b).

10.7 MULTIPLE PHASE-SHIFT KEYING (MPSK)

We now generalize upon M-ary phase-shift keying. The previous section considsered the case of $M = 4$. For the general case, the various signals are given by Eq. (10-24).

$$s_i(t) = A \cos 2\pi f_0 t + \theta_i) \qquad (10\text{-}24)$$

where the index, i, takes on values from 0 to $M - 1$. Assuming uniform spacing in a single-space representation, the angles are given by

$$\theta_i = \frac{(2i + 1)\pi}{M}$$

The signal-space diagram is shown in Fig. 10.29.

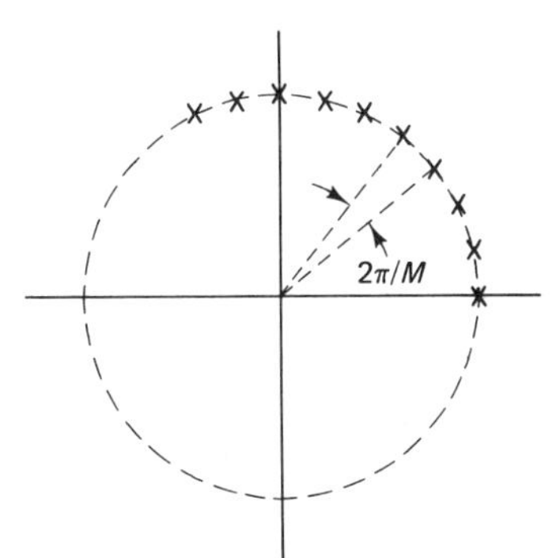

Figure 10.29 Signal-space diagram for MPSK

The transmitter can be viewed as a continuous phase modulator with baseband signal input. The baseband signal takes on one of M possible amplitude levels. Such a waveform can be generated from the binary baseband signal by combining groups of bits and then performing a digital-to-analog conversion. Thus, for example, if $M = 8$, we combine bits in triplets. A serial-to-parallel converter converts each three input bits to a 3-bit binary number which forms the input to the D/A converter. This is shown in Fig. 10.30.

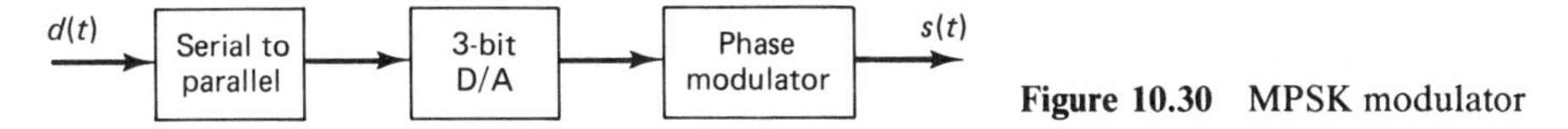

Figure 10.30 MPSK modulator

The corresponding receiver is shown in Fig. 10.31. This receiver forms two components of the point in signal-space representation (the *in-phase* and *quadrature* components of the received signal). These two numbers are processed by an analog-to-digital converter to reconstruct the data sequence. The detector requires a reconstructed version of the carrier in order to operate. Carrier recovery is performed by raising the received input signal to the Mth power and then filtering. This operation produces a sinusoid at M times the carrier frequency, much as the squaring loop provided a double-frequency carrier recovery for the binary case. The resulting sinusoid forms the input to a divide-by-M frequency-divider circuit.

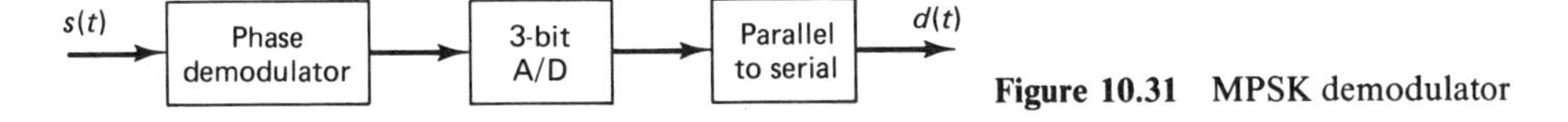

Figure 10.31 MPSK demodulator

10.7.1 Quadrature Amplitude Modulation

The points in the signal space for MPSK lie at equally spaced positions on the circumference of a circle. Improved performance results when these points are separated as widely as possible. *Quadrature amplitude modulation* (QAM) is one approach toward a different point distribution within the signal-space plane.

We present *16-QAM* as an example of this form of modulation. The signal-space diagram consists of 16 points in a uniform square array, as shown in Fig. 10.32. The individual signals are of the form

$$s_i(t) = A_i \cos(2\pi f_0 t + \theta_i)$$

The index, i, takes on values from 0 to 15.

This system combines 4 input bits to produce one signal burst. Both the phase and the amplitude of the sinusoidal burst are modulated. A block diagram of the modulator is shown in Fig. 10.33. The odd-numbered bits in the input data stream are combined in pairs to form one of four levels which modulate the sine term. The even-numbered bits are similarly combined to modulate the cosine term. The modulated sine and cosine terms are combined. The verification that this system creates the signal-space diagram of Fig. 10.32 is left to the problems at the end of this chapter.

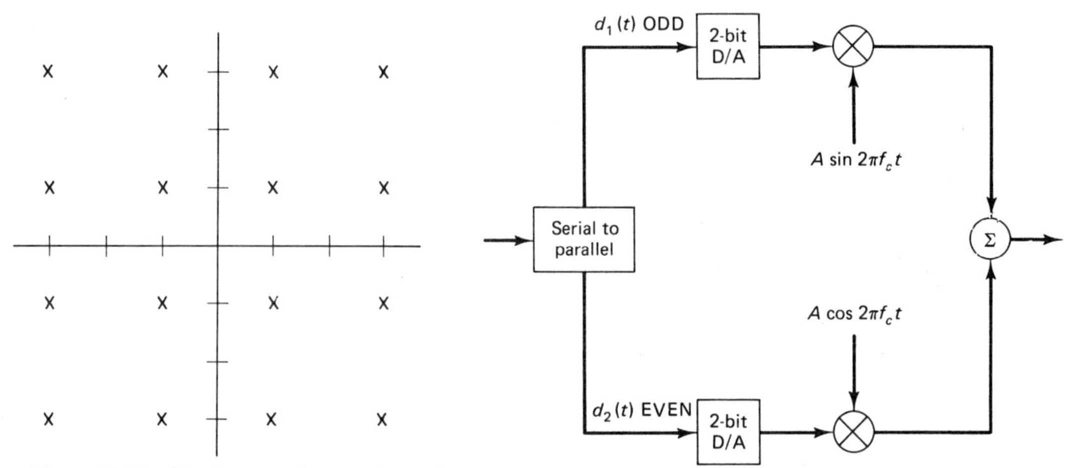

Figure 10.32 Signal-space diagram for 16-QAM

Figure 10.33 16-QAM modulator

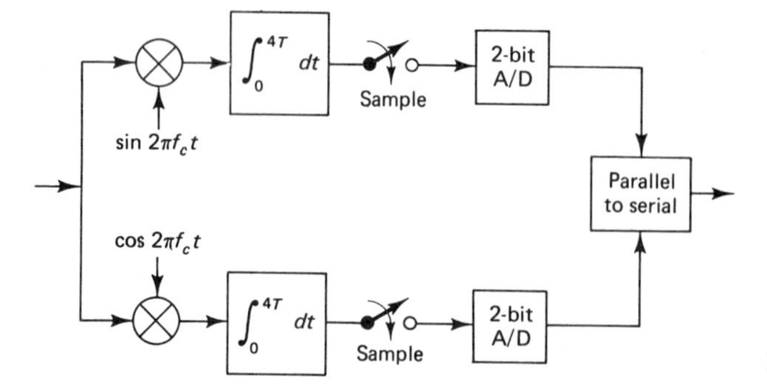

Figure 10.34 16-QAM demodulator

Figure 10.34 shows a demodulator for 16-QAM. This system should be compared to that of the modulator of Fig. 10.33 to verify that the original signal is recovered at the output.

While the 16-QAM system may appear complex as compared to simple binary systems, we shall show in the next chapter that it possesses the potential for providing better performance for a given bandwidth.

10.8 PERFORMANCE COMPARISONS

We now compare the various phase-shift keying systems with respect to bandwidth and bit error rate.

Let us first consider the bit error rate for MPSK systems. The signal set is of the form

$$s_i(t) = A \cos\left[2\pi f_0 t + \frac{2\pi i}{M}\right] \qquad 0 < t \leqslant T$$

where i ranges from 0 to $M - 1$. We can calculate the error probabilities by viewing this in the signal-space plane. The additive noise will cause an error if the phase angle of the received signal varies from the transmitted phase angle by more than π/M in either direction. Thus, in order to avoid errors, the received signal point must lie within the pie-shaped wedge of Fig. 10.35.

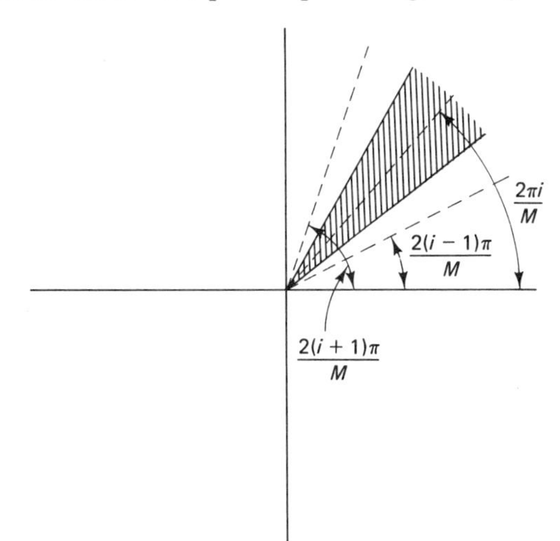

Figure 10.35 Phase region for error-free reception

The angle of the received waveform is found by taking the inverse tangent of the ratio of the quadrature term to the in-phase term. Thus,

$$\theta_r = \tan^{-1}\left(\frac{s_q}{s_i}\right)$$

where

$$s_i = \int_0^T \left\{ \left[A \cos\left(2\pi f_0 t + \frac{2\pi i}{M}\right) + n(t) \right] \cos 2\pi f_0 t \right\} dt$$

$$s_q = \int_0^T \left\{ \left[A \cos\left(2\pi f_0 t + \frac{2\pi i}{M}\right) + n(t) \right] \sin 2\pi f_0 t \right\} dt \qquad (10\text{-}25)$$

$n(t)$ is the additive bandlimited white noise.

The receiver performs nonlinear operations on the Gaussian bandlimited noise. Therefore, closed-form solutions are not possible. The symbol error rate is given by

$$P_s = \frac{M-1}{M} - \frac{1}{\sqrt{\pi}} \int_0^{\sqrt{E/N_0}} \exp(-y^2) \operatorname{erf}\left(y \cot \frac{\pi}{M}\right) dy$$

$$- \frac{1}{2} \operatorname{erf}\left(\sqrt{\frac{E_s}{N_0}} \sin \frac{\pi}{M}\right) \qquad (10\text{-}26)$$

We need to find the *bit* error rate so we can compare this to the result for the binary case. Let us assume that adjacent phase points correspond to binary numbers that differ by only one bit. That is, we assume that a *Gray code* is used in the serial-to-parallel conversion. We shall further assume that the signal-to-noise ratio is sufficiently large that symbol errors of more than one adjacent point are very unlikely. That is, when noise does cause an error, it causes a reception of a signal point which is separated from the correct point by an angle of $2\pi/M$. In that case, and with the Gray coding, one symbol error results in a single bit error. Since each symbol transmits $\log_2 M$ bits, the bit error rate is approximately related to the symbol error rate by

$$P_b = \frac{P_s}{\log_2 M}$$

In order to find the bit error rate for a given signal-to-noise ratio, we numerically solve Eq. (10-26) for the symbol error rate and divide this by $\log_2 M$. Values of the resulting bit error rate are shown in Fig. 10.36 for $M = 2, 4, 8,$ and 16. Note that as the number of bits combined into a single symbol increases, the probability of bit error also increases for a fixed bit signal-to-noise ratio. Alternatively, for a fixed bit error rate, the required signal-to-noise ratio increases as the number of phases increases. For a bit error rate of 10^{-5}, this "signal-to-noise-ratio penalty" approaches 6 dB for every doubling of M (beyond $M = 8$). This may seem undesirable, but remember that trade-off decisions are required in any design. The trade-off in this case is one of bandwidth. Each doubling of M represents a decrease in bandwidth by a factor of 2.

The bandwidth of the MPSK signal is shown in Fig. 10.37. Note that the bandwidth decreases inversely as M increases. Each doubling of M decreases the bandwidth by a factor of 2. This is in contrast with orthogonal MFSK, where an in-

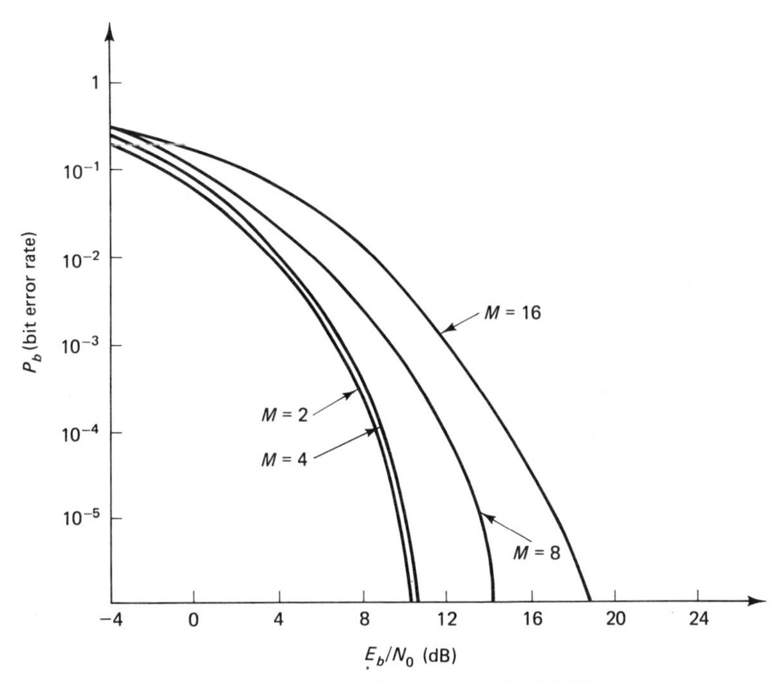

Figure 10.36 Bit error rate for MPSK

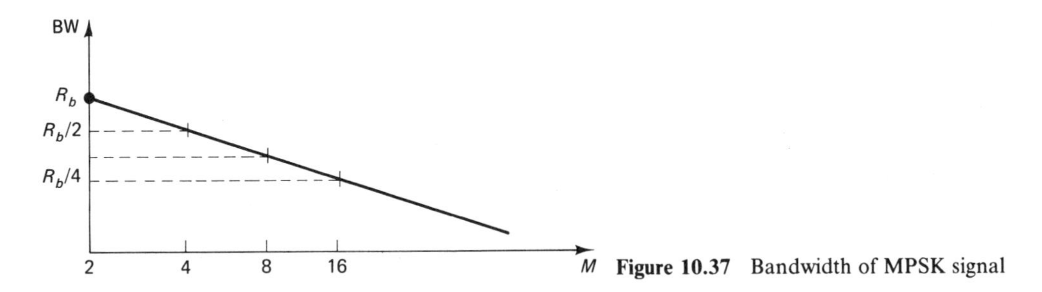

Figure 10.37 Bandwidth of MPSK signal

crease in M increases the bandwidth required for transmission. However, recall that in the MFSK system the bit error rate decreases with increasing M.

The next chapter attempts to compare the various signaling techniques in order to give guidance in the selection of the "best" system for a particular application.

10.9 DESIGN EXAMPLE

Problem

You are asked to design a binary communication system that will be used to transmit data at a rate of 6 kbps. The channel has a total bandwidth of 6 kHz. The additive noise has a power per hertz of $N_0 = 10^{-4}$. Design the system for a bit error rate no greater than 10^{-4}.

Discussion

The bandwidth of the channel poses a limitation in this design. Suppose, for example, you decide to use BPSK. We see (from Fig. 10.3) that a bandwidth equal to the bit rate is sufficient to pass only about 77% of the signal power.

Let us continue the BPSK design in order to arrive at a starting point for later comparison. If we assume that the BPSK signals still possess a correlation coefficient of -1 after bandwidth limiting (this is an approximation), we can account for channel bandlimiting by increasing the signal power at the source.

The next consideration is that of the detector design. If we can achieve coherence, it is not necessary to resort to differential signal formats, and the best possible performance is achieved. Coherence requires that the receiver be capable of reconstructing the carrier sinusoid and using this reconstructed signal in the detector. If the correlation coefficient is -1, a phase shift of 180 degrees is implied, and the transmission is suppressed-carrier. Carrier reconstruction is accomplished using a squaring (e.g., Costas) loop. We'll say more about this later; for now, let us assume that the loop can lock on to the sinusoid within an acceptable period of time. We are then talking about coherent BPSK, and the curves of Fig. 10.17 apply.

In order to achieve a bit error rate of 10^{-4}, the signal-to-noise ratio must be at least 8.4 dB. This translates to a received bit energy of 6.92×10^{-4} watt-sec. The signal power is found by multiplying this by the bit rate (i.e., divide by the bit period) to get

$$P_s = 4.15 \text{ watts}$$

Now since the channel is passing only 77% of the transmitted power, we need to transmit

$$P_s = 5.39 \text{ watt}$$

and the amplitude of the sinusoid must be

$$A = 3.28 \text{ V}$$

This completes the first cut at the design. We now begin to examine the approximations made in this process. The first relates to the bandwidth limiting. We noted that this limiting causes a 23% loss of signal energy, and that the correlation coefficient maintains a value close to -1. However, the more serious effect of the bandwidth limiting is a pulse spreading. This causes intersymbol interference and reduces the effectiveness of the detection scheme in the following intervals. That is, while the correlation coefficient might still be -1, the received pulses spread over more than one bit interval and reduce the difference between signals in the subsequent intervals. We could analyze this effect using the eye pattern or using the equations for peak intersymbol interference derived in Chapter 2. However, in this particular design we have a viable option.

Suppose we use QPSK instead of BPSK. The bandwidth of the channel is then matched to the symbol transmission rate, since we would be transmitting symbols at a rate of 3000/sec. Observation of Fig. 10.36 shows that we suffer a negligible signal-to-noise-ratio penalty in going from BPSK to QPSK. (This is not true as we go to higher levels of MPSK, where the curves start to spread by as much as 6 dB.) We therefore need a signal-to-noise ratio of about 6.9 dB, which translates to a sinusoidal voltage amplitude of 2.87 V. Note that this is an improvement over the 3.28 V needed for BPSK, but more importantly, the intersymbol interference is reduced by a considerable amount.

The above design assumes that we can build a coherent detector. This requires reconstruction of the carrier sinusoid using a squaring loop. These loops require a certain amount of acquisition time in order to lock on to the proper sinusoid. The statement of the problem did not include information regarding the length of the message. If, for example, this is a very short burst of information that is repeated at separated time intervals, we may not be able to tolerate the acquisition delay inherent in the control loops. On the other hand, if we wished to establish the channel and then send data for a long period of time, the setup time is no longer a serious consideration.

The other problem with squaring loops is the possibility of false lock, and the possibility of losing lock. These possibilities increase as we move from BPSK to QPSK. The sensitivity of the data and the length of the message indicate whether these problems are critical.

If the acquisition-time and false-lock possibilities of the loop are deemed unacceptable in a particular design, the alternative is to use a differential detector as shown in Fig. 10.12. We pay a signal-to-noise penalty in the use of this detector, as shown in the curves of Fig. 10.17.

In order to achieve a bit error rate of 10^{-4} using the differential system, we need a signal-to-noise ratio of 8.52. This translates to a sinusoidal voltage amplitude of 3.2 V, which compares to the 2.87 V needed for coherent QPSK. In many systems this may be an insignificant change, and we would opt for the simplicity of the differential system. However, if significant attenuation exists within the channel, this increased amplitude requirement may make it worthwhile to use the coherent system.

Note that in all of the above analysis we refer to signal power and amplitude *at the receiver.* We note also that, while the numbers may be realistic for a wire communication channel, they are orders of magnitude too high for a system which uses the air as a transmission medium.

10.10 SUMMARY

This chapter explores the modulation technique known as phase-shift keying. This technique is capable of providing lower error rates than are possible with FSK.

These lower error rates are achieved with a reduced bandwidth. However, we show that PSK detectors must be coherent, since the phase information must be preserved.

We begin the chapter with a general examination of the signal formats and with a derivation of PSK frequency characteristics. Section 10.3 explores the configuration of modulators and detectors. We present the concepts of squaring and Costas loops and explore their utility in reconstructing the carrier at the receiver.

Section 10.4 examines differential systems which take advantage of coherence between adjacent bit intervals.

The matched filter detector provides the best possible performance, but local reconstruction of the signal formats may not always be practical. This is explored in Section 10.5, where we present optimum performance and the performance degradation resulting from various mismatch errors. We also derive expressions for the performance of incoherent detectors.

The binary results are extended to quadrature phase-shift keying in Section 10.6. Minimum shift keying is presented as a technique that provides bandwidth efficiency by preventing abrupt phase changes. The following section, Section 10.7, generalizes the results to MPSK. We also examine quadrature amplitude modulation and the 16-QAM system.

Many tradeoff decisions must be made in the design of a PSK communication system. Section 10.8 explores some of these factors and presents performance comparisons among the various PSK configurations. We conclude the chapter with a design example. This example emphasizes design considerations involving the required signal-to-noise-ratio and the considerations inherent in choosing between BPSK and MPSK.

PROBLEMS

10.1. Find the power of a suppressed carrier BPSK signal which is bandlimited to a frequency between $f_c - \Delta f < f < f_c + \Delta f$. The cutoff frequencies are related to the bit rate by

(a) $\Delta f = R_b/2$.

(b) $\Delta f = R_b$.

(c) $\Delta f = 2R_b$.

10.2. A bipolar baseband signal is shaped by passing it through a low-pass filter with cutoff frequency $1/T$ (i.e., R_b). The filtered baseband signal phase-modulates a carrier with maximum phase deviation of $\Delta\theta = 180°$. The sampling period is 1 ms. Find the bandwidth of the resulting phase-modulated signal.

10.3. (a) Design a coherent detector for a binary transmission system which transmits the following two signals:

$$s_1(t) = A\cos(2\pi f_c t)$$

$$s_2(t) = A \cos(2\pi f_c t + 90°)$$

The two signals are equally likely.

(b) What factors affect the choice of integration time?

(c) Find the detector output under each hypothesis.

10.4. **(a)** Design a coherent detector for a binary transmission system which transmits the following two signals:

$$s_1(t) = A \cos(2\pi f_c t)$$

$$s_2(t) = A \cos(2\pi f_c t + 180°)$$

Use only one local oscillator and correlator. Assume that the signals are equally likely.

Now let the phase of the local oscillator vary from its correct setting by $\Delta\theta$. Find the output under each hypothesis as a function of $\Delta\theta$.

(b) What values of $\Delta\theta$ would cause decoding errors in the absence of noise?

(c) Show that the detector can be used to detect DPSK for any fixed value of $\Delta\theta$.

10.5. Demonstrate the performance of the DPSK detector of Fig. 10.12 for the data input

1 0 1 1 0 1 1 1 0 0

That is, discuss the form of the transmitted DPSK signal and find the detector output.

10.6. In the DPSK demodulator of Fig. 10.12, the delay path contains an error which causes the delay to vary between $0.9T$ and $1.1T$. Explore the effects of this "timing jitter" upon the demodulator and its performance.

10.7. A binary PSK system transmits the following two signals:

$$s_0(t) = 0.01 \cos(2\pi \times 1000t)$$

$$s_1(t) = 0.01 \cos(2\pi \times 1000t + \theta_1)$$

$$T = 10 \text{ ms}, \qquad N_0 = 2 \times 10^{-7}, \qquad P(H_0) = 0.5$$

(a) Plot the probability of error of a coherent detector as a function of θ_1.

(b) Repeat part (a) assuming that the local oscillators are mismatched in phase by 45 degrees.

(c) Assume that $\theta_1 = 180$ degrees. Plot the probability of false alarm, P_{FA}, as a function of $\Delta\theta$, the phase mismatch.

10.8. A discriminator detector is used to detect the following two signals:

$$s_1(t) = 0.05 \cos(2\pi f_c t + 90°)$$

$$s_2(t) = 0.05 \cos(2\pi f_c t - 90°)$$

$$T = 0.5 \text{ s}, \qquad N_0 = 10^{-4}$$

Assume that the signals are equally likely.

(a) Find the probability of error.

(b) Find the probability of error for a matched filter detector, and compare this to your answer to part (a).

10.9. Design a coherent DPSK system to transmit data at a rate of 10 kbps in AWGN with

$N_0 = 10^{-6}$. The system must achieve a bit error rate less than 10^{-5}. Find the bandwidth of the system and specify the minimum signal amplitude at the receiver.

10.10. Data are to be transmitted at a rate of 10 kbps, and the data occur in bursts of 30-bit length. Therefore, the receiver cannot use a carrier recovery loop because of the associated acquisition time. Design a system to achieve an error rate less than 10^{-5}.

10.11. Given the QPSK system with transmitted signals

$$s_1(t) = 10 \cos(1000t + 30°)$$

$$s_2(t) = 10 \cos(1000t + 120°)$$

$$s_3(t) = 10 \cos(1000t + 210°)$$

$$s_4(t) = 10 \cos(1000t + 300°)$$

The symbol period is 3 ms, and the additive noise has power $N_0 = 0.01$. Assume that all four signals are equally likely.

Design a matched filter detector and find the probability of error and the probability of correct transmission.

10.12. Verify the entries in the table showing the $s(t)$ for various $a_1(t)$ and $a_2(t)$ in Fig. 10.23.

10.13. Find the bandwidth of the OQPSK signal resulting from the signals of Fig. 10.24. Your answer should be given in terms of T, the sampling period.

10.14. Plot the instantaneous phase of an MSK waveform resulting from

(a) a train of static data (all 1's or all 0's).

(b) an alternating train of 1's and 0's.

10.15. Derive an expression for the signals at the inputs to the two A/D converters in the 16-QAM demodulator of Fig. 10.34.

10.16. Design a PSK system to transmit data at a rate of 10 kbps in AWGN with $N_0 = 10^{-6}$. The channel has a bandwidth of 2 kHz. Evaluate the performance of the system you design.

11

Design Considerations

11.1 INTRODUCTION

Chapters 7 through 10 of this text have outlined the four major signal formats used to send digital information. These are baseband, ASK, FSK, and PSK. Within each chapter we analyzed performance and presented curves illustrating the various trade-off decisions that must be made in parameter selection when applying a particular type of modulation.

In most real-life situations, the engineer must apply trade-off considerations not only to choose parameters within a particular modulating scheme but also to choose the modulating scheme itself. Thus, before even setting parameters, a decision must be made whether to use baseband, ASK, FSK, or PSK.

The current chapter combines results of the previous chapters in order to compare the various modulation schemes. Armed with this information and with the theory of the earlier chapters, you should be in a position to intelligently design a system from the bottom up.

11.2 PERFORMANCE COMPARISONS

We characterize performance by the *bit error rate,* and this has been plotted several times within the previous chapters as a function of signal-to-noise ratio.

Figure 11.1 combines results from Chapters 7 and 10 to allow comparison of ASK with PSK. Recall that suppressed-carrier PSK, where the phase shift is 180 degrees, is equivalent to coherent ASK with bipolar baseband.

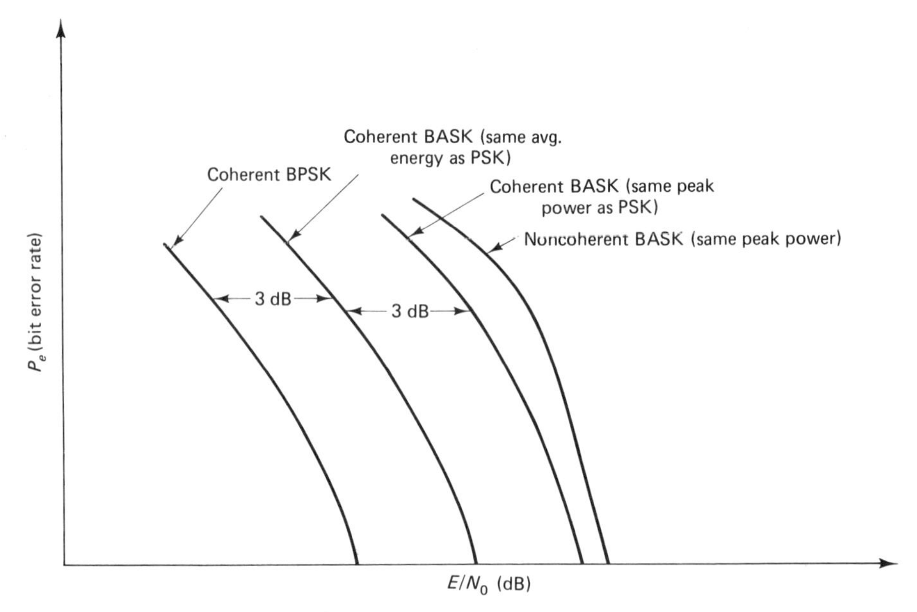

Figure 11.1 Bit error rate comparisons for ASK and PSK

The curve in Fig. 11.1 with the lowest bit error rate represents the performance of coherent BPSK with a phase shift of 180 degrees. Recall that this case was characterized by a correlation coefficient equal to -1, which represnts the best possible performance. The next best performance shown in the figure occurs for coherent binary ASK with the same *average* energy. This refers to the on-off keying case. The signal would have to have twice the peak power of the signals used for PSK in order to have the same *average* energy. Since the correlation coefficient for this case is 0 (as opposed to -1 for the previous case) the signal-to-noise ratio must be twice as large to realize the same bit error rate as obtained with coherent PSK. That is the reason why the first two curves are separated by 3 dB.

If the *peak* signal power in the binary ASK (on-off keying) case is the same as that of the PSK case, the *average* energy is one-half as large. Therefore, if the *peak* power of the ASK case matches the power of the PSK case, an additional 3 dB degradation occurs, as shown for the third curve of Fig. 11.1. This may seem to be an unfair comparison between the two systems, but it often represents the true practical situation where peak power may be limited by the transmitter electronics, or by FCC requirements.

If we transmit using on-off keying with the same peak power as for PSK, but now use an incoherent detector, the rightmost curve of Fig. 11.1 results. Although this represents yet another degradation in performance, the trade-off between performance and the difficulty of reconstructing the carrier at the receiver may make this degradation in performance a fair compromise. Incoherent detectors, such as the peak or envelope detector, are extremely simple to build.

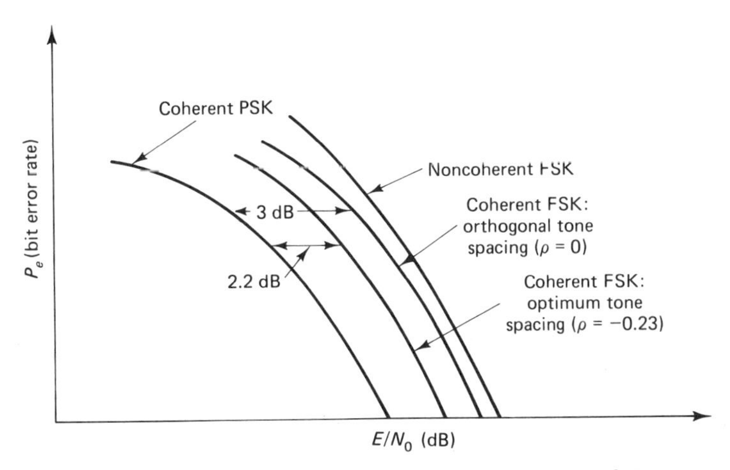

Figure 11.2 Bit error rate comparisons for PSK and FSK

Figure 11.2 illustrates another set of performance curves which allow a comparison between PSK and FSK. Once again, we show the coherent PSK performance curve as the best achievable performance for the binary case. If we use FSK with the optimum frequency separation, we can achieve a correlation coefficient of

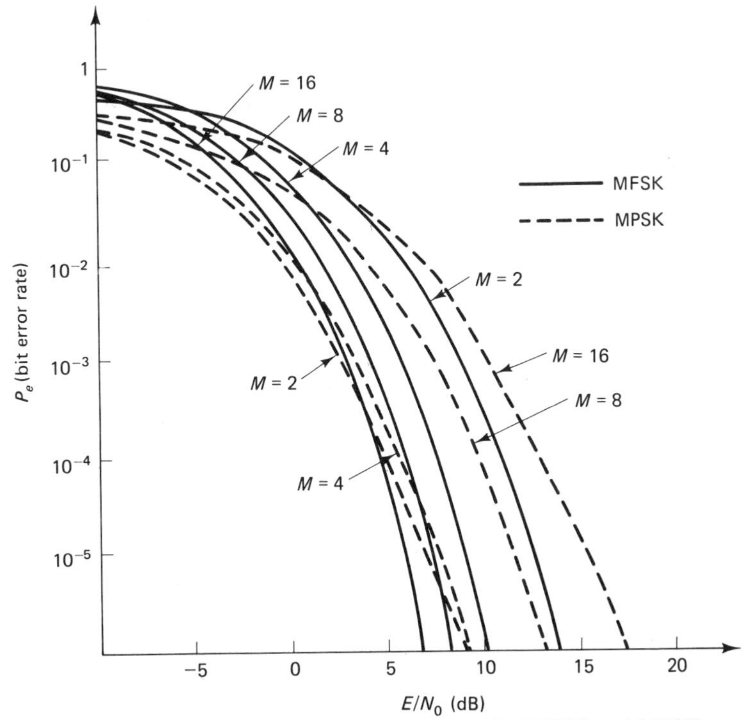

Figure 11.3 Bit error rate comparisons for MFSK and MPSK

−0.23. Recall that this requires perfect phase coherence, and if such coherence is achievable, one would probably opt for phase modulation. Nonetheless, the second curve shows the performance of this FSK system.

The third curve represents FSK with orthogonal tone spacing. This does not require phase coherence and achieves a correlation coefficient of 0. That is the reason why this curve is 3 dB to the right of the optimum PSK curve. In order to obtain the same performance with the FSK system, the signal-to-noise ratio must be twice that of the PSK system.

The final curve in Fig. 11.2 represents the performance of the FSK system where incoherent detection is used.

Figure 11.3 compares MFSK with MPSK. This is a superposition of Fig. 9.31 upon Fig. 10.36. Before reaching conclusions regarding the better performance of MFSK when compared to MPSK (at high signal-to-noise ratios), it is important to remember the bandwidth penalty that is paid with MFSK. This is expanded upon in the next section.

11.3 BANDWIDTH COMPARISONS

Figure 11.4 shows the ratio of bandwidth to the bit rate for various forms of modulation. Note that binary PSK has a bandwidth which is equal to the bit rate, so the ratio is plotted as unity. In the case of PSK and ASK, as the number of bits combined into a single symbol *increases,* the bandwidth decreases proportionately. Thus, for example, for 4-ary phase-modulation communications (4-PSK), the bandwidth is half as much as for binary (BPSK), provided that the bit rate remains

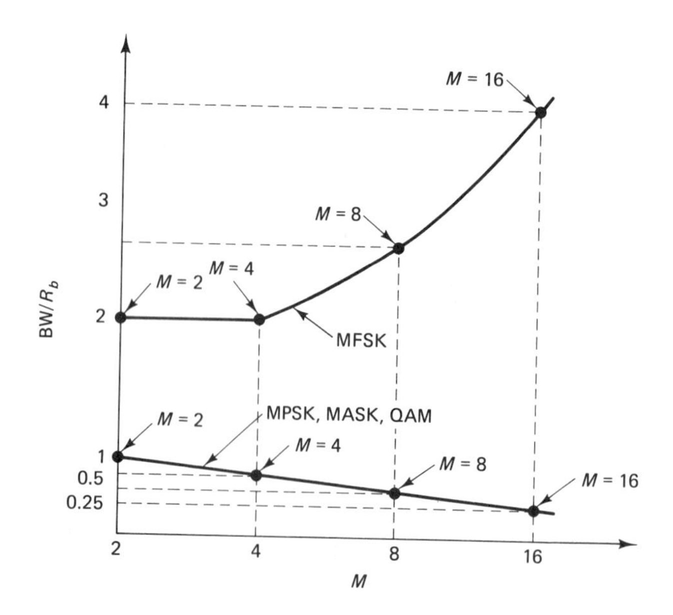

Figure 11.4 Bandwidth-per-bit-rate comparisons for various modulation techniques

constant. This is so because bits are combined in pairs, and the sinusoidal burst doubles in length. This cuts the bandwidth in half. The bandwidth reduces by an additional factor of 2 for every doubling in the number of bits combined into a single symbol.

In the case of FSK, we assume orthogonal tone spacing. For the binary case, the bandwidth is twice what it is for PSK. The bandwidth for the MFSK case depends upon two quantities: the total number of frequencies and the width of the frequency spectrum of each sinusoidal burst. As M increases, the number of frequencies increases, but the width of each individual spectrum decreases, since the sinusoidal bursts become longer. Between $M = 2$ and $M = 4$ these two effects cancel each other. For M greater than 4, the overall bandwidth increases, since it depends more heavily upon the number of frequencies (i.e., the spread beyond the highest and lowest frequencies becomes less significant than the spacing between these two frequencies).

11.4 BPS/HZ COMPARISONS

Until this point, we have been concentrating our system comparative evaluations upon bit error rate and upon system bandwidth. In one class of practical design situation, the bit error rate is specified and we must "optimize" the design for that given error rate. Thus, for example, a particular data communication system may specify a bit error rate of 10^{-7}, and the engineer is asked to design for the maximum transmission rate within a given bandwidth and at a given signal-to-noise ratio.

In situations such as those described above, a parameter of interest is the bit transmission rate per unit of system bandwidth.

Figure 11.5 illustrates the bit rate per unit of bandwidth as a function of the bit signal-to-noise ratio. The curves are shown for an assigned bit error rate of 10^{-4} which represents a typical value for voice transmission. Lower bit error rates are normally required for data transmission. For smaller bit error rates, the curves would move downward, indicating that a slower communication rate is required to reduce the error rate. The top curve on the figure represents the theoretical limit provided by Shannon's theorem. This is derived starting with the channel capacity expression,

$$C = \mathrm{BW} \log_2 \left(1 + \frac{S}{N}\right) \tag{11-1}$$

The signal-to-noise ratio is the ratio of transmitted power to the product of N_0 with the bandwidth. The power is the energy per bit divided by the time per bit. Alternatively, power is the energy per bit multiplied by the bit rate. We can therefore rewrite Shannon's theorem as follows:

$$C = \mathrm{BW} \log_2 \left(1 + \frac{ER_b}{N_0\,\mathrm{BW}}\right) \tag{11-2}$$

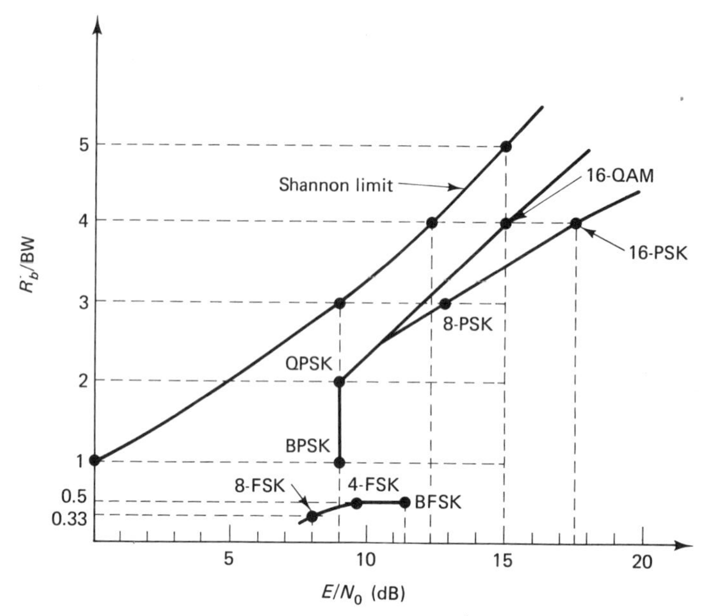

Figure 11.5 BPS/Hz comparisons for bit error rate of 10^{-4}

In Eq. (11-2), R_b is the bit transmission rate in bits/sec and E is the energy per bit. Finally, we let $C = R_b$ for the limiting case where we are communicating at the system capacity (an idealization which does not occur in real systems). Equation (11-2) then becomes Eq. (11-3):

$$\frac{R_b}{\text{BW}} = \log_2\left(1 + \frac{ER_b}{\text{BW}\,N_0}\right) \tag{11-3}$$

Equation (11-3) is solved for R_b/BW and plotted as the uppermost curve in Fig. 11.5. The other points on the figure are obtained from the appropriate curves in Chapters 9 and 10.

Note that PSK comes closer to the Shannon limit than does FSK systems, and that 16-QAM displays an improvement over PSK. Indeed, of the systems we have presented in this text, 16-QAM comes closest to the Shannon limit for bit signal-to-noise ratios above about 10 dB. We hasten to point out that the additional complexity of 16-QAM may cancel the benefits associated with this observation.

11.5 DESIGN CONSIDERATIONS

Another name for this section might be "*The Meaning of It All,*" but we hesitate to use that title for fear of being criticized as being too philosophical. Nonetheless, this section will attempt to tie the major pieces of this text together.

Some readers may be hoping that this section will give a recipe for system design—a type of "step-by-step" procedure. We shall resist this approach for several reasons. First, it reduces the approach to the subject from *education* to *training*. Second, there are no easy answers. The reason that communications is such an exciting field is that it is extremely challenging. The trade-off considerations are manifold, and the inputs to the design process are changing with time. In the past, hardware implementation considerations often provided the most important trade-off input. With advances in electronics—in particular, VLSI and VHSIC—implementation is becoming a less significant consideration in system design. Many systems which were previously of theoretical interest to provide an upper bound for system performance (i.e., they were viewed as being too difficult to implement) are now being used. In these circumstances, we can hope only to give the tools for communication system design, relegating detailed decisions to the ingenuity of the engineer. Many designs are "open-ended," and there may be more than one "correct answer."

In performing a system design, one begins with a set of requirements and constraints for the system. These requirements and constraints might include the following:

1. The required bit transmission rate.
2. The maximum allowable bit error rate.
3. The maximum system bandwidth.
4. The maximum transmitted signal power (and resulting maximum signal-to-noise ratio).
5. The maximum construction cost (i.e., complexity of detector).
6. The maximum power utilization of the detector.
7. The maximum acquisition time of the detector.

We now expand upon each of these requirements and constraints and relate them to the systems considered in this text.

11.5.1. Required Bit Transmission Rate

In a purely data communication system, the required bit transmission rate is a function of how the system is to be used. For example, if a corporation wishes to transmit financial data to its data processing division, it knows how much data must be sent and how soon that data must reach the receiver.

If the data contain redundancies, entropy coding can be used to reduce the number of information bits that must be transmitted. The required bit transmission rate should normally be reduced as much as possible prior to beginning the communication system design. Even this reduction contains trade-off considerations. There are costs involved in entropy coding. First, the actual hardware is not without costs—production, "real estate," and power. Special-purpose chips for entropy encoding are readily available, and while the cost is not high, this cost must be considered in the overall design. Such circuitry takes up space on a circuit board, and while this might seem negligible, it must be considered. Finally, any extra steps which must be performed at the transmitter require power and time. Some remote transmitters (e.g., space-based) have very limited power budgets, and the power used by entropy coders must be considered. Of course, any reduction in bit rate due to entropy coding will result in a reduction of the energy needed to transmit the encoded data.

The required bit transmission rate for a digitally encoded analog signal (e.g., speech) depends upon the frequency content of the analog waveform, the acceptable levels of signal-to-quantization-noise ratio, and the properties of the signal. The frequency content determines the minimum sampling rate. The signal-to-quantization-noise ratio determines the number of bits of quantization needed to achieve the required specifications. The properties of the signal determine whether more sophisticated compression techniques (e.g., ADM and DPCM) can be used to reduce the number of bits that must be transmitted.

We therefore see that the required bit transmission rate is a function of the information that must be transmitted and the amount of work we are willing to do prior to transmission. Although the latter of these two considerations often interacts with other decisions in the design process, we normally consider the transmission rate to be a specified quantity at the beginning of a design process.

11.5.2. Maximum Bit Error Rate

The sensitivity of the data will normally determine the maximum allowable bit error rate. In a speech or video signal, we know (from experience) the minimum signal-to-noise ratio for acceptable reception. For data, the requirements of the system will usually provide a maximum bit error rate. For example, military and financial applications would normally require a smaller bit error rate than would entertainment applications. If we find that this maximum rate cannot be achieved within the constraints to be discussed next, we can resort to "forward error correction" in the form of redundancy coding.

11.5.3. Maximum System Bandwidth

The channel through which the signal is transmitted will normally determine the maximum system bandwidth. This is set both by the transmission characteristics of the medium (e.g., the capacitance and inductance of a coaxial cable) and by regulations (e.g., the FCC allocates bandwidth in transmissions through the air). This constraint, when combined with the required bit transmission rate, will help in the choice of modulation scheme. For example, FSK requires more bandwidth than does PSK. Additional bandwidth savings can be accomplished by moving toward MPSK, but with a penalty being paid in signal-to-noise ratio. If bit error rate or transmitted power are not serious considerations, but the bandwidth is, we will probably move toward MPSK transmission or to 16-QAM. Figure 11.5 provides the basis for choosing a system to achieve the best possible transmission rate for a given bandwidth and signal-to-noise ratio. This assumes that the additional system complexity and acquisition time of control loops in the detector can be accepted as system trade-offs against bandwidth conservation.

We must also look at the signalling format and its relationships to system bandwidth. As an example, we saw that biphase baseband formats have frequency spectra that go to zero as the frequency approaches zero. This is in contrast to NRZ systems, where the frequency spectrum illustrates energy content down to DC. Some baseband channels do not transmit DC (i.e., they may contain AC coupling). Similarly, some carrier systems have problems with signals that have significant energy near the carrier frequency. In such cases, biphase formats would provide advantages over NRZ formats.

11.5.4. Maximum Transmitted Signal Power

The maximum transmitted signal power affects the received signal-to-noise ratio and therefore the bit error rate. The bit error rate increases as we move from binary to *M*-ary systems. Recall that we use *M*-ary systems to achieve bandwidth efficiency. If the bit error rate is providing the overriding constraint, we stay with the binary systems and attempt to use PSK instead of FSK or ASK. In this manner, we squeeze the minimum bit error rate from a given transmitted signal power. Of course, to do this we must be able to design a coherent detector and we must be able to accept the acquisition time, false lock, and drift associated with the control loops in these systems. This would be an ideal situation in which to concentrate upon entropy encoding to reduce the bit transmission rate as much as possible.

Noise reduction is sometimes possible as a means of raising signal-to-noise ratio. If we have control over the transmission path, it might sometimes be possible to shield the system from additive noise. If fading is causing selective reduction of signal-to-noise ratio, frequency or space diversity techniques might be used. That is, if one *signal path* is subject to strong noise or signal fading, we can change the transmission path. If one *frequency range* is subject to such problems, we can vary the frequency range.

11.5.5. Construction Cost

Incoherent detectors are inherently simpler than coherent detectors. Nonadaptive signal encoders are simpler than adaptive encoders. Incoherent ASK detectors are simpler than incoherent FSK detectors. Orthogonal tone spacing is easier to achieve in FSK than is optimum tone spacing. Binary systems are simpler than *M*-ary.

In all of the above observations, we could almost universally substitute the statement "systems with poorer performance are less expensive than systems with better performance." Alas, the more work we are willing to perform, the better job we can do in communicating information. Hardware implementation costs are decreasing rapidly, so it will be a rare design that will be dictated by such costs. Nonetheless, when designing a small number of systems, it is critical that "off-the-shelf" components be used. Major integrated electronics manufacturers can supply data sheets describing the systems which have been implemented in LSI and VLSI chips.

11.5.6. Maximum Power Utilization of Detector

We already discussed maximum transmitted signal power. We now turn our attention to the detector. The more complex the detector, the more power it needs to operate. Phase-lock loops require more power than do passive filters. Quadrature detectors must perform more operations than binary detectors. Some detectors are in remote locations, particularly those used for data acquisition. Some of these detectors rely upon solar power for their operation. In such cases, a careful power budget analysis may place constraints upon the system and may require one to live with a higher bit error rate in order to meet the power requirements.

11.5.7. Maximum Acquisition Time of Detector

Coherent detectors require time to acquire the carrier signal. In the case of transmitted carrier, the phase-lock loop in the detector takes time to lock on to the carrier term in the received signal. In the case of suppressed-carrier, squaring or Costas loops require even more time to lock on to the carrier. This acquisition time may not pose a serious problem in transmissions that are not time-sensitive. That is, we can afford to devote a number of bit transmissions at the beginning of the message to give the detector time to adjust. Other transmissions may be very short or may require transfer of data to begin very soon after transmission begins. In such cases, our options are to do a significant amount of work to provide coherence through external operations (e.g., derive local carriers from an external source locked to the transmitter), or to use incoherent detectors. The incoherent detectors require a higher signal-to-noise ratio to provide the same level of performance as do the coherent detectors. They therefore represent a viable alternative if the signal energy can be readily increased (through increase in signal power or decrease in transmission rate) or if the bit error rate is below specifications by a sufficient amount.

11.6 SUMMARY

This chapter presents overall considerations in system design. The previous chapters each focused upon a particular form of modulation. The current chapter presents information that would be used to choose the best modulation technique for a given application.

Section 11.1 concentrates upon comparisons of performance. The bit error rate is examined. We show that PSK is capable of better performance than ASK and FSK. We also show that coherent systems outperform incoherent systems.

Section 11.2 compares the various modulation techniques with respect to bandwidth. We see that a trade-off is possible between performance and bandwidth. That is, it is possible to use smaller bandwidths for transmission of digital information, but the performance of such systems is degraded relative to the higher-bandwidth systems.

Section 11.3 examines the relationships between bit rate per hertz of bandwidth and signal-to-noise ratio. The results of this section show the trade-off between signalling rate and signal-to-noise ratio. We also illustrate comparisons of the various modulation techniques, and we motivate the interest in more complex systems such as 16-QAM.

Section 11.4 contains a non-formula discussion of the various trade-off considerations involved in a total system design. We discuss various system constraints, including complexity of design, power budgets, and the matching of the system design to the communication requirements.

PROBLEMS

11.1. You are assigned the job of designing a PCM digital system to transmit voice signals which are limited to 3 kHz in frequency. The baseband channel has a bandwidth of 60 kHz, and the required signal-to-noise ratio at the receiver is 12 dB. The AWGN has a power of 10^{-6} watts/Hz. Your design should specify the signalling format (e.g., NRZ-L, biphase-M), the signal amplitude levels, and the block diagram of the receiver.

11.2. Repeat Problem 11.1, but use delta modulation instead of PCM.

11.3. Four messages are to be transmitted on a band-pass channel with bandwidth of 20 kHz. The messages have probabilities $\frac{1}{2}$, $\frac{1}{4}$, $\frac{1}{8}$, and $\frac{1}{8}$. The most probable message occurs 2000 times each second on the average.

(a) Design a system that could be used to transmit this information. Your design should include entropy coding and a specification of all signal formats and voltage levels. Find the probability of message error as a function of the additive noise power.

(b) You must reduce the error probability obtained in the design of part (a) by a factor of 2, and you are already using the maximum signal power. Explore the possibility of using error-correction coding to accomplish this task.

11.4. You design a system using MPSK where eight different phases are used. Theoretical calculations show that the system should achieve a bit error rate of 10^{-5}. After the system is built, the bit error rate is measured at 10^{-4}.

Discuss the steps you would take to determine the cause of the degradation in performance. How would you change the system in order to achieve the lower bit error rate?

11.5. Compare the maximum bit transmission rate of an OOK ASK system with that of BFSK and BPSK. Assume that the peak signal amplitude is 1 V, the AWGN has power of 10^{-6} watts/Hz, and the maximum allowable bit error rate is 10^{-6}.

11.6. Repeat Problem 11.5, but compare OOK with MFSK and MPSK instead of BFSK and BPSK.

11.7. Compare the performance of a BPSK system to that of an MFSK system with the same error rate. In particular, examine the signal-to-noise ratio required to maintain a specified bit error rate, and the complexity of detector design.

11.8. Compare the performance of a BPSK system that contains a phase mismatch to that of a BFSK system with orthogonal tone spacing and incoherent detection. Assume that the bit error rate is 10^{-5}, the transmission rate is 1000 bits/sec, the signal amplitude is 1 V, and the AWGN has power $N_0 = 10^{-5}$ watts/Hz.

11.9. Your task is to design a system to transmit information at 10 kbps. The AWGN has power $N_0 = 10^{-6}$ watts/Hz, and the maximum allowable bit error rate is 10^{-6}.

(a) Design a BPSK system to meet the specifications.

(b) Following the design of part (a), you find that phase coherence cannot be maintained in the detector. You decide to use either DPSK or FSK. Evaluate these two choices with respect to bandwidth and complexity of the detector.

12

Secure Communication

With the exception of a long history of military applications, *security* in communications is a relatively recent concern. In analog nonmilitary communication, people have been somewhat concerned with intrusion in wire communications (e.g., tapping of phones) and unauthorized interception of airwave transmissions (e.g., receiving the microwave signal in phone trnasmission or listening to proprietary satellite broadcasts such as scrambled pay TV signals).

The advent of data communications has extended security needs from the primarily military arena to the consumer market. Corporations are rapidly substituting electronic data transmission for what used to be transmitted via courier service or postal carriers. Data movement is important in diverse financial transactions from electronic money transfer to computer preparation of payroll checks. Security is important not only to prevent unauthorized listeners from gaining access to the signal, but also to prevent alteration of the data.

There are several generic classes of approach toward achieving message security. The oldest method is to confine the signal to a wire transmission channel and to carefully limit physical access to that channel. A second method is to disguise the signal in such a way that an unauthorized listener would not be able to tell the difference between the signal and background noise. The third class of method is to disguise the actual data using coding techniques, so that the message will be incomprehensible to an unauthorized listener even though that listener may correctly receive the signal.

Since the first of these three approaches is fairly obvious, the current chapter concentrates upon the second and third.

12.1 SPREAD SPECTRUM

One goal of the communication systems we have previously studied is to transmit the maximum amount of information with the minimum probability of error in the minimum channel bandwidth and using the minimum signal power. That is an ambitious set of goals—and, indeed, it is not possible to satisfy all of these requirements simultaneously.

Spread spectrum throws out the minimum-bandwidth requirement and intentionally uses a bandwidth that is at least 10 times, and typically over 100 times, the minimum required to send the signal information. If this spectrum spreading is done properly, the transmitted signal looks to the unauthorized listener like wideband noise. As an additional advantage, the spreading could decrease the probability of error (for certain types of noise), much as wideband FM yields higher signal-to-noise ratios than narrowband. We should point out that wideband FM would not be considered to be spread spectrum. To be known as spread spectrum, the frequency spreading must depend upon a signal other than the baseband signal.

We will confine our attention to binary communication. The extension to M-ary is conceptually straightforward and is left as an exercise to the student (a favorite trick when textbook authors become lethargic in the later chapters of a text!).

Suppose that we start with a wideband binary noise signal. This signal randomly varies between 1 and 0 and has a frequency spectrum whose width is proportional to the bit rate. Suppose that we now take a relatively narrowband information signal and modulate the noise with this. Figure 12.1 shows this modulation process for a representative signal and noise segment. Note that modulo-2 binary addition is used for the modulation, so the sum $1 + 1 = 0$ (this is an exclusive-OR logic operation). That is, when the high-frequency noise bit is a binary 1, the result of the modulation is a bit reversal in the information signal. When the noise is 0, no bit reversal occurs. The modulated signal is wideband with a bandwidth approximately equal to that of the noise. Actually, its bandwidth is the sum of the noise and signal bandwidths.

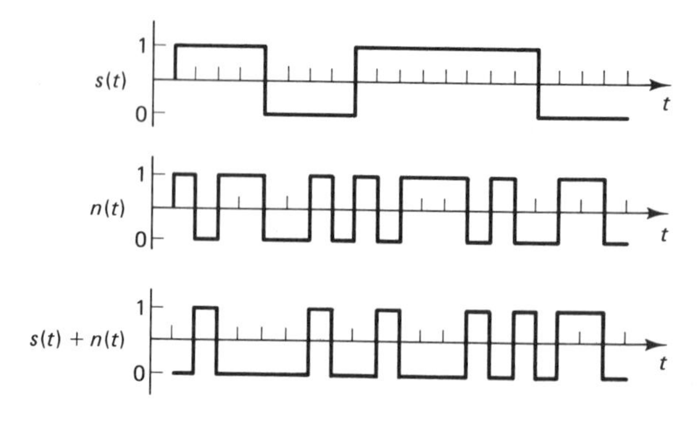

Figure 12.1 Generation of spread-spectrum signal

The effect of the modulation is to reverse blocks of the wideband noise bits. For example, if the wideband noise has a bit rate of 1 Mbps, and the signal has a bit rate of 1 kbps, we are effectively dividing the noise waveform into blocks of 1000 bits length. If the information bit is a 0, the noise block is transmitted without modification. If the information bit is a 1, every bit in the associated noise block is reversed. Since the noise frequency is much higher than the information frequency, the resulting bit sequence still appears to be random.

The signal can be recovered by modulating (exclusive-OR) a second time with the same noise waveform. That is, since binary addition of the information with the noise effectively reverses the information bit whenever the noise is a 1, a second reversal brings the signal back to where it started.

We noted that the additive noise is random. How, then, is it possible to repeat the operation at the receiver? Suppose that instead of being random the noise is *pseudorandom*. The PN sequence is a cyclic code that is generated using a relatively short initializing sequence and a tapped delay line (shift register). Although the PN sequence possesses many of the properties of wideband noise, it is exactly reproducible at remote locations. When the PN code is used for modulation, the process is known as *direct-sequence PN spreading.*

Any form of digital modulation can be spread, but PSK is usually chosen. Because of the constant amplitude of this type of modulation, it is generally more difficult for an unauthorized listener to detect the information signal.

We define a *spread spectrum process gain,* G_p, as the ratio of the signal-to-noise out of the demodulator to the signal-to-noise into the demodulator. It can be shown that this improvement ratio is approximately equal to the ratio of the radio-frequency bandwidth to the information rate.

Example 12-1

Find the process gain for a spread spectrum system using a 1-Mbps PN code for direct-sequence spreading and a 1-kbps information data signal.

Solution. The RF bandwidth of this spread-spectrum waveform is approximately

$$BW = 2 \times 10^6 \text{ Hz}$$

This is the bandwidth of the PN wave out to the first zero of the $(\sin f)/f$ characteristic. The information rate is 1 kbps, so the process gain is

$$G_p = \frac{2 \times 10^6}{10^3} = 2000 = 33 \text{ dB}$$

An explanation for this gain is that the receiver multiplies the received waveform by the reconstructed PN sequence. This spreads the received noise spectrum while it reconstructs (de-spreads) the original data signal. The de-spread signal is put through a low-pass filter, thereby cutting the widened noise by 33 dB.

Clearly, the receiver noise advantage is attributable to the spreading of the received noise by the multiplying wideband PN sequence. Therefore, an interference source (jammer) would be most effective if its power were concentrated in one frequency rather than spread. We see in the next section that this observation does not apply to all types of spread spectrum. In particular, the present result applies only to direct-sequence spread spectrum.

12.1.1 Code-Division Multiple Access (CDMA)

The concepts of direct-sequence spread spectrum can be extended to provide a multiplexing technique different from frequency- or time-division multiplexing. This technique is known as *code-division multiple access* (CDMA).

We begin with the two-channel system of Fig. 12.2. The noise signals, $n_1(t)$ and $n_2(t)$, represent two different PN noise sequences at the same bit rate. We assume that this bit rate is much higher than that of the signals, $s_1(t)$ and $s_2(t)$. This is the same assumption that was made for direct-sequence spread spectrum. Since the two modulated signals are added together prior to transmission through the channel, we would hope that there is some relatively simple scheme for separating the two signals at the receiver. The two modulated signals, $s_1(t)n_1(t)$ and $s_2(t)n_2(t)$, overlap in both time and frequency. Therefore, our usual techniques of time or frequency gating may not be used. However, we may take advantage of the property that the two noise sequences are approximately uncorrelated with each other.

The resulting demultiplexer is illustrated in Fig. 12.2(b). The signal at the output of the first multiplier is given by

$$s_{o1}(t) = s_1(t)n_1^2(t) + s_2(t)n_1(t)n_2(t)$$

Since $n_1^2(t)$ is unity, this can be reduced to

$$s_1(t) + s_2(t)n_1(t)n_2(t) \qquad (12\text{-}1)$$

Because the two noise functions are essentially uncorrelated, their product is a noise waveform at the frequency of each of the original noise waveforms. Therefore, the second term in Eq. (12-1) represents a spread-spectrum signal with a wide bandwidth, while the first term has a bandwidth that is much narrower. A low-pass filter can then be used to effect the same type of noise reduction as experi-

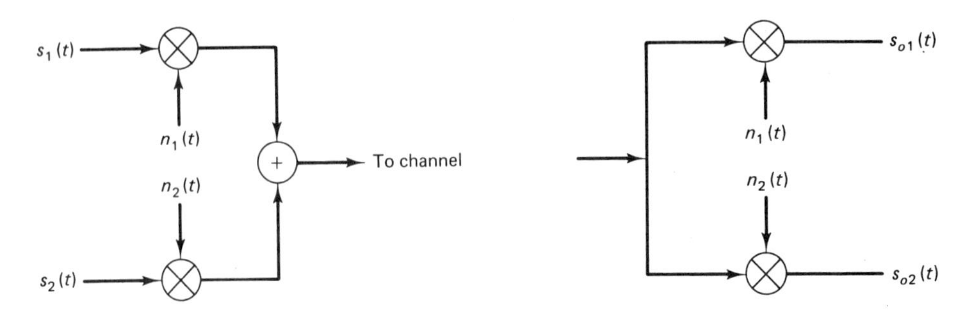

Figure 12.2 CDMA for two channels

enced with broadband jamming in spread spectrum. The amount of crosstalk [i.e., how much of $s_2(t)$ is present in the demultiplexed version of $s_1(t)$] is therefore a function of the ratio of the original signal bandwidth to the spread-spectrum bandwidth. The same analysis applies to the demultiplexed version of $s_2(t)$. We are discussing crosstalk in the context of Eq. (12-1) which illustrates time signals. If we now view data sequences, the receiver becomes a detection device and the consequences of crosstalk would be an increase in bit error rate. If a matched filter detector is used, the crosstalk does not affect performance since the spreading functions $[n_i(t)]$ are orthogonal (i.e., the cross products integrate to zero).

This CDMA system can be extended to more than two signals, but the amount of crosstalk increases with the number of signals. The crosstalk translates into bit errors in the detection of the individual signals. In spite of this increase in error rate, the simplicity of CDMA as a multiplex system applied to spread spectrum has made it attractive for communications in a relatively high signal-to-noise ratio environment where the bit error rate is very small to begin with.

12.2 FREQUENCY HOPPING

Frequency hopping is a second method for spreading the spectrum of an information signal. As the name implies, it is generated by hopping frequencies around. For example, if binary FSK is the chosen modulation technique, the two frequencies used to send a binary 1 and a binary 0 can be shifted over a wide band according to some well-defined rule. The rule must be sufficiently well defined such that the receiver can *dehop* the signal in order to demodulate the data signal. The hopping is often controlled by a PN noise sequence partitioned into groups of bits which define the frequencies.

In a typical system, the carrier (center) frequencies are selected from a set of frequencies equally spaced over the available channel bandwidth. The hopping pattern is a repetitive sequence of, perhaps, seven frequencies. Figure 12.3 shows a typical frequency-hopping system.

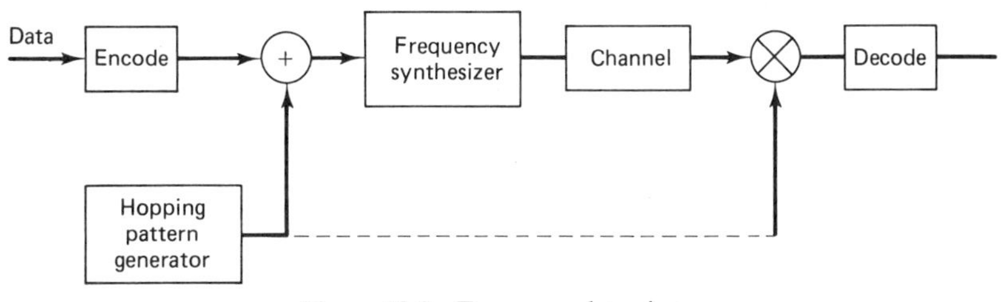

Figure 12.3 Frequency hopping

We start with the data that are configured using a particular form of modulation (e.g., FSK). A *hopping pattern* is selected and fed into a hopping-pattern generator. The output adds to the encoded signal, and then a frequency synthesizer (a

VCO in the simplest form) translates this into frequency bursts of the appropriate frequency. At the receiver, the same hopping pattern must be fed into a frequency synthesizer, which serves as a local oscillator. The output of this oscillator is multiplied by the incoming wave, thereby dehopping the signal, much as a synchronous demodulator demodulates AM.

The receiver often decodes the data for each hopping frequency rather than for each sampling period (each sampling period contains a number of hops). As an example, suppose that the hop rate were three times the bit rate. Then three frequencies would be used for every information bit. The decoder would (usually) use what is known as *majority logic* to decode the data signal. If the decoded information bit was a binary 1 for at least two out of the three frequencies in the period, the decoder would assume that a binary 1 was being sent. Thus, at least two individual bit errors would have to be made to cause an information-bit decoding error.

Example 12-2

Assume that the probability of bit error is 10^{-3} at each of the three frequency hops. Find the probability of information-bit error if majority logic is used.

Solution. The probability that two bit errors occur within a 3-bit interval is

$$P_{e2} = \binom{3}{2}(10^{-3})^2(1 - 10^{-3})$$

The probability of all three bits being in error is

$$P_{e3} = \binom{3}{3}(10^{-3})^3$$

The overall probability of information-bit error is then the sum of the above two probabilities,

$$P_e = P_{e2} + P_{e3} = 3 \times 10^{-6}$$

which is considerably less than 10^{-3}.

Of course, the probability of error during a single-frequency hop is a function of the length of the hop interval. Recall that in the coherent biphase PSK case, the probability of error using matched filter detectors was given by

$$P_e = \tfrac{1}{2}\,\text{erfc}\sqrt{\frac{E}{N_0}}$$

Now, since

$$E = \frac{A^2T}{2}$$

the length of the interval has a definite effect upon the probability of error.

Example 12-3

In a coherent biphase PSK system, the ratio of signal to noise power (N_0) is unity. A matched filter detector with integration time of 2 s is used. Compare the probability of error using this detector to the probability of error for (a) a system that divides the 2-s observation into three equal intervals and applies majority logic, (b) a system that divides the 2-s observation into five equal intervals and applies majority logic.

Solution. The basic matched filter detector has an error probability given by

$$P_e = \tfrac{1}{2}\,\text{erfc}\sqrt{\frac{E}{N_0}} = \tfrac{1}{2}\,\text{erfc}\,(1)$$

$$= 0.0786$$

(a) If three segments of $\frac{2}{3}$ s each are now used, the probability of error in each interval is

$$P_e = \tfrac{1}{2}\,\text{erfc}\sqrt{0.333} = 0.2115$$

The probability of at least two errors out of the three intervals is

$$P_e = \binom{3}{2}(0.2115)^2(0.7885) + \binom{3}{3}(0.2115)^3$$

$$= 0.1153$$

(b) If five segments are now used, the probability of error in a single $\frac{4}{10}$-s interval is

$$P_e = \tfrac{1}{2}\,\text{erfc}\sqrt{0.2} = 0.2623$$

The probability of three or more errors out of the five intervals is then

$$P_e = \binom{5}{3}(0.2623)^3(0.7377)^2 + \binom{5}{4}(0.2623)^4(0.7377) + \binom{5}{5}(0.2623)^5$$

$$= 0.1169$$

Although Example 12-3 would seem to indicate that performance is degraded by going to finer time divisions, it should be noted that this example specified an extremely low signal-to-noise ratio. Practical values would take us out on the tail of the Gaussian distribution, where the probability of single interval errors does not change very much with reductions in the interval length. This is explored more fully in the problems at the back of this chapter.

In reality, the hop rate is chosen to be as rapid as practically feasible to make the system as immune to interference as possible. For example, suppose that the hop time is longer than the time it takes for the signal to make a round trip to a jamming source. If PSK is used, the jammer can receive the hopped signal, modulate it with noise, and retransmit it. If the hop frequency has not changed by the time this altered signal is received, it will tend to cause serious errors.

The most annoying type of interference for a frequency-hopped spread spectrum is *wideband jamming.* Narrowband jamming would affect only one hop frequency and would have a negligible effect upon the overall signal reception. This should be contrasted with the result of the previous section for a direct-sequence spread-spectrum signal.

12.3 CRYPTOGRAPHY

The challenge of cryptography is to change a message into a form that only the intended receiver can comprehend. This must be done in a way that is economical both for the sender and for the intended receiver. At the same time, it must be very difficult (i.e., expensive in time and/or equipment) for an unauthorized receiver to comprehend the message.

Time-sensitive data is now almost exclusively distributed by electronic transmission, while in the past, courier service and hard copy were used. This makes privacy an increasingly critical consideration. Systems must become more and more complex as equipment available to unauthorized listeners becomes less expensive and more sophisticated. We are involved in a highly sophisticated technological "war" with no end in sight. For each major advance in providing secrecy in transmission, there are comparable advances in the art of "code breaking." We are even creating new problems with new transmission technologies. For example, while it used to require some expertise to listen in on wire transmissions, intercepting a satellite broadcast requires no sophisticated "tapping." One need simply locate a receiver within the "footprint" of the satellite transmission. These, plus a number of other, considerations are driving a transition to ever more complex crytographic systems.

Shortly after World War II, public interest and intrigue with elementary forms of cryptography led to children's games, where a child could join a *club.* Coded messages were sent among club members. These messages used a simple form of encryption where each letter of a word was changed to some other letter according to a secret *key.* For example, each time the letter A was encountered, it might be written instead as an M. This one-to-one correspondence or *substitution code* is still seen in puzzle books, where a mathematical operation is replaced by letters and the player must decode this by associating an integer with each letter. One disadvantage of this type of code is that the same input segment always codes into the same output segment. Thus, in coding English text, unauthorized decoding is relatively simple using the *relative-frequency* approach. For example, the most common letter is assumed to be E. Regular occurrences of a three-letter word are probably "THE," and so on. Repeated-letter words also help, such as "THAT" and "LOOK."

Some substitution codes are random. That is, the encypherer decides upon the substitute for each symbol (letter) in a haphazard manner. In such cases, the intended receiver needs the entire equivalence table in order to decypher the received code. Other codes are more systematic. For example, we can *add* 2 to each letter of the alphabet. That is, A becomes C, B becomes D, and so on. In this case, the receiver need only know the rule for substitution.

A second elementary coding technique is *transposition,* where the order of symbols is permuted. As an example, let us examine the phrase

AS A SIMPLE EXAMPLE

If this is permuted pairwise, we have

SA S AMILPE EAXPMEL

Longer permutations are usually used. Transpositions can also be combined with the previously described substitution technique. Experienced cryptanalysts (code breakers) have little trouble with these approaches, especially when a computer is used as an aid. In most cases, code breaking begins with a symbol-frequency analysis.

Higher levels of enciphering use several combinations of permutations in a well-defined pattern. For example, if we permute the foregoing message in alternating pairs and triplets, we get

SA I ASPMEL EAXLMPE

Similarly, we can cycle among various substitution algorithms. For example, for the first 100 symbols we might use a substitution alphabet that changes the letter A into the letter C, and for the next 100 symbols we might use a different alphabet, which changes A to Z. All of this is probably mind-boggling to the amateur puzzle player, but the experienced spy can still break all of these combined techniques, especially when having access to a computer.

Example 12-4

Given that coding is performed by one-to-one substitution. How many combinations would have to be tried to decode a received encyphered message?

Solution. Assuming that the alphabet constitutes the set of symbols, the number of possible substitution codes is 26!, or about 4×10^{26}. In its crudest form, the code breaker would have to try each of these combinations and perform tests on the result to see if the decyphered message "makes sense." This project would appear to require several lifetimes to perform.

If certain knowledge of the language is assumed, the task can be greatly simplified. For example, if 10-letter substitutions can be easily identified using such properties as vowel combinations, frequency of occurrence, letter combinations such as QU, and frequent words such as THE, the remaining 16 substitutions would require about 2×10^{13} combinations in order to try all possibilities. This is still an impractically large number, but it does represent a dramatic simplification from the original task.

The substitution-code approach is not very exciting when the message alphabet is binary (instead of 26-ary as it was for the alphabet), since only two choices exist. We either send the message unaltered, or send its complement changing 1's to 0's, and vice versa. As an alternative, we can combine bits and treat these as words using the techniques described above. For example, the eight possible 3-bit combinations can be permuted. A more interesting technique is to sometimes change a bit and at other times not change it. This can be envisioned as adding a binary sequence to the desired mesage, bit by bit. The added sequence is known as a *key,* since knowledge of it unlocks the code. When the added sequence is equal to 1, the effect of the addition is to change the message bit. When the added bit is zero, the message bit is not changed. For example, if the message is

$$1 \quad 1 \quad 0 \quad 1 \quad 0 \quad 1 \quad 1 \quad 0 \quad 1$$

and the key sequence to be added is

$$1 \quad 0 \quad 1 \quad 0 \quad 1 \quad 0 \quad 1 \quad 0 \quad 1$$

then the transmitted message is found by adding each pair of bits using modulo-2 addition to obtain

$$\begin{array}{r} 1 \quad 1 \quad 0 \quad 1 \quad 0 \quad 1 \quad 1 \quad 0 \quad 1 \\ +\,1 \quad \underline{0 \quad 1 \quad 0 \quad 1 \quad 0 \quad 1 \quad 0 \quad 1} \\ =0 \quad 1 \quad 1 \quad 1 \quad 1 \quad 1 \quad 0 \quad 0 \quad 0 \end{array}$$

The receiver would have to know the key sequence and perform another addition. When the received sequence is added to the key, the result is the original message. Figure 12.4 illustrates the overall encyphering and decyphering system using the key.

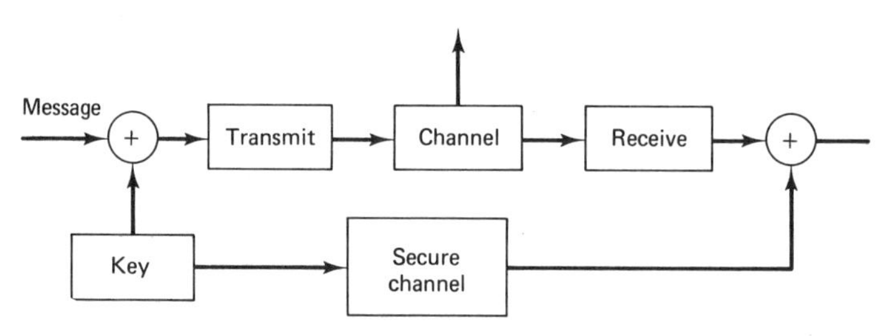

Figure 12.4 Key system for secure communication

Before continuing, we should introduce some terminology relating to the key system of Fig. 12.4. The unencrypted message is known as *plaintext.* Following transformation by the key, the plaintext becomes *cyphertext.* A *channel* can be either *public* or *private,* the latter being effectively restricted to authorized users only.

One way for a system to perform secure communication is to use a random key that is the same length as the message. Unfortunately, this would require that the key somehow be transmitted to the receiver for use in decoding. Perhaps this transmission could be done using the Postal Service if one had sufficient time available. Of course, if only the one message is to be sent and the time is available, one might as well send the message itself by post rather than sending just the key. Even when the same key is used for a number of messages, this random-key method is not very desirable, since for long messages the key must be long (perhaps infinite to cover all possible messages?). The key could, alternatively, be pseudorandom, in which case the receiver would have to know only the initiating sequence. We note that encryption of this type is similar to direct-sequence spread spectrum, except that the PN sequence and the information sequence have the same frequency instead of the noise being wideband, as it was in spread spectrum.

12.3.1 Key Management

It should be obvious that distribution of keys is a major consideration in a cryptographic system. This type of system can be compared to a combination lock. Most people understand the algorithm (e.g., right to first number, left 1+ turns to second number, right to third number), but the actual numbers are carefully guarded. If any unauthorized listener learns the key, security is destroyed.

Keys can be distributed via any very secure system. For example, we have already mentioned the Postal Service as one possibility. This secure channel is usually much too slow to allow use for the actual message and is therefore suited only to key distribution.

Key distribution can be enhanced by requiring multiple keys to operate a system. For example, suppose the key is 64 bits long and is derived by doing modulo-2 addition (exclusive-OR) of two 64-bit sequences. Each of these sequences is given to only one trustworthy employee. At the receiving end, one employee feeds in one of the sequences, and then vacates the premises. The second employee feeds in the second sequence, and the electronics is used to develop the key.

12.3.2 Originator Authentication

So far we have dealt with the aspect of *privacy*. A second important consideration is *authentication.* How can we assure that the receiver knows who is sending the message? If data communications is to effectively replace *hard-copy* communication, we need the digital equivalent of a signature to avoid the possibility of forgery. This signature must be unique and secure. It must be sufficiently protected to eliminate the possibility of disputes (e.g., What happens if I argue with my bank that I never instructed the machine to transfer funds from my checking to savings account?)

A simple illustration of a technique for originator authentication is electronic money transfer using bank machines. The user requests that money be transferred, or that money actually be spit out from the machine into the user's hand. Authentication in this case is usually accomplished by assigning a secret code word to each user. The user (transmitter) must give the correct code in order for the receiver to accept the message (instructions). For additional security, the code word is not sufficient to identify the originator. A coded magnetic card is also required. Thus, security is maintained by requiring two identifications, one of which is difficult to duplicate (the card) even if the other becomes known.

A second technique for secure originator authentication is used in *computer time sharing,* where users having active accounts are assigned *passwords* which are required to gain access to the computer. Security at the sending terminal is assured either by having the password not appear on the screen, or by effectively writing over the word so other users cannot distinguish it. A second concern is security at the receiving end. If passwords are received with no changes, a user might be concerned about unauthorized use of the account by those who process the received message. This shortcoming can be eliminated by having the sending terminal do a *transformation* on the password. It is the transformed word that the system operates upon. If the transformation function is difficult to invert, knowing the transformed password does not yield the actual word. Only the user knows that, so a measure of security is assured.

12.3.3 Data Scrambling

The term *data scrambling* is applied to systems which reorder, or permute, the data sequence. This operation often precedes encryption, since the randomized data stream possesses a number of desirable properties. If the original data stream contains long strings of 1's or 0's (static data), bit or symbol synchronizers can lose lock. Other types of synchronizers, particularly in biphase systems, may exhibit a false-lock condition with certain repetitive data sequences. Randomizing of the data reduces these problems.

The simplest form of randomizing consists of permutations of the incoming data. Figure 12.5 illustrates a simple system for permuting data in blocks of 8 bits. This system is exceedingly simple and, as might be expected, has a number of shortcomings. One is that it can be tricked into revealing the particular permuta-

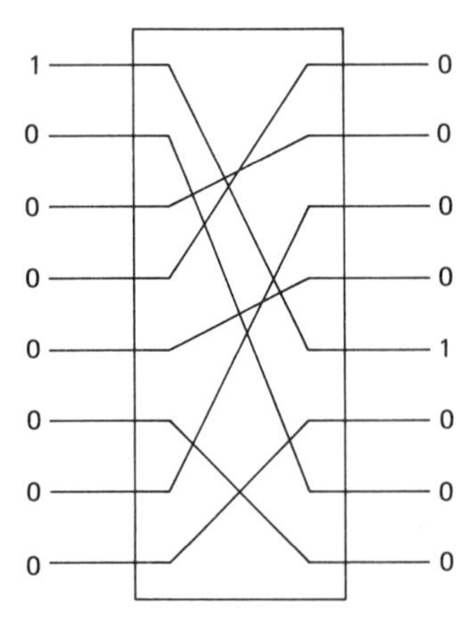

Figure 12.5 Data scrambling using permutations

tion by placing an input data stream with only one 1 in the sequence. This quickly reveals the interconnections to an unauthorized listener.

More complex scrambling systems use shift registers to perform convolutional operations upon the incoming data train. We studied such operations earlier in the context of cyclic encoding. The mathematical operation performed by the system is that of dividing the input information sequence by a generating polynomial. The bits in the output sequence are the coefficients of the quotient of this division.

There are advantages in standardizing the scrambling algorithm, since equipment can be purchased "off the shelf." The CCITT (Comité Consultatif Internationale de Telegraphique et Telephonique—we discuss the role of this body in detail in the next chapter) has issued a series of recommended standards for data scrambling within Section V of their standards. A representative standard is *Recommendation V.27,* which is illustrated in Fig. 12.6. The generating polynomial for this system is derived by viewing the taps on this delay line (each X^{-1} block represents a delay by one bit period). It is given by

$$1 + X^{-6} + X^{-7}$$

The result is that the transmitted data are randomized over a sequence length of $2^7 - 1$, or 127 bits. The adders in the figure are modulo-2 addition devices.

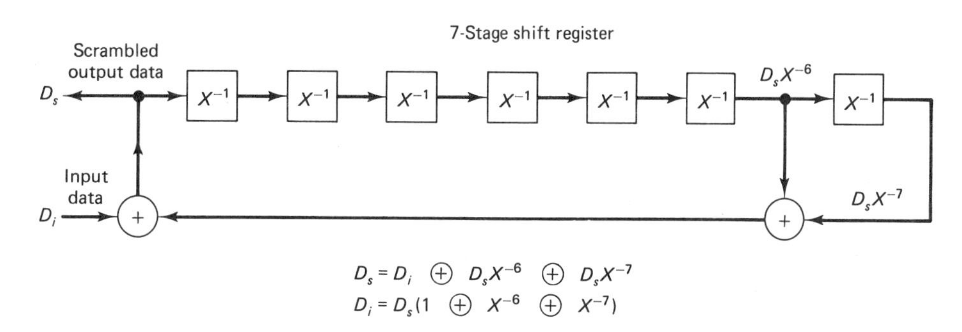

Figure 12.6 CCITT recommendation V.27 for data scrambling

12.3.4 Block vs. Stream Ciphers

Encryption techniques can be categorized into two broad areas: *block* encryption and *data-stream* encryption. This is somewhat analogous to block and convolutional codes, which we discussed in Chapter 4.

In block encryption, the plaintext data sequence is divided into blocks of a particular fixed length. Each of these blocks is encrypted separately. Thus, if a block is N bits long, there are 2^N possible ways to encipher it. Of course, the algorithm that maps a block of plaintext into a block of ciphertext must be unique. That is, we must be able to go backward from the ciphertext into the plaintext.

While it is not strictly required, the length of the ciphertext block is usually the same as that of the plaintext block.

Block ciphers can be viewed as digital substitution ciphers. We can view a dictionary with 2^N entries. We simply look up the replacement for the particular plaintext block.

Stream ciphers are generated bit by bit as the plaintext is generated. This is done by performing addition of the stream with a *keystream*. The keystream either can be independent of the plaintext stream (i.e., essentially random, but known at the transmitter and receiver) or can be derived from the plaintext.

12.3.5 Public Key Systems

The transmission of keys from party to party is somewhat awkward if one desires secure data communication to be as feasible some day as analog telephone communication is today. Wouldn't it be wonderful if there were a way to dial up anybody connected to the phone system and, without making any advance arrangements with that party, proceed to send secure data? *Public key cryptosystems* provide such a capability of random-access communication.

The difference between public key and nonpublic key is that in the public system, the original key is replaced by two keys, one used by the sender and one used by the receiver. Assume that the key used for encryption is denoted K_1. In order to decrypt, a different but related key, K_2, is needed. Although the two keys certainly must be related to each other, it is critical that an unauthorized party who obtains one of the keys will not find a practical method for deriving the second one. Thus, the keys can be published in a type of phone directory. If a source wishes to transmit to a receiver, it looks up the key for that receiver and chooses the complementary key for transmission, using some rule unique to that source. An unauthorized listener will know the rule for the source (this is published in the same directory), but without knowing which receiver is being sent to, the decoding key cannot be easily found.

Until we get to specific examples, it is probably difficult to envision an enciphering operation where, if an unauthorized intruder knows the transmission key, the deciphering key cannot be easily found.

Suppose that each person generates a pair of keys which are complementary. That is, if one of the keys is used for encyphering, the other can be used for deciphering. Further assume that knowing the enciphering key, one cannot easily find the deciphering key. The person now advertises the encyphering key in a phone-type directory, but keeps the deciphering key secret. Anyone wanting to send to that receiver simply uses the published code. We would call the enciphering key the *public key* and the deciphering key the *private key*.

Any number in any base system can be used as a key. For example, suppose that the key is 121 in base 10. If binary coding is being used, we can rewrite this as 1111001. The message can then be partitioned into 7-bit groupings, and this key added to each group. Alternatively, we can replace the scalar key number by a key *vector*. The first partitioned message segment would be added to the first element of the vector, the second segment to the second, and so on.

Having introduced the subject in broad terms, we will now illustrate two specific public key examples. In each of these, it will not be the actual key that is made public, but some function derived from that key.

Suppose that a number is generated from two random numbers, x_1 and x_2, and a constant, a, as follows:

$$K = a^{x_1 x_2}$$

Further suppose that communicator A knows only the random number x_1, and B knows only the random number, x_2. Both of these numbers are needed to derive the key, so if both are published in the directory, any unauthorized listener could derive the key. Suppose instead that a^{x_1} is published associated with person A, and a^{x_2} is listed for B. Then if A wants to send a message to B, B's code (a^{x_2}) is looked up in the directory, and A raises this to the power x_1, the secret number that only A knows. This forms the key, $a^{x_1 x_2}$. An unauthorized listener would have access to a^{x_1} and a^{x_2} from the directory, but to form the key the following operation would be required:

$$K = a^{x_1 \log_a(a^{x_2})}$$

This involves computing logs, which is not terribly difficult. However, if all operations are now modulo-q, where q is a prime number, the log then becomes more difficult to perform.

Functions of this type are known as *one-way functions*. A simple class of examples is the polynomials of the form

$$y = \sum_n a_n x^n$$

Knowing x, it is easy to find y, but knowing y, it is by no means easy to find x.

Example 12-5

In the public key example above, let $x_1 = 7$, $x_2 = 4$, $a = 2$, and $q = 5$. Evaluate the various keys.

Solution. The public key for x_1 is given by

$$a^{x_1} \bmod q = 2^7 \bmod 5 = 128 \bmod 5 = 3$$

The public key for x_2 is

$$a^{x_2} \bmod q = 2^4 \bmod 5 = 16 \bmod 5 = 1$$

The key for the code is then

$$K = (a^{x_1})^{x_2} \bmod 5 = (a^{x_2})^{x_1} \bmod 5 = 1$$

We note that with $q = 5$, there are only five possible keys. For that reason, a practical system would use a much higher value of q. q must, however, be a prime number to yield a unique solution.

12.3.6 Trapdoor-Knapsack

Another public key system uses a technique of the *trapdoor–knapsack* type to generate a vector key. The word *trapdoor* is used to refer to a complex computational situation wherein knowing some clue (e.g., the location of a trapdoor in an enclosed room) greatly simplifies the problem.

The (classic) *knapsack* problem gives 10 items, including food, clothing, and first aid, which might be placed in a knapsack. The weight of each item is also given. Then the problem is that given the exact weight of the knapsack contents, which of the possible sets of items is in the sack? An alternative statement of the knapsack problem specifies a one-dimensional knapsack of a certain length, and rods of differing lengths. The problem then asks for the correct combination of rods to exactly fill the knapsack.

Example 12-6

Given a collection of seven items weighing 1, 2, 5, 8, 11, 17, and 21 pounds, find the combination(s) of items that would exactly fill a knapsack with a capacity of 47 pounds.

Solution. Since the sum of all weights excluding the heaviest is less than 47 (it is 44), clearly the knapsack must contain the 21-pound item. The remaining items must add to a weight of 26 pounds. Since the sum of the 5 lightest items is 27, it is not necessary to include the 17-pound item. By successive comparisons, we find that the following two combinations of items solve the problem:

$$2, 5, 8, 11, 21$$

$$1, 8, 17, 21$$

It should be noted that not all weights are possible with this choice of items. For example, there is no combination of items that would add to 4 pounds.

If there were five possible items, there would be 32 possible combinations (i.e., 5 possibilities of one item, 10 of two items, and so on, including the choices of all five items and no items). Clearly, as the number of possible items in the bag increases, the number of combinations an organized search would have to try in order to solve the problem increases dramatically. Therefore, knowing the total weight of the knapsack does not really help someone to say much about individual items. But knowing some of the items, or secret rules for choosing sets of items, would help someone solve the problem.

One way to simplify the knapsack problem is to specify that each item's weight must be greater than the sum of the preceding weights. Thus, as an example (we will use this example throughout the discussion), consider six items where the weights are written in vector form as $(1, 2, 5, 10, 25, 51)$. Note that each weight is greater than the sum of the previous weights. A sequence of this type of sometimes called a *superincreasing function*. You are told that the total weight is 64. Starting with the heaviest item, you could work your way backward to find

$$64 = 51 + 10 + 2 + 1$$

Thus, in vector form, multiplication of the vector weights by (110101) yields 64. This binary vector could be used as the enciphering key, and the sum (64), could be published in a directory. Of course, this is much too simple and anyone can find the key.

We illustrate next a more practical scheme using the same weight vector. We start by choosing a base for modulo arithmetic. This base must be greater than the sum of the weights, so we shall choose 100 for this example (the sum of the six weights in this example is 94). It is then necessary to find any number and its inverse (in this base system). The numbers 11 and 91 are an inverse pair, since $11 \times 91 = 1$ modulo 100. We now multiply each of the weights by the first number, 11, doing the multiplication modulo-100 to yield

$$(11, 22, 55, 10, 75, 61)$$

This vector is published in the directory, whereas all of the other numbers (11, 91, 100) are kept secret.

Now if someone wants to send to you, your vector is checked in the directory, and a dot (inner) product is taken between the message vector and that vector. For example, if 101101 is to be sent, the dot product is given by

$$11 + 0 + 55 + 10 + 0 + 61 = 137$$

The transmitter then sends 137 to you. At the receiver, you multiply this by the inverse of 11 (91) to get

$$91 \times 137 = 12467 = 67 \text{ modulo } 100$$

You then examine your original weights, which nobody else knows,

$$(1, \quad 2, \quad 5, \quad 10, \quad 25, \quad 51)$$

to find that the required multiplication vector to achieve 67 is (101101), the message!

Of course, in the real system, instead of using a knapsack with six possible items, much larger numbers are used.

Example 12-7

Using a trapdoor–knapsack system with unmodified weights,

$$(1, \quad 2, \quad 6, \quad 11, \quad 23, \quad 50, \quad 94)$$

find the public key. Also demonstrate the process by sending 1010101.

Solution. It is first necessary to choose a base for the modulo arithmetic. Since the sum of the weights is 187, any number larger than this is acceptable. Choosing 188, we need find a complementary set of numbers. The numbers 7 and 403 are complementary, since

$$7 \times 403 \bmod 188 = 2821 \bmod 188 = (188 \times 15) + 1 \bmod 188$$
$$= 1$$

The public key is found by multiplying each weight by 7 and publishing these numbers modulo 188. The public key is then

$$(7, \quad 14, \quad 42, \quad 77, \quad 161, \quad 162, \quad 94)$$

If someone wishes to send the binary message 1010101 to you, they look up your public key and send the inner product of that key with the binary message. Thus, they transmit

$$7 + 42 + 161 + 94 = 304$$

To recover the message, you multiply this received message by 403, using modulo-188 arithmetic to get

$$304 \times 403 \bmod 188 = (651 \times 188) + 124 \bmod 188 = 124$$

The result, 124, represents the knapsack weight. Looking at the original weights to determine the items in the sack, you find that

$$1 + 6 + 23 + 94 = 124$$

yielding the vector

$$1 \quad 0 \quad 1 \quad 0 \quad 1 \quad 0 \quad 1$$

as the correct message.

12.3.7 The Data Encryption Standard

We have presented a number of tools for use in encrypting a message prior to transmission. In general, the simpler the system we use, the easier it is for an unauthorized listener to either intercept the message or to *spoof* the system. It is therefore desirable to use reasonably complex combinations of the various techniques in a particular application.

Unfortunately, the more complex the scheme, the more development time and resources must be devoted to configuring the system and implementing it in hardware. This provides the impetus for a standard which is generally adopted. The use of a standard allows hardware to be independently developed. Such hardware can be made available "off-the-shelf."

The *Data Encryption Standard* (*DES*) was developed by IBM and certified to be "secure" by the National Security Agency. It was adopted as a federal standard in 1976 and approved by the National Bureau of Standards in 1977. The standard has been reduced to readily available hardware. In fact, it has been implemented on integrated circuits (ICs) by many manufacturers.

The DES can be configured for both block and stream enciphering. The only thing that differs between users is the particular key, which must be carefully guarded to prevent unauthorized deciphering.

When the DES is used for block enciphering, the key is 64 bits long. Of these, 56 bits represent the secure key sequence, and the remaining 8 bits are developed as parity bits. Each of these parity bits provides odd parity within an 8-bit sequence.

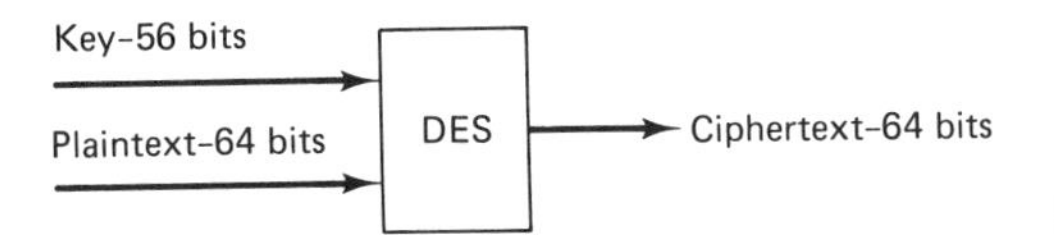

Figure 12.7 DES for block encyphering

The use of the DES for block enciphering is shown in Fig. 12.7. In this mode, the DES is known as an *electronic code book* (*ECB*), since a block code can be viewed as a substitution dictionary which replaces each 64-bit plaintext sequence by a particular 64-bit ciphertext sequence. The operation is configured such that a change of only one bit in the key will propagate throughout the ciphertext in a manner which causes each bit to have approximately a 50% change of changing. Thus, if the wrong key is used, an average of half of the deciphered bits will be in error.

In the block mode, the same plaintext always results in the exact same ciphertext, provided the key is not changed. The DES performs its operations by configuring a complex system out of building blocks. Each *standard building block* (SSB) consists of a 48-bit key (derived from the 56-bit input key) which operates upon the 64 plaintext bits. The 64 plaintext bits are partitioned into two halves, and the key subset operates upon each of these halves to provide the output sequences. This is illustrated in Fig. 12.8. Sixteen standard building blocks are cascaded to form the block enciphering system. A parallel operation takes place at the receiver. This cascading is illustrated in Fig. 12.9.

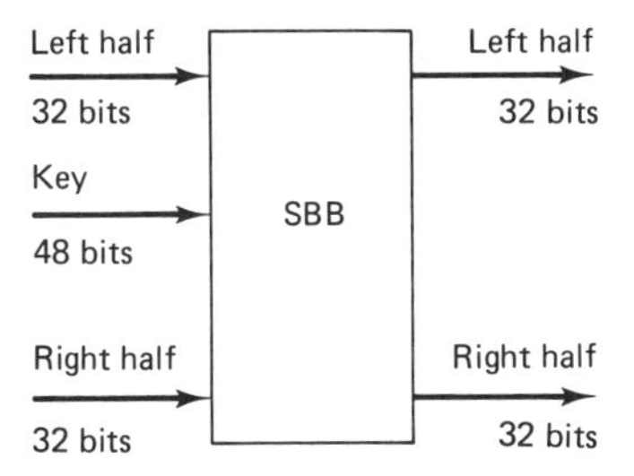

Figure 12.8 Standard building block of DES

The basic ECB operation can be improved by reconfiguring the system such that identical input blocks will not result in identical ciphertext blocks. This is done by providing a type of "interblock interference." That is, we have the various blocks interact with each other to provide a type of transient response, thus giving the system memory. This mode is known as *cipher block chaining* (*CBC*). It is implemented by providing for feedback so that we can add each plaintext block to the ciphertext block from the previous operation. Note that an error in transmission will now propagate through two blocks, possibly producing an unacceptably high probability of error. The CBC mode is often used in message authentication.

The DES can be configured to provide for a stream cipher. In this mode, the DES generates the bit stream which is added to the plaintext on a bit-by-bit basis. One particular configuration, known as *cipher feedback* (*CFB*), is illustrated in Fig. 12.10. In this figure, K is any number between 1 and 64. These K bits are added to K

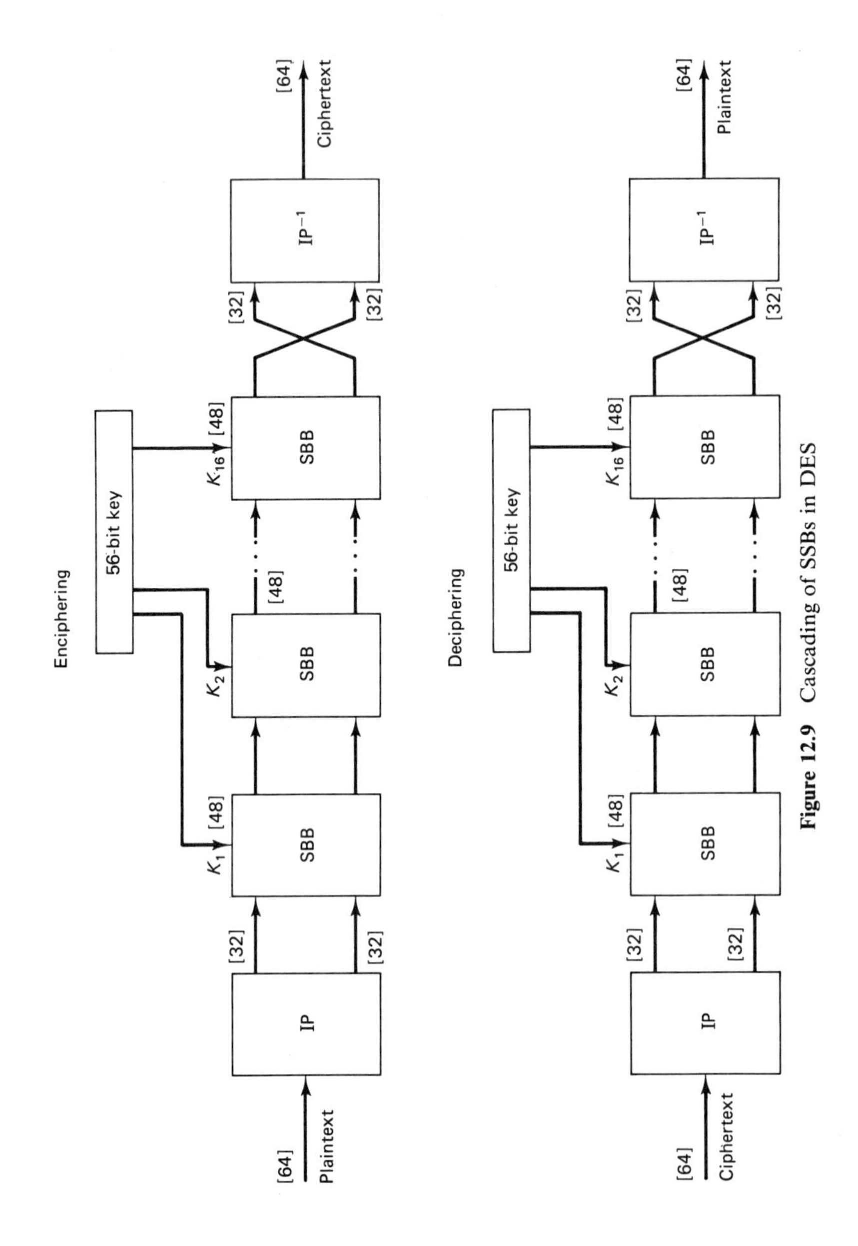

Figure 12.9 Cascading of SSBs in DES

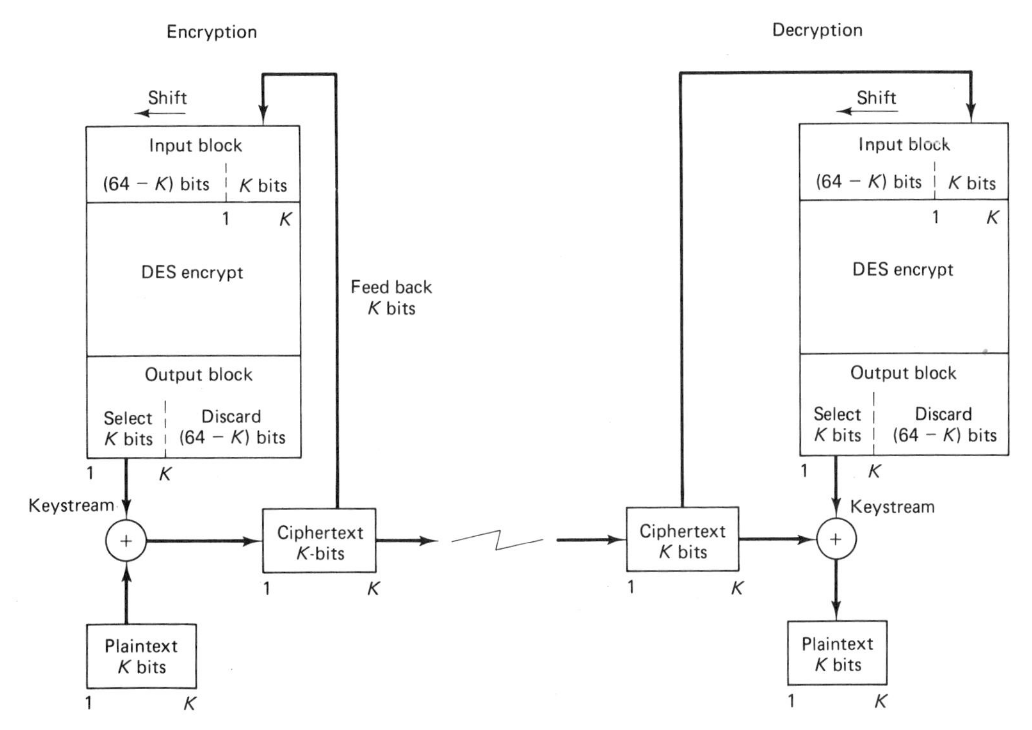

Figure 12.10 DES in cipher feedback mode

plaintext bits to produce K ciphertext bits. These ciphertext bits are fed into the input block (shift register) to form the K least significant bits. In this manner, the key sequence is generated in K-bit blocks and is dependent upon the data. If errors occur in transmission, they propagate across K bits, thereby causing errors in that number of output decrypted bits.

A second form of stream cipher configuration is *output feedback* (*OFB*). The difference between this and CFB is that the generated key stream is now independent of the data. The system is illustrated in Fig. 12.11. Note that it is the output block that is fed back. Since the keystream and not the ciphertext is fed back to generate the new bits of the keystream, this is also known as the *key-autokey* mode.

Regardless of the DES configuration, the operation depends upon a 56-bit key. There are thus 2^{56} or approximately 7×10^{16} possible keys. This appears to be an exceptionally high number, so that an exhaustive computer search by an unauthorized eavesdropper would take an inordinate amount of time. However, as computers become faster and more sophisticated, it may be necessary to increase this number. In fact, DES chips can be cascaded in series to increase the key length proportionately. If two systems are cascaded, the effective key length is 112 bits.

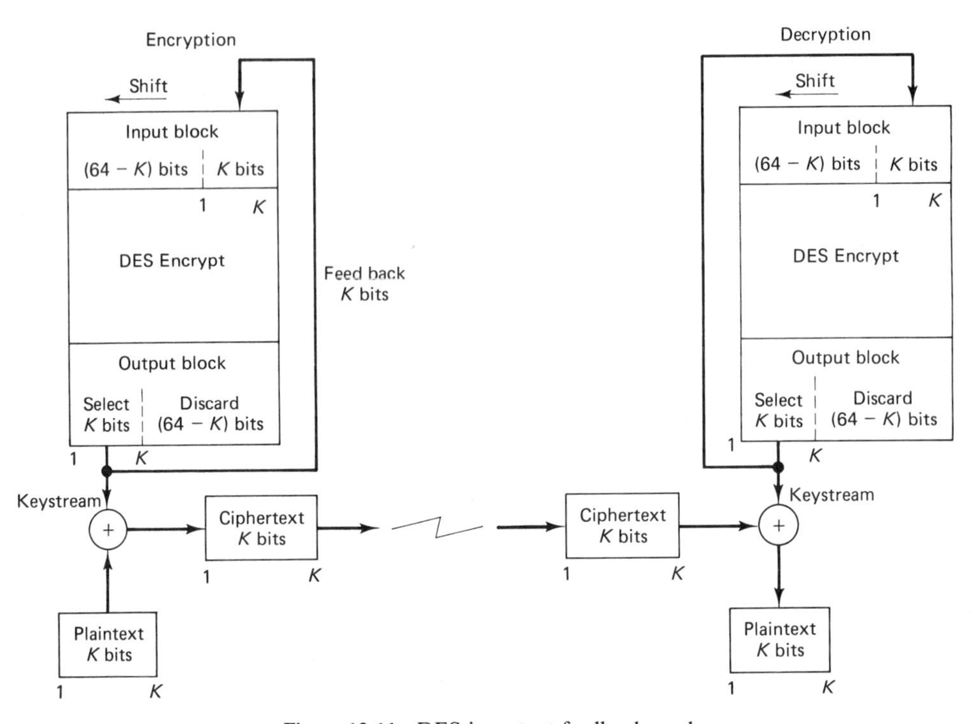

Figure 12.11 DES in output feedback mode

12.4 SUMMARY

This chapter explores techniques for providing communication security. Security has always been an important consideration for military systems. However, as data communication replaces hard copy in many sensitive consumer applications, security is moving out of the military arena into the public domain.

We begin with a discussion of spread spectrum, a technique that intentionally broadens the frequency spectrum of a signal in order to make it appear as broadband noise. As long as the intended receiver knows the algorithm used for spreading the spectrum, it can "refocus" the signal back to the narrowband original waveform. An unauthorized listener would observe little power in any narrow band and could not reconstruct the waveform.

Section 12.1 presents direct-sequence PN spreading, where a wideband pseudonoise signal is modulated by the narrowband information waveform in order to produce the spread signal. We examine code-division multiple access (CDMA) as a simple multiplexing scheme which uses the direct-sequence spreading concept.

Section 12.2 presents an alternate technique for achieving spread spectrum, known as frequency hopping. In this technique, the transmitted frequencies vary over a wide band according to a fixed pattern.

Cryptography is explored in Section 12.3. We examine traditional cryptography where a key is used to selectively reverse bits in the message (plaintext). We then present the related considerations of key management and originator authentication. Scrambling is explored, and the CCITT Recommendation V.27 is presented.

The exciting development of public key cryptosystems is examined, and the trapdoor–knapsack technique is analyzed. We end the chapter with a discussion of the data encryption standard (DES), where we examine the various configurations of this important building block.

PROBLEMS

12.1. Sketch the output of a direct-sequence spread-spectrum system where the input is the data signal

$$1 \quad 0 \quad 1 \quad 1 \quad 0 \quad 1$$

and the spreading sequence is 4-bit pseudonoise at three times the frequency of the data signal.

12.2. Find the process gain for a spread-spectrum system using a 5-Mbps PN code for direct-sequence spreading and a 2-kbps information data signal.

12.3. An FSK system with frequency hopping can be viewed as an MFSK system. Suppose the two frequencies of a BFSK system are hopped at a rate of four times the bit rate. Examine the difficulties that an unauthorized listener would have in intercepting this message. (*Hint:* Start by exploring the effects of the decrease in SNR due to the increase in symbol rate.)

12.4. If the probability of bit error is 10^{-2} for each of seven frequency hops and majority logic decoding is used, find the probability of error for a 4-bit message.

12.5. A puzzle writer takes a decimal arithmetic problem and substitutes one letter of the alphabet for each integer. The same integer always results in the same letter.

(a) If all ten integers appear in the original problem, how many substitution codes are possible?

(b) Given the arithmetic operation in letter form, how many combinations would have to be tried to assure one of hitting on the correct code?

(c) If one letter equivalent is found, how many combinations will then have to be tried to find the others?

12.6. You are given the following arithmetic problem, where a substitution code has been applied. Each letter corresponds to an integer.

$$\begin{array}{r} X\ X\ C\ X\ C \\ +\quad J\ X\ C \\ \hline = X\ C\ X\ C\ M \end{array}$$

Find the numerical equivalent of each letter.

12.7. A long message of greater than 1000 symbols is received where the symbols have been permuted in groups of three. How many combinations would have to be tried to be sure of arriving at the correct uncoded message?

12.8. Using trapdoor–knapsack with unmodified weights

$$(1,\quad 2,\quad 7,\quad 12,\quad 25,\quad 57,\quad 109)$$

find a public key and demonstrate the encryption process for sending the plaintext

$$1\quad 0\quad 1\quad 1\quad 0\quad 1\quad 1$$

12.9. When the DES is configured in the block mode, how many possible keys must an unauthorized listener try in order to be assured of using the correct message key?

13

Computer Communications Networks

This chapter presents a brief introduction to the field of computer communications. This field is vast and expanding, and whole books have been written about subsets of it. The interested reader is referred to the references for study beyond this introduction.

The earlier chapters of this text explored techniques for sending digital information. These techniques were *transparent* to the particular source of the data. Thus, for example, a QPSK system does not really care whether the data are derived by sampling and quantizing an analog voice channel or are directly generated by a computer. Thus, in some sense, this entire text has been addressing the field of computer communications.

As distributed computing becomes more and more common, the need for communications escalates. For example, the typical office contains word processing, electronic mail, databases, and sometimes mainframe computing. Many of these functions are accessed through desktop terminals that communicate with the various components of the network. Another driving force toward computer communications is the growth of home workstations and of public switching networks such as "Telenet" and "Tymnet."

The number of computer terminals in use in this country exceeded 10 million in the 1980s, and the growth rate is accelerating. The potential for growth is suggested by the fact that in the United States there are about 25 telephones and 10 televisions for every computer terminal [*IEEE Communications Magazine,* May 1984, p. 51].

Computer networks originated in the 1950s and were first used for defense. *Time-sharing* computer systems appeared in the early 1960s. The early 1960s also brought the first synchronous communication satellite (SYNCOM), which had a tremendous impact upon the growth of computer communication systems. The following decade brought the first major communications network (ARPANET), which we discuss in Section 13.8.

The 1980s brought an expanding interest in a communication network confined to a limited geographical area, the local area network (LAN). We discuss this important class of networks in Section 13.6.

Unless each organization wishes to design and build its own communication system from the ground up, a certain amount of uniformity and standards are needed. This need led to the development of network protocols and standards by five major standards groups—the *Comité Consultatif Internationale de Telegraphique et Telephonique* (*CCITT*), the *International Standards Organization* (*ISO*), the *European Computer Manufacturers Association* (*ECMA*), the *Electronic Industries Association* (*EIA*), and the *Institute of Electrical and Electronic Engineers* (*IEEE*). We discuss some of the standards developed by these groups in the later sections of this chapter.

13.1 SWITCHED SYSTEMS

Figure 13.1 shows one possible network architecture which connects six users. The lines joining the users are interconnections, which permit any user to communicate with any other user in the network. The number of lines shown in the figure is 15. In general, for N users we require

$$\frac{N(N-1)}{2}$$

interconnections in order to provide the capability of communication between any

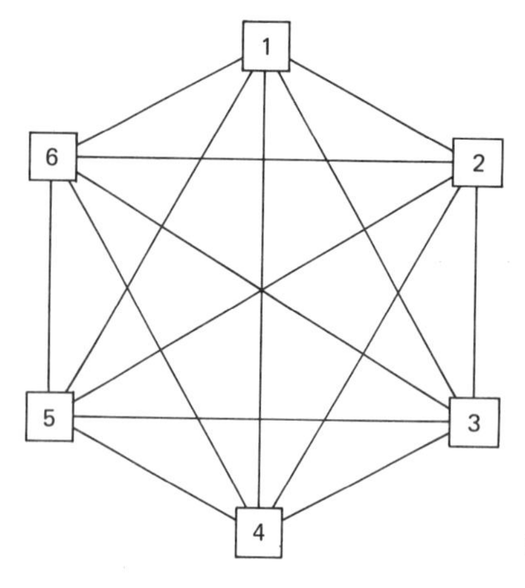

Figure 13.1 Simple network architecture

pair of users. The number of connections increases geometrically as the number of users increases.

The obvious alternative is to move to a switching architecture, as illustrated in Fig. 13.2. Here, the number of interconnections is equal to the number of users, thus resulting in a savings of

$$\frac{N(N-3)}{2}$$

connections.

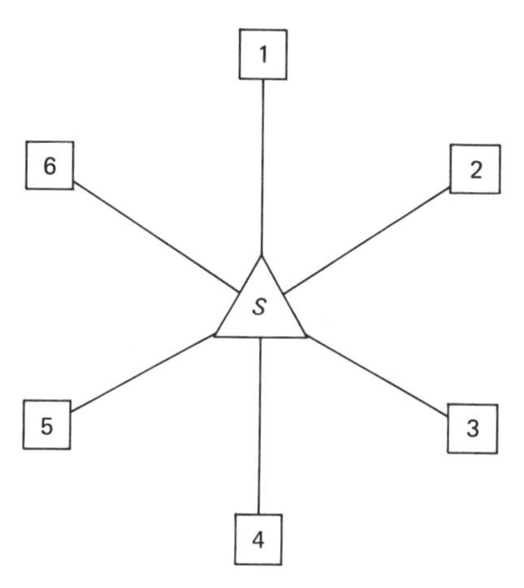

Figure 13.2 Switching network architecture

If the network becomes too large, or is distributed over a large geographical area, it may be necessary to increase the number of switching centers. Multiple switching centers may also be necessary to achieve a particular network *hierarchy,* which provides certain controls over the transmission of data. An example of a more complex interconnection of 30 users is shown in Fig. 13.3. In this system there are three major switching points, known as *tandem nodes,* and ten smaller switching points, known as *regional nodes.* Because of the interconnections between regional nodes, the architecture provides alternate paths for communications. This is a desirable feature in case failures occur in the network, or if use is so heavy that a simple structure would result in denying users access to the system at peak times.

There are three major categories of switched networks. Their differences involve the manner in which interconnections are established between users. The three categories are circuit switching, message switching, and packet switching. They are described in the next three sections.

13.2 CIRCUIT SWITCHING

Traditional telephone connections are an example of *circuit switching.* A hardware connection is established through the network, and it remains in place throughout

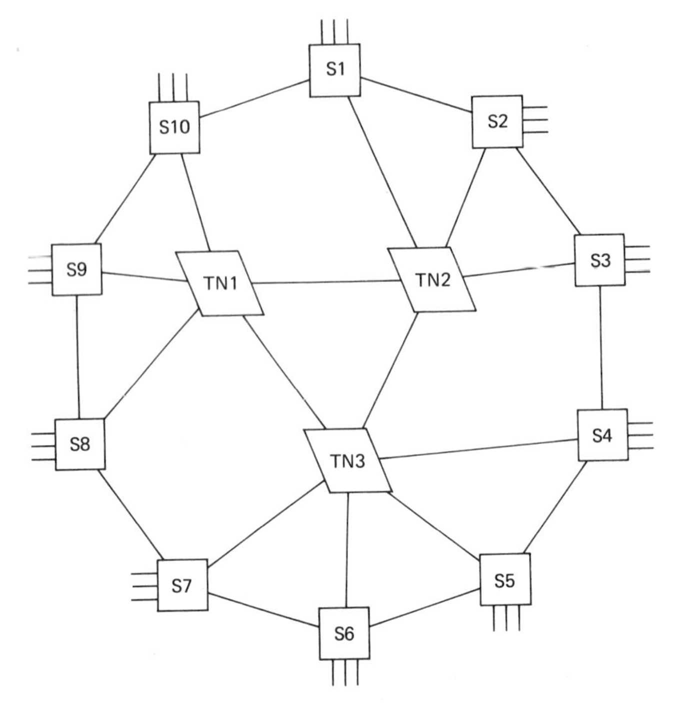

Figure 13.3 Multiple switching and hierarchy in network architecture

the duration of the transmission. The actual transmitted signals are not affected by the connections, so the system provides a *transparent* connection.

A switch scans each of the user connections in an established sequence looking for a *request for service*. This is indicated by a change in the state of the line. For example, in traditional telephone circuits, the user lifts the receiver off the hook, thus generating a dial tone. The period of the scanning operation must be acceptable to the user. That is, in the telephone example, the wait for a dial tone must not be annoyingly long. The amount of time one is willing to wait is a function of the length of the message. In a 10-minute phone call, a wait of a few seconds would not be unacceptable. In a 1-second data-burst message, a 0.1-second delay may prove unacceptable.

Among the advantages of circuit switching is low overhead. Control systems are needed to fix the routing of the message, but once this is done, the control equipment must simply monitor for the end of the transmission. The overhead is therefore not a function of the message length.

Once the connection is established, the timing of the message is preserved. That is, the spacing between transmitted message segments is identical to the spacing between the corresponding received message segments. The timing is thus *synchronous*.

A disadvantage of circuit switching is that the connection channel is dedicated to the pair of users throughout the time of the message transfer. This is often highly inefficient. For example, a keyboard terminal with a human operator may transmit data at a peak rate on the order of 100 bps, while the channel can handle rates considerably higher. There are also rest periods during which no data are being transmitted.

Circuit-switched networks are nomally full-duplex, allowing simultaneous transmissions in both directions. Since the network simply provides a nonintelligent interconnection between users, the terminal transmitting and receiving equipment of one user must be compatible with that of the other. This compatibility extends to both software and hardware. That is, the signal voltages, rates, and formats must be matched as well as the languages, coding, and transmission techniques.

13.3 MESSAGE SWITCHING

Modern-day society has been accused of being the most wasteful in history. We are guilty of very low efficiency in utilization. Let us consider some common examples. If you are the average engineer, you probably own at least one automobile. This vehicle probably sits in a parked position 85% to 95% of the time. If resources were as scarce as some of us believe, this situation would be unacceptable. Instead, as soon as you arrived at work and parked the car, someone else would pick it up and use it for shopping during the day. At night, when you retired, the neighbor down the street who works the night shift would pick up this same car. Indeed, the day may come when the cost of dedicated ownership is so high that society will be willing to consider such shared utilization.

Your author gained personal experience of the intermittent-use phenomenon as a newly hired engineer employed by the Bell Telephone Labs. New engineers were often given rotational assignments in various departments to learn about the scope of Bell System activities. A common assignment was to a telephone switching office. In those days most offices used *step-by-step* switches. As people dialed each digit of a phone number, a noisy switch clicked away. The new engineer, sitting in the New York switching office, heard this clicking every few seconds as a subscriber dialed a number, but it was not distracting. Every 15 minutes, however, the equipment went crazy with switching operations so thunderous that the young engineer was actually frightened. This phenomenon came to be known as the "toilet bowl effect," since the water pressure in a large city such as New York showed a similar behavior. A plot of the pressure resembled that of Fig. 13.4, showing a dip every 15 minutes. Both of these effects were due to the regular timing of commercial breaks on TV, during which people ran to use the bathroom or make a phone call.

Data communication experiences a related effect. Many communication systems involve a rapid data exchange, perhaps followed by a long rest period. If several terminals at the same location are in similar circumstance, time sharing of

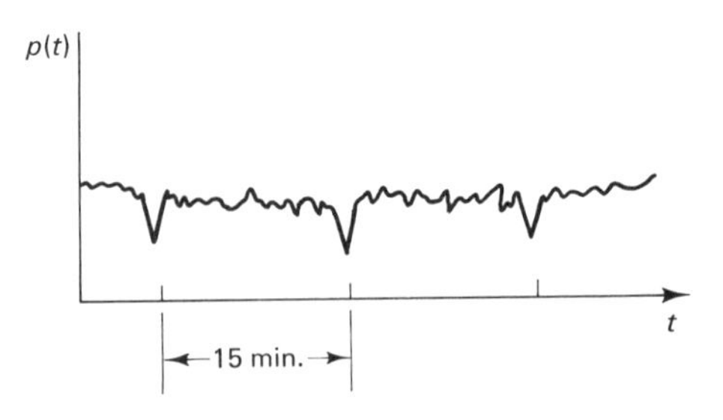

Figure 13.4 Water pressure vs. time

the channel is possible. But suppose that the terminals are spread over a wide geographic area? In such cases, the sharing must take place at points more internal to the communication network.

Because circuit-switched systems usually do not make efficient use of the network hardware, more sophisticated techniques have been developed to provide a sharing of resources. These techniques fall into the category of *store-and-forward* systems.

Among the simplest store-and-forward systems are *message-switched* networks. These provide for the entire message to enter the system, be stored, and be transmitted when links are available. The storing operation takes place at each switch within the network, and the message is routed through the network from switch to switch.

As long as the storage capability of the first level of switches closest to the source is not exceeded, the source user will not get a busy signal. Recall that in the circuit-switched network, messages can be blocked if no lines are available. In message-switching networks, blockages usually do not occur. If the system experiences a large number of messages, the various messages will be *delayed* rather than *blocked*.

Message timing is preserved. That is, even if a message contains pauses, these pauses form part of the information that is stored at each switch on its way through the network. Thus, although efficiency improves over that attainable using circuit switching, there is still a reasonable amount of lost capacity in this system.

Message-switching systems require considerable storage capability at each switch, and the variable delays may be unacceptable in an interactive communication application.

13.4 PACKET SWITCHING

Packet switching is a more sophisticated version of store-and-forward transmission where the information is broken into segments, or packets, and transmitted through the network using a store-and-forward technique. The packets typically consist of about 2000 bits or less. In addition to the desired data signal, each packet contains address information. Protocols are used to allocate network resources only during useful parts of the information transfer. Timing of the message is not preserved, since the various portions of the message may arrive at the user having taken different paths through the network. That is, the communication is *asynchronous.* Clearly, sufficient overhead information must be provided to allow reassembly of the pieces at the receiver.

Because this operation requires intelligent switches, it is possible to reassemble the message using a different set of rules than that used at the transmitter. For that reason, packet switching permits communication between users even if their basic equipment is incompatible.

Packet switching permits duplex operation, but the two directions of flow normally occur along different interconnection paths.

There are two distinct approaches to packet switching: the datagram and the virtual-circuit technique. In the *datagram* system, a packet enters and decisions are made at each node as to how to traverse the distance to the next node. The *virtual-circuit technique* differs in that a path is established between sender and receiver and the entire message usually uses this path. Although the same path may not be used for all packets in the message, the path is established before a packet enters the network, and packets arrive at the receiver in the same order as transmitted. The virtual circuit is therefore a compromise between dedicated circuits and the datagram.

13.5 PROTOCOLS

Protocols were introduced in Section 5.8 as rules that provide for the orderly exchange of data between users. We now expand upon that introduction and look at several popular protocols in some detail.

There are two approaches toward protocols—the *layered* approach and the *hierarchical* approach. The U.S. Department of Defense (DOD) has defined a protocol architecture (DPA). Although this hierarchical protocol provides specific advantages over the layered protocols, we shall focus our attention upon the traditional layered approach. This approach was suggested by the International Standards Organization (ISO), which defined seven levels of protocol, as discussed below. Many applications require only the first two.

Level 1, the *physical level,* deals with the mechanical, electrical, functional, and procedural access to the system. It includes standards for the pin connections, voltage levels, and physical signal formats used in data communications.

Level 2, the *data link level,* concerns itself with the data formats, the synchronization, error control, and flow control commands. Its purpose is to make the physical connections reliable.

Level 3, the *network level,* is responsible for establishing, maintaining, and terminating the connections. It is the internal network protocol which establishes the call for switched connections.

Level 4, the *transport level,* oversees the entire process from end to end and establishes priorities. In the case of packet-switched networks, this level provides the necessary reordering of packets at the receiver.

Level 5, the *session level,* provides for a particular control structure. It concerns itself with applications, log-on, authentication, and interaction at the software level.

Level 6, the *presentation level,* controls display formats and conversion of the various data codes. It can also include encryption operations. It is concerned with the syntax of the interchange of data.

Level 7, the *application level,* provides distributed information services. It deals directly with application software programs.

Entire books have been devoted to protocols, giving detailed descriptions of the most popular examples. We choose to present one example, the X.25. This is probably the most widely used standard for packet-switched network communication.

13.5.1 X.25 Packet-Switched Network Standard

X.25 is a standard which deals with interfaces between the data terminal equipment (DTE) and the data circuit-terminating equipment (DCE). The standard is not dependent upon internal structure of the network. We begin by tracing the history of this standard.

The *International Telecommunications Union* (*ITU*) is the major international organization responsible for setting standards. This organization has a consultative committee for international telegraphy and telephone (*CCITT*), which is charged with formulating standards for international communication. These standards are voluntary. The United States is represented on the CCITT through the National Bureau of Standards.

In 1968 the CCITT created a group to look at new data transmission services. It was not until 1975 when a draft of X.25 was developed. This draft eventually had to be ratified by 200 members from 20 countries. The formal adoption of recommendation X.25 came in September 1976, and several revisions followed at later dates.

X.25 is a layered protocol which spans three levels: the physical, the link, and the packet. The standard specifies parameters at the user and network level and not at the packet structure level. It applies to the virtual-circuit mode of operation and provides for either a permanent virtual circuit (PVC) or a switched virtual circuit (SVC).

Level 1 is the physical level and includes details of signal synchronization. Level 2 provides for error detection and correction (by retransmission). Level 3 involves flow control, multiplexing, and call set-up.

The *physical level* of X.25 specifies synchronous circuits which use duplex transmissions. It includes two other recommendations, the X.21 and the X.21 bis.

X.21 defines the digital circuits as switched synchronous systems. X.21 bis is for use of analog transmission lines in data transmission and is compatible with the RS-232 physical standard.

The *link level* provides for error-free exchange of data between the data terminal equipment (DTE) and the data circuit-terminating equipment (DCE). There are two possible protocols at the link level, these being the *link access procedure* (*LAP*) and the *link access protocol balanced procedure* (*LAPB*).

The *packet level* presents procedures to establish virtual circuits for transmission of data.

Figure 13.5 shows the typical X.25 packet format. The first byte of the packet consists of a general format identifier (4 bits) followed by the logical channel group number. The transmission system allows for up to 16 groups of 256 channels each to be sent on one virtual circuit. It therefore requires 4 bits to identify the group number and 8 bits to identify the channel number. The channel-number identifier is the second byte of the packet. A user would subscribe to a network and be assigned a range of channel numbers. The logical channel group number also identifies whether the packet is for switched or permanent virtual-circuit transmission.

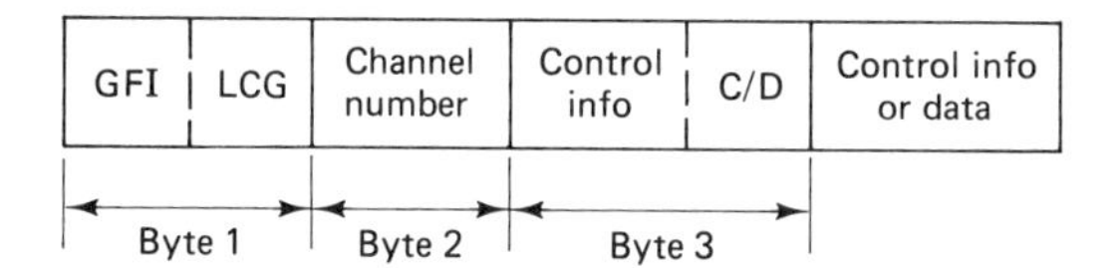

Figure 13.5 X.25 packet format

The third byte contains control information. This byte also contains a single bit, *C/D*, which identifies whether the packet is a data packet or a supervisory packet. The first three bytes are followed either by additional control information or by data.

Although it is not our intent to give a detailed treatise on X.25, it might prove instructive to trace a particular call set-up sequence. This should give some physical significance to the terms defined earlier.

Suppose a DTE wishes to initiate a call. It sends a *Call Request* packet, which contains the address of the DTE being called. The address is normally a twelve-digit decimal number. The calling DTE specifies a logical group and channel number in the Call Request packet, and these will identify the particular conversation for the following packets. The control information of the packet (the third byte) specifies this as a Call Request packet. The network inserts the address of the calling DTE, although this can be specified by that DTE in an optional field of the Call Request packet.

The called DTE will normally send a *Call Accept* packet, also known as a *Call Connected* packet. This DTE can also send a *Call Cleared* packet, which would indicate that the call request is being rejected.

After the call is set up, four types of packets are used: Data, Receiver Ready, Receiver Not Ready, and Reject. This provides for "handshaking."

Othet types of packets cover error monitoring, interrupt requests, resets, and restarts. The interested reader is referred to the references for additional details.

13.6 LOCAL AREA NETWORKS

Progress in digital communications and in microelectronics and microprocessors has led to a blending of technologies between computers and communications. This blending has greatly expanded the accessibility of data communications to many environments which were previously excluded owing to costs and considerations relating to efficiency of scale. Traffic profile studies have shown that an average of about 80% of the information transmitted within a local environment is generated within that same environment. Thus, only about 20% of an organization's information exchange takes place with terminals outside of that local environment.

Any medium-size company, set of buildings, college campus, or government organization can now configure its own communication system to provide high-speed channels within a limited geographic area.

There is no precise point at which a network becomes "local area." The term is used to define privately owned networks with transmission lengths of the order of 1 meter to about 5 km. At distances above approximately 5 km the network is known as "long haul," and for distances less than 1 meter we use the term "input/output bus."

Local area networks (LANs) normally contain "intelligent" switches. That is, store-and-forward techniques are usually used instead of circuit switching. Because these networks transmit over relatively short distances, they operate with high reliability and low error rates. They typically support several hundred independent devices and functions, such as word processing, television, sensors, terminals, minicomputers, voice, and FAX. Typical communication speeds range from less than 1 kbps for some terminals to about 56 kbps for communication with minicomputers. Data rates of the order of 100 Mbps may be required in some applications. LANs also provide for access to communication outside the geographical confines of the system through a 1544-kbps connection to long-haul networks.

Transmission technologies in LANs utilize twisted pairs of wires, coaxial cable, and fiber optics. Twisted pairs are used with maximum data rates of about 2 Mbps and maximum ranges of no more than several kilometers. Coaxial cable can be used at data rates up to about 50 Mbps (for 75-ohm cable) and ranges no greater than about one kilometer. Optical fiber technology is developing rapidly and is capable of exceeding the performance of coaxial cable in both maximum data rate and transmission distance.

Network *topology,* which deals with the method by which end points are connected together by the network, has three broad classes.

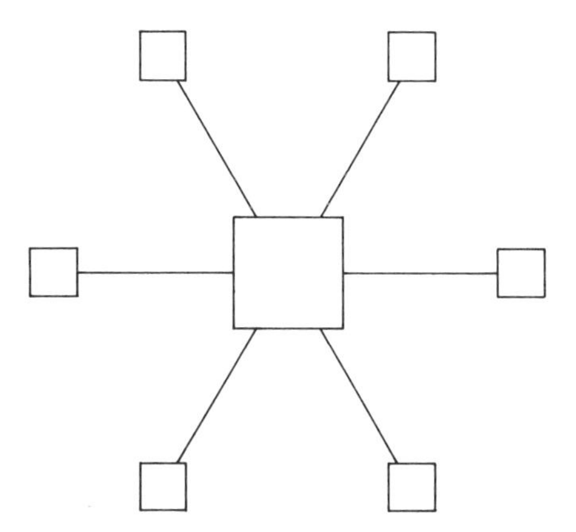

Figure 13.6 Star topology

The *star topology* is illustrated in Fig. 13.6. This topology includes a central switch which is common to all end points. Communication between any two end points requires routing through the common switch. This is a simple topology and requires fairly straightforward control signals. The central switch exerts complete control over the network.

The *bus topology* is illustrated in Fig. 13.7. This topology requires no switches, and all end points share the same transmission medium, or bus. A transmission from any end point can be received by any other end point. Unless some form of multiplexing is used, only one device can use the system at any time. This system is used with packet transmission, and each packet can be seen by every end point. The correct station recognizes its address in the packets intended for it.

The *tree topology* is a variation of the bus topology, and it is illustrated in Fig. 13.7.

The *ring topology* is illustrated in Fig. 13.8. The blocks with "R" in them are "repeaters" which receive data on one of two paths and transmit them on another path. Data circulate around the ring in only one direction (e.g., clockwise or counterclockwise). Packet transmission is usually used in the ring topology. Because the distance between repeaters is shorter than the distance between end points, improved error performance can be expected.

LANs can use either baseband or broadband transmission techniques. Broadband systems can employ frequency multiplexing to allow a particular communication channel to be used for transmission of more than one signal. Broadband systems can also transmit over greater distances than baseband systems. The *Ethernet* system, a baseband coaxial LAN, is discussed in detail in Section 13.8.

Protocols for LANs span layers 1 and 2 and (depending upon interpretation) layer 3. Layer 1, the physical layer, includes specifications of signal voltages, bit durations, and the mechanical and electrical characteristics establishing the link.

Layer 2, the data link layer, uses error detection and frame acknowledgement.

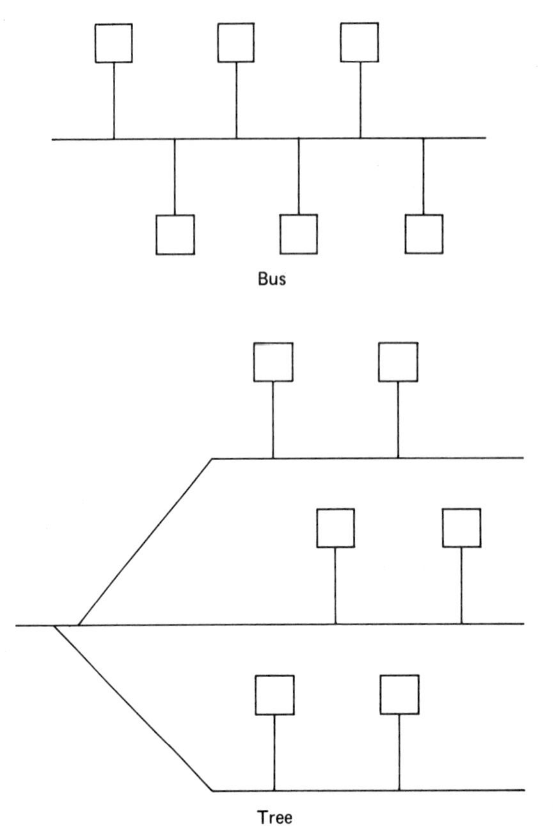

Figure 13.7 Bus and tree topologies

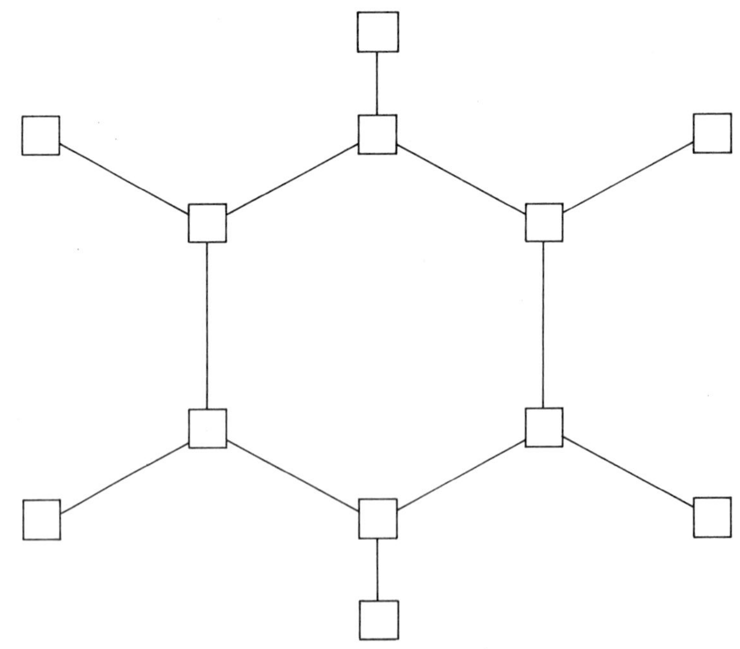

Figure 13.8 Ring topology

That is, the frames of data must be controlled in some manner during transmission through the network.

Layer 3, the network layer, is responsible for control of routing through the network. Since there are direct links between any two points, routing is normally not required. But layer 3 also concerns itself with the devices attached to the network and with assuring that the network can support multiple types of devices. Thus, portions of layer 3 are required by all but the simplest of LANs.

13.7 INTEGRATED SERVICES DIGITAL NETWORK (ISDN)

Integrated services digital network (*ISDN*) is not a single concept. It is a term used for an evolving technology of networks including both voice and nonvoice services. ISDN is a single network which supports a wide range of communication applications. It encompasses networks connecting with the home as well as with businesses, and includes both local area and long-haul networks. Services for the home include telephone, electronic mail, interactive video, information services including electronic banking, and monitoring (e.g., alarms). Business applications are essentially the same as those of LANs described in the previous section.

The goal of ISDN is to extend the success of voice phone, where about 600 million subscribers (worldwide) can establish a voice connection within seconds, to diverse forms of mixed voice/data applications. This is to be accomplished through access to a single wall plug, rather than through multiple networks as has been the case in the past.

The CCITT defines ISDN as follows: "A network evolved from the telephony Integrated Digital Network that provides end-to-end digital connectivity to support a wide range of services, including voice and nonvoice services, to which users have access by a limited set of standard multipurpose user-network interfaces."

Introduction of ISDN into the existing telephone network requires a phase-in over a period of one to two decades. The concepts and capabilities include capability to handle the wide variety of services efficiently, bandwidth allocation on a demand basis, low error rates, low message delays, and security. As if these demands were not challenging enough, it is critical that the ISDN provide for evolution as new technologies develop. That is, the system must be as immune as possible from obsolescence.

The CCITT ratified the first set of ISDN standards in October of 1984. This is the I.100 series which covers the general aspects of ISDN. This permitted implementation on a limited basis while other more detailed standards were still being developed and debated.

Typical user interconnections to ISDN are illustrated in Fig. 13.9. The blocks labeled TE1 are ISDN terminals which are compatible devices. The TE2 terminals are non-ISDN terminals, so reconfiguration of the signals is required. This is

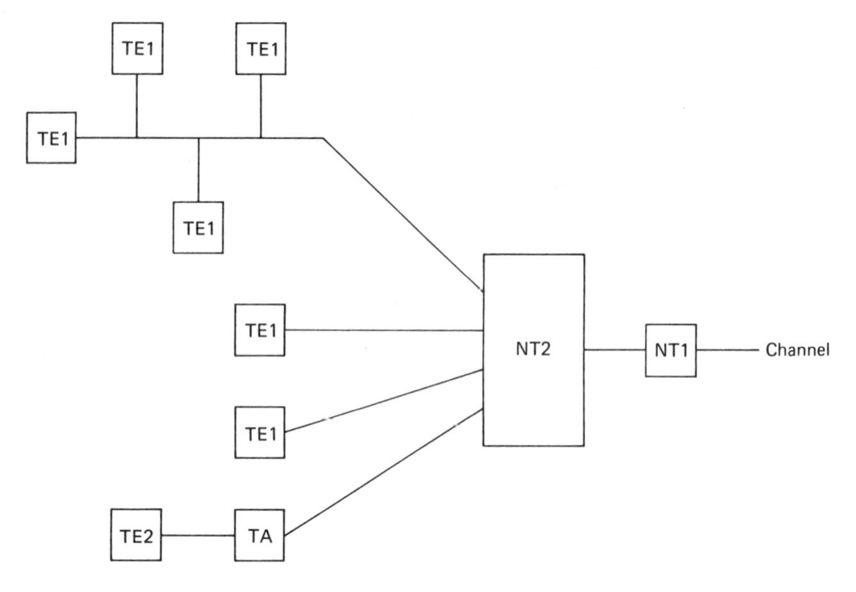

Figure 13.9 Typical user interconnections in ISDN

performed by terminal adapters (TA). NT1 and NT2 represent network terminators which multiplex the various signals. The use of ISDN for voice takes place through the TE2 path, and the terminal adaptor can perform a variety of functions. It digitizes the voice (PCM) and applies the appropriate companding relationship. It can also do audio processing to correct for distortion or modify the frequency characteristics. This provides a great deal of flexibility.

ISDN uses a basic transmission rate of 192 kbps. The system provides for full-duplex operations of three separate channels. Two of these provide for 64-kbps transmission and are known as bearer channels or *B-channels*. The B-channel is used to carry circuit-switched voice or data, including video and video conferencing. The third channel provides for 16-kbps transmission and is known as the *D-channel*. The D-channel is used for signaling and control. That is, it includes the functions of setting up of calls and terminating of calls. It can also contain a limited amount of data if the full capacity is not needed for signaling. The two B-channels and one D-channel total 144 kbps, leaving 48 kbps for housekeeping purposes, including framing and maintenance.

The transmission system used in ISDN may be either packet-switched or circuit-switched. The important thing is that the transmission technique be "transparent" to the user.

As ISDN evolves, a great deal of literature will certainly appear. Standards will receive attention, as will integration of LANs into the long-haul network. The interested reader is referred to the technical literature for up-to-date developments.

13.8 EXAMPLES OF NETWORKS

13.8.1 ARPANET

In the late 1960s the *Advanced Research Projects Agency* (*ARPA*) sponsored a project to establish a pilot packet-switching network. The goal was to interconnect various research centers so that their computer resources could be shared. The network was to be capable of sending data at 50 kbps. The system grew quite rapidly to a level of over 200 host computers in the United States (continental plus Hawaii), England, and Norway by the 1980s. Figure 13.10 shows a map of ARPANET in the United

Key to list of sites (not all of the current sites are illustrated)

1. Aberdeen R & D center
2. ALOHA, University of Hawaii
3. NASA Ames Research Center
4. Advanced Research Projects Agency
5. Bolt Beranek and Newman
6. USAMERDC, Fort Belvoir
7. CALTECH
8. Computer Corp of America
9. Carnegie-Mellon University
10. Department of Commerce
11. AF Global Weather Center
12. Harvard University
13. University of Illinois
14. Lawrence Berkeley Laboratory
15. Lawrence Livermore Laboratory
16. M. I. T. Lincoln Laboratory
17. M. I. T.
18. Mitre Corporation
19. Moffett field
20. National Bureau of Standards
21. National Security Agency
22. New York University
23. Pentagon
24. Rome Air Development Center
25. Rand Corporation
26. Stanford University
27. Tymshare
28. University of California, Los Angeles
29. University of California, Santa Barbara
30. University of California, San Diego
31. University of Southern California
32. University of Utah
33. Xerox

Figure 13.10 Map of ARPANET

States. Each computer accepts blocks of data, each of which is subdivided into 128-byte packets. Upon receiving a packet, a minicomputer acknowledges it and repeats the routing process to the next computer.

The ARPANET consists of subnetworks of specialized communication processors [known as *interface message processors* (*IMPs*)] interconnected by communication links, plus a collection of host computers, each attached to an IMP. Each IMP can accommodate up to four hosts. There are also *terminal interface processors* (*TIPs*) which provide direct access to the subnetworks from the user terminals. Each TIP can accommodate up to 63 terminals. The ARPANET topology is illustrated in Fig. 13.11.

Since diverse types of computers are employed, protocols are required to make sure the various computers are compatible. The *Network Control Program* first establishes a virtual channel between a pair of host computers. The *file transfer protocol* (FTP) provides for interhost transfer which can be in any one of a number of code forms, including binary and ASCII. Address information is added to the message.

Each message consists of a header plus text materials. The header contains labeling information, including the sender, intended receiver, data, subject, copies to, and so on, much as in the case of an office memo. *Message transmission protocols* (*MTPs*) are evolving to provide more sophisticated functions. For example, a user may wish to send a message to a group that is defined by some identifier (just think of the junk-mail potential). An important operational feature is that a copy of each packet is retained at each node until confirmation is received from the next node. This is a protective measure to decrease the probability of lost packets.

As the ARPANET develops, additional features will be added. In particular, the ARPANET is being used as a testing ground for packet-switching technology and performance evaluations. The future of electronic mail depends upon the sophistication of systems such as the ARPANET. Techniques are being devised, for example, for a receiver to survey its mail without having to get a complete copy of every letter.

13.8.2 ALOHA

ALOHA is perhaps the earliest example of a packet-switching system. It was originated at the University of Hawaii toward the end of the 1960s. Each station divides its message into frames. Whenever a frame is ready for transmission, it is sent into the network. The transmitter then listens for a specified length of time for a return acknowledgment. This length of time is the maximum round-trip excursion time expected in the network. If no acknowledgment is received, the transmitter resends the frame. If a receiver receives a frame with errors (checking is done), it ignores it. Such errors can occur either because of transmission errors or because two frames interact in what is known as a *collision*. If a transmitter makes a specified number of attempts to send a frame without acknowledgment, it simply "gives up" and tries later. Because of the lack of controls and intelligent switches, ALOHA is

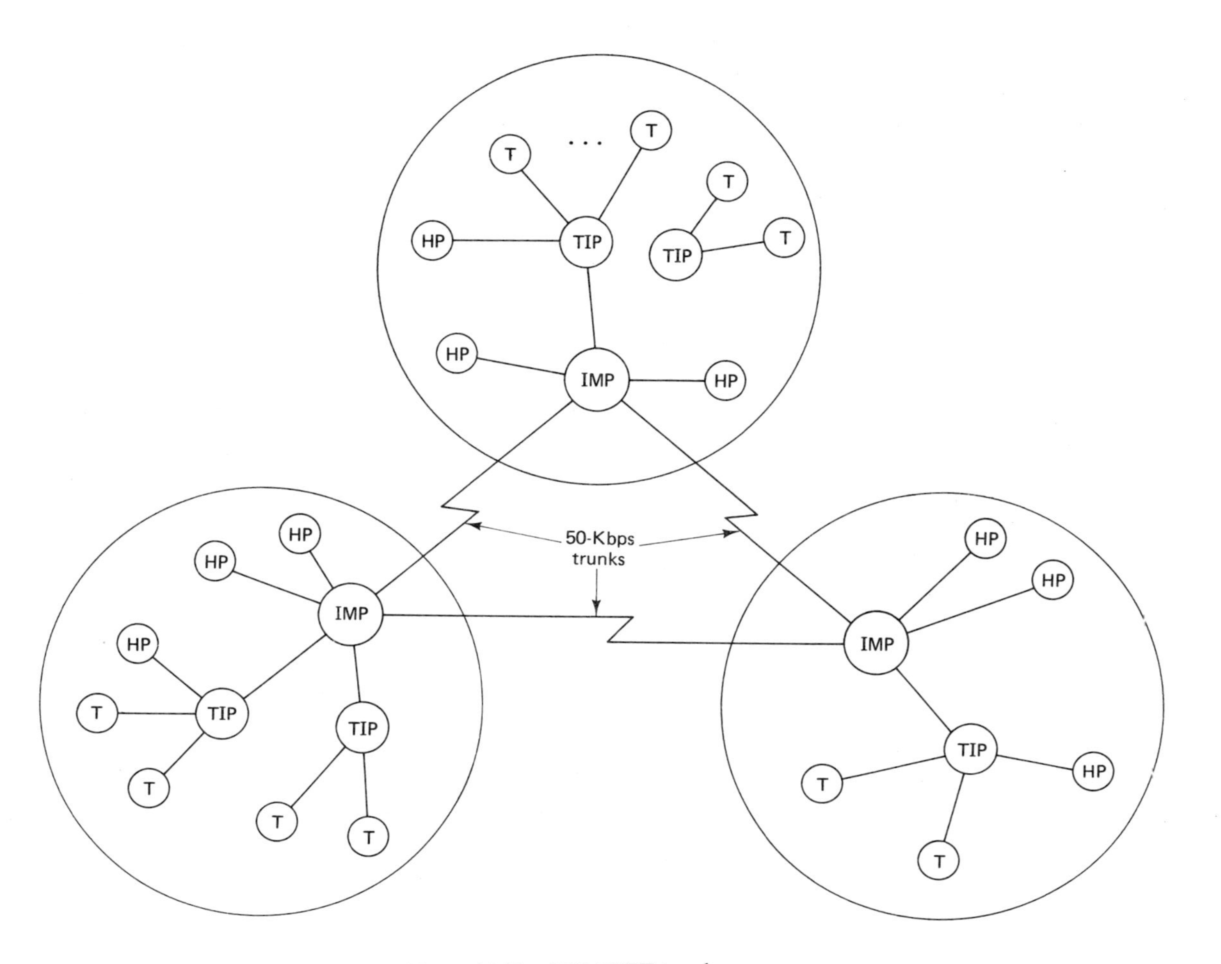

Figure 13.11 ARPANET topology

highly inefficient. The maximum utilization of the channel without experiencing unacceptable numbers of collisions is less than 20%. This figure arises from a simple probability analysis.

13.8.3 Ethernet

Ethernet was one of the first local networks in existence. It was developed by Xerox Corporation in conjunction with Digital Equipment Corp and Intel Corp. An early demonstration of Ethernet took place in 1982, when ten companies were linked for electronic mail and word processing.

Ethernet employs a baseband coaxial bus architecture. Each segment of coaxial cable is up to 500 m long and may connect to up to 100 transceivers. Each transceiver tests the line for the presence of a signal before injecting information.

Ethernet uses packets containing 72 to 1526 bytes of data. The packets include address information, error-correction capability, and the data. A biphase code is used to transmit the data at 10 Mbps. Both individual and group addresses are possible.

13.8.4 Time-Division Multiple Access (TDMA)

Time-division multiple access (TDMA) is a technique for sharing a communication channel among a number of users. Each transmitter is assigned a time slot (known as a *frame*), and the time slots are chosen to be nonoverlapping when the various signals arrive at the receiver. (Contrast this with ALOHA, where users transmit at will.) Since the distances from the various transmitters to the receiver are not necessarily the same, the time slots must be carefully chosen with this in mind. That is, although signals may be nonoverlapping at the time of transmission, if two signals take different amounts of time to reach the receiver, they may overlap upon reception. This consideration is not too critical with relatively short transmission paths. Small differences in transmission time can be accounted for by inserting dead bands between transmissions. It does, however, become a serious consideration in satellite communications systems, where transmission delays can become considerable.

Each transmitter is assigned what is known as a *frame*. The frames are specified such that if every sender were to transmit at the beginning of its assigned frame, all transmissions would be received simultaneously at the receiver. The frames are defined so there are a certain number of indexed frames per second, perhaps 1000 or more. A particular transmitter need not use a portion of every frame. For example, if there are 1000 frames per second and a user wants to send data at less than 1 kbps, some frames will be skipped. On the other hand, higher-speed users will need to send more than 1 bit per frame.

The frames are grouped in combinations known as *superframes,* each containing perhaps 64 frames. Figure 13.12 illustrates the hierarchy. Most users transmit a data burst during each frame. Each frame consists of a transmission that starts with

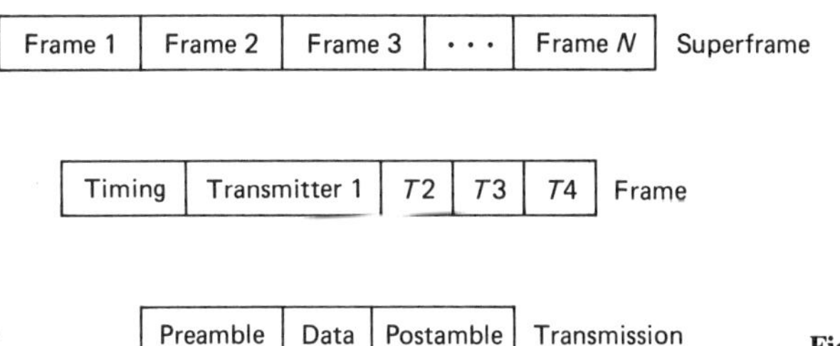

Figure 13.12 Hierarchy in TDMA system

timing information followed by signals from a number of transmitting stations, each having a portion of the frame for its data burst. To keep track of all this timing, the frame is divided into intervals, and the start of each interval is referenced with an index number. Each transmitter is given a portion of the frame by the designation of two indices, the start and stop of its assigned interval.

A burst from a particular transmitter starts with a *preamble* to establish timing and identification information. This preamble is followed by data, which might represent a PCM signal. The burst may be composed of several messages addressed to different receivers. The burst is concluded with a *postamble*.

Suppose, for example, that a TDMA system is set up with 10 transmitting stations sending to a communication satellite. The system has been divided into 1000 frames per second, and therefore each frame is 1 ms long. The frame is divided into 1-μs chunks, and each frame is therefore indexed from 1 to 1000. The absolute timing of the frames has been carefully set so that the ten transmitters have frames synchronized with respect to arrival time at the satellite. Suppose that you have applied as a user who needs to send data at a rate of 50 kbps. You are therefore assigned part of each frame, say for example from identification index 100 to identification index 150. This gives you 50 μs of each frame, with frames occurring every millisecond. During each frame you must therefore transmit 50 bits of information plus the preamble and postamble. Since you have 50 μs to transmit these 50 bits plus identification, you might send these bursts at a 2-Mbps rate, thereby having time for a 100-bit burst every frame.

13.8.5 Joint Tactical Information Distribution System

The *Joint Tactical Information Distribution System* (*JTIDS*) is an ambitious program developed for the Department of Defense (DOD). It represents a comprehensive military application of TDMA and *spread spectrum*.

JTIDS provides secure communication among a variety of users, including aircraft, airbases, ships, and ground stations. Communication can be point-to-point, or one subscriber can send a message simultaneously to multiple receivers. Messages can be either digitized voice or data. Although all users share the same channel, more than one channel can be provided in a geographic area through the use of multiplexing. Users can exceed the range of the system, which is line-of-sight (owing to the frequencies employed), through deployment of airborne relay stations.

The bandwidth of a single net is 57.6 kHz. The channel is divided into 128 time slots per second. Each terminal has a digital clock to set up the necessary time references.

The time slots are combined into 12-second frames, each containing $12 \times 128 = 1536$ time slots. Since some users might not even require one slot every 12 seconds, the frames are combined into groups of 64, giving an overall epoch of 12.8 minutes. This represents the maximum separation between time slots for any particular subscriber. That is, the bookkeeping requires that no subscriber be given less than one time slot every 12.8 min. Thus, the upper limit on the number of subscribers is about 98,300. Of course, many users (tactical surveillance, for example) will use far more than the minimum number of time slots.

Each time slot consists of a synchronization burst, data, and a guard band with no transmission. This guard band accounts for differing path lengths. These different lengths could not be accounted for by slot assignments, since so many subscribers are mobile. In addition to the use of spread spectrum, some users encrypt data for additional security.

13.9 SUMMARY

This entire text has dealt with the topic of digital communication. The digital information can result either from A/D conversion of an analog signal or from a signal which originates in a digital format. Data signals can originate from a variety of sources. These include data acquisition devices (e.g., transducers), digital keyboards (e.g., automatic teller machines), and computers. Thus, in some sense, a major portion of this text has dealt with computer communications.

As communication between computers becomes more common, attention must be given to the global aspects of this form of communication. For example, we must address the architecture of computer networks and the rules that must be followed to allow orderly interchange of information. We must also examine the techniques of interconnecting a large number of geographically dispersed users.

The current chapter concentrates upon the overall aspects of computer communications networks. It does not repeat the internal detailed analyses of the previous chapters, and so we do not deal with such considerations as establishment of synchronization and choice of signal formats. The results of the previous chapters are directly applicable to computer networks.

We begin with a general discussion of switched systems and of the advantages of switching as compared to direct interconnection between all possible terminals. Sections 13.2, 13.3, and 13.4 then expand upon the three major classifications of switching. The first is circuit switching, where a hardware connection is established between the sender and receiver for the entire duration of the information exchange. The other two types of switching rely upon smart switches, which use the store-and-forward technique. In message switching, the entire message is stored at each switch prior to transmission to the next switch and eventually to the receiver. In packet switching, the message is divided into groups of bits, known as packets, and these are transmitted through the network from switch to switch. Different packets may take different routes through the network, so effort is required to reassemble the pieces at the receiver. As each switching technique is presented, we examine its advantages and disadvantages relative to the other systems.

The orderly exchange of information requires an elaborate set of protocols. These are examined in Section 13.5, where we present the layered approach to protocols. The CCITT X.25 standard is examined as an example of a popular protocol for packet-switched networks.

Section 13.6 explores the subset of computer networks known as local area networks. The star, bus, and ring topologies are presented.

As the public demands increased access to data communication networks, the integrated services digital network (ISDN) is evolving. This is an ambitious system which would allow mixed voice and data use by consumers in a manner as convenient as in the traditional voice communication system. Thus, one need only plug a wire into a wall receptacle to have access to multiple data and voice channels. The concept of ISDN is presented in Section 13.7, where the terminology and network configuration are developed.

Several large computer communication networks have developed over the years. Among these are ARPANET, ALOHA, and Ethernet. The history and broad aspects of these systems are presented in Section 13.8. We also explore the concept of time-division multiple access (TDMA) and its application to the joint tactical information distribution system (JTIDS).

Appendix I

Answers to Selected Problems

This appendix contains answers to selected problems. The probability of an answer appearing in this appendix is, in some ways, inversely proportional to the significance of the problem. That is, only the simple problems lend themselves to brief numerical answers. The more complex design problems involve tradeoff decisions, and often have many "correct" solutions.

These answers are provided for purposes of reinforcement. Since the solutions are not presented, you are cautioned against spending an undue amount of time trying to match the answer given here. After all, there is always the probability that a typographical error crept into the printing. Indeed, there is even an extremely small probability that your author made a mistake in solving these problems.

Chapter 1

1.1. **(a)** 1001111110; **(c)** 11101100.
1.3. **(a)** 4575; **(c)** 44.
1.5. **(a)** 5555; **(c)** 5760.
1.9. $PR\{n > 2\} = PR\{2 + 3 + 4 + 5 + 6\} = 5/6$.
1.11. 1/8.
1.13. 5/32.
1.16. Mean = 1.125; Variance = 0.567.

1.20. **(a)** $p(x) = \dfrac{1}{\sqrt{2\pi}\,4} \exp\left[-(x-3)^2/32\right]$

(b) Probability is equal to 0.27.
1.24. $m = 1.182$.

1.28. **(b)** $E\{y\} = 2/\sqrt{2\pi}$.
1.34. $m_x = \sigma\sqrt{\pi/2}$; $\text{Var}\{x\} = 0.429\sigma^2$.
1.37. **(a)** $A = 1$; **(b)** $1 - e^6$; **(c)** 0.5.
1.44. $S(0) = 3$; $S(1) = -j$; $S(2) = 1$; $S(3) = j$.
1.47. Sample values of S are 1.23; $0.282 - j0.035$; $0.456 - j0.5$; $-0.083 + j0.144$; $0.097 - j0.137$; $0.334 + j0.396$; $0.103 + j0.175$; $0.193 - j0.026$; $0.359 + j0.115$; $0.309 - j0.003$; $0.843 - j0.071$; $0.177 - j0.133$; $0.427 - j0.766$; $0.022 + j0.15$.

Chapter 2

2.2. **(a)** Probability of bit error is 0.556. Probability of 2 errors out of 10 symbols is 0.021. **(b)** Probability of correct transmission is 0.017.
2.7. **(a)** $0.988 \cos(t - t_0)$.
(b) The output is given by

$$\frac{\sin(t + \pi/20)}{2(t + \pi/20)} + \frac{\sin(t - \pi/20)}{2(t - \pi/20)}$$

(c) The output is given by

$$\frac{\sin 10(t + \pi/20)}{2(t + \pi/20)} + \frac{\sin 10(t - \pi/20)}{2(t - \pi/20)}$$

2.9. Use an upper cutoff frequency of 1 Hz. Then $a_0 = 0.225$; $a_n = 0.45$. Output given by Eq. (2.28) with these values.
2.12. **(a)** $A(500) = -8$ dB; $A(1000) = 0$ dB; $\theta(500) = \pi$; $\theta(1000) = 2\pi$; $v_{out}(t) = -0.158 \cos(2\pi \times 500t) + \cos(2\pi \times 1000t)$

(b) $$v_{out}(t) = \frac{1}{\pi^2}\cos(2\pi \times 1000t) + \frac{0.1}{9\pi^2}\cos(2\pi \times 3000t)$$

2.16. $G_v(f) = 0.5A^2\pi[\delta(f - f_c) + \delta(f + f_c)]$.
2.18. **(a)** $SNR = 600$; **(b)** $SNR_o/SNR_i = 0.011$ dB; **(c)** $SNR = 15.18$ dB.

Chapter 3

3.2. **(a)** Function is unity; **(b)** $\delta(t)$; **(c)** yes; **(d)** use LPF
3.4. Aliasing error is $0.5a_1 \cos(2\pi/3)t + 0.5a_1 \cos(14\pi/3)t$.
3.8. 010; 101; 110; 111.
3.10. 1000; 1010; 0011; 1110; 0000; 1111.
3.18. mse $= 0.75\sigma^2$.
3.19.

$$A = \frac{\sin f_m T}{f_m T}$$

$$\text{mse} = \frac{N_0}{2\pi}\left[f_m - \frac{\sin^2 f_m T}{f_m T^2}\right]$$

3.23. mse $= 0.012\sigma^2$.

3.25. $N > 6.8$, so use 7 bits.
3.29. **(a)** mse = 0.016; mse(uniform) = 0.021.
3.32. **(a)** Improvement factor is 0; **(b)** improvement factor is 0.379; **(c)** improvement factor is 0.
3.36. $s(t_0 + T) = As(t_0) + Bs'(t_0)$; $A = R(T)/R(0)$; $B = -R'(T)/R''(0)$; mse $= R(0) - R^2(T)/R(0) + [R'(T)]^2/R''(0)$.

Chapter 4

4.4. **(a)** Information content is 633 kbits; **(b)** 19 Mbits/second; **(c)** 1000 words transmit 15.6 kbit of information; **(d)** picture is worth 40.58 kilowords.
4.5. **(a)** Entropy is 4.164 bits/letter; **(b)** 10.68 bits; **(c)** 18.36 bits; **(d)** not independent.
4.8. **(a)** Uniquely decipherable but not instantaneous; **(b)** not uniquely decipherable; **(c)** uniquely decipherable and instantaneous.
4.10. Detect up to three errors; correct one error.
4.14. **(a)**

$$[H] = \begin{bmatrix} 0 & 1 & 1 & 1 & 1 & 0 & 0 & 0 \\ 1 & 1 & 1 & 0 & 0 & 1 & 0 & 0 \\ 1 & 1 & 0 & 1 & 0 & 0 & 1 & 0 \\ 1 & 0 & 1 & 1 & 0 & 0 & 0 & 1 \end{bmatrix}$$

(b) Detect 33 errors, correct 1 error.
4.20.

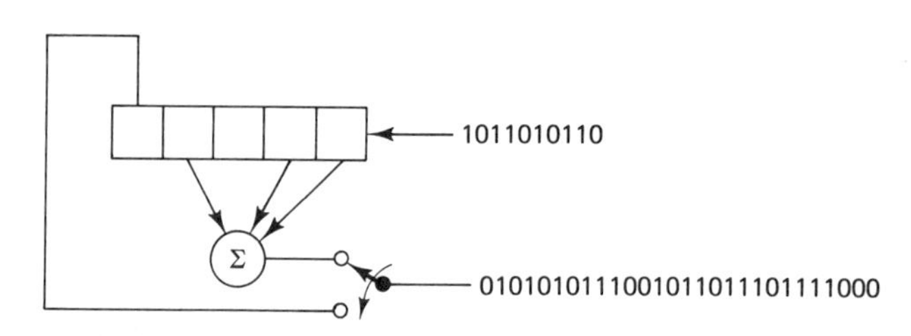

4.23. **(a)** $\lambda^4 + \lambda^2$ mod2; **(b)** $\lambda^4 + 1$ mod2; **(c)** $\lambda^4 + \lambda^3 + \lambda + 1$ mod2.
4.26. **(a)** 0 1 0 0 1 1 1 0 1 0 . . . ; **(b)** 7 bits; **(c)** 3; **(d)** detect three errors, correct one error.

Chapter 5

5.7. **(a)** $h(t) = 10 \cos(2\pi \times 1000[t - 0.01]) = s(t) \quad 0 < t < 0.01$; **(b)** $SNR = 0.5$
5.9. **(a)** Decision rule is given by the following diagram:

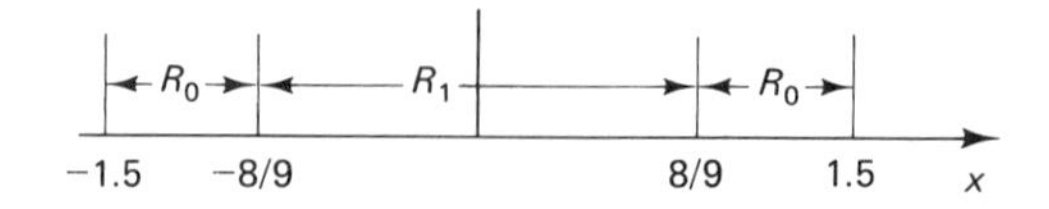

(b) Minimum error detector is the same as Bayes detector.
(c) The decision rule is given by the following diagram:

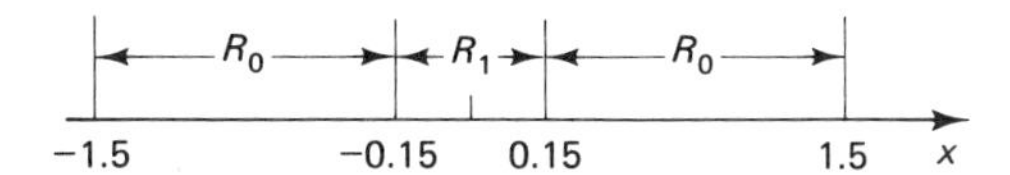

5.11. The decision rule is given by the following diagram:

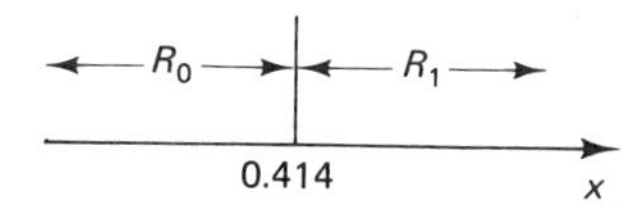

(b) $P_{FA} = 0.172$; $P_M = 0.086$; **(c)** $C = 0.172$; **(d)** $C = 0.167$.

5.14. If $\lambda_0 < 1/2$, always choose H_1. If $\lambda_0 > 1/2$, the decision rule is given by the following diagram:

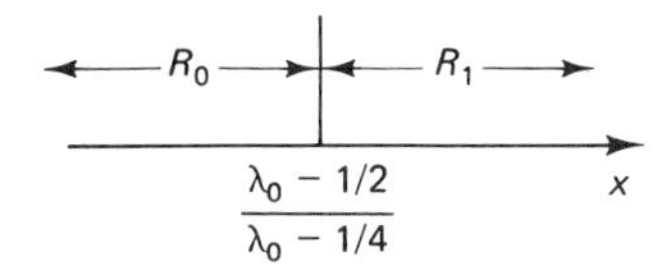

5.16. **(a)** The decision rule is given by the following diagram:

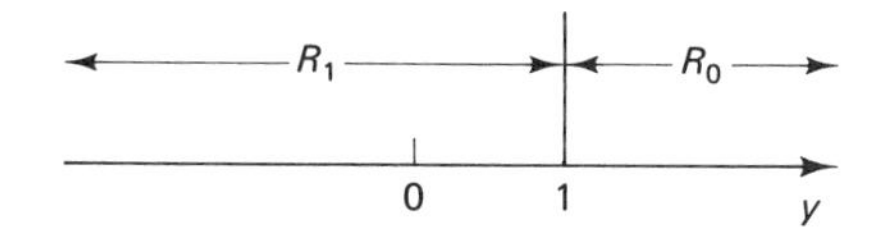

(c) The decision rule is given by the following diagram:

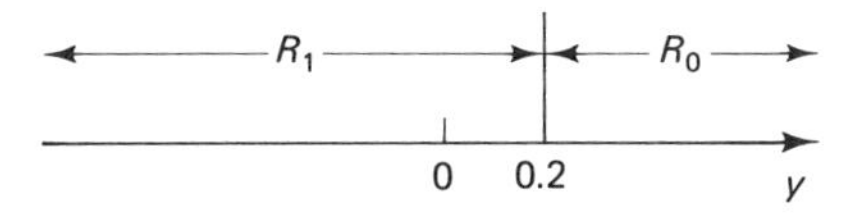

5.19. The decision rule is given by the following diagram:

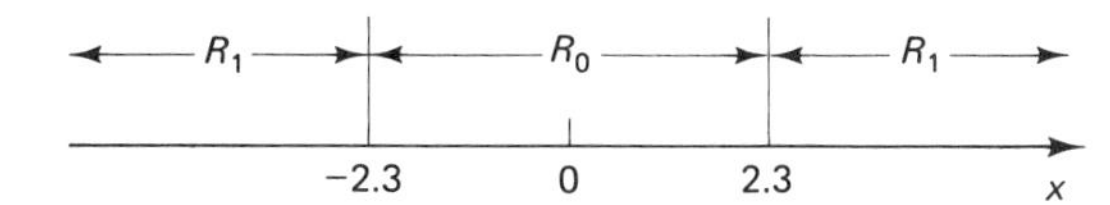

Chapter 6

6.4. $\sigma^2 = 0.5ABk_0N_0(BW)$ where BW is the equivalent bandwidth of the filter, $H(f)$.

6.6. Probability is 0.364.

6.10. Assuming a bit error rate of 10^{-3} and a correlation of 15 required, the error probability if 5×10^{-22}.

Chapter 7

7.3. **(a)** Bandwidth is 28 Hz out to the first zero of frequency spectrum.

7.6. Peak interference is 9.4×10^{-2}; mean square interference is 1.3×10^{-4}.

7.8. **(a)** $0.25 \operatorname{erfc}\left(\frac{A}{\sqrt{2}\sigma}\right) + 0.125 \operatorname{erfc}\left(\frac{A - \Delta}{\sqrt{2}\sigma}\right) + 0.125 \operatorname{erfc}\left(\frac{A + \Delta}{\sqrt{2}\sigma}\right)$

(b) $P_e = 8.9 \times 10^{-6}$.
(c) $P_e = 3.7 \times 10^{-7}$.

7.9. **(a)** Peak ISI = 0.036; **(b)** Mean square ISI = 0.00064; **(c)** Probability of error is zero.

7.12. $f(t + \Delta) = Af(T) + Bf'(t)$; $A = R(\Delta)/R(0)$; $B = R'(\Delta)/R''(0)$; mse $= R(0) - R^2(\Delta)/R(0) - [R'(\Delta)]^2/R''(0)$.

7.16. **(b)** Decision rule is to choose H_1 if $y > -0.55$; **(c)** $P_e = 0.0019$.

7.18. **(b)** $P_e = 0.5 \operatorname{erfc}(\sqrt{0.125/N_0})$

7.20. $P_e = 0.5 \operatorname{erfc}\left(\sqrt{\frac{T - 2\Delta}{4N_0}}\right)$

7.22. $P_c = A^2B$ where

$$A = \frac{1}{\pi N_0}\int_{-\infty}^{\infty}\int_{-\infty}^{y_1} \exp[-(y_1 - 1)^2/N_0] \exp[-y_3^2/N_0]\, dy_3\, dy_1$$

$$B = \frac{2/\sqrt{3}}{\pi N_0}\int_{-\infty}^{\infty}\int_{-\infty}^{y_1} \exp\left(\frac{-y_1^2 + y_1y_3 - y_3^2}{3N_0/4}\right) dy_3\, dy_1$$

7.23. Let y_i be the output of a filter matched to $s_i(t)$. The decision rule is then to choose the maximum among

y_1

y_2

$0.5y_1 + 0.5y_2 + 0.45y_3 + 0.166$

If the first of these three is maximum, choose H_1. If the second is maximum, choose H_2, and if the third is maximum, choose H_3.

Chapter 8

8.8. $P_e = 7 \times 10^{-7}$.
8.10. $P_e = 6.14 \times 10^{-6}$.
8.12. $P_e = 0.168$.
8.14. $P_{FA} = 0.0343$; $P_M = 0.05$.
8.16. Use detector of Fig. 7.23. Threshold is $z_0 = 76.8$.

Chapter 9

9.1. Bandwidth is $2(1 + k_f)$ Hz.
9.8. **(a)** $P_e = 3.7 \times 10^{-7}$; **(b)** $P_e = 1.86 \times 10^{-6}$.
9.9. **(a)** $P_e = 2.56 \times 10^{-7}$; **(b)** $P_e = 0.026$.
9.10. $P_{FA} = 3.72 \times 10^{-6}$.
9.15. $P_e = 0.387$.

Chapter 10

10.1. **(a)** Power out to half of the bit rate is 77.3% of $A^2/2$. **(b)** Power out to the bit rate is 90.2% of $A^2/2$. **(c)** Power out to twice the bit rate is 94.8% of $A^2/2$.
10.8. **(a)** $P_e = 1.15 \times 10^{-3}$; **(b)** $P_e = 2 \times 10^{-4}$.

Chapter 12

12.2. $BW = 10^7$ Hz; $G_p = 37$ dB.
12.4. Probability of a single bit error is 3.4×10^{-7}. Probability of a correct 4-bit message is $1 - (1.36 \times 10^{-6})$.
12.7. Since there is no sync, we must try nine possibilities.
12.8. If we choose 214 as base, then $7 \times 153 = 1 \bmod 214$. To send 1 0 1 1 0 1 1, transmit 446. Sum of original weights is 109.

Appendix II
The Error Function

x	erf (x)	erfc (x)	x	erf (x)	erfc (x)
0	0	1	0.05	0.056	0.944
0.10	0.112	0.888	0.15	0.168	0.832
0.20	0.223	0.777	0.25	0.276	0.724
0.30	0.329	0.671	0.35	0.379	0.621
0.40	0.428	0.572	0.45	0.475	0.525
0.50	0.521	0.479	0.55	0.563	0.437
0.60	0.604	0.396	0.65	0.642	0.358
0.70	0.678	0.322	0.75	0.711	0.289
0.80	0.742	0.258	0.85	0.771	0.229
0.90	0.797	0.203	0.95	0.821	0.179
1.00	0.843	0.157	1.05	0.862	0.138
1.10	0.880	0.120	1.15	0.896	0.104
1.20	0.910	0.0901	1.25	0.923	0.0768
1.30	0.934	0.0659	1.35	0.944	0.0564
1.40	0.952	0.0481	1.45	0.960	0.0400
1.50	0.966	0.0338	1.55	0.972	0.0284
1.60	0.976	0.0238	1.65	0.980	0.0199
1.70	0.984	0.0156	1.75	0.987	0.0128
1.80	0.989	0.0105	1.85	0.991	8.53×10^{-3}
1.90	0.993	6.91×10^{-3}	1.95	0.994	5.57×10^{-3}
2.00	0.995	4.59×10^{-3}	2.05	0.996	3.68×10^{-3}
2.10	0.997	2.93×10^{-3}	2.15	0.998	2.33×10^{-3}
2.20	0.998	1.84×10^{-3}	2.25	0.999	1.44×10^{-3}
2.30	0.999	1.13×10^{-3}	2.35	0.999	8.80×10^{-4}
2.40	0.999	6.82×10^{-4}	2.45	0.999	5.26×10^{-4}
2.50	1.000	4.03×10^{-4}	2.55	1.000	3.08×10^{-4}
2.60	1.000	2.34×10^{-4}	2.65	1.000	1.77×10^{-4}
2.70	1.000	1.33×10^{-4}	2.80	1.000	7.46×10^{-5}
2.90	1.000	4.09×10^{-5}	3.00	1.000	2.20×10^{-5}
3.10	1.000	1.16×10^{-5}	3.20	1.000	6.00×10^{-6}
3.30	1.000	3.06×10^{-6}	3.40	1.000	1.52×10^{-6}
3.50	1.000	7.43×10^{-7}	3.60	1.000	3.56×10^{-7}
3.70	1.000	1.67×10^{-7}	3.80	1.000	7.70×10^{-8}
3.90	1.000	3.48×10^{-8}	4.00	1.000	1.54×10^{-8}
4.10	1.000	6.70×10^{-9}	4.20	1.000	2.86×10^{-9}
4.30	1.000	1.19×10^{-9}	4.40	1.000	4.89×10^{-10}
4.50	1.000	1.97×10^{-10}	4.60	1.000	7.75×10^{-11}
4.70	1.000	3.00×10^{-11}	4.80	1.000	1.14×10^{-11}
4.90	1.000	4.22×10^{-12}	5.00	1.000	1.54×10^{-12}

Appendix III
The Marcum-Q Function

The following table lists approximate values of the Marcum-Q function for integer arguments. That is, it lists $Q(\alpha, \beta)$, where α is as shown in the left margin and β is given in the top row. More complete tables can be found in papers by Marcum.

α \ β	1	2	3	4	5	6	7	8	9	10
1	0.875	0.268	0.044	0.003	0	0	0	0	0	0
2	0.915	0.603	0.213	0.034	0.002	0	0	0	0	0
3	0.991	0.885	0.595	0.195	0.031	0.002	0	0	0	0
4	0.999	0.986	0.875	0.575	0.186	0.029	0.002	0	0	0
5	0.999	0.999	0.984	0.865	0.564	0.180	0.027	0.002	0	0
6	0.999	0.999	0.999	0.984	0.860	0.543	0.176	0.027	0.002	0
7	0.999	0.999	0.999	0.999	0.981	0.855	0.523	0.173	0.026	0.002
8	0.999	0.999	0.999	0.999	0.999	0.980	0.850	0.511	0.170	0.026
9	0.999	0.999	0.999	0.999	0.999	0.999	0.977	0.845	0.497	0.167
10	0.999	0.999	0.999	0.999	0.999	0.999	0.999	0.975	0.840	0.482

Appendix IV
References

CHAPTERS 1, 2, 3

Books

FEHER, KAMILO, *Digital Communications, Microwave Applications,* Englewood Cliffs, N.J.: Prentice-Hall, 1981.

FEHER, KAMILO, *Digital Communications, Satellite/Earth Station Engineering,* Englewood Cliffs, N.J.: Prentice-Hall, 1983.

GURRIE, MICHAEL L., and O'CONNOR, PATRICK J., *Voice/Data Telecommunications Systems—An Introduction to Technology,* Englewood Cliffs, N.J.: Prentice-Hall, 1986.

MCMENAMIN, J. MICHAEL, *Linear Integrated Circuits,* Englewood Cliffs, N.J.: Prentice-Hall, 1985.

PEEBLES, PEYTON Z., Jr., *Digital Communication Systems,* Englewood Cliffs, N.J.: Prentice-Hall, 1987.

RODEN, MARTIN S., *Analog and Digital Communication Systems,* 2nd ed., Englewood Cliffs, N.J.: Prentice-Hall, 1985.

RODEN, MARTIN S., *Digital and Data Communication Systems,* Englewood Cliffs, N.J.: Prentice-Hall, 1982.

SCHWARTZ, MISCHA, *Information Transmission, Modulation and Noise,* 4th ed., New York: McGraw-Hill, 1986.

TAUB, HERBERT, and SCHILLING, DONALD, *Principles of Communication Systems,* 2nd ed., New York: McGraw-Hill, 1986.

ZIEMER, RODGER E., and PETERSON, ROGER L., *Digital Communications and Spread Spectrum Systems,* New York: Macmillan, 1985.

ZIEMER, R. E., and TRANTER, W. H., *Principles of Communications,* 2nd ed., Boston, Mass.: Houghton Mifflin, 1985.

Papers

FISCHER, THOMAS, "Digital VLSI Breeds Next-Generation TV Receivers," *Electronics Magazine,* August 11, 1981, pp. 97–103.

KANEKO, HISASHI, and ISHIGURO, TATSUO, "Digital Television Transmission Using Bandwidth Compression Techniques," *IEEE Communications Magazine,* July 1980, pp. 14–22.

JURGEN, RONALD K., "The Problems and Promises of High-Definition Television," *IEEE Spectrum,* December 1983, pp. 46–51.

LERNER, ERIC J., "Digital TV: Makers Bet on VLSI," *IEEE Spectrum,* February 1983, pp. 39–44.

SKLAR, BERNARD, "A Structured Overview of Digital Communications—A Tutorial Review," *IEEE Communications Magazine:* Part I—August 1983, pp. 2–17; Part II—October 1983, pp. 6–21.

CHAPTER 4

Books

RODEN, MARTIN S., *Digital and Data Communication Systems,* Englewood Cliffs, N.J.: Prentice-Hall, 1982.

TAUB, HERBERT, and SCHILLING, DONALD, *Principles of Communication Systems,* 2nd ed., New York: McGraw-Hill, 1986.

VITERBI, ANDREW, J., and OMURA, JIM K., *Principles of Digital Communication and Coding,* New York: McGraw-Hill, 1979.

ZIEMER, R. E., and TRANTER, W. H., *Principles of Communications,* 2nd ed., Boston, Mass.: Houghton Mifflin, 1985.

Papers

SKLAR, BERNARD, "A Structured Overview of Digital Communications—A Tutorial Review," *IEEE Communications Magazine:* Part I—August 1983, pp. 2–17; Part II—October 1983, pp. 6–21.

CHAPTERS 5 and 6

Books

LINDSEY, WILLIAM C., and SIMON, MARVIN K., *Telecommunication Systems Engineering,* Englewood Cliffs, N.J.: Prentice-Hall, 1973.

ZIEMER, RODGER E., and PETERSON, ROGER L., *Digital Communications and Spread Spectrum Systems,* New York: Macmillan, 1985.

Papers

FRANKS, L. E., "Carrier and Bit Synchronization in Data Communication—A Tutorial Review," *IEEE Transactions on Communications,* Vol. COM-28, No. 8, August 1980, pp. 1107–1120.

SKLAR, BERNARD, "A Structured Overview of Digital Communications—A Tutorial Review," *IEEE Communications Magazine:* Part I—August 1983, pp. 2–17; Part II—October 1983, pp. 6–21.

CHAPTERS 7, 8, 9, and 10

Books

FEHER, KAMILO, *Digital Communications, Microwave Applications,* Englewood Cliffs, N.J.: Prentice-Hall, 1981.

FEHER, KAMILO, *Digital Communications, Satellite/Earth Station Engineering,* Englewood Cliffs, N.J.: Prentice-Hall, 1983.

PEEBLES, PEYTON Z. Jr., *Digital Communication Systems,* Englewood Cliffs, N.J.: Prentice-Hall, 1987.

RODEN, MARTIN S., *Digital and Data Communication Systems,* Englewood Cliffs, N.J.: Prentice-Hall, 1982.

SCHWARTZ, MISCHA, *Information Transmission, Modulation and Noise,* 4th ed., New York: McGraw-Hill, 1986.

SMITH, JACK, *Modern Communication Circuits,* New York: McGraw-Hill, 1986.

TAUB, HERBERT, and SCHILLING, DONALD, *Principles of Communication Systems,* 2nd ed., New York: McGraw-Hill, 1986.

Papers

SKLAR, BERNARD, "A Structured Overview of Digital Communications—A Tutorial Review," *IEEE Communications Magazine:* Part I—August 1983, pp. 2–17; Part II—October 1983, pp. 6–21.

SUNDBERG, CARL-ERIK, "Continuous Phase Modulation," *IEEE Communications Magazine,* April 1986, Vol. 24, No. 4, pp. 25–38.

CHAPTER 11

Papers

SKLAR, BERNARD, "A Structured Overview of Digital Communications—A Tutorial Review," *IEEE Communications Magazine:* Part I—August 1983, pp. 2–17; Part II—October 1983, pp. 6–21.

CHAPTER 12

Books

BEKER, HENRY J., and PIPER, FRED C., *Secure Speech Communications,* Orlando, Fla: Academic Press, 1985.

SIMON, MARVIN K., OMURA, JIM K., SCHOLTZ, ROBERT A., and LEVITT, BARRY K., *Spread Spectrum Communications,* Vol. 1, Rockville, Md.: Computer Science Press, 1985.

ZIEMER, RODGER E., and PETERSON, ROGER L., *Digital Communications and Spread Spectrum Systems,* New York: Macmillan, 1985.

Papers

ABBRUSCRATO, C. R., "Data Encryption Equipment," *IEEE Communications Magazine,* September 1984, Vol. 22., No. 9, pp. 15–21.

KAHN, DAVID, "Cryptology Goes Public," *IEEE Communications Magazine,* March 1980, pp. 19–28.

CHAPTER 13

Books

ALISOUSKAS, VINCENT F., and TOMASI, WAYNE, *Digital and Data Communications,* Englewood Cliffs, N.J.: Prentice-Hall, 1985.

CHORAFAS, DIMITRIS N., *The Handbook of Data Communication and Computer Networks,* Princeton, N.J.: Petrocelli Books, 1985.

COOPER, GEORGE R., and MCGILLEM, CLARE D., *Modern Communications and Spread Spectrum,* New York: McGraw-Hill, 1986.

HOUSLEY, TREVOR, *Data Communications and Teleprocessing Systems,* Englewood Cliffs, N.J.: Prentice-Hall, 1979.

KUO, FRANKLIN F., *Protocols and Techniques for Data Communication Networks,* Englewood Cliffs, N.J.: Prentice-Hall, 1981.

STALLING, WILLIAM, *Data and Computer Communications,* New York: MacMillan, 1985.

TAUB, HERBERT, and SCHILLING, DONALD, *Principles of Communication Systems,* 2nd ed., New York: McGraw-Hill, 1986.

Papers

CHATWANI, DILIP, "ISDN Line Card and Terminal Interfaces," WESCON 1986 Record, Electronic Conventions Management, Inc.

Telesis, Vol. 13, No. 2, 1986, Bell-Northern Research Ltd., Ottawa, Canada. All articles in this issue are relevant.

WONG, CHO LUN, and WOOD, ROB, *Implementation of ISDN,* Telesis, Vol. 13, No. 3, Ottawa, Canada: Bell-Northern Research Ltd., 1986.

Appendix V
Glossary of Terms

NOTE: This appendix presents an assortment of terms for which you may need brief definitions. It does not include major areas of the text material for which extensive study is required. As such, it should not be used to replace reference to text material, which can be done through the text index.

Aliasing. Error due to sampling at less than Nyquist rate.

ALOHA Network. Local area network which was originated at the University of Hawaii.

Amplitude Shift Keying (ASK). Technique for sending digital signals using sinusoidal bursts of differing amplitudes.

Applications Layer. Part of ISO OSI protocol model dealing with interface between software and presentation layer.

Baud. Rate of symbol transfer. If each symbol represents one bit, this baud is the same as the bit rate. If each symbol represents more than one bit, the baud is less than the bit rate.

Biphase Coding. Also known as Manchester. A scheme for sending binary information using transitions in the middle of bit intervals.

Bipolar. Digital transmission signal using both positive and negative voltages.

Bit Error Rate. Probability of bit error.

Burst Error. Error affecting a sequence of consecutive bits as contrasted with isolated bit errors.

Channel Capacity. Rate below which arbitrarily low error probabilities are possible.

Channel Encoder. Changes signal to increase efficiency or to reduce effects of transmission errors.

Cipher Text. Sequence resulting after encrypting a message.

Circuit Switching. Switching technique in which route through circuit is established and maintained for the entire exchange of data.

Comité Consultatif Internationale de Telegraphique et Telephonique (CCITT). Committee of the ITA that deals with voluntary worldwide standards for data communications.

Cyclic Redundancy Checking. An error checking code that deals with polynomials that have a given general polynomial as a factor. If division results in a zero remainder, it is assumed that no errors are present.

Data Compression. Bit reduction technique that takes advantage of redundancies in the data.

Data Encryption Standard. Encryption standard endorsed by the National Bureau of Standards for use in nondefense applications.

Delay Distortion. Distortion caused by different time delays of the various frequency components of a signal.

Encryption. Coding technique that makes data appear random to an unauthorized listener.

Entropy. Average information per message.

Envelope Delay. Also known as group delay; the delay experienced by the various frequency components of a signal.

Ethernet. Local area network developed by Xerox Corporation.

Forward Error Correction. Process to correct errors at the receiver without requiring retransmission of message segments.

Frequency Division Multiplexing (FDM). Technique for simultaneous transmission of multiple signals based upon separation of frequency content.

Frequency Shift Keying (FSK). Technique for transmitting digital signals by using sinusoidal bursts of differing frequencies.

Full-Duplex Circuit. Circuit capable of simultaneous transmission in both directions.

Half-Duplex Circuit. Circuit capable of transmission in both directions, but not simultaneously.

Host. The central computer which supports a number of terminals in a network.

Huffman Coding. Entropy coding technique that achieves data compression.

Intersymbol Interference. Adjacent symbols interfering with each other, usually due to pulse spreading by bandlimited channels.

Manchester Coding. See biphase.

Message Switching. Switching technique in which entire message is routed through the network from switch to switch.

MODEM. Device that modulates and demodulates digital signals.

Multiplexer. Device that combines multiple signals into a single waveform.

Network Layer. Part of ISO OSI protocol model concerned with routing of data from source to destination.

Nyquist Rate. Minimum sampling rate; equal to twice the highest signal frequency component.

ON-OFF Keying. Signaling technique that switches sinusoidal signal off and on to represent zeros and ones.

Packet Switching. Switching technique in which message is broken into packets which are routed through the network from switch to switch. Different packets may take different routes through the network.

Phase Shift Keying (PSK). Technique for transmitting digital signals by using sinusoidal bursts of differing phases.

Public Key Encryption. Encryption system that uses one key for encryption and a different key for decryption. Encryption keys are made public so messages can be sent without preestablishing secret keys.

Simplex Circuit. Circuit that can transmit in only one direction.

Source Encoder. Transforms one or more analog signal into a train of symbols.

Spread Spectrum. Intentional widening of frequency spectrum to reduce probability of unauthorized signal interception.

Star Topology. Local area network connecting stations with links that radiate from a central switch.

Store and Forward. Messages (or packets) are stored at switching centers while a search is made for a path to the next switching center.

T1 Transmission System. Early commercial digital system introduced by Bell System in 1962.

Telegraph-Grade Channel. Channel with frequency cutoff of about 300 Hz.

Time Division Multiplexing (TDM). Technique for simultaneous transmission of multiple signals based upon separation in time.

Transport Layer. Part of ISO OSI protocol model concerned with interface between hardware and software.

Voice-Grade Channel. Channel with bandwidth between 300 Hz and 4 kHz.

Wideband Channel. Channel with bandwidth greater than 4 kHz.

Index

D

E

U

V

W